AF352839

Industrial and Technological Applications of Power Electronics Systems

Industrial and Technological Applications of Power Electronics Systems

Editors

Ryszard Strzelecki
Galina Demidova
Dmitri Vinnikov

MDPI • Basel • Beijing • Wuhan • Barcelona • Belgrade • Manchester • Tokyo • Cluj • Tianjin

Editors
Ryszard Strzelecki
Gdansk University of Technology
Poland

Galina Demidova
ITMO University,
Russia

Dmitri Vinnikov
Tallinn University of Technology
Estonia

Editorial Office
MDPI
St. Alban-Anlage 66
4052 Basel, Switzerland

This is a reprint of articles from the Special Issue published online in the open access journal *Energies* (ISSN 1996-1073) (available at: https://www.mdpi.com/journal/energies/special_issues/ Applications_of_Power_Electronics_Systems).

For citation purposes, cite each article independently as indicated on the article page online and as indicated below:

LastName, A.A.; LastName, B.B.; LastName, C.C. Article Title. *Journal Name* **Year**, *Volume Number*, Page Range.

ISBN 978-3-0365-0822-1 (Hbk)
ISBN 978-3-0365-0823-8 (PDF)

Contents

About the Editors

Ryszard Strzelecki (M'97-SM'07) was born in Bydgoszcz, Poland. He received the M.Sc. and Ph.D. degrees in Electronic Engineering from the Department of Industrial Electronics, National Technical University of Ukraine "Kyiv Polytechnic Institute", Kyiv, Ukraine, in 1981 and 1984, respectively, and the Dr. Sc. degree in Electrical Engineering from the Institute of Electrodynamics, The National Academy of Sciences of Ukraine, Kyiv, in 1991. He is currently a full professor with the Gdańsk University of Technology (Gdańsk, Poland), co-head of the Laboratory of Power Electronics and Automated Electric Drive, ITMO University (St. Petersburg, Russia), and Scientific Consultant for Power Electronics to the Management Board of AREX Ltd. In 2020, he was elected to the Committee on Electrical Engineering of the Polish Academy of Sciences. His research activity is concentrated on the topology, control, and industry application of power electronic conditioners, particularly for power quality enhancement and power flow control on distributed electrical networks.

Galina Demidova received her B.Sc., engineer, and Ph.D. degrees from ITMO University, Saint Petersburg, Russia. At present she is an Associate Professor at the Faculty of Control Systems and Robotics, Engineer of the Industrial Department of Precision Electromechanical Systems, and researcher of the International Scientific Laboratory "Power electronics and Automated Electric Drive ", at ITMO University. She has performed more than 20 R&D projects, dealing with control systems in digital electric power drives for tracking telescopes. She has authored or coauthored more than 50 conference and journal papers. She is co-Guest Editor in several Special Issues, and an organizer of the IEEE Conference. Her current research interests include adaptive control, fuzzy logic control, motion control, automatic control, fuzzy neural networks, hybrid intelligent systems, genetic algorithms, electromechanical systems, wind turbines, and multi-agent control

Dmitri Vinnikov Guest Editor 3· Prof. Dr. Dmitri Vinnikov Dmitri Vinnikov was born in Tallinn, Estonia, in 1976. He received the Dipl.Eng., M.Sc., and Dr.Sc.techn. degrees in Electrical Engineering from Tallinn University of Technology, Tallinn, Estonia, in 1999, 2001, and 2005, respectively. He is currently a Research Professor and the Head of the Power Electronics Group, Department of Electrical Power Engineering and Mechatronics, Tallinn University of Technology (Estonia), and a visiting professor in the Faculty of Control System and Robotics, ITMO University, Saint Petersburg, Russia. He is the Head of R&D and co-founder of Ubik Solutions LLC.— an Estonian start-up company dedicated to innovative & smart power electronics for renewable energy systems. Moreover, he is one of the founders and leading researchers of ZEBE—the Estonian Centre of Excellence, performing research toward the development of zero energy and resource efficient smart buildings and districts. He has authored or co-authored two books, five monographs, and one book chapter, as well as more than 300 published papers on power converter design and development, and is the holder of numerous patents and utility models in this field. His research interests include applied designs of power electronic converters and control systems, renewable energy conversion systems (photovoltaic and wind), impedance-source power converters, and implementation of wide bandgap power semiconductors. D. Vinnikov is a Chair of the IEEE Estonia Section .

Preface to "Industrial and Technological Applications of Power Electronics Systems"

Since the turn of the century, interest in electrical power systems has been growing steadily, in part due to a tendency to move from directly controlled to intelligent autonomous energy systems. In particular, the increasing presence of renewable energy sources and the development of novel technologies, which demand active and often ultra-precise power supply systems, have generated extensive research in the area of advanced power electronics systems. The importance and scope of the application of regulated power sources in various technological systems are also growing, e.g., using plasma, ultrasounds, and superconductors. Furthermore, researchers pay great attention to loads in these systems, which are mostly represented by various types of electric drives that should be energy efficient. Hence, the main role in many modern technologies and industrial systems is to diversify power electronics converters by applying new topologies, components, and smart controls, where emphasis is placed on such merits as wide input voltage, load regulation range, improved quality of the input and output parameters, high control flexibility, and low cost. To promote research, and accelerate the transfer of knowledge and experience in the above areas, we propose a Special Issue of Energies on industrial and technological applications of power electronics systems. An important premise of this Special Issue would be the synergy effect derived from a combination of views and approaches from various power electronics application areas.

Ryszard Strzelecki, Galina Demidova , Dmitri Vinnikov
Editors

Article

A Parallel Estimation System of Stator Resistance and Rotor Speed for Active Disturbance Rejection Control of Six-Phase Induction Motor

Hamidreza Heidari [1,*], Anton Rassõlkin [1], Mohammad Hosein Holakooie [2], Toomas Vaimann [1], Ants Kallaste [1], Anouar Belahcen [1,3] and and Dmitry V. Lukichev [4]

[1] Department of Electrical Power Engineering and Mechatronics, Tallinn University of Technology, 19086 Tallinn, Estonia; anton.rassolkin@taltech.ee (A.R.); toomas.vaimann@taltech.ee (T.V.); ants.kallaste@taltech.ee (A.K.); anouar.belahcen@aalto.fi (A.B.)

[2] Department of Electrical Engineering, University of Zanjan, Zanjan 45371-38791, Iran; hosein.holakooie@znu.ac.ir

[3] Department of Electrical Engineering, Aalto University, 11000 Aalto, Finland

[4] Faculty of Control Systems and Robotics, ITMO University, 197101 St. Petersburg, Russia; lukichev@itmo.ru

* Correspondence: haheid@taltech.ee

Received: 3 February 2020; Accepted: 24 February 2020; Published: 2 March 2020

Abstract: In this paper, a parallel estimation system of the stator resistance and the rotor speed is proposed in speed sensorless six-phase induction motor (6PIM) drive. First, a full-order observer is presented to provide the stator current and the rotor flux. Then, an adaptive control law is designed using the Lyapunov stability theorem to estimate the rotor speed. In parallel, a stator resistance identification scheme is proposed using more degrees of freedom of the 6PIM, which is also based on the Lyapunov stability theorem. The main advantage of the proposed method is that the stator resistance adaptation is completely decoupled from the rotor speed estimation algorithm. To increase the robustness of the drive system against external disturbances, noises, and parameter uncertainties, an active disturbance rejection controller (ADRC) is introduced in direct torque control (DTC) of the 6PIM. The experimental results clarify the effectiveness of the proposed approaches.

Keywords: active disturbance rejection controller (ADRC); direct torque control (DTC); full-order observer; sensorless; six-phase induction motor (6PIM); stator resistance estimator

---

1. Introduction

Three-phase induction motor drives have become a mature technology in the last years, but investigations into concepts of multiphase induction motor drives are still taking place. Multiphase drive systems have a nearly 40-year history of research and study due to their promising advantages against the conventional three-phase systems. The phase redundancy of the multiphase drives provides extra merits such as fault-tolerant operation, series-connected multimotor drive systems, asymmetry and braking systems. Six-phase induction motors (6PIMs) are known for its fault-tolerant capability, low rate of inverter switches, and low DC-link voltage utilization compared with its three-phase one [1–3]. On the other hand, the modular three-phase structure of the 6PIM allows the use of well-known three-phase technologies. The 6PIM is successfully used in special applications, such as electric ships, electric aircrafts, electric vehicles, and melt pumps, where the high reliability and continuity of the operation are critical factors for the system [4]. The phase redundancy of the 6PIM provides the ability of the open-phase fault-tolerant operation without any extra electronic components [5,6].

Among different structures of the 6PIM [4], the asymmetrical 6PIM with double isolated neutral points, which consists of two sets of three-phase windings spatially shifted by 30 electrical degrees, has attracted the interest of many researchers [7–10]. The traditional three-phase control strategies, including switching table-based direct torque control (ST-DTC) [7], modulation-based DTC [8], the field-oriented control (FOC) [9], and finite control set-model predictive control (FCS-MPC) [10], can be extended to 6PIM (or other multi-phase machines) with some modifications to use more freedom degrees that exist in multi-phase machines. DTC is a well-accepted technique due to its simplicity, quick dynamics, and robustness [11]. The modulation-based DTC strategy offers better phase current, torque, and flux response. On the contrary, this method has more complexity against conventional ST-DTC. The ST-DTC approach has straightforward and simple structure, but it is completely overshadowed by low-order harmonics due to unused voltage vectors in the losses subspaces. To overcome this restriction, the idea of duty cycle control is introduced by several researchers [12,13].

The rapid development of intelligent and high-performance control technologies has also brought about changes in the adjustable speed drive system for different industrial applications [14,15]. To operate safely and reliably under different conditions, there is a lot of debate nowadays about the main control strategy of the system [16,17]. Among different high-performance control strategies of drive systems, the DTC strategy has a straightforward algorithm. The DTC technique is inherently speed sensorless. Nevertheless, if an outer speed loop is added to the DTC, the speed value is also necessary. Sensorless three/multi-phase induction machine drives are widely addressed in the technical literature due to multiple shortcomings of shaft encoders [18–23]. To investigate the instability problem of the traditional rotor flux-based model reference adaptive system (MRAS) speed estimators in the regenerating-mode low-speed operation, a stator current-based and back electromotive force-based MRASs are addressed in [19,20], respectively. In [21], two modified adaptation mechanisms are proposed to replace the classical proportional-integral (PI) regulator. The full-order Luenberger and Kalman filter observers are discussed in [22,23], respectively. Providing a DTC drive system with parallel identification of the rotor speed and the stator resistance is a challenging task because the operation of the DTC scheme is severely dependent on the stator resistance. This problem is sporadically reported for three-phase induction machines (3PIMs) [24,25], where the rotor speed and the stator resistance estimators encounter an overlap due to limited freedom degrees of 3PIM. In this paper, the problem of parallel estimation is investigated using more freedom degrees of 6PIM.

The outer speed control loop of the DTC scheme conventionally contains the PI regulator to obtain torque command from speed error. In general, the control law of a PID regulator is a linear combination of proportional-integral-derivative terms, which is suitable for linear systems. For nonlinear systems, such as the 6PIM drive system, the PI regulator has been given a lot of attention due to its simplicity. However, it suffers from multiple problems including: (1) tuning of its parameters; (2) high sensitivity against noise and external disturbances; and (3) loss of efficiency due to oversimplified control law [26,27]. One promising technique to relatively get rid of the drawbacks of PI regulator is active disturbance rejection controller (ADRC) [26,28]. The ADRC is a nonlinear control scheme, which provides a robust control against noises, external disturbances, and parameter uncertainties. For these reasons, the ADRC technique has recently attracted more attention for electric drive systems. To address this issue, a modified FOC scheme based on first-order ADRCs for current and speed control loops is proposed in [29]. A combined active disturbance rejection and sliding-mode controller for an induction motor is presented to achieve total robustness [30].

The aim of this paper is to present an ADRC-based DTC scheme for sensorless 6PIM drives. The speed estimator is based on adaptive full-order observer, and its control law is designed using Lyapunov stability theorem. Besides the speed estimation system, a stator resistance estimator is proposed using additional degrees of freedom of the 6PIM to enhance the robustness of the sensorless DTC strategy against stator resistance uncertainties. The adaptation law for the stator resistance estimator is derived using the Lyapunov stability theorem to ensure its overall convergence.

The rest of this paper is organized as follows. Section 2 introduces the mathematical model of the 6PIM. Section 3 presents the design procedure of the adaptive full-order observer, the speed estimator, and the stator resistance estimator. The DTC scheme of the 6PIM is discussed in Section 4, which includes the ST-DTC scheme, and ADRC in DTC. The experimental results are presented in Section 5. Finally, Section 6 summarizes the findings and concludes the paper.

2. Dynamic Model of 6PIM

There are two popular approaches for modeling of the multi-phase machines: (1) multiple d–q approach [9]; (2) vector space decomposition (VSD) approach [31]. The first method is exclusively used for modular three-phase structures-based multi-phase machines such as six-phase and nine-phase machines. However, the second method can be used for all types of multi-phase machines. In this research, the VSD approach is used, where a 6PIM with distributed windings is modeled in the three orthogonal subspaces, i.e., the $\alpha - \beta$, $z_1 - z_2$ and $o_1 - o_2$. Among them, only the $\alpha - \beta$ variables are in relation with electromechanical energy conversion, while $z_1 - z_2$ and $o_1 - o_2$ variables do not actively contribute to the torque production.

The schematic diagram of a six-phase voltage source inverter (VSI)-fed an 6PIM with two isolated neutral points is shown in Figure 1. The transfer between the normal $a - x - b - y - c - z$ variables and $\alpha - \beta - z_1 - z_2 - o_1 - o_2$ variables is performed by T_6 transformation matrix as follows [31]:

$$
T_6 = \frac{1}{3}
\begin{bmatrix}
1 & \frac{\sqrt{3}}{2} & -\frac{1}{2} & -\frac{\sqrt{3}}{2} & -\frac{1}{2} & 0 \\
0 & \frac{1}{2} & \frac{\sqrt{3}}{2} & \frac{1}{2} & -\frac{\sqrt{3}}{2} & -1 \\
1 & -\frac{\sqrt{3}}{2} & -\frac{1}{2} & \frac{\sqrt{3}}{2} & -\frac{1}{2} & 0 \\
0 & \frac{1}{2} & -\frac{\sqrt{3}}{2} & \frac{1}{2} & \frac{\sqrt{3}}{2} & -1 \\
1 & 0 & 1 & 0 & 1 & 0 \\
0 & 1 & 0 & 1 & 0 & 1
\end{bmatrix}
\tag{1}
$$

By applying T_6 matrix to the voltage equations in the original six-dimensional system, the 6PIM model can be represented in the three orthogonal submodels, identified as $\alpha - \beta$, $z_1 - z_2$, and $o_1 - o_2$. The voltage space vector equations of the 6PIM in the $\alpha - \beta$ subspace are written as follows:

$$
v_s = R_s i_s + p \mathbf{\Psi}_s \tag{2}
$$

$$
0 = R_r i_r + p \mathbf{\Psi}_r - j \omega_r \mathbf{\Psi}_r \tag{3}
$$

The flux linkages are

$$
\mathbf{\Psi}_s = L_s i_s + L_m i_r \tag{4}
$$

$$
\mathbf{\Psi}_r = L_m i_s + L_r i_r \tag{5}
$$

where v, i, $\mathbf{\Psi}$, R, and L represent voltage, current, flux linkage, resistance, and inductance, respectively, for stator (s subscript) and rotor (r subscript) quantities, and p denotes derivative operator. The electromagnetic torque produced by the 6PIM is expressed as

$$
T_e = 3P \mathbf{\Psi}_s \otimes i_s \tag{6}
$$

where P is pole pairs and $\otimes$ denotes the cross product.

The 6PIM voltage equations in the $z_1 - z_2$ subspace are the same as a passive R-L circuit as follows:

$$
v_{sz1} = R_s i_{sz1} + L_{ls} p i_{sz1} \tag{7}
$$

$$
v_{sz2} = R_s i_{sz2} + L_{ls} p i_{sz2} \tag{8}
$$

where L_{ls} is stator leakage inductance.

On the presumption that the stator mutual leakage inductances can be neglected, the 6PIM model in the $o_1 - o_2$ subspace has the same form of the $z_1 - z_2$ subspace. However, the applied 6PIM with two isolated neutral points avoids zero-sequence currents because it contains two sets of balanced three-phase windings.

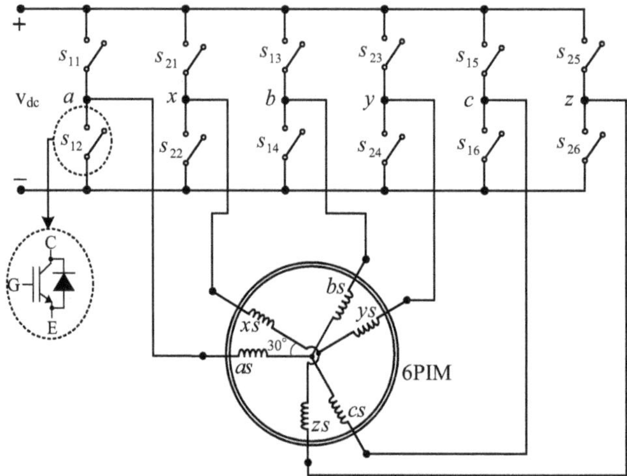

Figure 1. Six-phase two-level VSI-fed 6PIM.

3. Adaptive Full-Order Observer

The block diagram of the proposed R_s and ω_r estimators based on the adaptive state observer is shown in Figure 2. It contains the stator current and rotor flux observers, the stator resistance identifier, and the rotor speed estimator, which are discussed below.

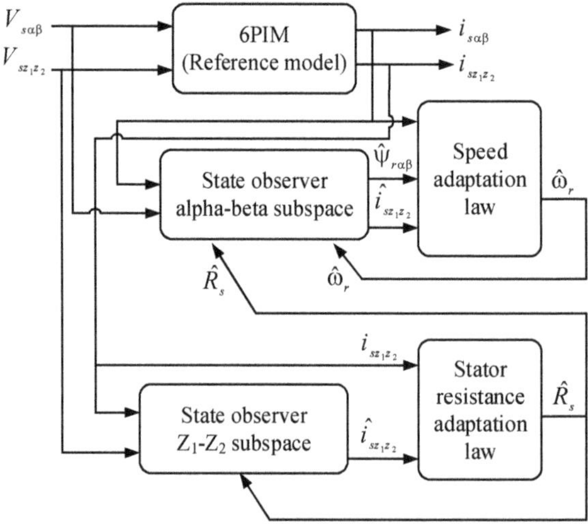

Figure 2. The block diagram of the proposed parallel estimation system of the stator resistance and the rotor speed based on an adaptive full-order observer.

3.1. Stator Current and Rotor Flux Observers

The general form of state-space model of the 6PIM in the $\alpha - \beta$ subspace is

$$\begin{cases} \dot{x}_1 = A_1 x_1 + B_1 u_1 \\ y_1 = C_1 x + D_1 u_1 \end{cases} \tag{9}$$

Assuming stator current and rotor flux as state variables and using Equations (2) and (3), the elements of state-space representation in $\alpha - \beta$ subspace will be

$$x_1 = \begin{bmatrix} i_{s\alpha} & i_{s\beta} & \psi_{r\alpha} & \psi_{r\beta} \end{bmatrix}^T \tag{10}$$

$$A_1 = \begin{bmatrix} (-\frac{R_s}{\sigma L_s} - \frac{1-\sigma}{\sigma T_r})I & \frac{L_m}{\sigma L_s L_r}(\frac{1}{T_r}I - \omega_r J) \\ \frac{L_m}{T_r}I & -\frac{1}{T_r}I + \omega_r J \end{bmatrix} \tag{11}$$

$$B_1 = \begin{bmatrix} \frac{1}{\sigma L_s}I & O \end{bmatrix}^T \tag{12}$$

$$u_1 = \begin{bmatrix} v_{s\alpha} & v_{s\beta} \end{bmatrix}^T \tag{13}$$

$$y_1 = \begin{bmatrix} i_{s\alpha} & i_{s\beta} \end{bmatrix}^T \tag{14}$$

$$C_1 = \begin{bmatrix} I & O \end{bmatrix} \tag{15}$$

with

$$I = \begin{bmatrix} 1 & 0 \\ 0 & 1 \end{bmatrix}, J = \begin{bmatrix} 0 & -1 \\ 1 & 0 \end{bmatrix}, O = \begin{bmatrix} 0 & 0 \\ 0 & 0 \end{bmatrix} \tag{16}$$

where $T_r = L_r/R_r$ is rotor time constant and $\sigma = 1 - L_m^2/L_s L_r$ is leakage coefficient.

The state observer of the 6PIM has a similar form of state-space representation except that an additional compensation term based on error of measurable states and observer gain matrix is added to it. The state observer can be written as

$$\begin{cases} \dot{\hat{x}}_1 = \hat{A}_1 \hat{x}_1 + B_1 u_1 + G_1(i_s - \hat{i}_s) \\ \hat{y}_1 = C_1 \hat{x}_1 \end{cases} \tag{17}$$

where the marker $^\wedge$ indicates the estimated values, and G_1 is the observer gain matrix. The matrix A_1 contains unknown parameters of the 6PIM such as the rotor speed and the stator resistance. These parameters can be estimated by the designing of a suitable adaptation control law with a nonlinear theorem such as a Lyapunov stability theorem. It is worth mentioning here that the matrix A_1 also contains the rotor time constant. However, simultaneous estimation of the rotor speed, the rotor time constant, and the stator resistance is challenging because of persistency of excitation conditions problem [32]. Some techniques have recently been developed based on signal injection to provide persistent excitation [33], which suffer from steady-state torque and speed ripples. In this paper, the stator resistance is estimated from additional degrees of freedom of the 6PIM, while the rotor speed is provided using the 6PIM equations in $\alpha - \beta$ subspace. This procedure provides the stator resistance independent from the rotor speed.

The observer gain matrix G_1 must be designed to ensure stability and good dynamic response of the observer at a wide range of the speeds. Using pole-placement method, the elements of matrix G_1 is provided as [22,34]

$$G_1 = \begin{bmatrix} g_1 & g_2 & g_3 & g_4 \\ -g_2 & g_1 & -g_4 & g_3 \end{bmatrix}^T \tag{18}$$

where

$$\begin{cases} g_1 = (1 - K_{po})(R_s L_r^2 + R_r L_m^2)/\sigma L_s L_r^2 \\ g_2 = (K_{po} - 1)\hat{\omega}_r \\ g_3 = (K_{po} - 1)(R_s L_s - K_{po} R_s L_r)/L_m \\ g_4 = (1 - K_{po})\hat{\omega}_r \sigma L_s L_r/L_m \end{cases} \tag{19}$$

where $K_{po} > 0$ is observer constant gain.

3.2. Stator Resistance Identification

In this paper, a stator resistance adaptation system is proposed using the machine model in the $z_1 - z_2$ subspace. This method can be utilized for any multi-phase machines. It is completely decoupled from the rotor speed and the rotor time constant, whereas most of the conventional stator resistance estimators, developed for three-phase machines, are related to these parameters. The proposed R_s estimator only depends on the stator leakage inductance L_{ls}, which can be approximately assumed to be constant.

The state-space model of 6PIM in the $z_1 - z_2$ subspace, with consideration of i_{sz1} and i_{sz2} as the state variables, can be derived from Equations (7) and (8) as follows:

$$
\begin{bmatrix} \dot{i}_{sz1} \\ \dot{i}_{sz2} \end{bmatrix} = \begin{bmatrix} -\frac{R_s}{L_{ls}} & 0 \\ 0 & -\frac{R_s}{L_{ls}} \end{bmatrix} \begin{bmatrix} i_{sz1} \\ i_{sz2} \end{bmatrix} + \frac{1}{L_{ls}} \begin{bmatrix} v_{sz1} \\ v_{sz2} \end{bmatrix}
$$

$$
\begin{bmatrix} i_{sz1} \\ i_{sz2} \end{bmatrix} = \begin{bmatrix} 1 & 0 \\ 0 & 1 \end{bmatrix} \begin{bmatrix} i_{sz1} \\ i_{sz2} \end{bmatrix} \tag{20}
$$

In this case, the proposed states observer is given by

$$
\begin{cases} \dot{\hat{x}}_2 = \hat{A}_2 \hat{x}_2 + B_2 u_2 \\ \hat{y}_2 = C_2 \hat{x}_2 \end{cases} \tag{21}
$$

It should be noted that a correction term $G_2(x_2 - \hat{x}_2)$ is neglected in Equation (21) due to the inherent stability of the observer.

The proposed adaptation law for the stator resistance estimation is

$$
\hat{R}_s = K_{pr}\epsilon_{R_s} + K_{ir}\int \epsilon_{R_s} dt \tag{22}
$$

where K_{ir} and K_{pr} are the integral and proportional gains, respectively, and ϵ_{R_s} is the stator resistance error signal

$$
\epsilon_{R_s} = \hat{i}_{sz1}(i_{sz1} - \hat{i}_{sz1}) + \hat{i}_{sz2}(i_{sz2} - \hat{i}_{sz2}) \tag{23}
$$

The proof for the stator resistance adaptation law is presented in Appendix A.

3.3. Rotor Speed Estimation

In order to design the speed adaptation law, it is considered as an unknown parameter. First, an appropriate positive definite function is chosen as the Lyapunov candidate. Then, the adaptation law is obtained using the Lyapunov criterion to ensure asymptotic stability of the system. The speed adaptation law is

$$
\hat{\omega}_r = K_{p\omega}\epsilon_\omega + K_{i\omega}\int \epsilon_\omega dt \tag{24}
$$

where $K_{p\omega}$ and $K_{i\omega}$ are proportional and integral gains, respectively, and ϵ_ω is the speed error signal as follows:

$$
\epsilon_\omega = (i_{s\alpha} - \hat{i}_{s\alpha})\hat{\psi}_{r\beta} - (i_{s\beta} - \hat{i}_{s\beta})\hat{\psi}_{r\alpha} \tag{25}
$$

The proof for the speed adaptation law is presented in Appendix B.

4. DTC of 6PIM

4.1. ST-DTC Scheme

A six-phase VSI contains overall $2^6 = 64$ different voltage space vectors, 60 active, and four zero vectors, where the active voltage vectors are distributed in four non-zero levels depicted in Figure 3. The electrical angle of each sectors is $30°$. The 6PIM phase-to-neutral voltages can be calculated as

$$\begin{bmatrix} V_a \\ V_b \\ V_c \\ V_x \\ V_y \\ V_z \end{bmatrix} = \frac{V_{dc}}{3} \begin{bmatrix} 2 & -1 & -1 & 0 & 0 & 0 \\ -1 & 2 & -1 & 0 & 0 & 0 \\ -1 & -1 & 2 & 0 & 0 & 0 \\ 0 & 0 & 0 & 2 & -1 & -1 \\ 0 & 0 & 0 & -1 & 2 & -1 \\ 0 & 0 & 0 & -1 & -1 & 2 \end{bmatrix} \begin{bmatrix} S_a \\ S_b \\ S_c \\ S_x \\ S_y \\ S_z \end{bmatrix} \tag{26}$$

where $S_i = \{0, 1\}, i = \{a, x, b, y, c, z\}$ is the switching state. When $S_i = 1$ ($S_i = 0$), the corresponding stator terminal is connected to positive (negative) DC-link rail. The voltage space vectors are given by

$$v_s = \frac{1}{3}[V_a + aV_x + a^4 V_b + a^5 V_y + a^8 V_c + a^9 V_z] \tag{27}$$

$$v_z = \frac{1}{3}[V_a + a^5 V_x + a^8 V_b + aV_y + a^4 V_c + a^9 V_z] \tag{28}$$

where $v_z = v_{sz1} + jv_{sz2}$ and $a = e^{j\pi/6}$.

The flux estimator is obtained from

$$\psi_{s\alpha} = \int (v_{s\alpha} - \hat{R}_s i_{s\alpha}) dt \tag{29}$$

$$\psi_{s\beta} = \int (v_{s\beta} - \hat{R}_s i_{s\beta}) dt \tag{30}$$

and the toque estimator is obtained from (6). In the traditional ST-DTC, the torque and stator flux errors are applied to hysteresis regulators to provide the sign of torque (ϵ_T) and stator flux (ϵ_ψ). According to gained signals and also the position of stator flux, a proper large voltage vector is selected based on Table 1 during each sampling period. From Figure 3, the corresponding voltage vectors in the $z_1 - z_2$ subspace will produce large current harmonics, when only large voltage vectors are used to control the torque and flux. Hence, it can alleviate the current harmonics through reduction of the $z_1 - z_2$ components by applying a combined voltage vector during each sampling period. This technique is referred to as duty cycle control, where a virtual vector (synthesized by large and medium voltage space vectors) is applied to the inverter in each sampling period because the large and medium voltage vectors are in the opposite direction in the $z_1 - z_2$ subspace (see Figure 3). The duration of the applied vectors is calculated in order to reduce the average volt-seconds in the $z_1 - z_2$ subspace [4]. The block diagram of the proposed sensorless DTC strategy with the adaptive full-order observer is shown in Figure 4a. In this figure, the speed control loop is based on the ADRC strategy, which will be discussed in the next subsection.

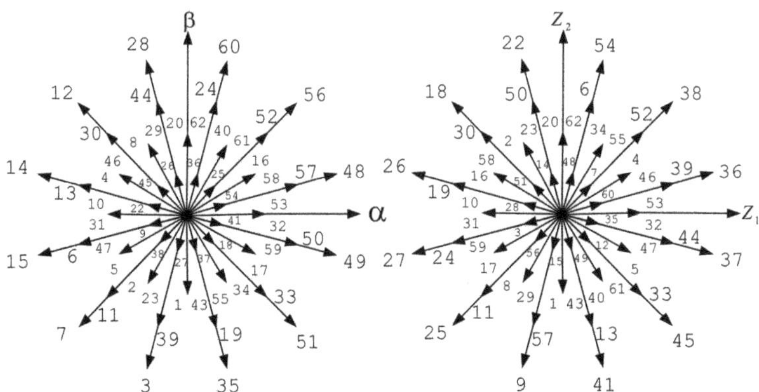

Figure 3. The $\alpha - \beta$ (top side) and the $z_1 - z_2$ (down side) vector subspaces for a six-phase VSI.

Table 1. Switching table of DTC strategy.

ϵ_T	ϵ_ψ	Selected Voltage *
1	1	V_{m+1}
1	0	V_{m+4}
0	1	V_0
0	0	V_0
-1	1	V_{m-2}
-1	0	V_{m-5}

* m is sector number.

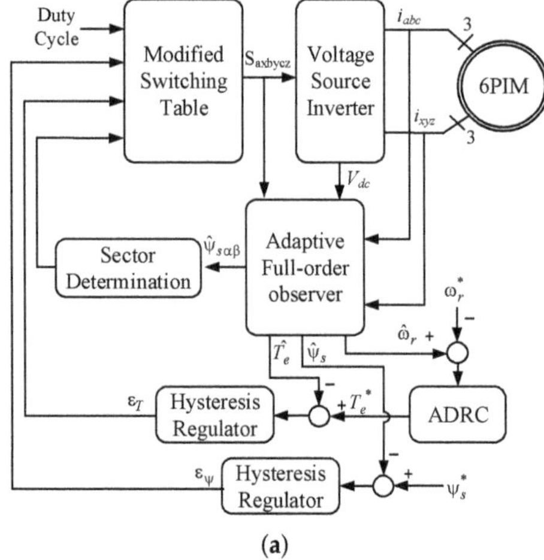

(a)

Figure 4. *Cont.*

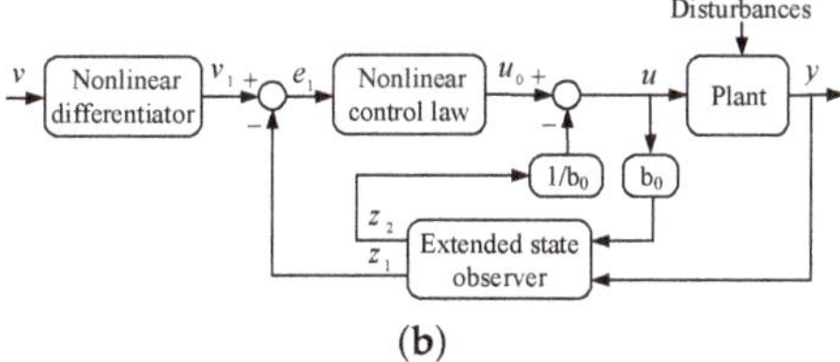

Figure 4. Block diagram of (**a**) the proposed sensorless DTC strategy; (**b**) ADRC.

4.2. ADRC in DTC

To enhance the robustness of the DTC technique against external disturbances and measurement noises, the ADRC is proposed to replace with the conventional PI regulator in the outer speed control loop. The block diagram of ADRC is shown in Figure 4b. It consists of three main elements: (1) nonlinear differentiator; (2) extended state observer; (3) nonlinear control law.

In some industrial applications, the command values are changed as step function, which is not suitable for the control system because of a sudden jump of output and control signals. To solve this problem, the nonlinear differentiator is used, which makes a reasonable transient profile from command signals for tracking [26]. The nonlinear differentiator can be expressed by

$$\begin{cases} v_1(k+1) = v_1(k) + hv_2(k) \\ v_2(k+1) = v_2(k) + hf_1(v_1(k) - v(k), v_2(k), r_0, h_0) \end{cases} \tag{31}$$

where f_1 is a nonlinear function as

$$f_1(v_1(k), v_2(k), r_0, h_0) = - \begin{cases} a(k)/h_0 & |a(k)| \le r_0 h_0 \\ r_0 \text{sign}(a(k)) & |a(k)| > r_0 h_0 \end{cases} \tag{32}$$

with

$$a(k) = \begin{cases} v_2(k) + y_0(k)/h_0 & |a(k)| \le r_0 h_0^2 \\ v_2(k) + (a_0(k) - r_0 h_0)/2 & |a(k)| > r_0 h_0^2 \end{cases}$$

$$y_0(k) = v_1(k) + h_0 v_2(k)$$

$$a_0(k) = \sqrt{(r_0 h_0)^2 + 8 r_0 |y_0(k)|}$$

where r_0 and h_0 are the parameters of the nonlinear differentiator, and h is sampling period.

The extended state observer is an enhanced version of feedback linearization method to compensate the total disturbances of the system. Using this observer, the state feedback term can be estimated online; hence, it is an adaptive robust observer against model uncertainties and external disturbances. The extended state observer is represented as follows:

$$\begin{cases} z_1(k+1) = z_1(k) + h[z_2(k) - \beta_1 f_2(e(k), \alpha_1, \delta_1) + b_0 u(k)] \\ z_2(k+1) = z_2(k) - h\beta_2 f_2(e(k), \alpha_1, \delta_1) \\ e(k) = z_1(k) - y(k) \end{cases} \tag{33}$$

where the nonlinear function f_2 is defined as

$$f_2(e(k), \alpha, \delta) = \begin{cases} e(k)/\delta^{1-\alpha} & |e(k)| \le \delta \\ |e(k)|^\alpha \text{sign}(e(k)) & |e(k)| > \delta \end{cases} \tag{34}$$

where $\alpha_1, \delta_1, \beta_1, \beta_2$, and b_0 are the parameters of the extended state observer.

The conventional PI controller is based on the linear combination of proportional and integral terms of error, which may degrade the performance of the DTC scheme. Different nonlinear combination of error can be presented to overcome this problem. In this paper, the following nonlinear control law is used:

$$\begin{cases} e_1(k) = v_1(k) - z_1(k) \\ u_0(k) = \beta_3 f_2(e_1(k), \alpha_2, \delta_2) \\ u(k) = u_0(k) - z_2(k)/b_0 \end{cases} \tag{35}$$

where α_2, β_3, and δ_2 are the parameters of nonlinear control law.

5. Experimental Validation

5.1. Description of Experimental Setup

The schematic and photograph of the experimental setup are shown in Figure 5a,b, respectively. The principal elements are

- a TMS320F28335-based digital signal processor (DSP) board.
- two custom-made two-level three-phase VSIs based on BUP 314D IGBTs and LEM LTS 6-NP current transducers.
- an LEM LV25-P voltage transducer.
- an Autonics incremental shaft encoder.
- a magnetic powder brake mechanically coupled to the 6PIM.
- a bridge rectifier.
- a 1-hp three-phase induction motor, which has been rewound to provide an asymmetrical 6PIM. The specifications of the 6PIM are shown in Table 2.

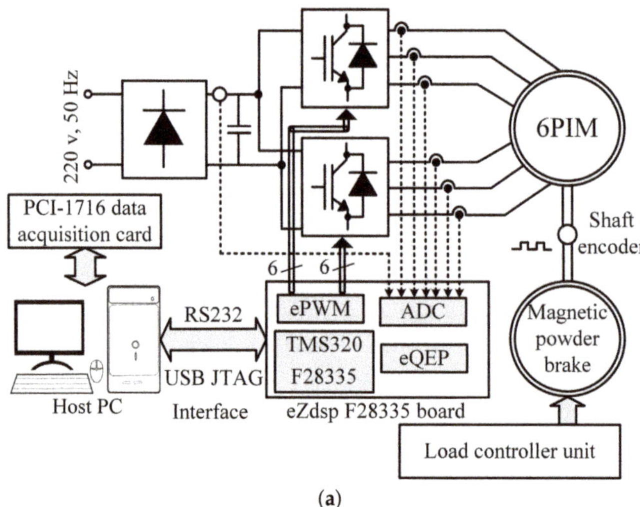

(a)

Figure 5. *Cont.*

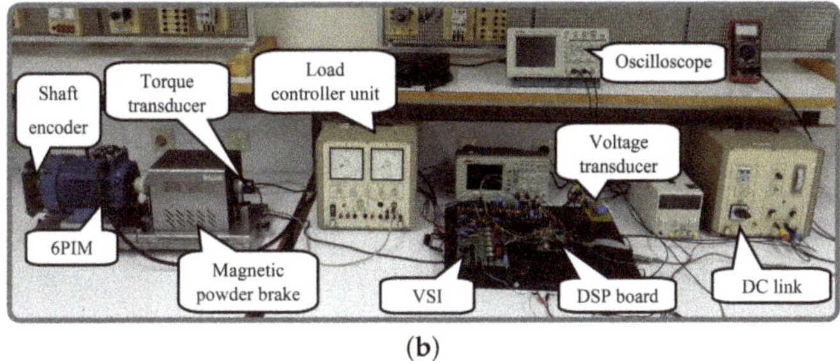

(**b**)

Figure 5. Experimental setup (**a**) schematic (**b**) photograph.

Table 2. The parameters of 6PIM.

Symbol	Quantity	Value
T_n	Nominal torque	2 Nm
P	Pole pairs	1
R_s	Stator resistance	4.08 Ω
R_r	Rotor resistance	3.73 Ω
L_s	Stator inductance	443.6 mH
L_r	Rotor inductance	443.6 mH
L_m	Magnetizing inductance	429.8 mH
J	Moment of inertia	0.000718 kg·m^2

5.2. Experimental Results

The performance of the proposed sensorless DTC strategy has been experimentally surveyed using DSP platform, programmed through Code Composer Studio (CCS v.3.3) and MATLAB. The IQmath and digital motor control (DMC) libraries have been used to provide optimized code. A 10 kHz sampling frequency with a 2 μs dead-band has been adopted. The experimental results have been captured using an Advantech PCI-1716 data acquisition card (DAQ) and serial port with LABVIEW and MATLAB, respectively. The serial communications interface (SCI) module has been employed to provide a serial connection between host PC and DSP. An incremental shaft encoder has been used to verify the performance of the speed estimation algorithm. All of the experiments have been carried out in sensorless mode as well as closed-loop adaptation of the stator resistance under various test scenarios, emphasizing on the low-speed region.

The experimental results of the proposed parallel estimation system of stator resistance and rotor speed under 50% initial stator resistance mismatch are shown in Figure 6. The speed command is 7% rated speed under rated load torque. In this test, the electric drive is allowed to start with a wrong stator resistance. This causes an error in estimated electromagnetic torque and actual speed. However, the estimated speed and the stator flux follow their reference values because of the controller action. It can be seen that the estimation error of the speed and the electromagnetic torque due to detuned stator resistance are removed within short seconds after activation of the stator resistance estimator at $t = 5$ s.

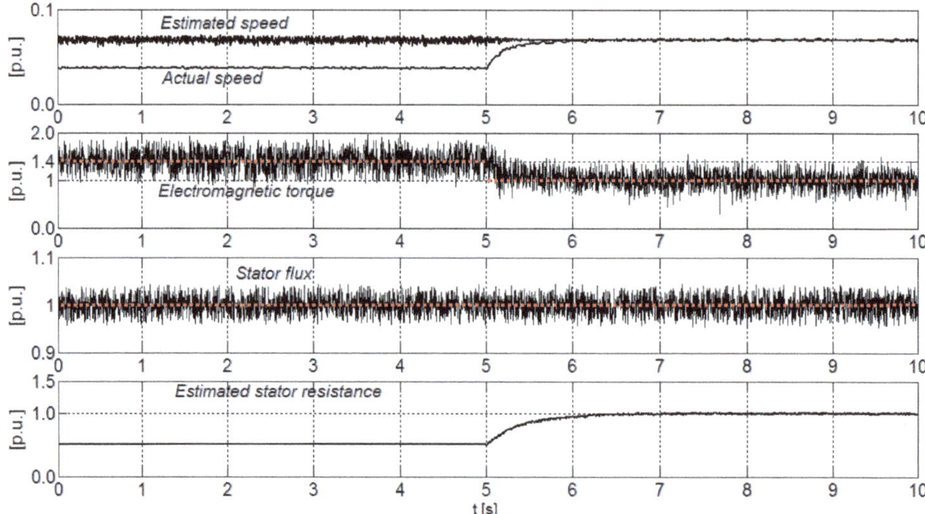

Figure 6. Experimental results of the proposed parallel estimation system under initial mismatch of stator resistance.

As already mentioned, the proposed parallel estimation system has the merit of avoiding overlap between stator resistance and rotor speed estimators, whereby the stator resistance is independently estimated from rotor speed using additional freedom degrees of 6PIM. The experimental results of estimated stator resistance under speed changes and load change are shown in Figure 7a,b, respectively. In Figure 7a, the speed command is changed as a step function from a very low speed to 17% rated speed, and, in Figure 7b, a load torque is suddenly applied to the motor at $t = 2$ s. It can be clearly adjudged that the adaptation process of stator resistance is independent of speed and load torque changes.

Disturbance-free operation of the ADRC-based speed controller is evaluated through a comparative study of its performance and the conventional PI regulator. The experimental results for the estimated speed under sudden load torque changes at 7% rated speed when the conventional PI and introduced ADRC are utilized as speed controllers are shown in Figure 8. As can be seen, applying the external load torque to the 6PIM leads to a larger overshoot (undershoot), when the conventional PI regulator is employed. The ADRC properly improves the disturbance rejecting capability, which in turn provides a robust performance against load torque changes.

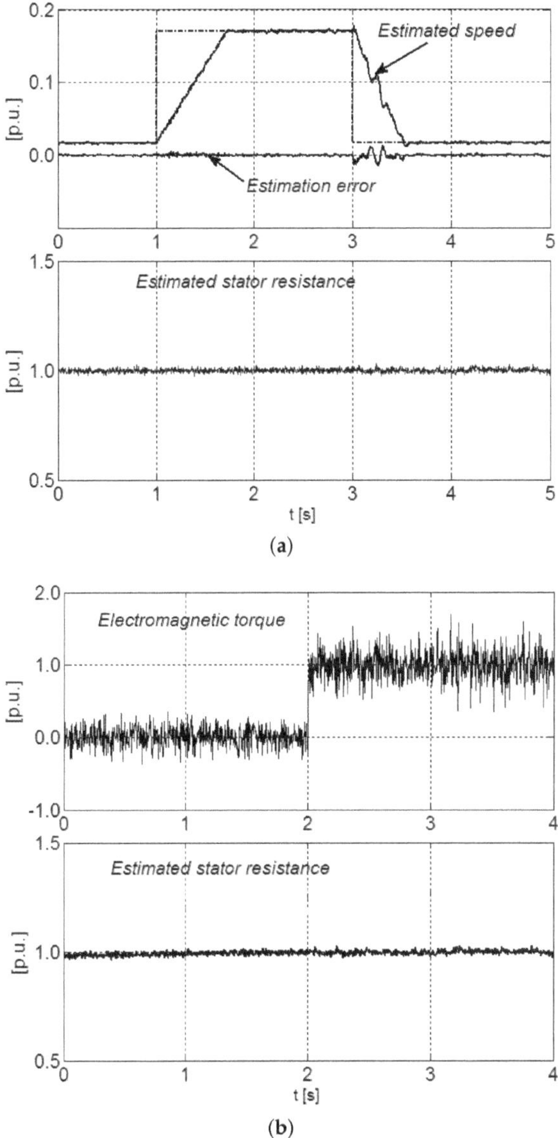

Figure 7. Experimental results of the estimated stator resistance under (**a**) speed changes (**b**) load torque change.

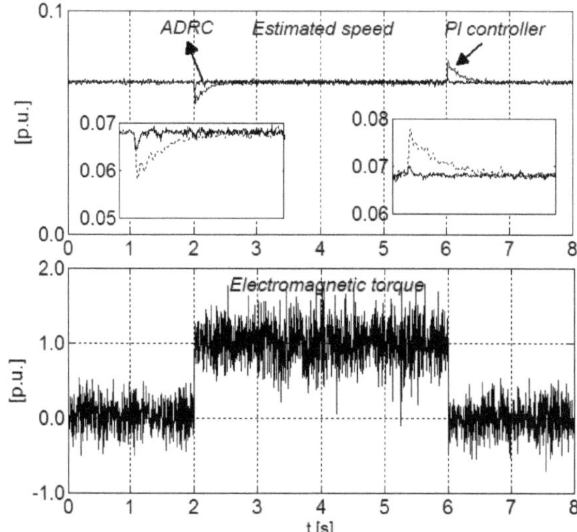

Figure 8. Experimental results of the estimated speed with PI and ADRC-based speed controllers under load changes.

6. Conclusions

Multiphase electrical machines and drives have different advantages over their traditional three phase counterparts. In recent years, multiple research works have been published to explore the specific advantages of multiphase machines and drives. In this regard, a parallel estimation system of the stator resistance and the rotor speed for direct torque-controlled 6PIM was proposed in this paper. The speed estimator is based on an adaptive full-order observer, which estimates the speed signal using the 6PIM model in the $\alpha - \beta$ subspace, while the stator resistance estimator employs the 6PIM model in the $z_1 - z_2$ subspace. Hence, the stator resistance is identified independently of the rotor speed. The rotor speed- and the stator resistance-adaptation laws were derived using the Lyapunov stability theorem. The performance of the proposed sensorless DTC was experimentally investigated, where the obtained results confirmed its capabilities in terms of accuracy as well as no overlap between the stator resistance and the rotor speed estimators. In order to provide a robust performance for the DTC technique against external load torques, the PI regulator was replaced by an ADRC, as a well-known disturbance-free controller. The better performance of the DTC scheme based on ADRC was verified through a comparative study with the conventional PI regulator.

Author Contributions: Conceptualization, methodology, validation, formal analysis, writing–original draft preparation, H.H., M.H.H.; writing—review and editing, resources, A.R.; project administration, T.V.; investigation, A.K.; funding acquisition, D.V.L.; supervision, A.B. All authors have read and agreed to the published version of the manuscript.

Funding: The research has been supported by the Estonian Research Council under grant PSG453 "Digital twin for propulsion drive of autonomous electric vehicle" and was financially supported by Government of Russian Federation (Grant 08-08).

Conflicts of Interest: The authors declare no conflict of interest.

Appendix A. The Design of Adaption Law for Stator Resistance Estimation

The quadratic Lyapunov function for asymptotic stability of the proposed stator resistance estimation system is defined as

$$V_r = e_r{}^T e_r + \frac{\Delta R_s^2}{\lambda_r} \tag{A1}$$

where λ_r is a positive constant, $\Delta R_s = \hat{R}_s - R_s$, $\hat{R}_s$ is the estimated stator resistance, R_s is the real stator resistance, and e_r is the error matrix of the state variables in the $z_1 - z_2$ subspace as

$$e_r = x_2 - \hat{x}_2 = \begin{bmatrix} i_{sz1} - \hat{i}_{sz1} & i_{sz2} - \hat{i}_{sz2} \end{bmatrix}^T \tag{A2}$$

The asymptotic stability of the stator resistance estimator is assured when the Lyapunov candidate function V_r is positive definite as well as its time derivative pV_r is negative definite. The time derivative of the Lyapunov candidate function is calculated as

$$pV_r = e_r^T pe_r + pe_r^T e_r + \frac{2}{\lambda_r}\Delta R_s p\hat{R}_s \tag{A3}$$

With some mathematical manipulation, Equation (A3) can be written as

$$pV_r = e_r^T(A_2 + A_2^T)e_r - [e_r^T \Delta A_2 \hat{x} + \hat{x}_2^T \Delta A_2^T e_r]$$
$$+ \frac{2}{\lambda_r}\Delta R_s p\hat{R}_s \tag{A4}$$

The first term of Equation (A4) is inherently negative definite. The stability of the system is eventually assured, when the sum of the last two terms of Equation (A4) is zero as

$$\frac{2}{\lambda_r}\Delta R_s p\hat{R}_s - [e_r^T \Delta A_2 \hat{x}_2 + \hat{x}_2^T \Delta A_2^T e_r] = 0 \tag{A5}$$

which leads to

$$\hat{R}_s = -\frac{\lambda_r}{2}\int \epsilon_{R_s} dt \tag{A6}$$

where the tuning signal ϵ_{R_s} is

$$\epsilon_{R_s} = \hat{i}_{sz1}(i_{sz1} - \hat{i}_{sz1}) + \hat{i}_{sz2}(i_{sz2} - \hat{i}_{sz2}) \tag{A7}$$

A PI regulator is employed to enhance the dynamic behaviour of the proposed estimator, instead of Equation (A6) as

$$\hat{R}_s = K_{pr}\epsilon_{R_s} + K_{ir}\int \epsilon_{R_s} dl \tag{A8}$$

where K_{ir} and K_{pr} are the integral and proportional constants.

Appendix B. The Design of Adaption Law for Speed Estimation

The Lyapunov candidate function for asymptotic stability of the speed estimation system is

$$V_\omega = e_\omega^T e_\omega + \frac{\Delta\omega_r^2}{\lambda_\omega} \tag{A9}$$

where λ_ω is a positive constant, $\Delta\omega_r = \hat{\omega}_r - \omega_r$, and e_ω is the error matrix of the estimated and real values in $\alpha - \beta$ subspace as

$$e_\omega = x_1 - \hat{x}_1 \tag{A10}$$
$$= \begin{bmatrix} i_{s\alpha} - \hat{i}_{s\alpha} & i_{s\beta} - \hat{i}_{s\beta} & \psi_{r\alpha} - \hat{\psi}_{r\alpha} & \psi_{r\beta} - \hat{\psi}_{r\beta} \end{bmatrix}^T$$

In this case, the first-order time derivative of Lyapunov function can be deduced as

$$pV_\omega = e_\omega^T[(A_1 - G_1C_1) + (A_1 - G_1C_1)^T]e_\omega \tag{A11}$$
$$+ (e_\omega \Delta A_2 \hat{x}_2 + \hat{x}_2 \Delta A^T e_\omega) + \frac{2}{\lambda_\omega} \Delta \omega_r p\hat{\omega}_r$$

The first term of Equation (A11) is guaranteed to be negative definite by suitable adopting of observer gain matrix G_1. The Lyapunov stability criterion is satisfied, if the sum of second and third terms of Equation (A11) is zero. With some calculations, the adaptation law for speed estimator is acquired as

$$\hat{\omega}_r = K_{p\omega}\epsilon_\omega + K_{i\omega}\int \epsilon_\omega dt \tag{A12}$$

where the tuning signal ϵ_ω is

$$\epsilon_\omega = (i_{s\alpha} - \hat{i}_{s\alpha})\hat{\psi}_{r\beta} - (i_{s\beta} - \hat{i}_{s\beta})\hat{\psi}_{r\alpha} \tag{A13}$$

References

1. Levi, E.; Bojoi, R.; Profumo, F.; Toliyat, H.A.; Williamson, S. Multiphase induction motor drives—A technology status review. *IET Electr. Power Appl.* **2007**, *1*, 489–516. [CrossRef]
2. Heidari, H.; Rassõlkin, A.; Vaimann, T.; Kallaste, A.; Taheri, A.; Holakooie, M.H.; Belahcen, A. A novel vector control strategy for a six-phase induction motor with low torque ripples and harmonic currents. *Energies* **2019**, *12*, 1102. [CrossRef]
3. Levi, E. Multiphase Electric Machines for Variable-Speed Applications. *IEEE Trans. Ind. Electron.* **2008**, *55*, 1893–1909. [CrossRef]
4. Holakooie, M.H.; Ojaghi, M.; Taheri, A. Direct Torque Control of Six-Phase Induction Motor With a Novel MRAS-Based Stator Resistance Estimator. *IEEE Trans. Ind. Electron.* **2018**, *65*, 7685–7696. [CrossRef]
5. Che, H.S.; Duran, M.J.; Levi, E.; Jones, M.; Hew, W.P.; Rahim, N.A. Postfault Operation of an Asymmetrical Six-Phase Induction Machine With Single and Two Isolated Neutral Points. *IEEE Trans. Power Electron.* **2014**, *29*, 5406–5416. [CrossRef]
6. Lu, H.; Li, J.; Qu, R.; Ye, D. Fault-tolerant predictive current control with two-vector modulation for six-phase permanent magnet synchronous machine drives. *IET Electr. Power Appl.* **2018**, *12*, 169–178. [CrossRef]
7. Holakooie, M.H.; Ojaghi, M.; Taheri, A. Modified DTC of a Six-Phase Induction Motor With a Second-Order Sliding-Mode MRAS-Based Speed Estimator. *IEEE Trans. Power Electron.* **2019**, *34*, 600–611. [CrossRef]
8. Bojoi, R.; Farina, F.; Griva, G.; Profumo, F.; Tenconi, A. Direct torque control for dual three-phase induction motor drives. *IEEE Trans. Ind. Appl.* **2005**, *41*, 1627–1636. [CrossRef]
9. Singh, G.K.; Nam, K.; Lim, S.K. A simple indirect field-oriented control scheme for multiphase induction machine. *IEEE Trans. Ind. Electron.* **2005**, *52*, 1177–1184. [CrossRef]
10. Gregor, R.; Barrero, F.; Toral, S.L.; Duran, M.J.; Arahal, M.R.; Prieto, J.; Mora, J.L. Predictive-space vector PWM current control method for asymmetrical dual three-phase induction motor drives. *IET Electr. Power Appl.* **2010**, *4*, 26–34. [CrossRef]
11. Zaid, S.A.; Mahgoub, O.A.; El-Metwally, K.A. Implementation of a new fast direct torque control algorithm for induction motor drives. *IET Electr. Power Appl.* **2010**, *4*, 305–313. [CrossRef]
12. Hoang, K.D.; Ren, Y.; Zhu, Z.Q.; Foster, M. Modified switching-table strategy for reduction of current harmonics in direct torque controlled dual-three-phase permanent magnet synchronous machine drives. *IET Electr. Power Appl.* **2015**, *9*, 10–19. [CrossRef]
13. Tatte, Y.N.; Aware, M.V. Torque Ripple and Harmonic Current Reduction in a Three-Level Inverter-Fed Direct-Torque-Controlled Five-Phase Induction Motor. *IEEE Trans. Ind. Electron.* **2017**, *64*, 5265–5275. [CrossRef]
14. Rassõlkin, A.; Vaimann, T.; Kallaste, A.; Kuts, V. Digital twin for propulsion drive of autonomous electric vehicle. In Proceedings of the 2019 IEEE 60th International Scientific Conference on Power and Electrical Engineering of Riga Technical University (RTUCON), Riga, Latvia, 7–9 October 2019; pp. 1–4.

15. Kuts, V.; Tahemaa, T.; Otto, T.; Sarkans, M.; Lend, H. Robot manipulator usage for measurement in production areas. *J. Mach. Eng.* **2016**, *16*, 57–67.

16. Sell, R.; Leier, M.; Rassõlkin, A.; Ernits, J.P. Self-driving car ISEAUTO for research and education. In Proceedings of the 2018 19th International Conference on Research and Education in Mechatronics (REM), Delft, The Netherlands, 7–8 June 2018; pp. 111–116.

17. Rassõlkin, A.; Sell, R.; Leier, M. Development case study of the first estonian self-driving car, iseauto. *Electr. Control Commun. Eng.* **2018**, *14*, 81–88. [CrossRef]

18. Kumar, R.; Das, S.; Syam, P.; Chattopadhyay, A.K. Review on model reference adaptive system for sensorless vector control of induction motor drives. *IET Electr. Power Appl.* **2015**, *9*, 496–511. [CrossRef]

19. Gadoue, S.M.; Giaouris, D.; Finch, J.W. Stator current model reference adaptive systems speed estimator for regenerating-mode low-speed operation of sensorless induction motor drives. *IET Electr. Power Appl.* **2013**, *7*, 597–606. [CrossRef]

20. Rashed, M.; Stronach, A.F. A stable back-EMF MRAS-based sensorless low-speed induction motor drive insensitive to stator resistance variation. *IEE Proc. Electr. Power Appl.* **2004**, *151*, 685–693. [CrossRef]

21. Holakooie, M.H.; Taheri, A.; Sharifian, M.B. MRAS Based Speed Estimator for Sensorless Vector Control of a Linear Induction Motor with Improved Adaptation Mechanisms. *J. Power Electron.* **2015**, *15*, 1274–1285. [CrossRef]

22. Holakooie, M.H.; Ojaghi, M.; Taheri, A. Full-order Luenberger observer based on fuzzy-logic control for sensorless field-oriented control of a single-sided linear induction motor. *ISA Trans.* **2016**, *60*, 96–108. [CrossRef]

23. Habibullah, M.; Lu, D.D.C. A Speed-Sensorless FS-PTC of Induction Motors Using Extended Kalman Filters. *IEEE Trans. Ind. Electron.* **2015**, *62*, 6765–6778. [CrossRef]

24. Lee, K.B.; Blaabjerg, F. Sensorless DTC-SVM for Induction Motor Driven by a Matrix Converter Using a Parameter Estimation Strategy. *IEEE Trans. Ind. Electron.* **2008**, *55*, 512–521. [CrossRef]

25. Lascu, C.; Boldea, I.; Blaabjerg, F. Direct torque control of sensorless induction motor drives: A sliding-mode approach. *IEEE Trans. Ind. Appl.* **2004**, *40*, 582–590. [CrossRef]

26. Han, J. From PID to Active Disturbance Rejection Control. *IEEE Trans. Ind. Electron.* **2009**, *56*, 900–906. [CrossRef]

27. Mahmudizad, M.; Ahangar, R.A. Improving load frequency control of multi-area power system by considering uncertainty by using optimized type 2 fuzzy pid controller with the harmony search algorithm. *World Acad. Sci. Eng. Technol.* **2016**, *10*, 1051–1061.

28. Wang, G.; Wang, B.; Li, C.; Xu, D. Weight-transducerless control strategy based on active disturbance rejection theory for gearless elevator drives. *IET Electr. Power Appl.* **2017**, *11*, 289–299. [CrossRef]

29. Li, J.; Ren, H.P.; Zhong, Y.R. Robust Speed Control of Induction Motor Drives Using First-Order Auto-Disturbance Rejection Controllers. *IEEE Trans. Ind. Appl.* **2015**, *51*, 712–720. [CrossRef]

30. Alonge, F.; Cirrincione, M.; D'Ippolito, F.; Pucci, M.; Sferlazza, A. Robust Active Disturbance Rejection Control of Induction Motor Systems Based on Additional Sliding-Mode Component. *IEEE Trans. Ind. Electron.* **2017**, *64*, 5608–5621. [CrossRef]

31. Zhao, Y.; Lipo, T.A. Space vector PWM control of dual three-phase induction machine using vector space decomposition. *IEEE Trans. Ind. Appl.* **1995**, *31*, 1100–1109. [CrossRef]

32. Kubota, H.; Matsuse, K. Speed sensorless field-oriented control of induction motor with rotor resistance adaptation. *IEEE Trans. Ind. Appl.* **1994**, *30*, 1219–1224. [CrossRef]

33. Wang, K.; Chen, B.; Shen, G.; Yao, W.; Lee, K.; Lu, Z. Online updating of rotor time constant based on combined voltage and current mode flux observer for speed-sensorless AC drives. *IEEE Trans. Ind. Electron.* **2014**, *61*, 4583–4593. [CrossRef]

34. Yin, Z.; Zhang, Y.; Du, C.; Liu, J.; Sun, X.; Zhong, Y. Research on Anti-Error Performance of Speed and Flux Estimation for Induction Motors Based on Robust Adaptive State Observer. *IEEE Trans. Ind. Electron.* **2016**, *63*, 3499–3510. [CrossRef]

Article

Adaptive Maximum Torque per Ampere Control of Sensorless Permanent Magnet Motor Drives

Anton Dianov [1] and Alecksey Anuchin [2,*]

[1] Home Appliances Division, Samsung Electronics, Suwon 16677, Korea; anton.dianov@samsung.com
[2] Electric Drives Department, Moscow Power Engineering Institute, 111250 Moscow, Russia
* Correspondence: anuchinas@mpei.ru

Received: 16 August 2020; Accepted: 24 September 2020; Published: 27 September 2020

Abstract: Interior permanent magnet synchronous motor (IPMSM) efficiency can be improved by using maximum torque per ampere control (MTPA). MTPA control utilizes both alignment and reluctance torques and usually requires information about the magnetization map of the electrical machine. This paper proposes an adaptive MTPA algorithm for sensorless control systems of IPMSM drives, which is applicable in industrial and commercial drives. This algorithm enhances conventional control schemes, where the output of the speed controller is the commanded stator current and the direct current is calculated using an MTPA equation; therefore, it can be easily implemented in the previously developed drives. The proposed algorithm does not use any motor parameters for the calculation of the MTPA trajectory, which is important for systems operating in changing environmental conditions, because motor inductances and flux linkage strongly depend on the stator current and the rotor temperature, respectively. The proposed algorithm continuously varies the current phase and in such a way it tries to minimize the magnitude of the stator current at the applied load torque. The main contribution of this paper is the development of a technique to overcome the main disadvantage of seeking algorithms–the necessity of a precision information about the rotor position. The proposed method was verified experimentally.

Keywords: interior permanent magnet motors; maximum torque per ampere; sensorless control; adaptive control

---

1. Introduction

Interior permanent magnet synchronous motors (IPMSM), compared with machines of other types, have higher torque to weight ratios, higher efficiency, output power per volume and mass per volume values, which make them attractive for use in compact drives, high-efficient drives, drives with high dynamics, etc. At the same time, the high price of rare-earth metals, which are necessary for producing strong magnets, restricts the popularity of permanent magnet (PM) motors. However, over the past decade, the price of rare-earth magnets has decreased; therefore, the area usage of PM motors is widening. As a result, they attract more attention, and many researchers have investigated the control systems of these machines.

The main feature of IPMSMs is their asymmetry along direct and quadrature axes, which creates reluctance torque. At the same time, permanent magnet synchronous machines (PMSM), which have equal direct and quadrature inductance and idle load conditions, demonstrate magnetic asymmetry at load; thus, they may also produce reluctance torque. As a result, modern efficient control systems must consider these facts and utilize the reluctance torque of the PM motors by employing one of the maximum torque per ampere (MTPA) techniques.

Information on rotor position is required to control PM motors; therefore, precision and high-performance drives are equipped with position encoders, whose resolution depends on the desired dynamic and precision of control. At the same time, low-cost systems and motor drives with higher reliability have a tendency to eliminate additional parts, especially moving parts, such as speed and position encoders. Therefore, sensorless control algorithms have almost become a standard in these applications [1,2]. Therefore, modern control systems of PM motors, in order to be used in a variety of applications, have to be sensorless [3] and must be able to implement MTPA techniques [4,5].

An analysis of the operating conditions of PM motors and their impact on the motor parameters showed that motor direct and quadrature inductances strongly depend on the motor stator current and may decrease due to steel saturation by more than 50% [6]. At the same time, the temperature of the rotor impacts the flux linkage of magnets and may decrease it by 10% [7]. Furthermore, magnet degradation during the lifetime of the motor may also decrease the flux linkage by 15%. As a result, it would be beneficial to develop an MTPA algorithm, which can adapt to the variations in motor parameter and provide efficient control of the motor, despite its environment.

The conventional MTPA approach involves the calculation of one of the MTPA equations—e.g., Equation (1) obtained from motor equations:

$$i_d = -\frac{\psi_m}{4(L_d - L_q)} - \sqrt{\frac{\psi_m^2}{16(L_d - L_q)^2} + \frac{I_s^2}{2}},\tag{1}$$

where I_s represents the stator current, ψ_m represents the permanent magnet flux linkage, L_d and L_q represent the d-axis and q-axis inductances, respectively, and i_d stands for the direct current component providing the MTPA. This approach is simple and can be easily implemented in a sensorless control systems and is discussed and studied in [8–12]; however, such techniques are sensitive to the variation in motor parameter due to operating conditions. For example, the accurate knowledge of motor parameters was required in [10], but the change in the motor inductances due to the saturation effect was not taken into account.

To solve this problem, different adaptive MTPA techniques were proposed. The authors of [13,14] proposed to enhance the conventional MTPA algorithm with on-line estimation techniques of motor inductances; however, these papers do not suggest a solution for the flux-linkage estimation. Furthermore, these methods need fine-tuning, and the control system has to be equipped with a high-speed processor capable of executing additional calculations at every calculation step, together with basic control routines.

The authors of [15] proposed a method with a fast dynamic response, which uses a recursive least squares (RLS) parameters estimator to track the MTPA trajectory. However, this method calculates many square roots, which significantly load a microcontroller unit (MCU), even with optimizations; therefore, the use of this method is limited.

A group of methods described in [16,17] proposes several similar MTPA techniques which are based on the high-frequency signal injection and the analysis of response. These methods do not need motor parameters, but high-frequency signals cause noises and vibrations, which are undesirable in many drives.

To overcome this problem, the authors of [18–21] proposed an interesting technique called the virtual signal injection (VSI). This method detects the MTPA trajectory analytically by the injection of a virtual signal into a motor model. It does not use motor parameters and does not inject real signals into the system; therefore, undesired noise and vibrations are excluded. However, despite perfect reported results of VSI methods, we do not share the optimism of the authors. We found that these algorithms were very sensitive to the variation in stator resistance, which is not a problem in other algorithms, including the conventional one.

Another approach used for tracking the MTPA trajectory is seeking algorithms, which do not use motor parameters and can effectively operate in a changing environment. An example of this technique is described in [22], where the authors continuously varied the phases of the stator current and tracked the minimum of the current magnitude. The main disadvantages of this idea are lower dynamics and the necessity of a position encoder.

After a detailed analysis of the pros and cons of the existing techniques, the authors found that the seeking algorithm reported in [22] is the best candidate for developing a motor drive, provided that it can be adapted to the operation without a position encoder.

2. MTPA Seeking Algorithm

The seeking algorithm reported in [22], which was selected for further improvements, continuously varies the phase of the stator current γ to provide the minimum stator current I_s for the given torque. The flowchart of this method is shown in Figure 1. It can be clearly seen that in each calculation step, the motor phase is modified by a small disturbance angle $\Delta\gamma$, and the resulting value γ is checked to be inside the limits. After that, the new value of the phase of the stator current is applied, and the control system waits until the end of transient. After that, the tuning algorithm measures the average magnitude of the stator current over the calculation step and compares it to the value measured at the previous step.

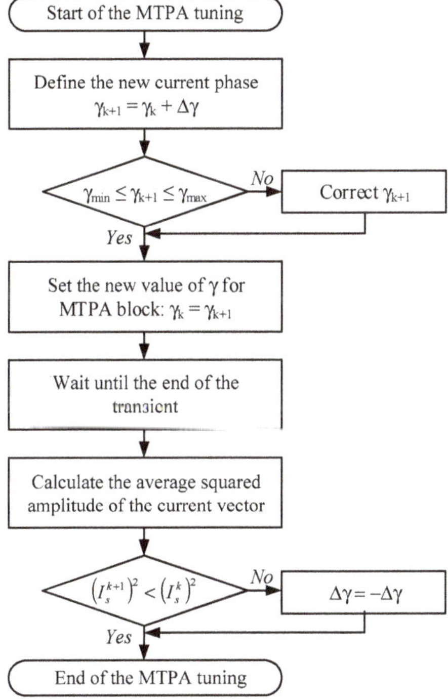

Figure 1. Flowchart of the MTPA seeking algorithm.

If the current value obtained at the current calculation step is less than the same value from the previous step, the stator current has been rotated in the correct direction, and vice versa. If the stator current has been rotated to the incorrect direction, the sign of disturbance value $\Delta\gamma$ is reversed, and in the next step the stator vector will be rotated in the proper direction. This process is illustrated in Figure 2, where the current vector rotates to track the constant torque loci.

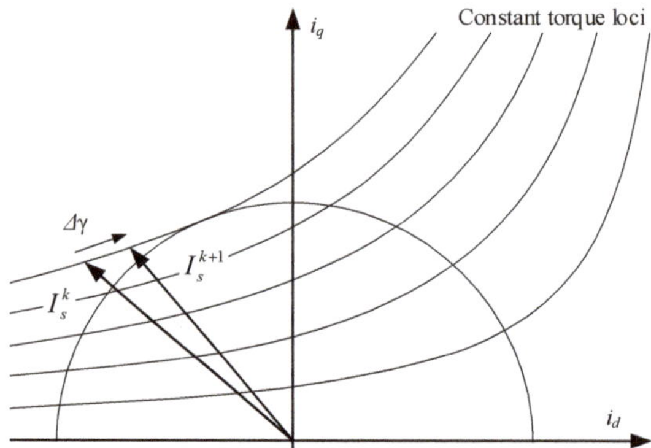

Figure 2. Variation of stator current phase.

This seeking algorithm has the advantages of being insensitive to motor parameter variation and the possibility of being easily implemented in the previously developed control schemes.

At the same time, this technique demonstrates excellent results only when the rotor position is measured precisely, and may fail when a significant error appear. As can be seen from Figure 2, the minimum current for the constant torque curve is not strongly pronounced, and the variation of the stator current angle causes only minor changes to the current magnitude. This problem is illustrated by the data in Table 1, calculated for the test motor, the parameters of which are given in the section below. The table illustrates the increase in the magnitude of the stator current when its angle varies with the step of one degree. As can be seen, the variation in stator current is quite small and lies below one percent for a range of ±5°. Therefore, to detect the minimum stator current, it is very important to know the rotor position precisely. Unfortunately, sensorless drives contain a position error with a typical value of 5°–10°, which varies over revolution and restricts the operation of the seeking algorithm in sensorless systems. The original algorithm [22] at the calculation step k applies the stator current with a phase γ_k. In the next calculation step $k + 1$, the algorithm applies the stator current with a phase γ_{k+1}, which differs from the γ_k at the fixed disturbance angle $\Delta\gamma$. During each calculation step, the algorithm measures (integrates) the magnitude of the stator current and then compares these magnitudes. The lower current magnitude corresponds to the phase angle being closer to the true MTPA angle. Thus, if the position error is not constant, the stator current is applied at different phase angles during each calculation step; therefore, its magnitude varies, producing incorrect measurements of the current. In order to overcome this problem and use the advantages of the seeking technique, an advanced method was proposed, which makes the operation of the seeking algorithm in sensorless drives possible.

Table 1. Increase in Stator Current due to Angle Variation.

Angle	Increase, %	Angle	Increase, %
$\gamma + 1°$	0.02	$\gamma - 1°$	0.02
$\gamma + 2°$	0.09	$\gamma - 2°$	0.10
$\gamma + 3°$	0.19	$\gamma - 3°$	0.22
$\gamma + 4°$	0.33	$\gamma - 4°$	0.39
$\gamma + 5°$	0.56	$\gamma - 5°$	0.62

3. Proposed Enhanced Algorithm

As mentioned earlier, the main problem with the implementation of the seeking algorithms in the sensorless systems is the absence of precision of information on the rotor position. The typical position estimation error of the back-EMF-based estimator is shown in Figure 3. This picture demonstrates that the estimation error is significant for the seeking technique described. The proposed algorithm belongs to the perturb and observe methods, which involves modifying one parameter of the system and analyzing its response by measuring another parameter. Algorithms such as these may fail if another disturbance appears in the system and impacts the measured parameter. At the same time, our experiments showed that the average value of the position estimation error is stable and mainly depends on the variation in the relationship between the direct and quadrature inductances, while instant error depends on disturbance factors, such as cyclic mechanical load, non-sinusoidal back-emf, etc. Therefore, the previously developed seeking algorithm may operate properly if its calculation step contains an integer number of electrical revolutions. In that case, the average position error at consequent calculation steps will be the same, and the average current magnitudes may be compared. Stator resistance variation due to temperature change affects the average error value, but the seeking algorithm compensates for this error.

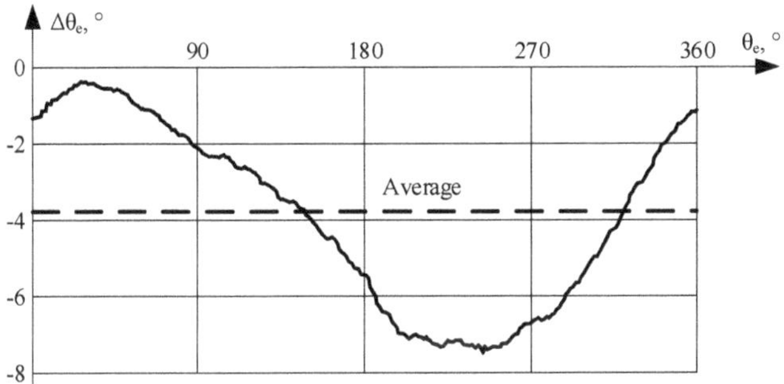

Figure 3. Rotor position estimation error.

The number of electrical revolutions that are contained in one calculation step is denoted as N. Then, the maximum calculation time is denoted as T_{max}. These parameters define the minimum motor speed n_{min}, where the proposed algorithm can operate. If the motor speed n is less than the minimum speed n_{min}, the tuning algorithm must be stopped. Then, the length of the current calculation step T_{cs}^k is defined and compared to the length of the previous calculation step T_{cs}^{k-1}. If they are the same, the calculations at these intervals may be compared, otherwise they may not. If the consecutive calculation steps are different, the tuning algorithm calculates the squared amplitude of the stator current for the current measurement interval $\left(I_s^k\right)^2$ and proceeds to the next step. If the length of the previous calculation step and the length of the current calculation step are the same, the tuning algorithm defines the new MTPA angle γ in the same manner as a basic algorithm, measures the squared amplitude of stator current $\left(I_s^k\right)^2$, and compares it to the same value from the previous iteration $\left(I_s^{k-1}\right)^2$. If the squared amplitude of the stator current at the current step is less, it means that the MTPA angle γ was modified in the correct direction, and the same disturbance value will be applied in the next step. If the MTPA angle γ was modified in the wrong direction, then the sign of disturbance value $\Delta\gamma$ in the next step will be reversed. A flowchart of the proposed algorithm is shown in Figure 4.

This algorithm is quite simple, and the most important things are the proper selection of N and T_{max}, which define the errors, the dynamic response of the algorithm, and its minimum operating speed. The higher the number of electrical revolutions in the calculation step, the more reliable and stable the algorithm operates; however, at the same time, its dynamic response decreases. These parameters are suggested to be selected experimentally by monitoring the performance of the tuning algorithm.

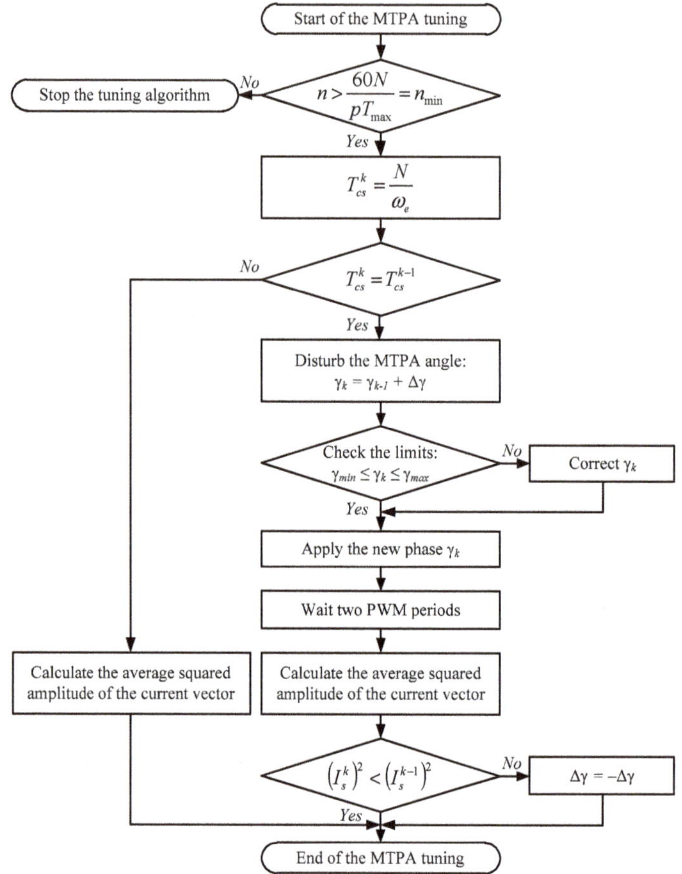

Figure 4. Flowchart of the proposed algorithm.

4. Experimental Setup

The experimental motor used in the experiments is the mass production (MP) device, the parameters of which are given in Table 2. However, these inductances strongly depend on the motor current and vary, as shown in Figure 5.

Table 2. Motor Rated Parameters.

Parameter	Value	Units
Number of poles	2P = 6	-
Rated speed	2000	rpm
Phase resistance	1.5	Ohm
d-axis inductance	54	mH
q-axis inductance	95	mH
Back-EMF constant	0.15	V·s/rad

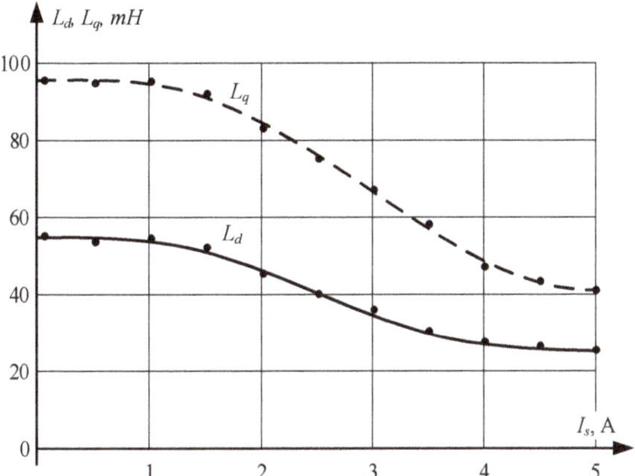

Figure 5. Motor inductances vs. current.

The control system used in the experiment is the same as in MP versions described in [23] (Figure 6). It drives the IPM motor, using the sensorless control, which nowadays can be considered to be a standard for many appliances. This control is based on the back-EMF estimation methods discussed in [24], which is enhanced by the initial position estimation necessary for excluding the reverse rotation while starting. The performance of the implemented estimation algorithm was verified using a quadrature encoder, which proved that the algorithm perfectly operates in the speed range over 10 Hz, with the estimation error being not more than several electrical degrees.

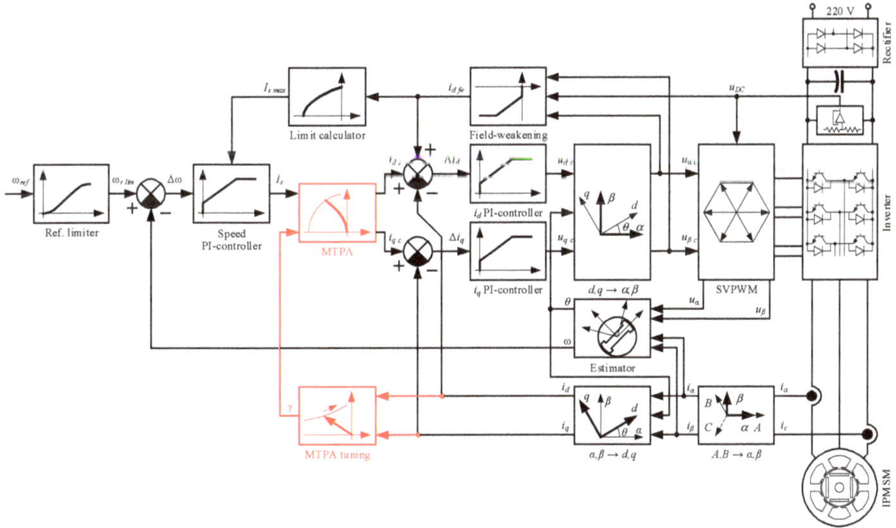

Figure 6. Structure of the sensorless control scheme of the IPM motor drive.

The inverter used for driving the motor is based on the smart power module FSAM10SH60 from "Fairchild" (10 A/600 V), which contains six IGBTs and embedded gate drivers. This drive was developed for a standard 220–240 V, 50/60 Hz supply source. The control system of the drive under test is based on a 60 MIPS Cortex-M3 microcontroller, which operates the inverter at 10 kHz PWM.

This system is equipped with two current sensors and a DC-link voltage sensor, whose signals are processed by a 12-bit ADC of the microcontroller, with a sampling time of 100 μs.

The control system of the experimental drive is a conventional vector control system without the position encoder used in MP devices. It involves an outer speed loop and two inner current loops implemented in the *dq* reference frame, where the electrical position and speed are provided by the estimator. The experimental drive implements open-loop starting and acceleration with immediate closing and the reinitialization of the controllers.

The control system measures two phase currents and DC-link voltage, which are then transformed into phase currents and voltages, respectively. After that, the three phase values, *abc*, are converted into two axis stationary reference frames $\alpha\beta$, using the Clarke transformation:

$$\begin{bmatrix} X_\alpha \\ X_\alpha \end{bmatrix} = \frac{2}{3} \begin{bmatrix} 1 & -\frac{1}{2} & -\frac{1}{2} \\ 0 & \frac{\sqrt{3}}{2} & -\frac{\sqrt{3}}{2} \end{bmatrix} \begin{bmatrix} X_a \\ X_b \\ X_c \end{bmatrix} \tag{2}$$

where X denotes any converted value. The conversion gain of 2/3 provides equality of amplitudes in *abc* and $\alpha\beta$ reference frames, which is easier for tuning. Then, the values are transformed into a synchronous *dq* reference frame using the Park transformation:

$$\begin{bmatrix} X_d \\ X_q \end{bmatrix} = \begin{bmatrix} \cos\theta & \sin\theta \\ -\sin\theta & \cos\theta \end{bmatrix} \begin{bmatrix} X_\alpha \\ X_\alpha \end{bmatrix} \tag{3}$$

where θ represents the angle of angular displacement.

The control system uses a field-weakening controller, which increases the maximum speed by up to +50% of the rated velocity by weakening the field of the rotor with i_d current. The drive under test also includes an MTPA block for increasing efficiency and decreasing stator current. This MTPA block receives the stator current from the speed controller and then converts it into direct and quadrature components.

The only difference between the experimental and the conventional systems is the presence of the MTPA tuning block with the proposed algorithm, which outputs the MTPA angle of the decomposition of the commanded stator current. These changes, including additional block and corresponding connections, are shown in red in Figure 6.

5. Experimental Results

5.1. Experimental Setup and Load Motor

The test jig used in our experiments included a load motor, represented by an HG-SR202, 2 kW AC servomotor from Mitsubishi Electric, which was equipped with an incremental position encoder. This motor was controlled using the MR-J4-200, an AC servo amplifier from Mitsubishi Electric, operated in the torque control mode. The AC servo amplifier and the inverter were connected to the PC, which was used to control the experiment and monitoring the data. This experimental setup is shown in Figure 7.

5.2. Motor Characteristics

At the beginning of our experimental work, the real characteristics of the motor were found. This experiment was performed using the test jig shown in Figure 7, where an incremental position encoder with a resolution of 4096 pulses was used. Motor inductances were found in several points for different values of stator current at the MTPA condition. They are shown in Figure 5, which demonstrates that saturation significantly impacts their values.

The motor MTPA characteristics were detected with a step of 1°, and they are demonstrated in Figure 8. It can be clearly seen that the experimental MTPA curve deviates from the theoretical curve at higher currents due to the saturation effect. Therefore, this fact must be taken into account when developing an efficient control system.

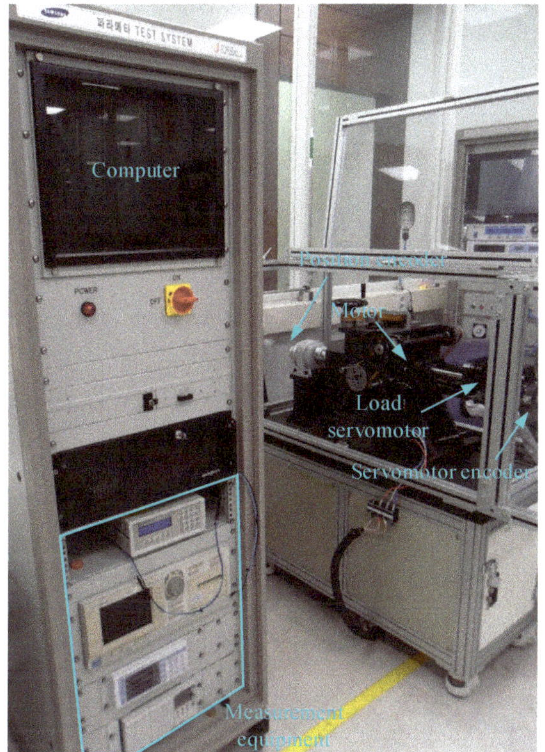

Figure 7. Experimental setup.

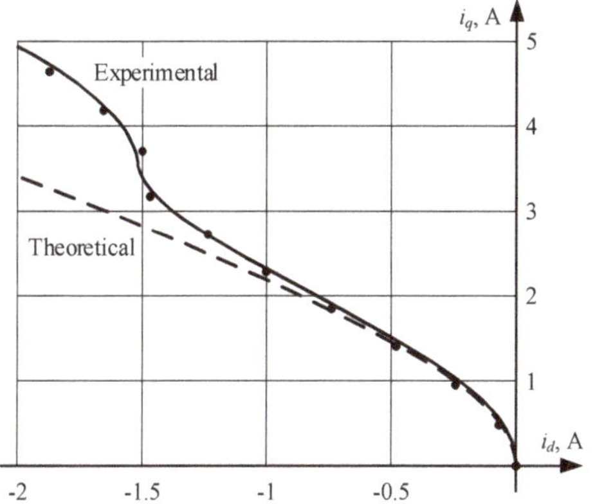

Figure 8. Theoretical and experimental MTPA curves.

5.3. The Performance of the Developed Algorithm

The proposed MTPA tuning algorithm was tested using the test bench described above. The load torque was programmed as a function with several steps, shown in Figure 9 so that the detection of the MTPA angle at different conditions can be dynamically monitored. A step-changing function with the following steps was used: 1, 2, 2.5, 3, 3.5, and 4 Nm. The values of the steps at lower currents are higher because the MTPA angle in this region changes faster, and it is easier to track it. At the same time, at a higher load, the steps are lower to check the behavior of the proposed algorithm in that region in more detail.

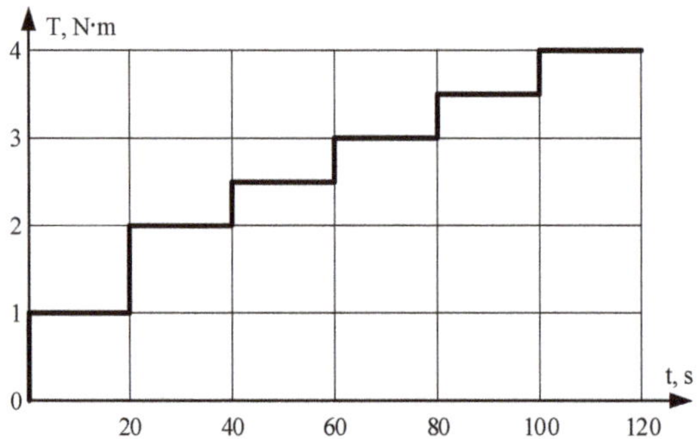

Figure 9. Commanded load torque.

When selecting the parameters of the algorithm, it was assumed that the MP sensorless drives rarely operate at a speed below 900 rpm, as they are focused on the total efficiency more than on a fast dynamic. At the same time, the higher number of electrical revolutions N used as a calculation step provides better stability and avoids side effects. Therefore, the maximum calculation time, as $T_{max} = 0.5$ s, and minimum operating speed as $n = 600$ rpm, were selected, which results in N from 15 revolutions or less. After several experiments, the disturbance angle $\Delta\gamma$ equal to 3° was selected, which was a compromise between precision and the algorithm's stability. The lower value of $\Delta\gamma$ makes it difficult to detect the current changes in our system; therefore, 3 degrees is a tradeoff value between the tolerance and quality of control. At the same time, in other systems, especially with motors of higher magnetic asymmetry, the lower values of $\Delta\gamma$ can be used.

The operation of a tuning algorithm at 900 rpm with $N = 15$ is shown in Figure 10, which demonstrates the proper detection of the MTPA angle. However, the defined value of γ contains some spikes. The results of the same experiment at 1500 rpm are presented in Figure 11, which proves the correct operation of the developed algorithm and demonstrates a lower number of spikes—i.e., it has higher stability. In the next experiment, the calculation step was increased to 30 electrical revolutions, and the results of this test are provided in Figure 12. It is observed that the stability of the proposed algorithm increased, and the defined MTPA angle almost did not contain significant deviations.

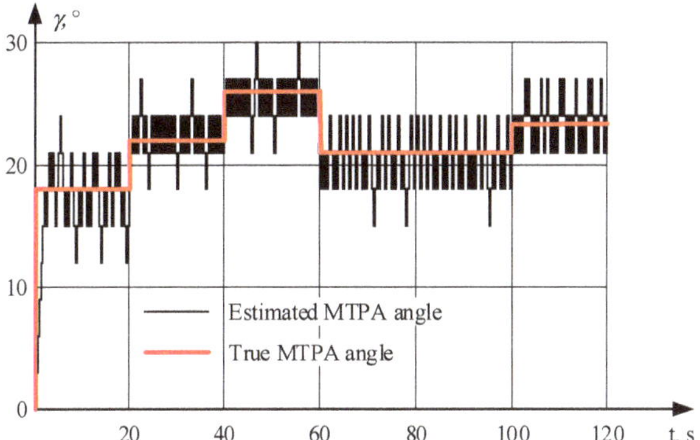

Figure 10. MTPA angle defined at 900 rpm with $N = 15$.

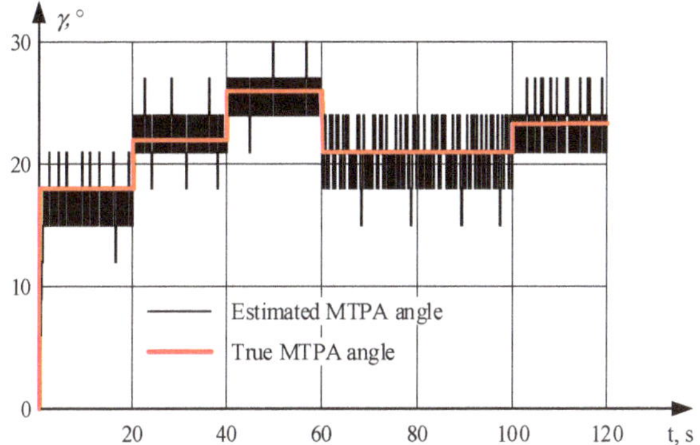

Figure 11. MTPA angle defined at 1500 rpm with $N = 15$.

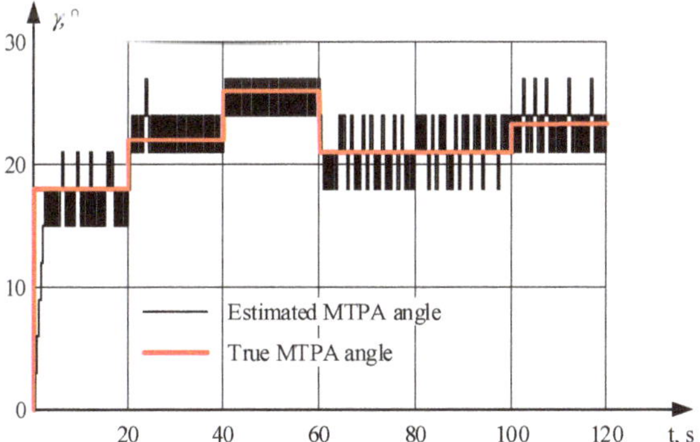

Figure 12. MTPA angle defined at 1500 rpm with $N = 30$.

6. Conclusions

This paper proposes the adaptive MTPA control algorithm capable of operating in sensorless drives. This algorithm does not use any motor parameters and conventional equations; therefore, it is insensitive to the motor parameter variation due to the operating conditions. The proposed method uses a seeking technique, which continuously varies the phase of the stator current and tracks the minimum of its magnitude. It is designed to be used in conventional control systems; therefore, it can be easily embedded in previously developed motor drives. Experimental results provided in this paper prove the feasibility of the developed algorithm and its perfect operation, despite the motor operating conditions. The proposed algorithm was implemented in the drives with reciprocating compressors and put into mass production.

Author Contributions: General idea, A.D.; Simulation software, A.D. and A.A.; Software implementation and experimental verification, A.D.; Methodology, A.A.; Data analysis, A.D. and A.A.; Writing original draft, A.D.; Writing—review and editing, A.A. Both authors have read and agreed to the published version of the manuscript.

Funding: This research is supported by the Russian Science Foundation grant (Project № 16-19-10618).

Conflicts of Interest: The authors declare no conflict of interest.

Nomenclature

AC	Alternating Current
ADC	Analog to Digital Converter
DC	Direct Current
EMF	Electromotive Force
IGBT	Insulated-Gate Bipolar Transistor
IPMSM	Interior Permanent Magnet Synchronous Motor
MCU	Microcontroller Unit
MIPS	Million Instructions Per Second
MP	Mass Production
MTPA	Maximum Torque Per Ampere
PC	Personal Computer
PM	Permanent Magnet
PMSM	Permanent Magnet Synchronous Machine
PWM	Pulse-Width Modulation
RLS	Recursive Least Squares

References

1. Lin, F.J.; Hung, Y.C.; Chen, J.M.; Yeh, C.M. Sensorless IPMSM Drive System Using Saliency Back-EMF-Based Intelligent Torque Observer with MTPA Control. *IEEE Trans. Ind. Inform.* **2014**, *10*, 1226–1241.
2. Yang, Z.; Shang, F.; Brown, I.P.; Krishnamurthy, M. Comparative Study of Interior Permanent Magnet, Induction, and Switched Reluctance Motor Drives for EV and HEV Applications. *IEEE Trans. Transp. Electron.* **2015**, *1*, 245–254. [CrossRef]
3. Sul, S.; Kwon, Y.; Lee, Y. Sensorless control of IPMSM for last 10 years and next 5 years. *CES Trans. Electr. Mach. Syst.* **2017**, *1*, 91–99. [CrossRef]
4. Sun, J.; Lin, C.; Xing, J.; Jiang, X. Online MTPA Trajectory Tracking of IPMSM Based on a Novel Torque Control Strategy. *Energies* **2019**, *12*, 3261. [CrossRef]
5. Ye, M.; Shi, T.; Wang, H.; Li, X.; Xia, C. Sensorless-MTPA Control of Permanent Magnet Synchronous Motor Based on an Adaptive Sliding Mode Observer. *Energies* **2019**, *12*, 3773. [CrossRef]
6. Wu, X.; Fu, X.; Lin, M.; Jia, L. Offline Inductance Identification of IPMSM with Sequence-Pulse Injection. *IEEE Trans. Ind. Inform.* **2019**, *15*, 6127–6135. [CrossRef]
7. Cho, S.; Shin, W.; Park, J.; Kim, W. A Torque Compensation Control Scheme of PMSM Considering Wide Variation of Permanent Magnet Temperature. *IEEE Trans. Magn.* **2019**, *55*, 1–5. [CrossRef]
8. Halder, S.; Agarwal, P.; Srivastava, S.P. MTPA based Sensorless Control of PMSM using position and speed estimation by Back-EMF method. In Proceedings of the IEEE 6th International Conference on Power Systems (ICPS), New Delhi, India, 4–6 March 2016.
9. Dianov, A.; Kim, N.S.; Lim, S.M. Sensorless starting of horizontal axis washing machines with direct drive. In Proceedings of the International Conference on Electrical Machines and Systems (ICEMS), Busan, Korea, 26–29 October 2013.
10. Joo, K.J.; Park, J.S.; Lee, J. Study on Reduced Cost of Non-Salient Machine System Using MTPA Angle Pre-Compensation Method Based on EEMF Sensorless Control. *Energies* **2018**, *11*, 1425. [CrossRef]

11. Guo, Q.; Zhang, C.; Li, L.; Zhang, J.; Wang, M. Maximum Efficiency per Torque Control of Permanent-Magnet Synchronous Machines. *Appl. Sci.* **2016**, *6*, 425. [CrossRef]
12. Baek, S.-K.; Oh, H.-K.; Park, J.-H.; Shin, Y.-J.; Kim, S.-W. Evaluation of Efficient Operation for Electromechanical Brake Using Maximum Torque per Ampere Control. *Energies* **2019**, *12*, 1869. [CrossRef]
13. Kim, H.; Hartwig, J.; Lorenz, R.D. Using on-line parameter estimation to improve efficiency of IPM machine drives. In Proceedings of the IEEE 33rd Annual IEEE Power Electronics Specialists Conference, Cairns, Australia, 23–27 June 2002.
14. Phowanna, P.; Boonto, S.; Konghirun, M. Online parameter identification method for IPMSM drive with MTPA. In Proceedings of the 18th International Conference on Electrical Machines and Systems (ICEMS), Pattaya, Thailand, 25–28 October 2015; pp. 1775–1780.
15. Liu, Q.; Hameyer, K. High-Performance Adaptive Torque Control for an IPMSM with Real-Time MTPA Operation. *IEEE Trans. Energy Conv.* **2017**, *32*, 571–581. [CrossRef]
16. Chen, Q.; Gu, L.; Lin, Z.; Liu, G. Extension of Space-Vector-Signal-Injection-Based MTPA Control into SVPWM Fault-Tolerant Operation for Five-Phase IPMSM. *IEEE Trans. Ind. Electron.* **2020**, *67*, 7321–7333. [CrossRef]
17. Bolognani, S.; Petrella, R.; Prearo, A.; Sgarbossa, L. Automatic Tracking of MTPA Trajectory in IPM Motor Drives Based on AC Current Injection. *IEEE Trans. Ind. Appl.* **2011**, *47*, 105–114. [CrossRef]
18. Sun, T.; Wang, J.; Koc, M. On Accuracy of Virtual Signal Injection Based MTPA Operation of Interior Permanent Magnet Synchronous Machine Drives. *IEEE Trans. Power Electron.* **2017**, *32*, 7405–7408. [CrossRef]
19. Sun, T.; Koç, M.; Wang, J. MTPA Control of IPMSM Drives Based on Virtual Signal Injection Considering Machine Parameter Variations. *IEEE Trans. Ind. Electron.* **2018**, *65*, 6089–6098.
20. Tang, Q.; Shen, A.; Luo, P.; Shen, H.; Li, W.; He, X. IPMSMs Sensorless MTPA Control Based on Virtual q-Axis Inductance by Using Virtual High-Frequency Signal Injection. *IEEE Trans. Ind. Electron.* **2020**, *67*, 136–146. [CrossRef]
21. Bedetti, N.; Calligaro, S.; Olsen, C.; Petrella, R. Automatic MTPA Tracking in IPMSM Drives: Loop Dynamics, Design, and Auto-Tuning. *IEEE Trans. Ind. Appl.* **2017**, *53*, 4547–4558. [CrossRef]
22. Dianov, A.; Young-Kwan, K.; Sang-Joon, L.; Sang-Taek, L. Robust self-tuning MTPA algorithm for IPMSM drives. In Proceedings of the 34th Annual Conference of IEEE Industrial Electronics, Orlando, FL, USA, 10–13 November 2008; pp. 1355–1360.
23. Dianov, A. Estimation of the Mechanical Position of Reciprocating Compressor for Silent Stoppage. *IEEE Open J. Power Electron.* **2020**, *1*, 64–73. [CrossRef]
24. Dianov, A.; Kim, N.-S.; Kim, S.-M. Sensorless starting of direct drive horizontal axis washing machines. *J. Int. Conf. Electr. Mach. Syst.* **2014**, *3*, 148–154. [CrossRef]

Article

Fault Diagnosis of PMSG Stator Inter-Turn Fault Using Extended Kalman Filter and Unscented Kalman Filter

Waseem El Sayed [1,2,*], Mostafa Abd El Geliel [1] and Ahmed Lotfy [1]

[1]   Electrical and control Department, College of Engineering and Technology, Arab Academy For Science and Technology and Maritime Transport, Abou Keer Campus, P.O. Box 1029, Alexandria 21500, Egypt; mostafa.geliel@aast.edu (M.A.E.G.); alotfy@aast.edu (A.L.)
[2]   Institute of Automatics, Electronics and Electrical Engineering, University of Zielona Gora, 65-417 Zielona Gora, Poland
*   Correspondence: waseem.elsayed@ieee.org

Received: 28 April 2020; Accepted: 4 June 2020; Published: 9 June 2020

Abstract: Since the permeant magnet synchronous generator (PMSG) has many applications in particular safety-critical applications, enhancing PMSG availability has become essential. An effective tool for enhancing PMSG availability and reliability is continuous monitoring and diagnosis of the machine. Therefore, designing a robust fault diagnosis (FD) and fault tolerant system (FTS) of PMSG is essential for such applications. This paper describes an FD method that monitors online stator winding partial inter-turn faults in PMSGs. The fault appears in the direct and quadrature (dq)-frame equations of the machine. The extended Kalman filter (EKF) and unscented Kalman filter (UKF) were used to detect the percentage and the place of the fault. The proposed techniques have been simulated for different fault scenarios using Matlab®/Simulink®. The results of the EKF estimation responses simulation were validated with the practical implementation results of tests that were performed with a prototype PMSG used in the Arab Academy For Science and Technology (AAST) machine lab. The results showed impressive responses with different operating conditions when exposed to different fault states to prevent the development of complete failure.

Keywords: extended Kalman filter (EKF); permanent magnet synchronous generator (PMSG); fault diagnosis (FD); stator inter-turn short circuit

---

1. Introduction

In the last decade, the permeant magnet synchronous generator (PMSG) has been used in many industries, especially, for renewable energy applications [1–3], aircraft [4,5], and propulsion systems [6]. Consequently, this has generated growing concern about the operation reliability of the PMSG, especially in safety critical applications like the shaft generators (SG) in marine applications.

The PMSG faults inexorably decrease the reliability of the system, which may lead to malfunction or a failure in the system. Moreover, most PMSG applications are safety-critical, which makes the presence of fault an unwanted option. Mechanical, magnetic, and electrical faults are the major types of faults that may occur in a PMSG [7]. Extensive research has examined the detection of mechanical faults, which is the most usual fault in the PMSG; these faults can be divided into eccentricity faults [8–10], and bearing faults [11–14], based on [15], the bearing faults represent from 40 to 50% of the total faults while the eccentricity fault represents from 5 to 10% in the machines. Further research has considered the detection of demagnetization faults [16–18]. Both types of faults cause torque to unbalance; followed by an increase in the overall temperature of the machine. The high temperature may cause the deterioration of the stator winding insulation, which may lead to the presence of a stator inter-turn fault [19–22], based on [21], the stator electrical faults represent 38% of the total faults in the machines. All these papers have focused on the stator winding inter-turn fault in any phase, which is a particular case fault that, if not addressed, affects the machine's voltage magnitude and balance, and may lead to other catastrophic failures.

Fault diagnosis (FD) techniques were used to detect the place and severity of the fault, followed by isolation with minimal losses. This can be divided into three main approaches: signal-based, artificial intelligence-based, and model-based techniques [7,23]. First are the signal-based techniques; they emphasize the analysis of the measured signal to detect the presence of specific frequency components relating to the fault. Moreover, it requires knowledge of the fault signatures, this knowledge can be acquired from the stator voltage and current, torque signal, and similar variables [7]. The advantages of these methods are the non-dependency on a specific model [24]. However, if the signal contains many harmonics, it may give an erratic estimation for fault. Furthermore, it needs a batch set of samples to analyze the signal; this causes a delay in time in determining a fault estimation. Wavelet transform (WT) is one of these methods that is presented in [25] and [26]. Additionally, Hilbert Hang transform (HHT) and Wigner–Ville have been shown to produce considerable results [17,27]. Also, the vibroacoustic techniques are used in condition monitoring for the machines in [28] and [29].

Secondly, artificial intelligence (AI) methods have been extensively studied in the fault diagnosis of electrical machines. These techniques require a deep understanding of fault signatures under several faulty conditions. However, it needs a set of logged data for the definite fault, which may be undetermined. In addition, some of these techniques do not cope with the online monitoring required for inter-turn short circuit detection due to the computational burden taking time for these techniques to fulfill the FD. Neural networks (NN) [30,31], particle swarm optimization [32], and fuzzy logic [33,34] are AI methods that have been used in stator windings FD of PMSG. A lot of researchers have used a combination of them, such as using the neuro-fuzzy technique [35] or using the AI technique with the signal base technique, such as using the wavelet transform (WT) with the adaptive neuro-fuzzy inference system (ANFIS) in [25].

The third choice is model-based FD techniques, which require the use of a system model. These techniques give the precise estimation of the fault if the mathematical model used is accurate, so they can estimate parameters that are hard to measure [36,37]. Moreover, these techniques offer online parameter identification with the required fast response for taking action. However, these techniques require an accurate model for the system to make a robust estimation in all operating conditions, which is so rare to find, this means that the model-based technique is not used in a lot of complex systems. In [38], the recursive least square (RLS) method is used to estimate the stator inter-turn faults, and the technique provides good response and early detection for the fault. The extended Kalman filter (EKF) has been used in [20,39–41], for the detection of the fault in PMSG and the induction motor (IM). Other researches take into account the use of unscented Kalman filter (UKF) in parameters estimation of PMSG, as an enhancement tool for the control system [42]. The model-based technique is also used in the industrial process control fault diagnoses in [37,43]. In [44], the research presented uses the graph of the process to find an accurate model for the system.

In this contribution, a comparison between the use of the EKF and the UKF is presented in the fault diagnosis of the stator inter-turn faults for PMSG, which has not been addressed before in any other research work. The mathematical model and the equivalent circuit in both healthy and faulty states were implemented based on the model in [21,40]. The procedures of fault percentage and location estimation using EKF and UKF are presented, and the simulation results of parameter estimation in both healthy and faulty conditions, showing the response of both techniques in the case of inter-turn short circuits through several operating conditions and scenarios, are discussed. Moreover, two scenarios were proposed for the decision-making process based on the severity of the fault. The results were validated by applying a practical emulation for the fault in a laboratory prototype machine and discussed.

2. The Faulty PMSG Model

The model was implemented in the direct and quadrature (dq-frame) in [39], in both states, healthy and faulty.

2.1. PMSG Healthy State Model

The healthy state represents the machine in the case of no fault; in this case, the internal current outgoing from the machine is the same current consumed from the load. Figure 1 shows the equivalent circuit of the machine in the abc-frame, R_s and L_s are the stator resistance and inductance. E_a, E_b, E_c, are the induced voltages, and the output current from the generator is represented by I_a, I_b, I_c.

To simplify the model, the equations of the machine should be converted to the dq-frame. Figure 2 shows the equivalent circuit of the machine in the dq-frame. L_d and L_q are the direct and quadrature inductance, I'_{sd} and I'_{sq} is the internal direct and quadrature current of the generator respectively, I_{sd} and I_{sq} are the terminal direct and quadrature current of the generator respectively, the V_{sd} and V_{sq} are the direct and quadrature stator terminal voltages, and ω_e is the electrical angular speed that can be related to rotor mechanical angular speed ω_m. All the equations representing the machine on the dq-frame are given in [39].

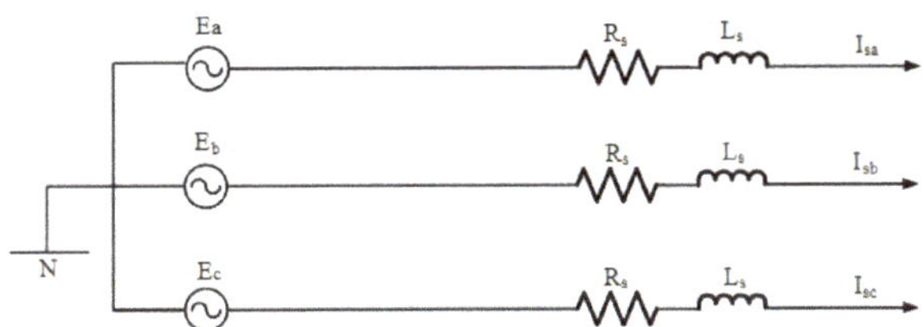

Figure 1. The equivalent circuit of healthy permeant magnet synchronous generator (PMSG) in the abc-frame.

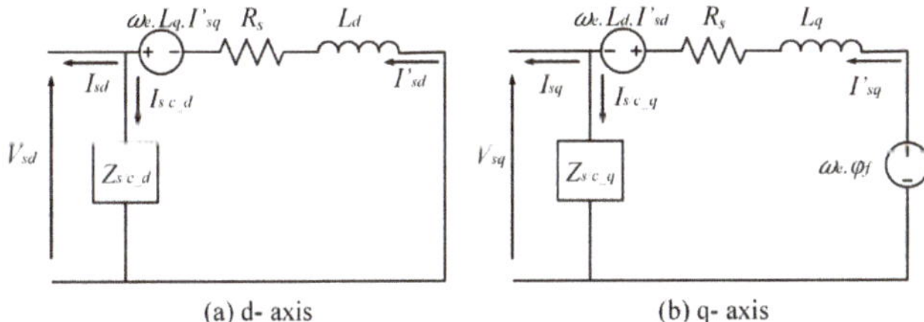

(a) d- axis   (b) q- axis

Figure 2. The equivalent circuit of healthy PMSG in the dq-frame.

2.2. PMSG Faulty State Model

In the case of a PMSG stator winding fault, the number of turns in a certain phase is reduced due to the deterioration in the stator winding insulation, which causes a short circuit in this phase. Most stator winding insulation failures are caused by high temperatures and overloading. When a stator winding short circuit happens, the value of generator impedance changes, consequently the amplitude of stator current harmonics will increase, the torque will drag, and potential overheating will appear, and so on, this symptom may lead to complete failure if the fault was not addressed.

The short circuit current $[I_{s/c}]_{dq}$ is generated inside the machine as shown in Figure 3 due to the presence of the short circuit impedance $Z_{s/c}$ in any phase; this impedance value changes according to the ratio between the number of inter-turn short-circuit windings and the total number of turns in one healthy phase. Figure 4 shows the equivalent circuit of the faulty machine in the dq-frame. The mathematical equations representing the faulty state model of the PMSG in dq-frame are given in [39].

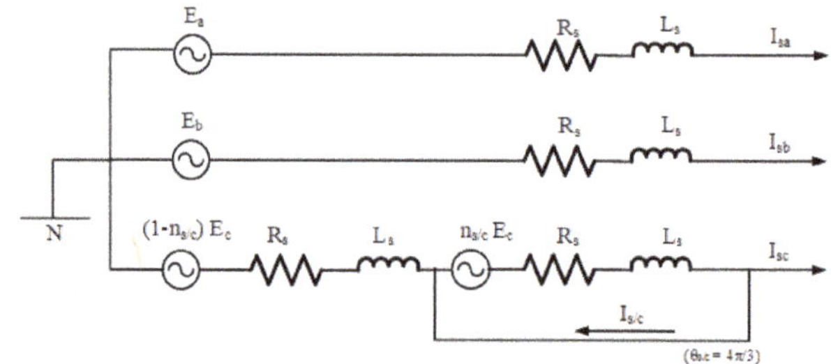

Figure 3. The short circuit turns ratio representation.

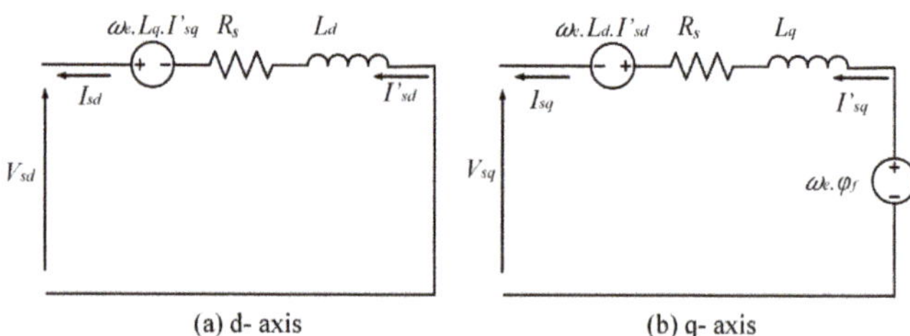

(a) d- axis   (b) q- axis

Figure 4. The equivalent circuit of the faulty state PMSG in the dq-frame.

3. Parameter Estimation Procedures

The general faulty PMSG state-space model and EKF algorithm are presented in this section.

3.1. General PMSG State-Space Model

To use the EKF and UKF techniques to estimate the presence of the fault, the precise state-space model of the machine should be highlighted. Based on [39], the state-space model equation of the faulty machine can be written as:

$$\begin{cases} \dot{x}_m(t) = A_m{\cdot}x_m(t) + B_m{\cdot}u_m(t) + W_m(t) \\ y_m(t) = C_m{\cdot}x_m(t) + D_m{\cdot}u_m(t) + V_m(t) \end{cases} \tag{1}$$

where:

$$x_m(t) = \begin{bmatrix} I'_{sd} & I'_{sq} \end{bmatrix}^T \quad u_m(t) = \begin{bmatrix} V_{sd} & V_{sq} & \varphi_f \end{bmatrix}^T \quad A_m = \begin{bmatrix} -\dfrac{R_s}{L_d} & \omega_e{\cdot}\dfrac{L_q}{L_d} \\ -\omega_e{\cdot}\dfrac{L_d}{L_q} & -\dfrac{R_s}{L_q} \end{bmatrix}$$

$$B_m = \begin{bmatrix} -\dfrac{1}{L_d} & 0 & 0 \\ 0 & -\dfrac{1}{L_q} & \dfrac{\omega_e}{L_q} \end{bmatrix} \quad C_m = \begin{bmatrix} 1 & 0 \\ 0 & 1 \end{bmatrix}$$

$$D_m = \begin{bmatrix} D_1 & 0 \end{bmatrix}^T \quad D_1 = -\sum_{k=1}^{3} \frac{2{\cdot}n_{s/c\,k}}{(3 - 2{\cdot}n_{s/c\,k}){\cdot}R_s} {\cdot}P(\theta)^T{\cdot}Q(\theta_{s/c\,k}){\cdot}P(\theta)$$

where

$$P(\theta) = \begin{bmatrix} cos\theta & -sin\theta \\ sin\theta & cos\theta \end{bmatrix} \tag{2}$$

$$Q(\theta_{S/C}) = \begin{bmatrix} cos^2\theta_{S/C} & sin\theta_{S/C}{\cdot}cos\theta_{S/C} \\ sin\theta_{S/C}{\cdot}cos\theta_{S/C} & sin^2\theta_{S/C} \end{bmatrix} \tag{3}$$

The extension of the model states to estimate the presence of a fault in any phase is compulsory for the estimation process of EKF and UKF, the new states of the model become as follows:

$$\widetilde{X}_e(t) = \begin{bmatrix} X(t) \\ \lambda(t) \end{bmatrix} = \begin{bmatrix} I'_{sd} & I'_{sq} & n_{A\,s/c} & n_{B\,s/c} & n_{C\,s/c} \end{bmatrix}^T \tag{4}$$

where $\widetilde{X}_e(t)$ is the estimated state; After that, the model equations are linearized around a definite operating point followed by discretization at a sampling time Ts, the model expressed as:

$$\begin{bmatrix} \dot{I}'_{sd} \\ \dot{I}'_{sq} \\ \dot{n}_{A\,s/c} \\ \dot{n}_{B\,s/c} \\ \dot{n}_{C\,s/c} \end{bmatrix} = \begin{bmatrix} 1-T_s\frac{R_s}{L_d} & T_s\cdot w_e\cdot\frac{L_q}{L_d} & 0 & 0 & 0 \\ -T_s\cdot w_e\cdot\frac{L_d}{L_q} & 1-T_s\frac{R_s}{L_q} & 0 & 0 & 0 \\ 0 & 0 & 1 & 0 & 0 \\ 0 & 0 & 0 & 1 & 0 \\ 0 & 0 & 0 & 0 & 1 \end{bmatrix} \begin{bmatrix} I'_{sd} \\ I'_{sq} \\ n_{A\,s/c} \\ n_{B\,s/c} \\ n_{C\,s/c} \end{bmatrix} + \begin{bmatrix} -\frac{T_s}{L_d} & 0 & 0 \\ 0 & -\frac{T_s}{L_q} & \frac{w_e}{L_q} \\ 0 & 0 & 0 \\ 0 & 0 & 0 \\ 0 & 0 & 0 \end{bmatrix} \begin{bmatrix} V_{sd} \\ V_{sq} \\ \varphi_f \end{bmatrix} \tag{5}$$

3.2. Extended Kalman Filter Algorithm

The EKF gives an approximation of the optimal estimate. The non-linearity of the system's dynamics is approximated by a linearized version of the non-linear system model around the last state estimate. As in many cases, if the nonlinear system is approximately linearized, the EKF may not perform well [20]. If there is a bad initial guess regarding the underlying system's state, then this may cause a bad estimation. The first step in the EKF algorithm is the prediction step equations, which consist of state prediction and error covariance matrix update, and the second step is the correction step which corrects the predicted state estimate and it's covariance matrix as in Figure 5. Consider applying EKF to estimate the parameter λ_k in the PMSG system, the discrete linearized state-space model of PMSG is expressed as:

$$\begin{cases} \widetilde{X}_{e_{k+1}} = \widetilde{F}_k\,\widetilde{X}_{e_k} + W_k \\ \widetilde{Y}_k = \widetilde{H}_k\,\widetilde{X}_{e_k} + V_k \end{cases} \tag{6}$$

where

$$\begin{cases} \widetilde{F}_K = \begin{bmatrix} 1+T_sA(\lambda_k) & T_s\left(\frac{\partial A(\lambda_k)}{\partial\lambda_k}X_k + \frac{\partial B(\lambda_k)}{\partial\lambda_k}U_k\right) \\ 0 & I \end{bmatrix} \\ \widetilde{H}_K = \begin{bmatrix} C(\lambda_k)\left(\frac{\partial C(\lambda_k)}{\partial\lambda_k}X_k + \frac{\partial D(\lambda_k)}{\partial\lambda_k}U_k\right) \end{bmatrix} \end{cases} \tag{7}$$

$\widetilde{F}_K$ and $\widetilde{H}_K$ represent the state and output equations of the discrete linearized model. By substituting matrix A and B in (4) into (6), $\widetilde{F}_K$ and $\widetilde{H}_K$ in case of, $n_{A\,s/c}$, $n_{B\,s/c}$, $n_{C\,s/c}$, as an estimated parameter will be:

$$\widetilde{F}_K = \begin{bmatrix} 1-T_s\frac{R_s}{L_d} & T_s\cdot w_e\cdot\frac{L_q}{L_d} & 0 & 0 & 0 \\ -T_s\cdot w_e\cdot\frac{L_d}{L_q} & 1-T_s\frac{R_s}{L_d} & 0 & 0 & 0 \\ 0 & 0 & 1 & 0 & 0 \\ 0 & 0 & 0 & 1 & 0 \\ 0 & 0 & 0 & 0 & 1 \end{bmatrix} \quad \widetilde{H}_K = \begin{bmatrix} 1 & 0 & S_A\times q_1 & S_B\times q_1 & S_C\times q_1 \\ 0 & 1 & S_A\times q_2 & S_B\times q_2 & S_C\times q_2 \end{bmatrix} \tag{8}$$

where

$$S_A = -6/\left((3-2n_{A\,s/c})^2\times R_s\right) \quad S_B = -6/\left((3-2n_{B\,s/c})^2\times R_s\right)$$

$$S_C = -6/\left((3-2n_{C\,s/c})^2\times R_s\right) \quad \begin{bmatrix} q_1 \\ q_2 \end{bmatrix} = \left[P(\theta)^T\cdot Q(\theta_{\frac{s}{c}\,k})\cdot P(\theta)\right]_{2x2}\cdot\begin{bmatrix} V_{sd} \\ V_{sq} \end{bmatrix}$$

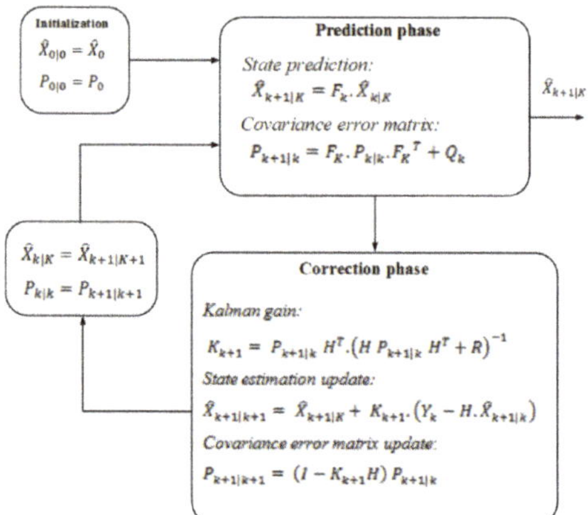

Figure 5. Extended Kalman filter (EKF) algorithm equations.

3.3. Unscented Kalman Filter Algorithm

Instead of using linearized equations using the Jacobin matrix to approximate the nonlinear model as the EKF approach, the UKF generates a finite set of sigma points to compute the predicted states and measurements and the associated covariance matrices [45]. Mathematically, the UKF process can be described as in Figure 6.

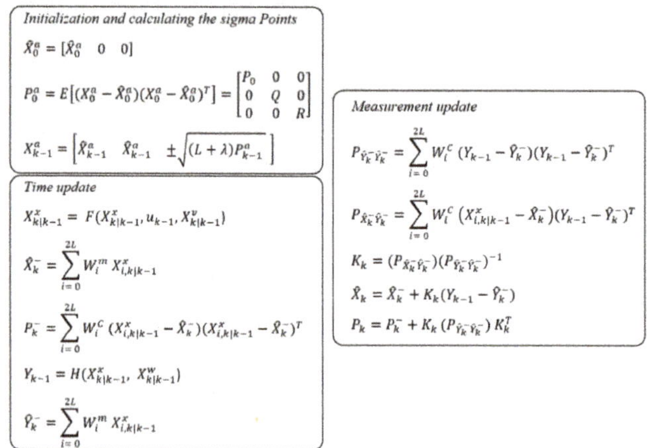

Figure 6. Unscented Kalman filter (UKF) algorithm equations.

Where W_i^m and W_i^c are weighting factors and they are equal to.

$$W_i^m = W_i^C = 1/2(L + \lambda) \tag{9}$$

where L is the state dimension and $\lambda = \alpha^2((L + k) - L)$, α can be tuned from 10^{-4} to 1 and k usually was chosen to be 0.

Figure 7 shows the block diagram of the fault diagnosis online monitoring for the PMSG.

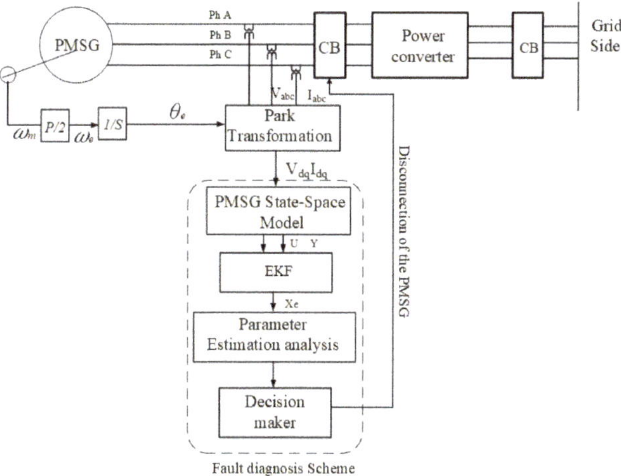

Figure 7. PMSG fault diagnosis on-line monitoring block diagram.

3.4. The Covariance Matrices Tuning

The noises covariance matrices are diagonals, Q can be divided into two matrices: q_x (for measured states) and q_λ (for the estimated parameters). Thus, Q and R can be expressed as:

$$\begin{cases} Q = q_x \cdot \begin{bmatrix} I_m & 0 \\ 0 & \frac{q_\lambda}{q_x} \cdot I_n \end{bmatrix} \\ R = r \cdot I_m \end{cases} \tag{10}$$

where m is the state's x numbers and n is estimated parameters λ numbers, q_x and r could be determined by measuring the variance of noises on input σ_u^2 and output signals σ_y^2 [20], they are expressed as Equation (9) and the ratio q_λ / q_x is set by the evolution time constant of the estimated parameters (τ) as expressed in Equation (10).

$$\begin{cases} q_x = \left(\frac{\partial f_k}{\partial u_k}\right)^2 \cdot \sigma_u^2 \\ r = \left(\frac{\partial H_k}{\partial X_k}\right)^2 \cdot \sigma_y^2 \end{cases} \tag{11}$$

$$\sqrt{\frac{q_\lambda}{q_x}} \approx \frac{T_s}{\tau \cdot \sqrt{\sum_{i=1}^{n}\left(\left|\frac{\partial f_{k_i}}{\partial u_k}\right| \cdot \left|\frac{\partial h_{k_i}}{\partial X_k}\right|\right)^2}} \tag{12}$$

4. Simulation Results

The machine parameters were taken from the nameplate of the generator, as shown in Table 1, the equations of the generator were used to simulate the output of the machine with different operating points, and the simulation run at sampling time $T_s = 100$ us.

Table 1. The PMSG parameters used.

Parameter	Symbol	Value
Nominal Power	P	1500 W
Nominal current	I_s	5 A
Nominal Voltage	V_s	100 v
Nominal Frequency	f	50 Hz
Stator resistance	R_s	1.2 Ω
Direct *axis* magnetizing inductance	L_d	4 mH
Quadrature *axis* magnetizing inductance	L_q	3 mH
Nominal Torque	T_m	9.7 Nm
Rotation speed	N_m	1500 rpm
Number of pole pairs	p	2
Total moment of system inertia	J	0.11 kgm^2

4.1. EKF VS. UKF Response

Figure 8 shows the instantaneous internal current of phase A of the machine at RMS load current of 0.75 A and frequency of 30 Hz; a simulated inter-turn fault was implemented at $t = 0.5$ s, this fault caused an increase in the current inside the machine, respectively, the voltage in the faulty phase decreased by a small amount and the machine started to become hotter. The current reached an RMS value of 1.63 in the case of $n_{A\ s/c} = 4\%$, which is more than double the used load current. Also, the current reached an RMS value of 4.6 A in the case of $n_{A\ s/c} = 16\%$, which is more than 6 times the load current (0.75 A). This implies the importance of taking fast action to save the machine from damage.

Figure 9 shows the estimation response of EKF and UKF in either a healthy or faulty state; it was noticed that the UKF technique gives more precise values for the fault estimation than the EKF. As the PMSG model used is a linearized and discretized model around a specified operating point in the case of EKF, the error in the estimation varies non-linearly with the value of the short circuit turns ratio. Besides, the covariance matrices (Q and R) were chosen, which play an important role in the quality of the estimation. Also, the presence of sensor errors and the use of a phased locked loop (PLL) in the estimation of the angular position θ cause error in the estimated parameters. Figure 10 shows the error-index, which indicates the values of the inter-turn short circuit that the EKF and UKF techniques will estimate varies the percentage of error. It was noticed that the UKF had much less error than the EKF, especially for short circuit turns ratios greater than 20%, the highest error detected in case of using UKF was at $n_{A\ s/c} = 4\%$ and reached 0.3%, however, the EKF estimation error reach 23.72% at $n_{A\ s/c} = 100\%$, the lowest estimation error detected by EKF was at $n_{s/c} = 16\%$.

The dynamic time response of detecting the fault was 0.02 s, which is very fast (approximately equal to 1.5 periodic cycles related to the used frequency). The covariance matrix Q, in this case, was tuned by time constant $\tau = 5$ ms, which increased the dynamic response; however, it increased the presence of noise in the estimation action. Figure 11 illustrates the effect of τ on the estimation time, and it was noticed that when the $\tau = 10$ ms, the estimation response reached a steady-state after 0.02 s with the presence of noise, however, when the $\tau = 80$ ms, the estimation response reached a steady-state after 0.2 s but with filtering action. Figure 12 shows the dynamic estimation responses versus the parameter estimation evolution time constant (τ); the detection time increased linearly with the increase of the τ value.

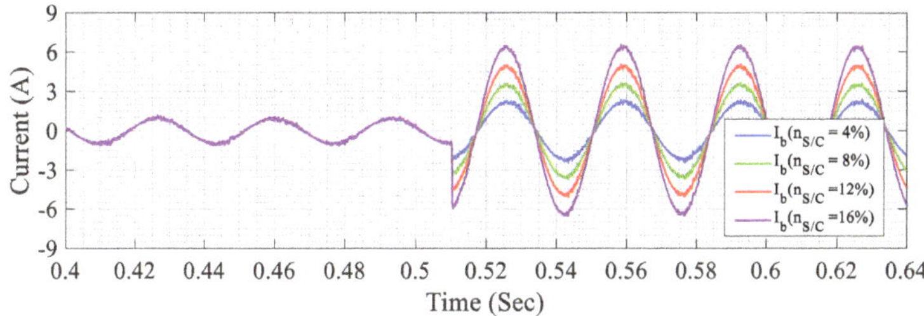

Figure 8. The instantaneous currents of phase A in case of inter-turn fault at 0.72 A load and 30 Hz frequency.

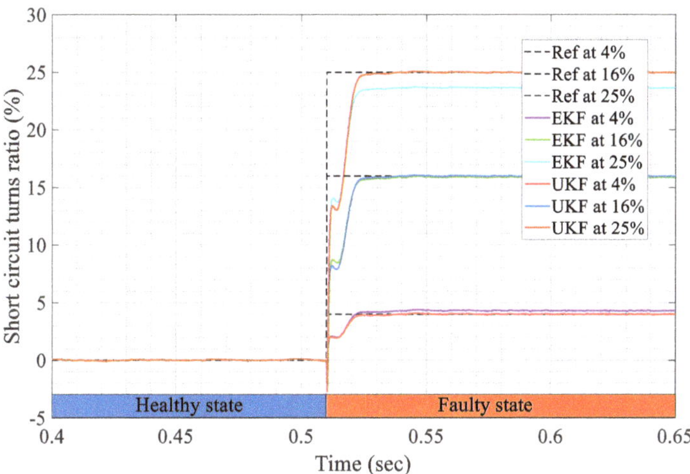

Figure 9. EKF estimation response in phase A at a load of 0.72 A and a frequency of 30 Hz.

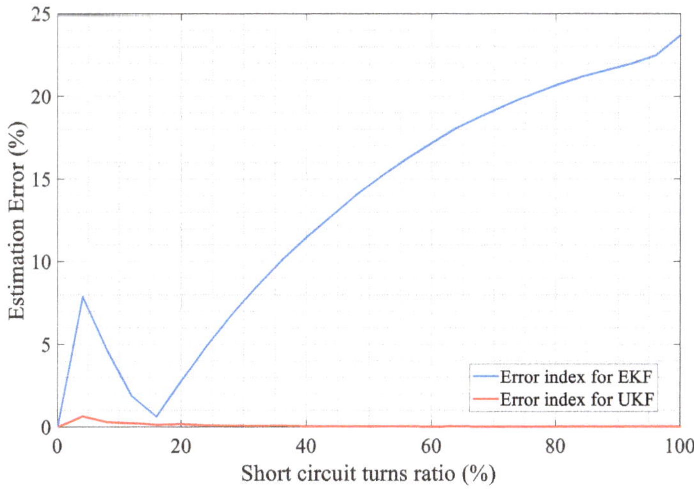

Figure 10. Error Index of the estimated parameter $n_{s/c}$.

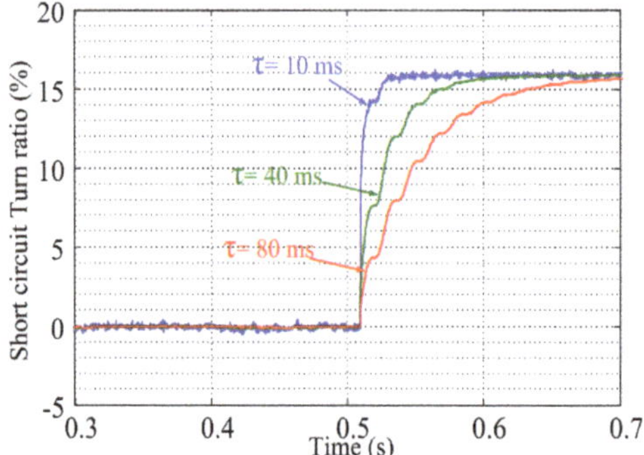

Figure 11. Estimation of 16% short circuit turn ratio in phase B in case of different τ.

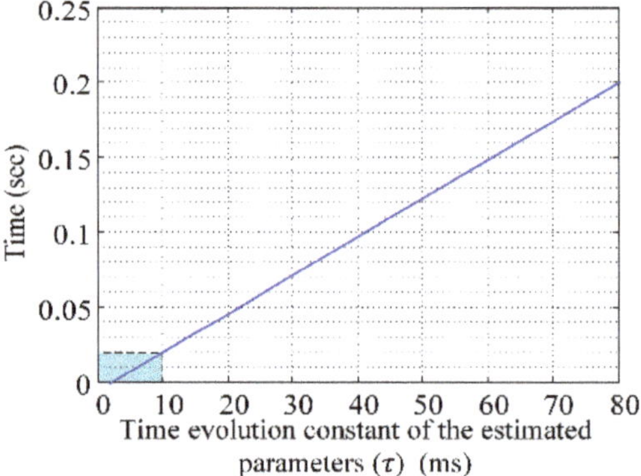

Figure 12. Index of the estimated parameters $n_{s/c}$.

4.2. Robustness Tests

The technique was tested in different operating conditions and showed a robust response; the same tests are done in cases of practical implementation, and are listed as the following:

Test 1: Variation of load current variation from 0.75 A to 3 A by 0.75 A step at a constant frequency of 30 Hz.

Test 2: Variation of frequency from 20 Hz to 50 Hz, with a 10 Hz step.

4.3. Load Variation Test

Figure 13 shows the estimated internal instantaneous currents in the presence of a 16% stator inter-turn short circuit in phase A in case of load variation from 0.75 to 3 A with a rate of 1 Hz. In the case of a 16% stator inter-turn fault, the current reached an RMS value of 5.3 A when I_{Load} = 1.5 A, and it reached an RMS value of 6 A when the I_{Load} = 2.25 A. This confirms the increase of fault severity as load current increases; this form of the fault requires fast action.

Figures 14 and 15 show the estimation response of EKF and UKF in load current variation from I_{Load} = 0.75 A to 3 A condition in the presence of 4%, 8%, 12%, and 16% stator inter-turn short circuit by a rate of 1 Hz. The time constant of the estimated parameters (τ) was chosen to be 10 ms based on the measured value of the input signal noises variance ($\acute{o}_x$). The estimation for both techniques show a constant response with the load variation with different short circuit values.

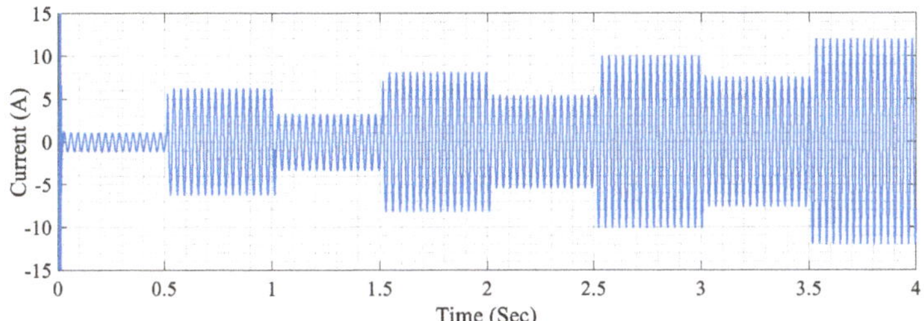

Figure 13. The instantaneous currents of phase A in case of fault at constant frequency of 30 Hz and current variation from 0.75 A to 3 A with step of 0.75 A.

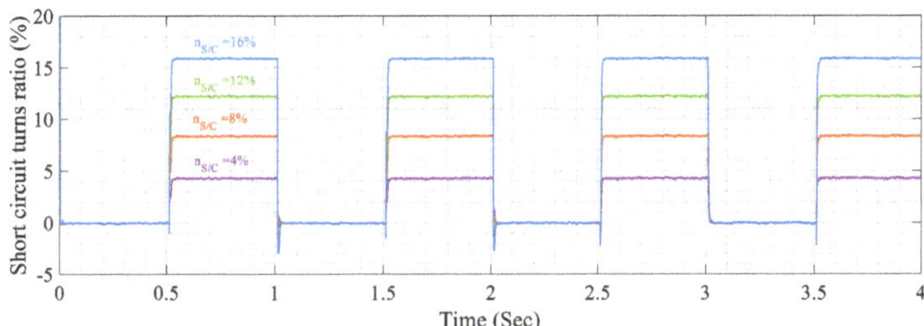

Figure 14. Estimation response in phase A at a frequency of 50 Hz with load variation from 0.75 to 3 A and 7.5 A step.

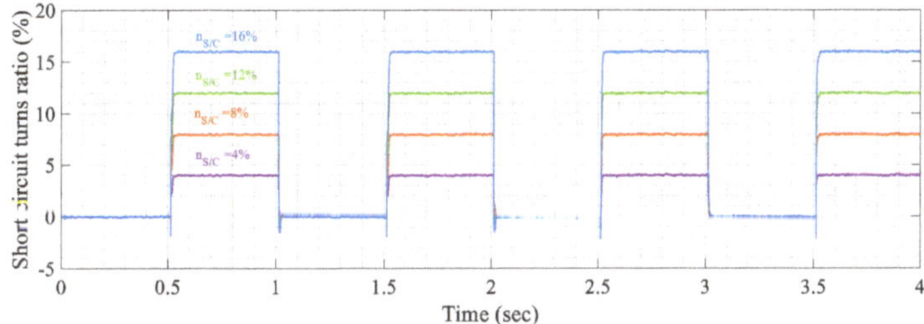

Figure 15. Estimation response in phase A at a frequency of 50 Hz with load variation from 0.75 to 3 A and 7.5 A step.

4.4. Frequency Variation Test

Figure 16 shows the estimated internal instantaneous currents in the case of faults in various frequencies, ranging from 20 Hz to 50 Hz with 10Hz step frequency and rate of change of 1 Hz. Figures 17 and 18 show the EKF the UKF estimation response in the presence of 4%, 8%, 12%, and 16% stator inter-turn short circuit in phase A at constant $I_{Load} = 0.75$ A and frequencies of 20 Hz, 30 Hz, 40 Hz and 50 Hz. The results show a constant response for both techniques with the frequency variation condition.

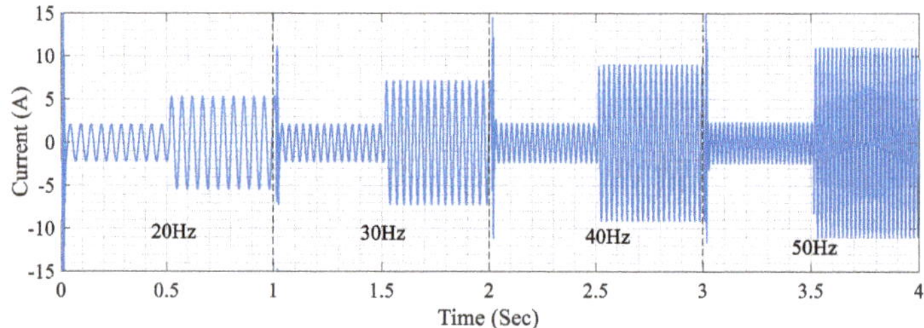

Figure 16. The instantaneous currents of phase A in case of fault at 0.75 A load and 50 Hz frequency.

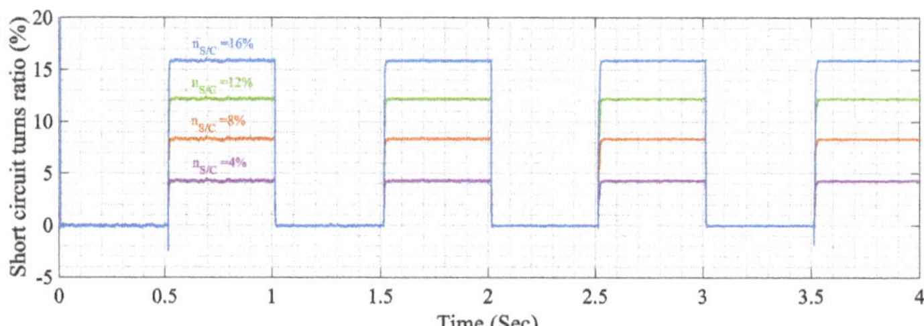

Figure 17. Estimation response in phase A at a load of 0.75 A with frequency variation from 20 to 50 Hz and 10 Hz step.

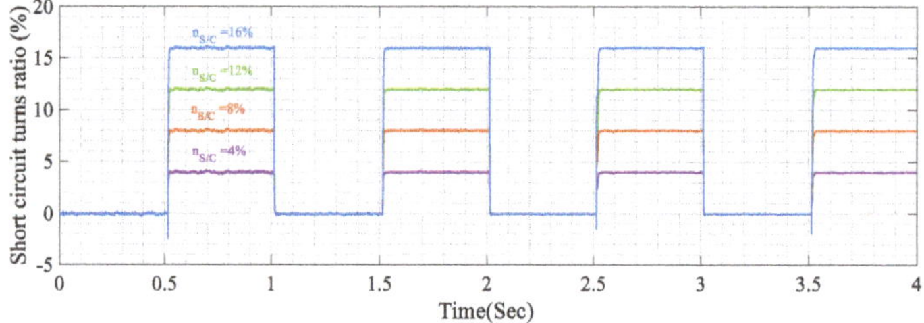

Figure 18. Estimation response in phase A at a load of 0.75 A with frequency variation from 20 to 50 Hz and 10 Hz step.

5. Experimental Results

5.1. Test Bench

For safety conditions, to prevent the used PMSG from being damaged, it is not possible to make an actual stator inter-turn fault. However, it is possible to validate this detection method by adding a shunt resistance $R_{S/C}$ between the needed phase and the neutral, to increase the current in this phase and make the machine unbalanced by a percentage equal to that of an inter-turn fault.

The generator used rotates by means of a separately excited DC motor as a prime mover; the shaft of the motor is coupled directly to the shaft of the PMSG. The power pack supplies the DC motor field with a constant DC supply, and the armature is supplied with a variable DC supply to control the speed of the generator. The load used is a three-phase variable load with a maximum RMS value of 5 A; the shunt resistance $R_{S/C}$ is variable resistance, which will be added to any phase of the three phases using a circuit breaker. Figure 19 shows all the power components of the test bench.

The three-phase voltages were measured by three voltage transformers. The transformers used were typical 220 v/12 v single-phase transformers. Hence, the voltages measured were connected to analog signal conditioning boards to manipulate the voltage to be level with the digital signal processor (DSP) voltage (from 0 to 3.3 v).

On the other hand, the currents were measured using three CTs at a ratio of 10000:5. The current signals measured were connected to the signal conditioning circuit board to convert the current into a manipulated voltage which was compatible with a DSP analog to digital (A/D) input. The DSP used (Texas Instrument TMS320F280) had all measured signals connected to the A/D port in the DSP. The EKF algorithm was implemented online with a sampling period of $T_s = 200$ μs. The relay board was used to take the action of disconnection of the faulty phase to prevent fault propagation leading to severe failure. Figure 20 shows the connection diagram for the whole circuit.

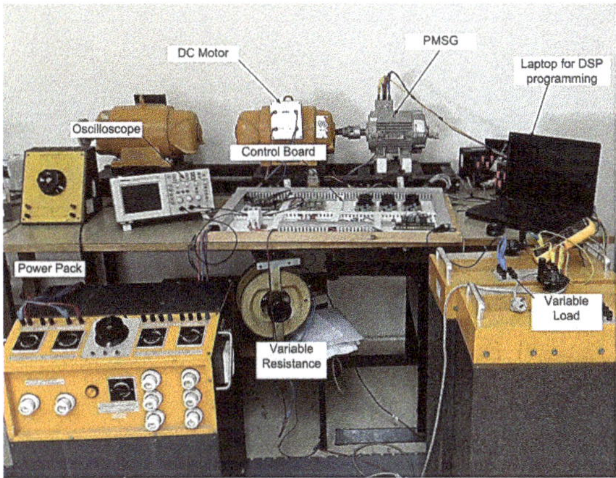

Figure 19. Test bench.

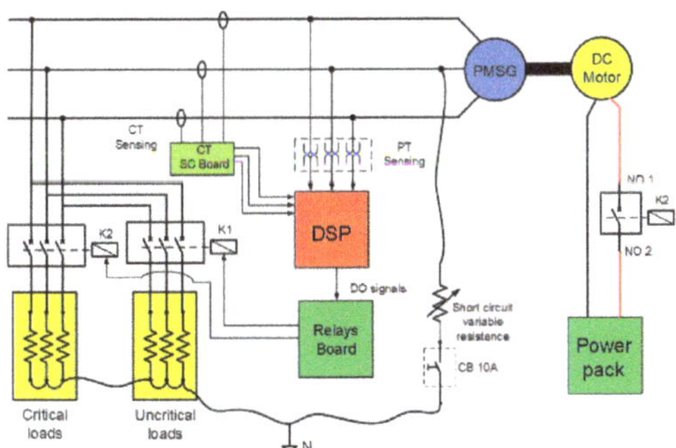

Figure 20. Connection diagram.

5.2. PMSG Test Output

The next step was to compare the measured output voltages and currents of the PMSG used in a simulation of a healthy state. The value of the measured currents and voltages was found to be approximately the same as that in simulation, but with more measurement noise around a mean of 0.4; this will affect the dynamic response of EKF estimation in case of a fault. The output was measured in different load and frequency operating conditions and showed the same output as the simulation. Figure 21a,b show the instantaneous three-phase currents and voltages in the case of a healthy state of PMSG with a load current 0.72 A and frequency of 30 Hz.

Figure 22 shows the difference between the three-phase instantaneous currents and voltages in simulation and practical implementations. It can be seen that the experimental results appear noisier than that of the simulation. Accordingly, the parameter estimation responses will require more filtering action, which will cause a delay in the dynamic response.

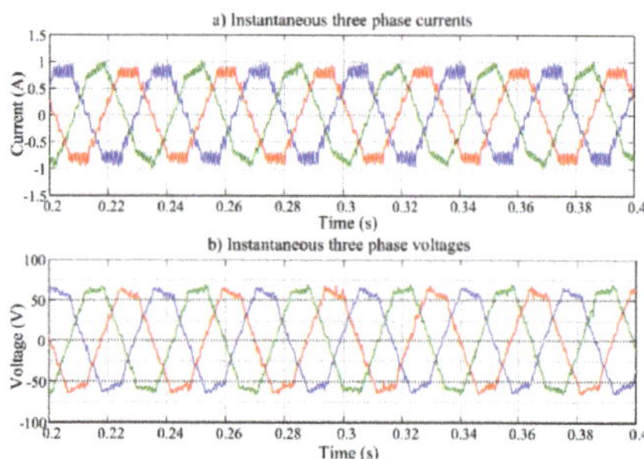

Figure 21. Instantaneous 3phase currents and voltages.

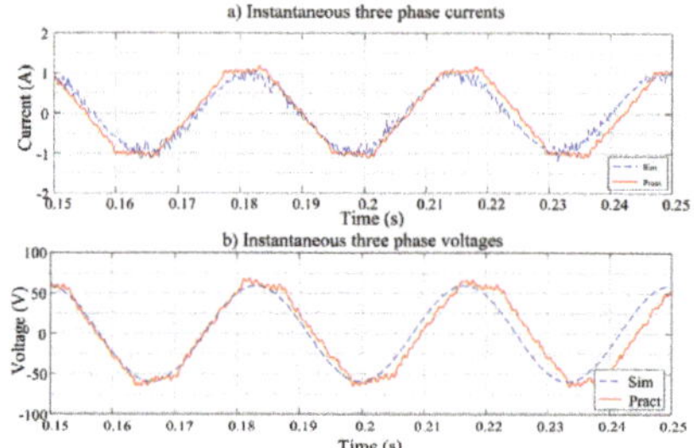

Figure 22. Simulation vs. Practical implementation 3phase currents and voltages.

5.3. EKF Response

The model of a faulty machine and the EKF model were implemented in the DSP, the input to the machine state-space model are the measured stator voltages in the dq-frame. The short circuit current was calculated from the measured dq stator load currents and is presented in the model as the feed-forward matrix D_m. Therefore, the measured three-phase voltages and currents must be converted in dq-frame to make the EKF estimator work probably.

Indeed, the detection of electrical angular position θ is essential to use it in the abc to dq0 transformation. There are two suitable solutions for the detection of electrical angular position θ; the first is to use an encoder sensor coupled directly to the machine shaft and uses its counts to calculate the mechanical angular position, and then calculates the electrical angular position. Nevertheless, this solution requires the addition of new hardware to the system. The second solution is to generate the electrical angular position θ from voltage signals through the three phases of the phased locked loop (PLL), this solution is more economical as extra sensing devices are not needed. Nevertheless, the angular position generated will be dependent on the nature of the measured voltage.

The machine works at a load current of 0.72 and a frequency of 30 Hz in a healthy state. The practical experiments tested the EKF estimation responses in different values of short circuit inter-turn to turn the ratio in all phases ($n_{A\ s/c}$, $n_{B\ s/c}$ & $n_{C\ s/c}$) and in different operating points. The Q and R were tuned at $\tau = 20$ ms to achieve the required fast response with a good filtering action.

Figure 23 display the response of EKF to estimate $n_{A\ s/c} = 4\%$ using these conditions. The parameter estimation showed an excellent response to this case when compared to the results of the simulation. Figure 24 display the estimated internal instantaneous currents in the presence of 4%, stator inter-turn short circuit in phase A at $t = 0.5$ s, respectively. The same response was noticed on the estimation of $n_{A\ s/c} = 8\%$ in Figure 25 followed by the estimated internal instantaneous currents in phase A in Figure 26, respectively. Also, The same response was noticed on the estimation of $n_{A\,s/c} = 12\%$ in Figure 27 followed by the estimated internal instantaneous currents in phase A in Figure 28, respectively. In addition, the estimation of the fault in $n_{A\,s/c} = 16\%$ casein Figure 29 and it's etimated internal current in Figure 30. It was noted that the current reached higher values when compared to the rated current of 0.72 A.

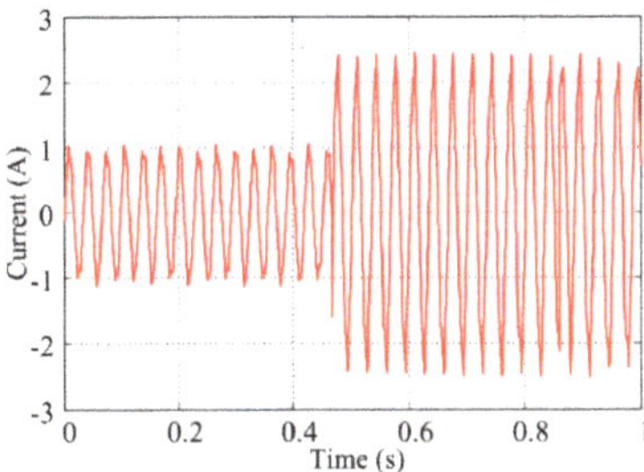

Figure 23. Current in phase A at $n_{A\,s/c} = 4\%$.

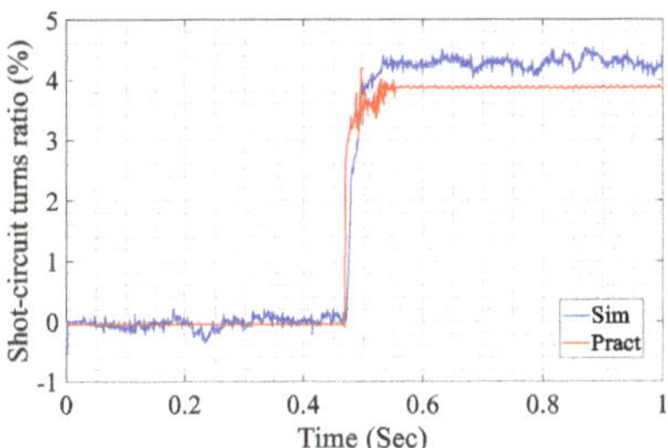

Figure 24. Estimation of 4% short circuit turns ratio in phase A.

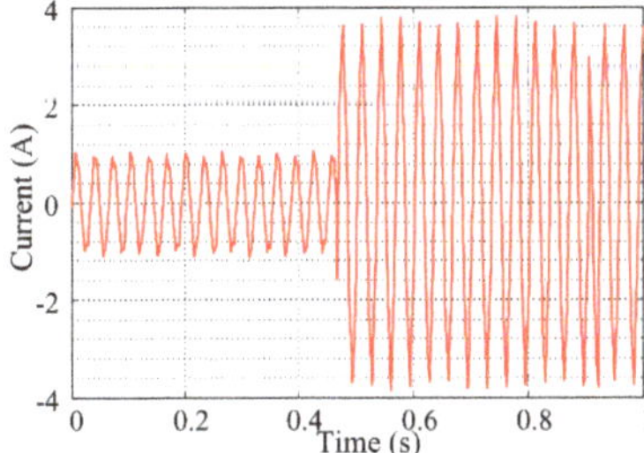

Figure 25. Current in phase A at $n_{A\,s/c} = 8\%$.

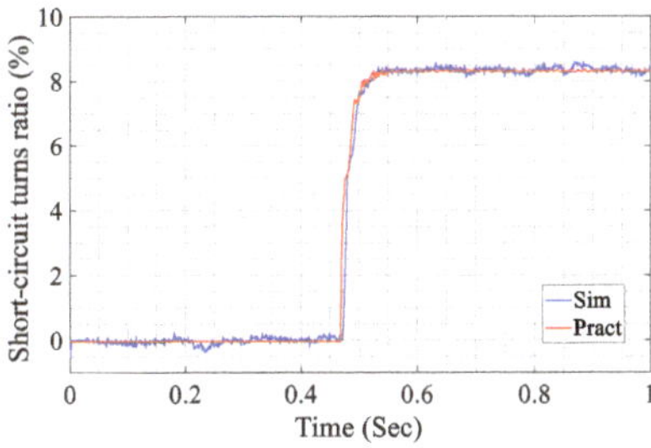

Figure 26. Estimation of 8% short circuit turns ratio in phase A.

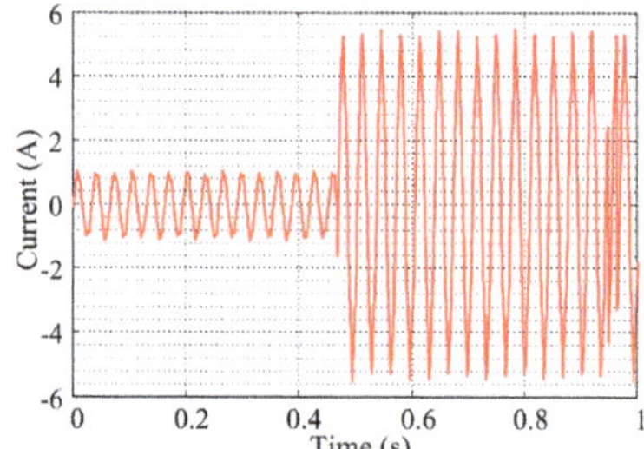

Figure 27. Current in phase A at $n_{A\,s/c} = 12\%$.

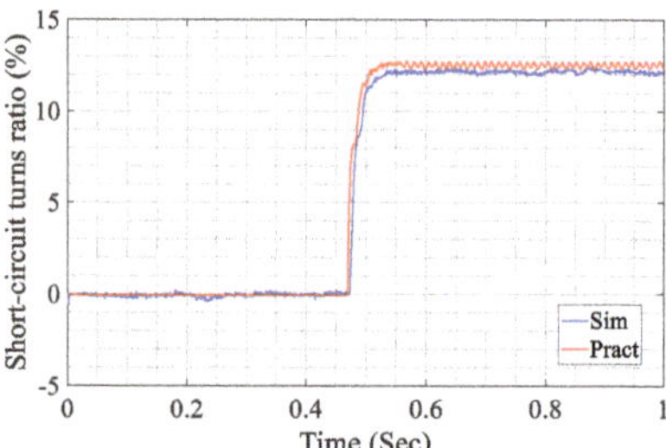

Figure 28. Estimation of 12% short circuit turns ratio in phase A.

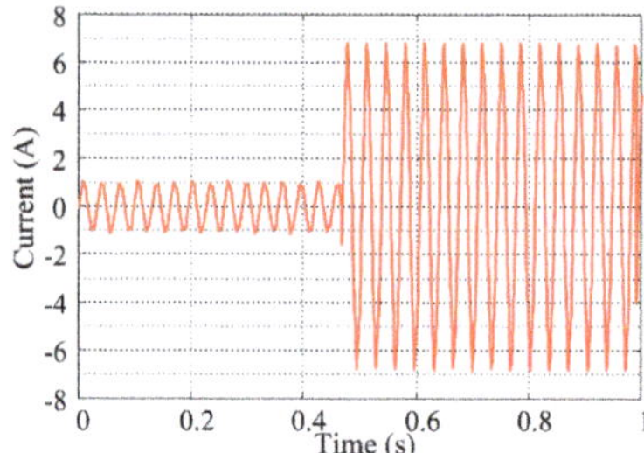

Figure 29. Current in phase A at $n_{A\,s/c} = 16\%$.

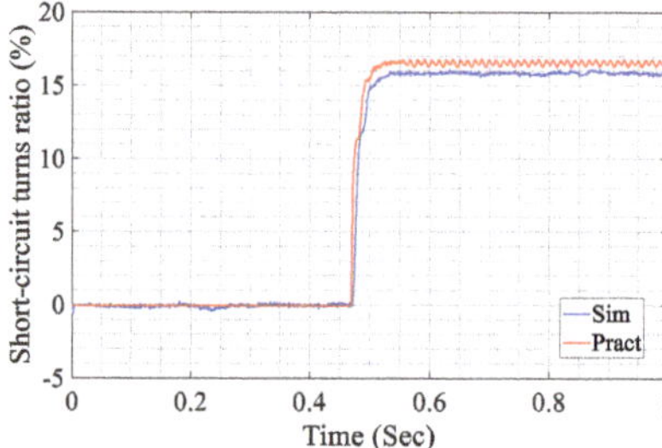

Figure 30. Estimation of 16% short circuit turns ratio in phase A.

5.4. Tuning of Covariance Matrices

Figure 31 shows the dynamic estimation response with different values of evolution time constant of the estimated parameter (τ) in the presence of a 16% fault in phase A. The weighting matrices (Q and R) were chosen based on measuring the variance of input noise and the variance of output noise to achieve the required fast dynamic response for the estimation of the parameters at different operating conditions. However, the change in the weighting matrices caused changes in the nature of the estimation response.

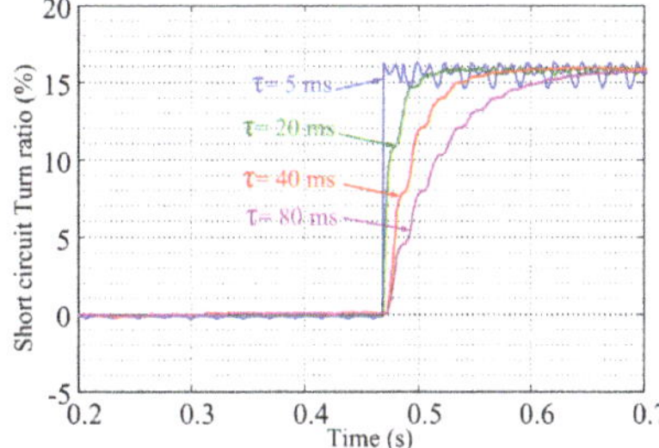

Figure 31. Estimation response with different τ at a constant frequency of 30 Hz.

5.5. Robustness Test

As in the simulation, the machine was tested using different load conditions; this approach tested the parameter estimation response to various load conditions. The tests are listed as following:

Test 1: Variation of frequency from 20 Hz to 50 Hz with a 10 Hz step.

Test 2: Variation of load Current variation from 0.72 A to 2.25 A by 0.75 A step at a constant frequency of 30 Hz.

The EKF showed a constant response in assays with different frequencies (20 Hz, 30 Hz, 40 Hz and 50 Hz) in the presence of 4%, 8%, 12%, and 16% stator inter-turn fault in phase A, and a load current of 0.72 A in a healthy state (Table 2). This emphasized the robustness of this technique when there was a variation in frequency. Moreover, the results confirmed the simulation results for the same machine during the same operating and fault conditions.

Table 2. EKF estimation response with different frequencies at constant load current in phase A.

Case	Freq (Hz)	Exact $n_{A\ s/c}$ (%)	Simulation $n_{A\ s/c}$ (%)	Practical $n_{A\ s/c}$ (%)
1	20	2%	2.15	1.94
2	20	4%	4.3	3.62
3	20	8%	8.3	7.52
4	20	10%	10.3	9.77
5	20	12%	12.22	12.1
6	20	16%	15.9	16.5
7	30	2%	2.15	1.97
8	30	4%	4.3	3.8
9	30	8%	8.3	7.85
10	30	10%	10.3	9.81
11	30	12%	12.22	11.8
12	30	16%	15.9	16.1
13	40	2%	2.2	1.97
14	40	4%	4.3	3.7
15	40	8%	8.5	7.53
16	40	10%	10.2	9.9
17	40	12%	12.3	12.3
18	50	2%	2.2	2.1
19	50	4%	4.1	4.1
20	50	8%	8.4	7.94
21	50	10%	10.2	10
22	50	12%	12.2	11.8

Besides, the estimation response was tested when exposed to variations in the current (1.5 A and 2.25 A) and at a constant frequency of 30 Hz. Again, the response of the EKF technique showed a robust estimation in load current variation at a constant frequency, (Table 3). For safety conditions, it was not able to emulate short circuit inter turns fault more than 12% as the current in the faulty state went over 5 A; 5 A being the maximum load current for this machine.

Table 3. EKF estimation response with different load currents in phase A at a constant frequency.

Case	Load Current (A)	Exact $n_{A\,s/c}$ (%)	Simulation $n_{A\,s/c}$ (%)	Practical $n_{A\,s/c}$ (%)
1	0.72	2%	2.15	1.97
2	0.72	4%	4.3	3.8
3	0.72	8%	8.3	7.85
4	0.72	10%	10.3	9.81
5	0.72	12%	12.22	11.8
6	0.72	16%	15.9	16.1
7	1.5	2%	2.15	2
8	1.5	4%	4.3	3.9
9	1.5	8%	8.3	8.2
10	1.5	10%	10.3	9.9
11	1.5	12%	12.22	12.1
12	2.25	2%	2.2	1.97
13	2.25	4%	4.3	3.85
14	2.25	8%	8.5	7.9
15	2.25	10%	10.2	9.5
16	2.25	12%	12.3	12

5.6. Decision-Making Process

The decision was taken based on the estimated total internal current in all three phases of the machine. The loads were divided into two groups: critical loads and uncritical loads were connected through contactors K2 and K1, respectively. To prevent the propagation of internal inter-turn faults inside the machine, there are two proposed scenarios:

- the disconnection of the machine;
- load shedding.

Figure 32 shows the flowchart presenting the FDS EKF technique and the proposed scenarios based on the operator's choice.

5.6.1. Scenario 1: The Disconnection of the Machine

This solution provides for the safety of the machine and prevents the propagation of the fault to other turns and phases. However, this solution affects the reliability of the operation, and it requires a backup for the disconnected generator.

This scenario is presented in the experimental work at RMS load current of 1.5 and 30 Hz frequency, in a healthy state the machine gives $n_{A\,s/c} = 0$, $n_{B\,s/c} = 0$ and $n_{C\,s/c} = 0$. At $t = 0.5$ s. A 4% inter-turn fault exists in phase A, the estimated parameter $n_{A\,s/c} = 4.1\%$ and the total estimated internal current reached an RMS value of 2.8 A. As the FDS works in parallel with the protection system of the machine, the disconnection of the machine will be based on the extremely inverse time (EIT) thermal characteristics curve of overcurrent relay based on IEEE standard [46], the expected time to disconnect the machine is 17 s, at TDS = 0.1 s. Figure 33b,c show the detection time and disconnection time of contactor K1 and contactor K2 after detecting the presence of a fault. An LCD was used to monitor the situation of the machine in both the healthy and faulty states; it also shows the expected disconnection time of both contactors and the position of the loads' contactor to inform the operator about the situation.

5.6.2. Scenario 2: Load Shedding

The second scenario is the disconnecting of some uncritical loads (contactor K1) to decrease the total current of the machine allowing it to run under the fault condition. This solution offers the reliability for the process; the machine can continue running in the presence of a fault but with partial loading. This solution does not solve the main problem of internal fault, but it gives the operator a suitable time to take corrective action; the fault may propagate for other turns or phases, respectively, increasing the internal short circuit current, causing a severe fault.

This scenario was implemented at RMS load current of 1.5 A and 30 Hz frequency; at $t = 0.5$ s, 2% inter-turn fault was emulated in phase C, which caused an increase in the total estimated internal current to 2 A. The fault was indicated, and the first group of loads (the uncritical loads) connected through contactor K1 was disconnected (Figure 34a). The disconnection of K1 decreased the current in the machine, and the total current became 1.25 A, allowing the machine to return to its normal state for a definite time and consequently canceling the alarm indication. After a time, the fault percentage increased to 8%, causing an increase in the internal current. In this case, the right decision was to disconnect the machine to solve the internal fault problem. The machine was disconnected after the estimated time based on the EIT characteristics of the overcurrent relay.

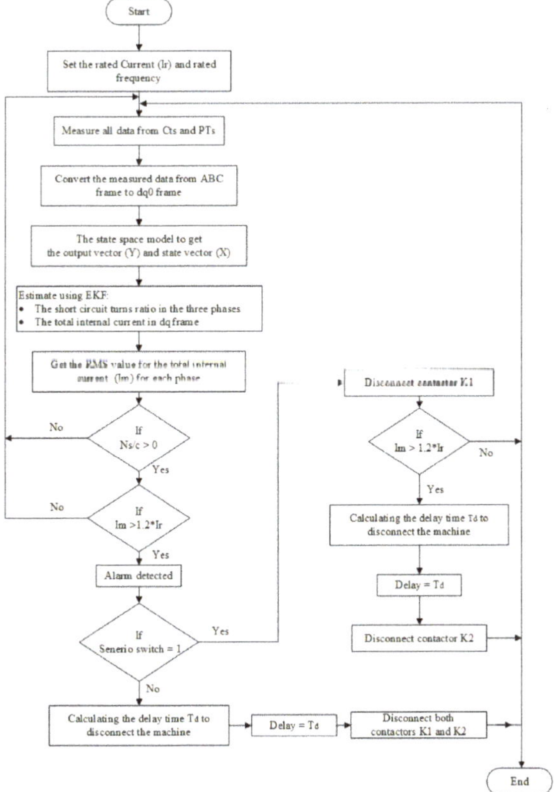

Figure 32. Decision-making process scenarios flowchart.

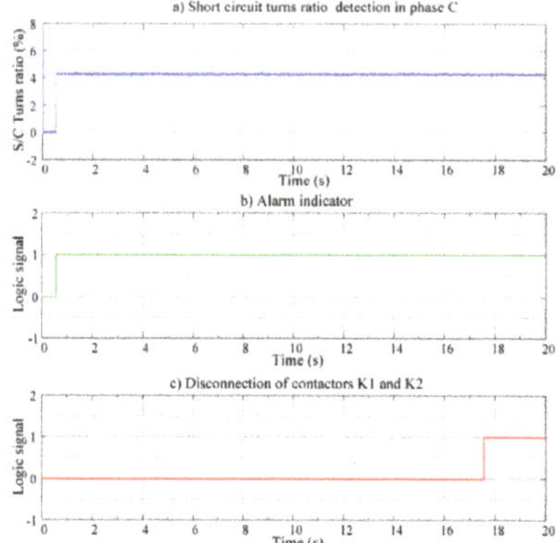

Figure 33. Scenario 1 Alarm indication and machine disconnection.

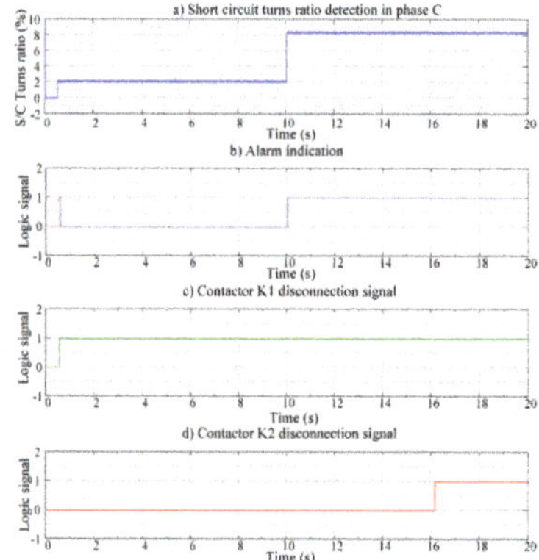

Figure 34. Scenario 2 Alarm indication and contactors disconnection.

6. Conclusions

The paper presents the detection and isolation of PMSG stator windings faults using the EKF and the UKF, which are model-based techniques. The model of the faulty machine was implemented in the state-space model using the machine equations in the dq-frame. The estimated states of the EKF and the UKF techniques were the short circuit turns ratio in each phase. It was noted that the proposed techniques have the following advantages:

- a fast and accurate response in relation to the time needed to take action in real time;
- a robust estimation, in the presence of process and measurement noises, in addition to load and frequency variations.

On the other hand, the UKF technique overcomes the EKF technique drawback of the inaccuracy of the technique in case of severe faults, as it is a nonlinear system and it was linearized around a definite operating point, and so the error of estimation increased as the value of the short circuit turn ratio increased.

Also, the tuning of the weighting matrices (Q and R) has a great impact on the estimated parameters. As indicated in the result, an increase in Q implies an acceleration of the dynamic response of the fault indicator with an increase in noise sensitivity, however, decreasing Q implies better filtering with a decrease in the dynamic response.

The results of this paper point to several exciting directions for future research work. The proposed technique can be used on FD of different types of faults such as bearing, eccentricity, and demagnetizations faults in machines. Moreover, other types of FD techniques may be used, such as artificial intelligence-based techniques and signal-based techniques, and comparing their results with the results of the EKF Technique. This result raises the ability to implement the fault tolerant control (FTC) technique in case of faults such as using the model predictive control (MPC) [47], which would increase the reliability of the machine safety-critical applications.

Author Contributions: Conceptualization, W.E.S., M.A.E.G. and A.L.; methodology, W.E.S. and M.A.E.G.; software, W.E.S.; validation, W.E.S., M.A.E.G. and A.L.; formal analysis, W.E.S., M.A.E.G. and A.L.; investigation, W.E.S., M.A.E.G. and A.L.; resources, W.E.S. and M.A.E.G.; data curation, W.E.S. and M.A.E.G.; writing—original draft preparation, W.E.S. and M.A.E.G.; writing—review and editing, W.E.S. and M.A.E.G.; visualization, W.E.S.; supervision, M.A.E.G. and A.L.; All authors have read and agreed to the published version of the manuscript.

Funding: This research received no external funding.

Conflicts of Interest: The authors declare no conflict of interest.

Nomenclature

L_d	Direct stator synchronous inductance.	mH		$y_m(t)$	Output vector.	
L_q	Quadrature stator synchronous inductance.	mH		T_s	Sampling period.	Sec
$[V_s]_{dq}$	Direct and quadrature stator voltages.	Volts		$I_{S/C}$	Short-circuit current.	Ampere
$P(\theta)$	dq transformation matrix.			$R_{S/C}$	Short-circuit resistance.	Ω
θ	Electrical angular position.	rad		$n_{S/C}$	Short-circuited turns ratio.	%
ω_e	Electrical angular velocity	rad/s		$\widetilde{F}_k$	State equation of the discrete model.	
$[E]$	Electromotive forces vector.	Volt		A_m	State matrix.	
$[Z_{s/c}]$	Equivalent fault impedance.	Ω		Q	State noises covariance matrix.	
$\widetilde{X}_{e_k}$	Extended state vector.			$W_m(t)$	State noises vector.	
$\hat{X}_{k|k}$	Extended state vector.			$x_m(t)$	State vector.	
$\theta_{\frac{s}{c}}$	Fault localization angle.			$[\mathrm{I_s}]_{dq}$	Stator currents vector after variable change in dq-frame.	Ampere
$Q\!\left(\theta_{\frac{s}{c}\,k}\right)$	Fault localization matrix.			$[\mathrm{I'_s}]_{dq}$	Stator currents vector in dq-frame.	Ampere
D_m	Feed forward matrix.			R_S	Stator resistance.	Ω
J	Inertia	Kg.m^2		L_S	Stator synchronous inductance.	mH
B_m	Input matrix.			P	The electromechanical power	Watts
$u_m(t)$	Input vector.	-		$P_{k|k}$	The error covariance matrix at time k	
K'_k	Kalman gain			$\widetilde{H}_k$	The output equations of the discrete linearized model.	
T_m	Load torque	Nm		$P_{k|k-1}$	The prior estimate of P_k	
R	Measurement noises covariance matrix			$\widetilde{F}_K$	The state equations of the discrete linearized model.	
$V_m(t)$	Measurement noises vector.			τ	The time constant of the estimated parameters.Sec	
H_k	Output equation of the discrete model.			σ_u	The variance of input signals noises.	
C_m	Output matrix.			σ_y	The variance of output signals noises.	

References

1. Zhang, Y.; Ji, T.Y.; Li, M.S.; Wu, Q.H.; Wu, Q.H. Application of Discrete Wavelet Transform for Identification of Induction Motor Stator Inter-turn Short Circuit. In Proceedings of the 2015 IEEE Innovative Smart Grid Technologies—Asia, ISGT ASIA 2015, Bangkok, Thailand, 3–6 November 2016. no. 51207058.
2. Wang, H.; Yang, J.; Chen, Z.; Ge, W.; Ma, Y.; Xing, Z.; Yang, L. Model Predictive Control of PMSG—Based Wind Turbines for Frequency Regulation in an Isolated Grid. *IEEE Trans. Ind. Appl.* **2018**, *54*, 3077–3089. [CrossRef]
3. Fakhari, M.; Arani, M. Assessment and Enhancement of a Full-Scale PMSG-Based Wind Power Generator Performance Under Faults. *IEEE Trans. Energy Convers.* **2016**, *31*, 728–739.
4. Jiang, Y.; Zhang, Z.; Jiang, W.; Geng, W.; Huang, J. Three-phase current injection method for mitigating turn-to-turn short-circuit fault in concentrated-winding permanent magnet aircraft starter generator. *IET Electr. Power Appl.* **2018**, *12*, 566–574. [CrossRef]

5. Gao, F.; Zheng, X.; Bozhko, S.; Hill, C.I.; Asher, G. Modal Analysis of a PMSG-Based DC Electrical Power System in the More Electric Aircraft Using Eigenvalues Sensitivity. *IEEE Trans. Transp. Electrif.* **2015**, *1*, 65–76.

6. Sarigiannidis, A.; Kladas, A. High Efficiency Shaft Generator Drive System Design for Ro-Ro Trailer-passenger Ship Application. In Proceedings of the 2015 International Conference on Electrical Systems for Aircraft, Railway, Aachen, Germany, 3–5 March 2015. Ship Propulsion and Road Vehicles (ESARS).

7. Wang, Z.; Yang, J.; Ye, H.; Zhou, W. A Review of Permanent Magnet Synchronous Motor Fault Diagnosis. In Proceedings of the 2014 IEEE Conference and Expo. Transportation Electrification Asia-Pacific (ITEC Asia-Pacific), Beijing, China, 31 August–3 September 2014; pp. 1–5.

8. Ebrahimi, B.M.; Roshtkhari, M.J.; Faiz, J.; Khatami, S.V. Advanced eccentricity fault recognition in permanent magnet synchronous motors using stator current signature analysis. *IEEE Trans. Ind. Electron.* **2014**, *61*, 2041–2052. [CrossRef]

9. Rosero, J.; Romeral, L.; Rosero, E.; Urresty, J. Fault Detection in dynamic conditions by means of Discrete Wavelet Decomposition for PMSM running under Bearing Damage. In Proceedings of the 2009 Twenty-Fourth Annual IEEE Applied Power Electronics Conference and Exposition, Washington, DC, USA, 15–19 Febuary 2009; pp. 951–956.

10. Ebrahimi, B.M.; Faiz, J.; Roshtkhari, M.J. Static-, dynamic-, and mixed-eccentricity fault diagnoses in permanent-magnet synchronous motors. *IEEE Trans. Ind. Electron.* **2009**, *56*, 4727–4739. [CrossRef]

11. Pacas, M.; Villwock, S.; Dietrich, R. Bearing Damage Detection in Permanent Magnet Synchronous Machines. In Proceedings of the 2009 IEEE Energy Conversion Congress and Exposition, San Jose, CA, USA, 20–24 September 2009; pp. 1098–1103.

12. Soualhi, A.; Jean-monnet, U.; Lyon, U. Early Detection of Bearing Faults by the Hilbert-Huang Transform. In Proceedings of the 2016 4th International Conference on Control. Engineering & Information Technology (CEIT-2016), Hammamet, Tunisia, 16–18 December 2016; pp. 16–18.

13. Rosero, J.; Romeral, J.L.; Cusido, J.; Ortega, J.A.; Garcia, A. Fault Detection of Eccentricity and Bearing Damage in a PMSM by Means of Wavelet Transforms Decomposition of the Stator Current. In Proceedings of the 2008 Twenty-Third Annual IEEE Applied Power Electronics Conference and Exposition, Austin, TX, USA, 24–28 Febuary 2008; pp. 111–116.

14. Huang, T.; Fu, S.; Feng, H.; Kuang, J. Bearing Fault Diagnosis Based on Shallow Multi-Scale Convolutional Neural Network with Attention. *Energies* **2019**, *12*, 3937. [CrossRef]

15. Zhang, H.; Zhang, S.; Yin, Y. A Novel Improved ELM Algorithm for a Real. *Math. Probl. Eng.* **2014**, *2014*. [CrossRef]

16. Vinson, G.; Combacau, M.; Prado, T.; Ribot, P. Permanent Magnets Synchronous Machines Faults Detection and Identification. In Proceedings of the IECON 2012—38th Annual Conference on IEEE Industrial Electronics Society, Montreal, QC, Canada, 25–28 October 2012; pp. 3925–3930.

17. Espinosa, A.G.; Rosero, J.A.; Cusidó, J.; Romeral, L.; Ortega, J.A. Fault detection by means of Hilbert-Huang transform of the stator current in a PMSM with demagnetization. *IEEE Trans. Energy Convers.* **2010**, *25*, 312–318. [CrossRef]

18. Ruiz, J.R.R.; Rosero, J.A.; Espinosa, A.G.; Romeral, L. Detection of demagnetization faults in permanent-Magnet synchronous motors under nonstationary conditions. *IEEE Trans. Magn.* **2009**, *45*, 2961–2969. [CrossRef]

19. Obeid, N.H.; Boileau, T.; Nahid-Mobarakeh, B. Modeling and diagnostic of incipient interturn faults for a three-phase permanent magnet synchronous motor. *IEEE Trans. Ind. Appl.* **2016**, *52*, 4426–4434. [CrossRef]

20. Aubert, B.; Regnier, J.; Caux, S.; Alejo, D. Kalman-Filter-Based Indicator for Online Interturn Short Circuits Detection in Permanent-Magnet Synchronous Generators. *Ind. Electron. IEEE Trans.* **2015**, *62*, 1921–1930. [CrossRef]

21. Chen, Y.; Liang, S.; Li, W.; Liang, H.; Wang, C. Faults and Diagnosis Methods of Permanent Magnet Synchronous Motors: A Review. *Appl. Sci.* **2019**, *9*, 2116. [CrossRef]

22. Liang, H.; Chen, Y.; Liang, S.; Wang, C. Fault Detection of Stator Inter-Turn Short-Circuit in PMSM on Stator Current and Vibration Signal. *Appl. Sci.* **2018**, *8*, 1677. [CrossRef]

23. Kościelny, J.M.; Syfert, M.; Sztyber, A. *Advanced Solutions in Diagnostics and Fault Tolerant Control*; Springer: New York, NY, USA, 2017.

24. Duque-Perez, O.; del Pozo-Gallego, C.; Morinigo-Sotelo, D.; Godoy, W.F. Condition monitoring of bearing faults using the stator current and shrinkage methods. *Energies* **2019**, *12*, 3392. [CrossRef]

25. Helal, A.H.; Badran, E.F.; Ashour, H.A. Synchronous Generator Stator Earth fault Classification and Location Using Wavelet Transform and ANFIS. In Proceedings of the 2017 Nineteenth International Middle East. Power Systems Conference (MEPCON), Menoufia University, Cairo, Egypt, 19–21 December 2017; pp. 19–21.

26. Devi, N.R. Diagnosis and Classification of Stator Winding Insulation Faults on a Three-phase Induction Motor using Wavelet and MNN. *IEEE Trans. Dielectr. Electr. Insul.* **2016**, *23*. [CrossRef]

27. Rosero, J.A.; Romeral, L.; Ortega, J.A.; Rosero, E. Short-circuit detection by means of empirical mode decomposition and Wigner-Ville distribution for PMSM running under dynamic condition. *IEEE Trans. Ind. Electron.* **2009**, *56*, 4534–4547. [CrossRef]

28. Cempel, C. Descriptive parameters and contradictions in TRIZ methodology for vibration condition monitoring of machines. *Diagnostyka* **2014**, *15*, 51–59.

29. Cempel, C. Generalized singular value decomposition in multidimensional condition monitoring of machines-A proposal of comparative diagnostics. *Mech. Syst. Signal. Process.* **2009**, *23*, 701–711. [CrossRef]

30. Patan, K.; Witczak, M.; Korbicz, J. Towards robustness in neural network based fault diagnosis. *Int. J. Appl. Math. Comput. Sci.* **2008**, *18*, 443–454. [CrossRef]

31. Skowron, M.; Orlowska-Kowalska, T.; Wolkiewicz, M.; Kowalski, C.T. Convolutional neural network-based stator current data-driven incipient stator fault diagnosis of inverter-fed induction motor. *Energies* **2020**, *13*, 1475. [CrossRef]

32. Medoued, A.; Lebaroud, A.; Laifa, A.; Sayad, D. Feature form Extraction and Optimization of Induction Machine Faults Using PSO Technique. In Proceedings of the 3rd International Conference on Electric Power and Energy Conversion Systems, Istanbul, Turkey, 2–4 October 2013; pp. 1–5.

33. Quiroga, J.; Liu, L.; Cartes, D.A. Fuzzy Logic Based Fault Detection of PMSM Stator Winding Short under Load Fluctuation Using Negative Sequence Analysis. In Proceedings of the American Control, Seattle, WA, USA, 11–13 June 2008; pp. 4262–4267.

34. Mini, V.P.; Sivakotaiah, S.; Ushakumari, S. Fault Detection and Diagnosis of an Induction Motor Using Fuzzy Logic. In Proceedings of the IEEE Region 8 SIBIRCON-2010, Irkutsk Listvyanka, Russia, 11–15 July 2010; pp. 459–464.

35. Korbicz, J.; Kowal, M. Neuro-fuzzy networks and their application to fault detection of dynamical systems. *Eng. Appl. Artif. Intell.* **2007**, *20*, 609–617. [CrossRef]

36. Pazera, M.; Korbicz, J. A Process Fault Estimation Strategy for Non-linear Dynamic Systems Marcin. In Proceedings of the International Conference on Recent Trends in Physics (ICRTP 2016), Lille, France, 17–18 November 2016; Volume 755.

37. Pazera, M.; Witczak, M.; Korbicz, J. Combined Estimation of Actuator and Sensor Faults for Non-linear Dynamic Systems. In Proceedings of the 2017 22nd International Conference on Methods and Models in Automation and Robotics, Miedzyzdroje, Poland, 28–31 August 2017; pp. 933–938.

38. Khov, M.; Regnier, J.; Faucher, J. Monitoring of Turn Short-circuit Faults in Stator of PMSM in Closed Loop by On-line Parameter Estimation. In Proceedings of the 2009 IEEE Int. Symp. Diagnostics Electr. Mach. Power Electron. Drives, Cargese, France, 31 August–3 September 2009.

39. Wafik, W.; El-geliel, M.A.; Lotfy, A. PMSG Fault Diagnosis in Marine Application. In Proceedings of the 2016 20th International Conference on System Theory, Control. and Computing (ICSTCC), Sinaia, Romania, 13–15 October 2016; pp. 626–631.

40. Saad, W.W.; El-geliel, M.A.; Lotfy, A. IM Stator Winding Faults Siagnosis Using EKF. In Proceedings of the 2017 fifth Advanced Control Circuits and Systems (ACCS'017), Alexandria, Egypt, 5–8 November 2017; pp. 34–39.

41. Lončarek, T.; Lešić, V.; Vašak, M. Increasing Accuracy of Kalman Filter-based Sensorless Control of Wind Turbine PM Synchronous Generator. In Proceedings of the 2015 IEEE International Conference on Industrial Technology (ICIT), Seville, Spain, 17–19 March 2015; pp. 745–750.

42. Zawirski, K.; Janiszewski, D.; Muszynski, R. Unscented and extended Kalman filters study for sensorless control of PM synchronous motors with load torque estimation. *Bull. POLISH Acad. Sci. Tech. Sci.* **2013**, *61*, 793–801. [CrossRef]

43. Kościelny, J.M.; Syfert, M.; Tabor, Ł. Application of Knowledge AboutResidual Dynamics for Fault Isolation and Identification. In Proceedings of the Conference on Control. and Fault-Tolerant Systems, SysTol, Nice, France, 9–11 October 2013; pp. 275–280.

44. Sztyber, A.; Ostasz, A.; Koscielny, J.M. Graph of a process—A new tool for finding model structures in a model-based diagnosis. *IEEE Trans. Syst. Man, Cybern. Syst.* **2015**, *45*, 1004–1017. [CrossRef]
45. Wan, E.A.; van der Merwe, R. The Unscented Kalman Filter for Nonlinear Estimation. In Proceedings of the IEEE 2000 Adaptive Systems for Signal. Processing, Communications, and Control. Symposium (Cat. No.00EX373), Lake Louise, AB, Canada, 4 October 2002; pp. 153–158.
46. Electric, S. *Extremely Inverse Time (EIT) Curves Current I/I s*; Schneider Electric: Rueil-Malmaison, France, 2008; p. 3000.
47. Witczak, P.; Luzar, M.; Witczak, M.; Korbicz, J. A Robust Fault-tolerant Model Predictive Control for Linear Parameter-varying Systems. In Proceedings of the 2014 19th International Conference on Methods and Models in Automation and Robotics (MMAR), Miedzyzdroje, Poland, 2–5 September 2014; pp. 462–467.

Article

GaN-Based DC-DC Resonant Boost Converter with Very High Efficiency and Voltage Gain Control

Zbigniew Waradzyn *, Robert Stala, Andrzej Mondzik, Aleksander Skała and Adam Penczek

Department of Power Electronics and Energy Control Systems, Faculty of Electrical Engineering, Automatics, Computer Science and Biomedical Engineering, AGH University of Science and Technology, al. Mickiewicza 30, 30-059 Krakow, Poland; stala@agh.edu.pl (R.S.); mondzik@agh.edu.pl (A.M.); skala@agh.edu.pl (A.S.); penczek@agh.edu.pl (A.P.)
* Correspondence: waradzyn@agh.edu.pl; Tel.: +48-12-617-2811

Received: 29 October 2020; Accepted: 30 November 2020; Published: 3 December 2020

Abstract: This paper presents a concept for the operation of a resonant DC–DC switched-capacitor converter with very high efficiency and output voltage regulation. In its basic concept, such a converter operates as a switched-capacitor voltage doubler (SCVD) in the Zero Current Switching (ZCS) mode with a constant output voltage. The proposed methods of switching allow for the switched-capacitor (SC) converter output voltage regulation, and improve its efficiency by the operation with Zero Voltage Switching (ZVS). In this paper, various switching patterns are proposed to achieve high efficiency and the output voltage control by frequency or duty cycle regulation. Some examples of the application of the proposed switching patterns are presented: in current control at the start-up of the converter, in a bi-directional converter, and in a modular cascaded system. The paper also presents an analytical model as well as the relationships between the switching frequency, voltage ratio and efficiency. Further, it demonstrates the experimental verification of the waveforms, voltage ratios, as well as efficiency. The proposed experimental setup achieved a maximum efficiency of 99.228%. The implementation of the proposed switching patterns with the ZVS operation along with the GaN-based (Gallium Nitride) design, with a planar choke, leads to a high-efficiency and low-volume solution for the SCVD converter and is competitive with the switch-mode step-up converters.

Keywords: boost converters; DC–DC power converters; GaN switch; resonant power conversion; zero current switching (ZCS); zero-voltage switching (ZVS)

---

1. Introduction

The switched-capacitor (SC) conversion principle applied in power electronic converters is a promising technology [1]. Switched-capacitor topologies are often proposed for the DC–DC conversion, which can achieve favorable features such as a high voltage ratio or light weight [1–27]. In some concepts, the SC converters can achieve continuous voltage regulation, as in references [2–4]. The SC converters, such as multipliers, can operate with a fixed voltage gain [5,6] or a discrete voltage ratio variation [7]. A switched-capacitor power converter can operate as a zero-current-switching (ZCS) circuit with limited current stress of its components by the application of oscillatory circuits for the recharging of the capacitors. Zero current switching (ZCS) converters can achieve high efficiency, which is demonstrated in [5–7]. A substantial part of losses in the SC converters results from conduction losses. This is reported in [4] for a resonant converter, discussed in detail in [5] for a voltage multiplier, and analyzed in [8] in a generic losses model of losses in the SC converters. However, the switching loss is still an important subject of an investigation into SC converters. Previous works related to the efficiency of the SC multipliers (SCVMs) [5–7] show that the losses associated with the discharging of the transistor output capacitance (C_{OSS} loss) can be significant. The reason for this is that the transistors of an SC converter operating in the ZCS mode are turned-on with the output capacitance charged. However, in the SC resonant doubler (SCVD—such as that presented in Figure 1), their output charge can be reduced by the reverse current flow in the transistor before its turn-on. Therefore, this paper introduces a method of switching off the SCVD (Figure 1), where the switching pulse is shorter than half the period of the oscillations in the resonant circuit. The method of efficiency improvement, by phase-shift control in a resonant switched capacitor converter (RSCC), has been demonstrated in [2] and developed in [3] with the use of the (Gallium Nitride) GaN switches. In the classic switched-capacitor voltage multiplier [5], the Zero Voltage Switching (ZVS) operation (with hard turn-off of the switches) is not applicable, as the topology does not allow for the current termination in the resonant circuits.

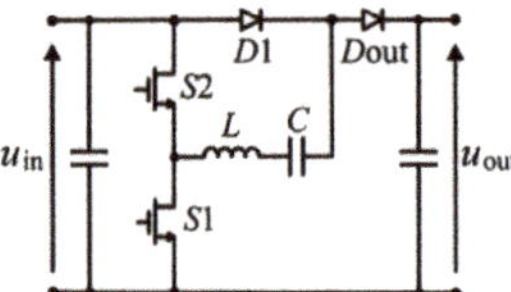

Figure 1. GaN-based resonant switched-capacitor voltage doubler (SCVD) with output voltage range from U_{in} to $2U_{in}$.

For the highest-efficiency operation, the switch needs to be turned off shortly before the zero-crossing of the oscillating current, and the other switch is turned on during its reverse conduction. In this mode, the ZVS and the low-current switching (LCS) is achieved when sufficiently fast switches are used. This can be accomplished by applying GaN switches. Other features of a GaN switch can also be favorable for the resonant SCVD. The linear function of the output capacitance C_{OSS} versus drain-source voltage V_{ds} is meant to improve the shape of the currents in the oscillatory circuits. The drain-source on resistance $R_{ds(on)}$ increase versus temperature is lower than in the case of silicon MOSFETs, which allows for operating at a higher temperature with high efficiency. A smaller area of GaN devices versus Si devices is beneficial for high power density design as well. Furthermore, a low gate charge of the GaN switch is important in this converter. Other features of GaN switches can be found in [28]. In [5], an impact of the dead-time between the input current pulses on the efficiency is demonstrated. The method of the reduction in Q_{oss} losses (the losses resulting from the output charge Q_{oss} discharge at transistor turn-ons) that is proposed in this paper improves the input current shape and allows for a decrease in both the root mean square (RMS) value of the currents in the SC converter, and its resistive losses. In [29], a comparison of power dissipation under the soft switching operation is presented between Si MOSFET 650 V, SiC MOSFET 900 V, GaN E-HEMT 600 V and GaN GIT 600 V. The lowest losses are reported there for the GaN GIT transistors (such type of switches is used in the experimental tests presented in this paper). In an SCVD design, this requires the application of switches with blocking voltage above 650 V (voltage limit of majority superjunction MOSFETs and GaN commercial devices); therefore, a SiC switch can be the most favorable option. In [29], the SiC power dissipation is located between those of GaN and Si superjunction MOSFET, and in [30], it has been concluded that 1200 V SiC devices have a better switching performance than those of 600 V. The application of SiC switches in a bidirectional MRSCC (multilevel resonant switched capacitor converter) converter with voltage ratio 0.5/2 kV is demonstrated in [31], where efficiency of 98.5% was achieved.

The amount of the dissipated energy associated with C_{OSS} losses may be difficult to predict on the basis of C_{OSS} datasheet parameter. In [32], a problem of underestimation of the energy stored in the output capacitance for a large signal operation of superjunction Si MOSFETs is presented. Therefore, the experimental results presented in this paper allow a credible characterization of the parameters of the switches associated with C_{OSS} losses: they present a comparison of efficiencies obtained in the ZCS and ZVS operation. This is one of the important contributions of the paper.

The major disadvantage of SC converters is that they have a limited regulation capability. In the vast majority of cases, the SC converter is a constant-voltage-ratio converter. However, the output voltage regulation is achievable by the use of a suitable topology and control. In this paper, an output voltage regulation capability of the SC doubler is investigated using suitable switching patterns. The method is based on the control of the switching frequency f_S in the range above the frequency of the oscillations in the resonant circuit of the converter and/or the turn-on time of the transistors. This method introduces a hard termination of the current in a transistor, which involves the reverse conduction of the second transistor and its turn-on at zero voltage.

A number of methods for the output voltage regulation in SC converters are analyzed in [2–4,7,9–24]. The phase-shift control in the RSCC presented in [2,3], aside from the improvement of efficiency, introduces the output voltage regulation. However, the methods proposed in this paper use the phase-shift concept as well, but introduce a number of switching patterns, present selected applications of the switching patterns and the analysis of voltage gain and efficiency. The proposed research brings an important contribution to the area of voltage regulation when compared to that presented in [2,3]. Further improvement in the SC resonant converter control by the use of phase shift method, switching frequency and dead-time control depending on the load conditions can be found in [23]. The reference [24] proposes modified Dickson RSCC step-down converters with ZVS operation and full-range regulation via modulation.

In [4], the voltage gain is controlled by the time delay introduced between the switching cycles. In [7], a method of regulation by reducing the number of active switching cells in the multiplier is proposed. This method enables the operation of the converter with various output voltage values in a steady state. The composition of the converter with switched capacitors and switched inductors [9–14] allows for the output voltage regulation by the duty cycle control. However, it requires additional passive magnetic components of significant values when compared with those used in the SCVD analyzed in this paper. A decrease in the volume and weight of converters by reducing the values of their inductive components is an important contemporary trend. Similarly, the elimination of ferrite components in the converter allows for an operation at higher temperatures, which is also favorable in many applications. This is achieved in the proposed SCVD. In [9], a 95 µH choke is used in the system combining SC circuits and a switch-inductor circuit that transfers the energy to the output. In [10], a converter integrating an SC converter and a cell that stores energy in an inductor (430 µH) to achieve high voltage gain with the output voltage regulation is presented. In [11], a converter utilizing a series/parallel connection of inductors (200 µH) and capacitors is proposed for high voltage gain with regulation. A converter operating on the principle of integration of a switch-mode DC–DC regulated converter (with the input inductance of 1.33 mH) with a simple SC circuit composed of diodes and capacitors is presented in [12]. In comparison to the inductors used in [9–12], with values that could be also typical for a DC–DC boost converter, the proposed resonant DC–DC converter uses an inductance of 10.4 µH in the experimental setup.

Some voltage gain control methods utilize the switching frequency variation. Such concepts can be found in [13] for a ladder resonant SC converter (RSC), and in [14,15] for two-switch SC converters.

A method for a variable number of voltage gains is presented in [7], and in [16–20]. In [16], a multilevel output voltage is generated by the appropriate connection of SC components, which gives the input voltage of a half-bridge inverter in the multilevel inverter. In [17], a multilevel DC–DC converter utilizes multiple DC sources. The methods and topologies presented in [18–20] demonstrate the ability for fractional voltage gain control. However, the method analyzed in this paper allows for continuous voltage regulation. Aside from the output voltage control, the proposed switching methods allow for controlling overload states or the start-up of the converter that may lead to an overcurrent.

Another issue, which is novel and presented in this paper, is associated with the bi-directional operation of the SCVD. This example is presented in Section 5, where the reverse conduction of the switch is used. The application of a GaN switch, in this case, makes it possible to avoid the losses connected with a reverse recovery charge (Qrr losses).

All in all, the major contributions of this paper related to efficiency improvement include: the proposition of various switching concepts for power loss reduction, analysis and the development of a model of efficiency, an implementation of the ZVS operation in an SCVD converter, experimental research with a GaN-based SCVD setup, and the demonstration of results related to operation and efficiency of the converter. Some issues such as analysis of various methods of switching and power losses modeling in the SCVD are novel in relation to previous works. This research is a follow-up to the contemporary trends of efficiency improvement and the analyses of prospective topologies for GaN switches favorable implementations.

The major contributions of this paper related to the output voltage regulation in the SCVD converter include the proposition and analysis of various switching methods with a model of voltage gain, and the presentation of examples of their capabilities. Some experimental results of steady-state voltage gain of the converter as well as the dynamical states of the output voltage control are also presented. The application of GaN switches [33,34] makes it possible to implement the proposed switching concepts in high-frequency converters.

The SC converters can be attractive in photovoltaic or fuel-cell low-power systems, where the ability for a high voltage gain is required. The demonstration, in this paper, of bi-directional DC–DC conversion suggests the possibility of implementing it in battery-powered systems. Low weight and volume, achieved in an inductiveless design, can be very favorable in such applications as well. When ferrites are not used in an SC-based converter, it can operate at a higher temperature, which allows the optimization of the converter towards a low volume of heat sink or operation in a higher temperature environment.

The paper is organized in the following way. Section 2 introduces the principle of operation of the SCVD and various switching patterns. The most advantageous patterns are selected and their operation, efficiency, voltage ratio and rated power are presented. Section 3 contains an analysis of the voltage gain control in the SCVD and demonstrates models of output voltage versus switching frequency and power for two switching patterns. Section 4 shows the model of efficiency of the SCVD. In Section 5, we present selected examples of the use of mixed switching patterns for the start-p control of an SCVD, in a bi-directional SCVD, and in a modular SC system. Section 6 presents the laboratory model of the SCVD converter as well as the experimental results which confirm the proper operation, regulation ability and high efficiency of the converter.

2. Operation Principle of the Resonant Power SCVD

According to the basic principle of the operation of a Switch-Capacitor Voltage Multiplier (SCVM) or an SCVD described in [5,6], the converter operates in the ZCS mode. The topology presented in Figure 1 is explained in [6] and in [2,3] (in the case of a converter equipped with four transistors). Both in the charging and discharging cycle of the switched capacitor, the current oscillates and reaches zero value (Table 1, pattern P1). In such a switching method, the theoretical voltage gain of the SCVD equals

$$G_U = U_{\text{out}}/U_{\text{in}} = 2 \tag{1}$$

In the ZCS mode, conduction losses and switching losses of the SCVD are caused by the discharging of the transistor output capacitance during the turn-on transitions (C_{OSS} loss). Using that soft-switching mode, the efficiency of the SCVD has the following analytical model (based on [5]):

$$\eta = 1 - \frac{\pi^2}{16} \frac{P_{\text{in}} r}{U_{\text{in}}^2} \frac{T_S}{T_1} - \frac{\Delta U_D}{U_{\text{in}}} - \frac{\Delta W_{\text{Sn}} f_S}{P_{\text{in}}} \tag{2}$$

where U_{in} is the input voltage, P_{in} is the input power, r denotes the total resistance, both of the circuit of charging and discharging the switched capacitor, including the resistance of the switch, and ΔU_D is the voltage drop across each diode, assumed to remain constant in the forward-conducting state. T_1 is the conduction time of the transistors, T_S is the switching period, f_S is the switching frequency, and ΔW_{Sn} is the energy lost at turn-ons in the resistances of both the switches in a single switching period.

Table 1. Switching patterns of SCVD and their basic features.

Pattern		
P1	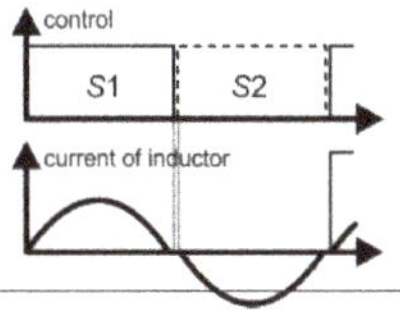	Operation without voltage regulation. Basic ZCS switching [5–7] Features: • problem with C_{OSS} losses.
P2	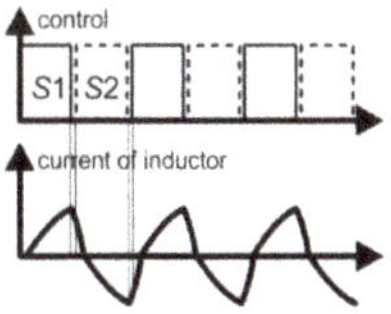	Regulation by switching frequency variation Operation above the resonant frequency with continuous inductor current. Features: • ZVS-low conduction losses (no blanking times between pulses, low RMS currents) [5], • output voltage regulation is possible by switching frequency regulation, • the output voltage range is limited by the maximal applicable switching frequency. Therefore, $U_{outmin} > U_{in}$.
P3	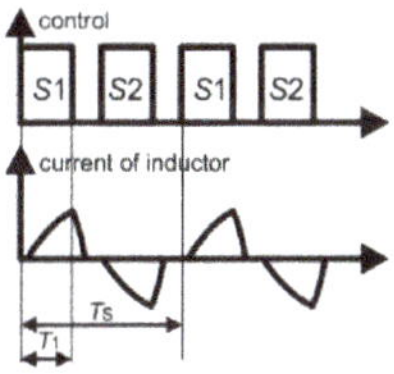	Regulation by duty cycle control Operation with short pulses and discontinuous inductor current. Features: • output voltage regulation is possible with the constant switching frequency, • the minimum output voltage is not limited: $U_{outmin} = U_{in}$, • hard turn-on.
P4	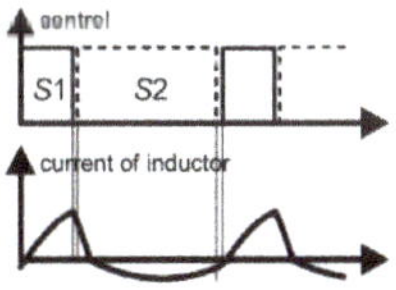	Regulation by duty cycle control Features: • ZVS, • output voltage regulation is possible with variable switching frequency, • the minimum output voltage is not limited: $U_{outmin} = U_{in}$, • control design similar to typical procedures for switch-mode DC–DC converters with a regulated duty cycle (switching frequency regulation required).

Figure 1 shows that the voltage on the switches of an SCVD equals the input voltage. However, in an SCVM, the voltages on the switches exceed the input voltage, and they increase in the switching cells nearer to the output [5].

In the case of the application of GaN switches, it is assumed that the converter can operate with very high efficiency under modes when the switches are turned-off while conducting. Therefore, each switching period can consist of four states that are presented in Figure 2.

• State 1: transistor $S2$ is switched-off and transistor $S1$ conducts the source current that charges the switched capacitor (SC);
• State 2: transistor $S1$ is switched-off and transistor $S2$ conducts reversely until the inductance current reaches zero; the charging of the SC is continued in this state;

- State 3: transistor *S1* is switched-off and transistor *S2* conducts the current that is forced by the source and the switched capacitor to flow to the output;
- State 4: transistor *S2* is switched-off and transistor *S1* conducts reversely until the inductance current reaches zero; the charging of the output capacitor is continued.

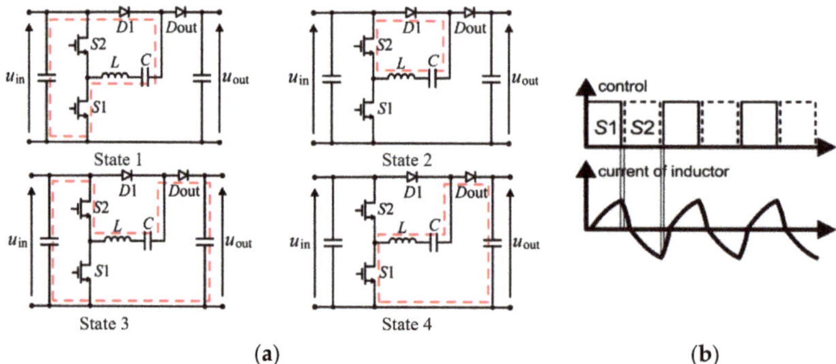

Figure 2. (**a**) Operating states of SCVD in charging and discharging cycles of the switched capacitor. (**b**) Idealized control signals and inductor current waveforms. Pattern P2.

Assuming a hard termination of the transistor current, various switching patterns can be proposed (Table 1). Pattern P1 with the ZCS switching is also shown for comparison.

Under the ZVS operation, the inductor current can discharge the capacitance of a switch before it starts flowing in the reverse direction (Figure 3). This can occur when the switch output capacitance is low, because a high-frequency SC converter is designed with a very low parasitic inductance. Therefore, a GaN or a superjunction MOSFET switch is very favorable in such an operating mode, due to short transition times. As the ZVS operation is more efficient than the ZCS switching, pattern P2 appears to be the most attractive. Furthermore, using pattern P2, switching can be achieved in the ZVS and nearly ZCS mode (LCS–low current switching), which is discussed in more detail in Section 4.

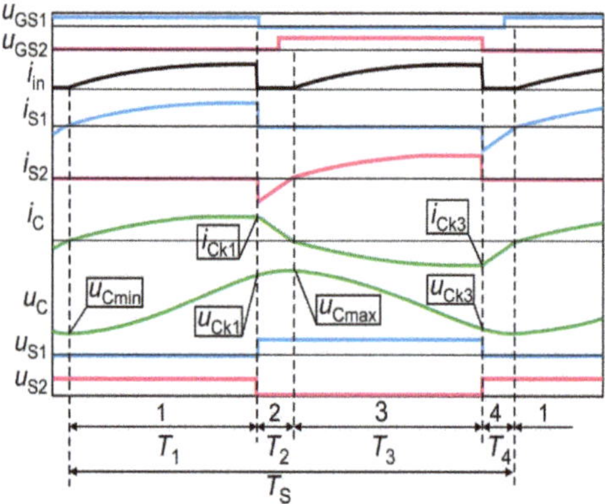

Figure 3. Theoretical time waveforms of currents and voltages in SCVD useful for the analysis–switching pattern P2.

3. Output Voltage Control of SCVD

An idea of the output voltage regulation of the SCVD assumes the shortening of the charging and/or discharging process of the switched capacitor. Thus, a lower amount of energy is transferred through this component. At the same time, high efficiency of operation can be achieved by reducing the RMS current and eliminating turn-on losses. Thus, when a wide range of switching frequency is assumed on the stage of the design, pattern P2 (Table 1) seems to be the most favorable, and the operation under this pattern will be further analyzed in more detail.

The analysis below refers both to the switching patterns P2 ($T_1 + T_2 + T_3 + T_4 = T_S$, Figure 3) and P3 ($T_1 + T_2 + T_3 + T_4 < T_S$) (Table 1). It assumes ideal power electronic switches, fixed values of input voltage U_{in} and power P_{in}, equal values of resonant frequency f_0 and characteristic impedance ρ of each current path in Figure 2, as well as neglecting parasitic resistances and voltage drops across the power electronic devices, where

$$\omega_0 = 2\pi f_0 = 1/\sqrt{(LC)}, \; \rho = \omega_0 L = \sqrt{L/C} \tag{3}$$

The capacitor is being charged in states 1 and 2 (Figures 2 and 3). The capacitor current and voltage are described by Equations (4)–(11) (time is counted from zero from the beginning of each state)). They present the current of a typical series LC circuit supplied from a voltage source, and the voltage across its capacitor, taking into account the initial values of the currents and voltages (Figure 3). The capacitor current and the voltage across it are given by

$$i_C(t) = \frac{U_{in} - U_{Cmin}}{\rho} \sin \omega_0 t = I_m \sin \omega_0 t \tag{4}$$

$$u_C(t) = U_{in} - (U_{in} - U_{Cmin}) \cos \omega_0 t \tag{5}$$

in cycle 1, with $i_C(T_1) = I_{Ck1}$, $u_C(T_1) = U_{Ck1}$, and by (6) and (7)

$$i_C(t) = I_{Ck1} \cos \omega_0 t - \frac{U_{Ck1}}{\rho} \sin \omega_0 t \tag{6}$$

$$u_C(t) = U_{Ck1} \cos \omega_0 t + \rho I_{Ck1} \sin \omega_0 t \tag{7}$$

in cycle 2, with $i_C(T_2) = 0$, $u_C(T_2) = U_{Cmax}$.

The capacitor is being discharged in states 3 and 4 (Figures 2 and 3). The capacitor current and voltage are as follows:

$$i_C(t) = -(U_{in} - U_{out} + U_{Cmax})/\rho \sin \omega_0 t \tag{8}$$

$$u_C(t) = U_{out} - U_{in} + (U_{in} - U_{out} + U_{Cmax}) \cos \omega_0 t \tag{9}$$

in cycle T_3, with $i_C(T_3) = I_{Ck3}$, $u_C(T_3) = U_{Ck3}$, and

$$i_C(t) = I_{Ck3} \cos \omega_0 t + (U_{out} - U_{Ck3})/\rho \sin \omega_0 t \tag{10}$$

$$u_C(t) = U_{out} - (U_{out} - U_{Ck3}) \cos \omega_0 t + \rho I_{Ck3} \sin \omega_0 t \tag{11}$$

in cycle 4, with $i_C(T_4) = 0$, $u_C(T_4) = U_{Cmin}$.

For further analysis, it is assumed that

$$T_1 = T_3, \qquad T_2 = T_4 \tag{12}$$

$$I_{Ck3} = -I_{Ck1} \tag{13}$$

I_m (4) is the amplitude of the input current and the switched capacitor current, and can be calculated based on the expression

$$I_{inav} = \frac{P_{in}}{U_{in}} = \frac{2}{T_S} \int_0^{T_1} I_m \sin \omega_0 t \, dt = \frac{1}{\pi} \frac{f_S}{f_0} I_m \left[1 - \cos\left(2\pi \frac{T_1}{T_0}\right) \right] \tag{14}$$

All the currents and voltages in the SCVD can be computed based on (4)–(14). Introducing normalized quantities

$$I_{mn} = \frac{I_m}{U_{in}/\rho}, \; I_{Ck1n} = \frac{I_{Ck1}}{U_{in}/\rho}, \; P_{in\_n} = \frac{P_{in}}{U_{in}^2/\rho}$$
$$U_{Ck1n} = U_{Ck1}/U_{in}, \; U_{Cmin-n} = U_{Cmin}/U_{in}, \; U_{Cmax-n} = U_{Cmax}/U_{in}, \; U_{outn} = U_{out}/U_{in}$$
$$U_{outn} = U_{out}/U_{in}$$
$$f_{Sn} = f_S/f_0, \; T_{1n} = T_1/T_S, \; T_{2n} = T_2/T_S \tag{15}$$

we obtain

$$I_{mn} = \frac{\pi P_{in-n}/f_{Sn}}{1 - \cos\!\left(2\pi T_{1n}/f_{Sn}\right)}, \; I_{Ck1n} = I_{mn}\sin\!\left(2\pi T_{1n}/f_{Sn}\right) \tag{16}$$

$$U_{Cmin-n} = 1 - I_{mn}, \quad U_{Ck1n} = 1 - I_{mn}\cos\!\left(2\pi T_{1n}/f_{Sn}\right) \tag{17}$$

$$T_{2n} = f_{Sn}\,\text{arctg}\!\left(I_{Ck1n}/U_{Ck1n}\right)/(2\pi) \tag{18}$$

$$U_{Cmax-n} = U_{Ck1n}\cos\!\left(2\pi T_{2n}/f_{Sn}\right) + I_{Ck1n}\sin\!\left(2\pi T_{2n}/f_{Sn}\right) \tag{19}$$

$$U_{outn} = U_{Cmin-n} + U_{Cmax-n}, \quad U_{Ck3n} = U_{outn} - U_{Ck1n} \tag{20}$$

3.1. Pattern P2—Continuous Capacitor Current Mode

The capacitor current is continuous if $T_1 + T_2 + T_3 + T_4 = T_S$ (pattern P2—Table 1 and Figure 3), which corresponds to (12)

$$T_S = 2(T_1 + T_2) \tag{21}$$

This switching pattern can be easily obtained by varying T_S and using long gate pulses, as shown in Table 1 for pattern P2–times T_1 and T_2 (Figure 3) will be set automatically.

Using (16)–(20) and taking into account that (21) $T_{2n} = 1/2 - T_{1n}$ yields

$$\pi\!\left(1 - 2T_{1n}\right) = f_{Sn}\,\text{arctg}\!\left[\frac{\pi P_{in-n}/f_{Sn}\,\sin\!\left(2\pi T_{1n}/f_{Sn}\right)}{1 - \left(1 + \pi P_{in-n}/f_{Sn}\right)\cos\!\left(2\pi T_{1n}/f_{Sn}\right)}\right] \tag{22}$$

From (22), normalized conduction time $T_{1n} = T_1/T_S$ of the transistor can be computed numerically. Figure 4 presents T_{1n} (a), and normalized output voltage U_{outn} (b) as a function of f_{Sn} for three values of $P_{in-n} = P_{in}/(U_{in}^2/\rho)$: 0.0344, 0.0688, and 0.1031. The value of $P_{in-n} = 0.0688$ corresponds to, e.g.,: $U_{in} = 200$ V, $P_{in} = 400$ W, and $\rho = 6.876$, which can be obtained for, e.g., $L = 10.4$ µH and $C = 0.22$ µF. These parameters correspond to those of the experimental setup presented in Section 6.

In switching pattern P2, the theoretical lower limit of the normalized switching frequency f_{Sn} is 1, which corresponds to switching pattern P1 (Table 1) with zero dead-times. The upper limit of f_{Sn} results from (21), and it depends on power, which can be seen in Figure 4.

The normalized conduction time T_1/T_S of the transistors decreases with increasing frequency, and the power is larger as the decrease rate is higher. Moreover, it is very important that varying f_S affects the output voltage. In switching pattern P2, it is possible to control the output voltage in the range of ca. $1.45\,U_{in}$ to $2\,U_{in}$. An increase in f_S results in decreasing voltage gain [see Figure 4b]. As in the case of T_1/T_S, the larger the power, the higher the decrease rate.

In the discussed switching pattern P2, the capacitor is never discharged to zero volts.

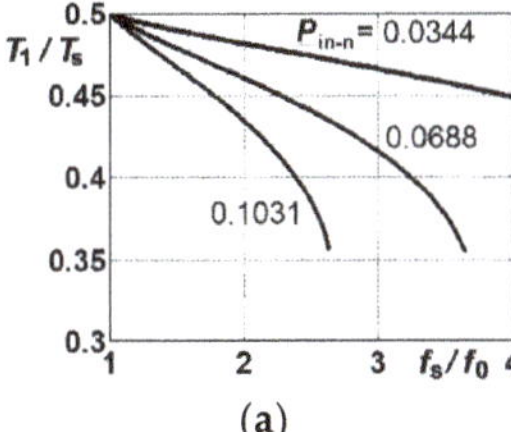 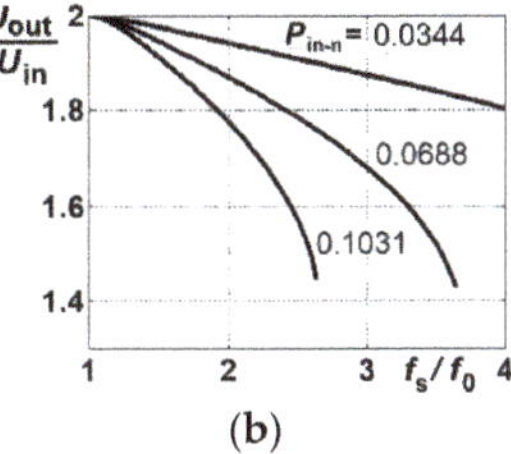

(a) (b)

Figure 4. (a) Normalized conduction time $T_{1n} = T_1/T_S$ of transistors, (b) normalized output voltage $U_{outn} = U_{out}/U_{in}$: as a function of $f_{Sn} = f_S/f_0$ for three values of normalized power $P_{in\text{-}n} = P_{in}/(U_{in}^2/\rho)$: 0.0344, 0.0688, and 0.1031. Switching pattern P2.

3.2. Pattern P3—Discontinuous Capacitor Current Mode

The continuous current mode is advantageous in terms of optimizing the converter's efficiency. However, the converter can also be operated at a fixed frequency f_S by varying the conduction time T_1 of the transistors. If $T_{1n} = T_1/T_S$ is lower than that in Figure 4a, the capacitor current becomes discontinuous–pattern P3 in Table 1. This operating mode offers a wider range of output voltage control.

The conduction time T_1 of the transistors is limited. Its maximum normalized value is equal to that shown in Figure 4a and its minimum value is limited by two factors. The first is the condition $U_{Cmin} \geq 0$, leading to

$$T_{1min\text{-}n} = T_{1min}/T_S = f_{Sn}/(2\pi)\,\arccos\!\left(1 - \pi P_{in\text{-}n}/f_{Sn}\right) \tag{23}$$

The second factor is the requirement for $U_{outn} \geq 1$, which results from the topology (Figure 1).

The output voltage regulation can be done by varying transistor conduction time T_1 at a given frequency f_S, where a decrease in T_1 leads to a decrease in the voltage gain ratio (Figure 5). This ratio falls with the rise of frequency f_S. Moreover, U_{out} is lower at higher powers.

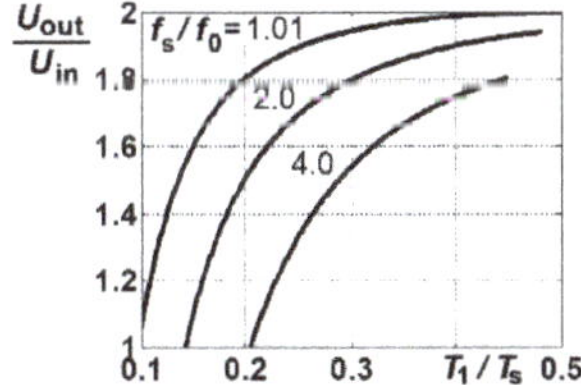

Figure 5. Normalized output voltage U_{out}/U_{in} of the converter as a function of $T_{1n} = T_1/T_S$ for three values of $f_{Sn} = f_S/f_0$: 1.01, 2.0, and 4.0. $P_{in\text{-}n} = P_{in}/(U_{in}^2/\rho) = 0.0344$. Switching pattern P3.

4. Efficiency of the SCVD

4.1. Model of Efficiency of the SCVD–Maximum Efficiency of the Converter without Switching Losses

The efficiency of the SCVD is determined by the resistance of its components, voltage drops on the diodes, input voltage, power, operating frequency, and the switching pattern. The following calculations have been performed for the SCVD with GaN switches using pattern P2 (Table 1). The assumptions for the analysis given in Section 3 remain valid, except for taking into account the circuit parasitic resistances, which are now added to the transistor resistances, and voltage drops on the diodes. Moreover, the conduction losses in the GaN transistors are computed taking into account its $R_{DS(on)}$, both for forward and reverse conduction. This can be done as the gate signals are applied during nearly the half-period $T_S/2$, except for dead-time, which is very short.

The current $i_{S1}(t)$ in transistor $S1$ is equal to the capacitor current $i_C(t)$ in state 4 (reverse conduction) and state 1 (forward conduction), and equal to zero in states 2 and 3 (Figures 2 and 3). The current $i_{S2}(t)$ in transistor $S2$ is phase-shifted, having the same shape and values. The RMS value of both currents is the same. It can be calculated from

$$I_S = \sqrt{\frac{1}{T_S}\int_0^{T_S} i_{S1}^2(t)dt} = \sqrt{\frac{1}{T_S}\left[\int_0^{T_1} i_{S1(1)}^2(t)dt + \int_0^{T_4} i_{S1(4)}^2(t)dt\right]} \tag{24}$$

where $i_{S1(1)}(t) = i_C(t)$ in state 1 (4), $i_{S1(4)}(t) = i_C(t)$ in state 4 (10), and $T_4 = T_2$ (12). After calculating the integrals using the equations mentioned above, and taking into account relationship (13), current I_S can be presented in the form

$$I_S = \sqrt{I_{SA}^2 + I_{SB}^2 + I_{SC}^2 + I_{SD}^2}. \tag{25}$$

where

$$
\begin{aligned}
I_{SA}^2 &= \frac{I_{Ck1}^2}{2}\left[T_{2n} + \frac{f_{Sn}}{4\pi}\sin\left(4\pi T_{2n}/f_{Sn}\right)\right]\\
I_{SB}^2 &= \frac{f_{Sn}}{4\pi}I_{Ck1}\frac{U_{Ck3}- U_{out}}{\rho}\left[1 - \cos\left(4\pi T_{2n}/f_{Sn}\right)\right]\\
I_{SC}^2 &= \frac{(U_{Ck3}- U_{out})^2}{2\rho^2}\left[T_{2n} - \frac{f_{Sn}}{4\pi}\sin\left(4\pi T_{2n}/f_{Sn}\right)\right]\\
I_{SD}^2 &= \frac{(U_{in}- U_{Cmin})^2}{2\rho^2}\left[T_{1n} - \frac{f_{Sn}}{4\pi}\sin\left(4\pi T_{1n}/f_{Sn}\right)\right].
\end{aligned}
\tag{26}
$$

The current $i_{D1}(t)$ in diode $D1$ is equal to the capacitor current $i_C(t)$ in states 1 and 2, and equal to zero in states 3 and 4 (Figures 2 and 3). The current $i_{Dout}(t)$ in diode $Dout$ is phase-shifted, having the same shape and values as $i_{D1}(t)$. The average value of both currents is the same and equal to

$$I_{Dav} = \frac{1}{T_S}\int_0^{T_S} i_{D1}(t)dt = \frac{1}{T_S}\left[\int_0^{T_1} i_{D(1)}(t)dt + \int_0^{T_2} i_{D(2)}(t)dt\right] \tag{27}$$

where $i_{D1(1)} = i_C(t)$ in state 1 (4) and $i_{D1(2)} = i_C(t)$ in state 2 (6).

$$I_{Dav} = \frac{f_{Sn}}{2\pi}\left\{I_m\left[1 - \cos\left(2\pi T_{1n}/f_{Sn}\right)\right] + I_{Ck1}\sin\left(2\pi T_{2n}/f_{Sn}\right) - \frac{U_{Ck1}}{\rho}\left[1 - \cos\left(2\pi T_{2n}/f_{Sn}\right)\right]\right\} \tag{28}$$

Conduction losses ΔP_C are the sum of losses in the transistors and the diodes

$$\Delta P_C = (r_1 + r_2)I_S^2 + (\Delta U_{D1} + \Delta U_{Dout})I_{Dav} \tag{29}$$

where r_1 and r_2 denote the total resistance, including the resistance of the transistor, in the circuits with $S1$ and $S2$, respectively; ΔU_{D1} is the voltage drop across diode $D1$, and ΔU_{Dout} is the voltage drop across diode $Dout$ (Figure 1). It is assumed that the voltage drops across the diodes remain constant in the conducting state. Therefore, the efficiency is

$$\eta = 1 - \frac{\Delta P_C}{P_{in}} = 1 - \frac{(r_1 + r_2)I_S^2}{P_{in}} - \frac{(\Delta U_{D1} + \Delta U_{Dout})I_{Dav}}{P_{in}} \tag{30}$$

If the resistances and diode voltage drops are the same, i.e.;

$$r_1 = r_2 = r,\ \Delta U_{D1} = \Delta U_{Dout} = \Delta U_D \tag{31}$$

we can rewrite the efficiency formula in the form

$$\eta = 1 - 2\left(r_n I_{Sn}^2 + \Delta U_{Dn}I_{Davn}\right)/P_{in-n} \tag{32}$$

where

$$I_{Sn} = \frac{I_S}{U_{in}/\rho}, \quad I_{Davn} = \frac{I_{Dav}}{U_{in}/\rho}, \quad r_n = r/\rho, \quad \Delta U_{Dn} = \Delta U_D/U_{in} \tag{33}$$

Figure 6 presents the model of efficiency created on the basis of (32). The peak efficiency achieves a maximum above the resonant frequency. It is assumed that the switching losses are reduced in this area as well.

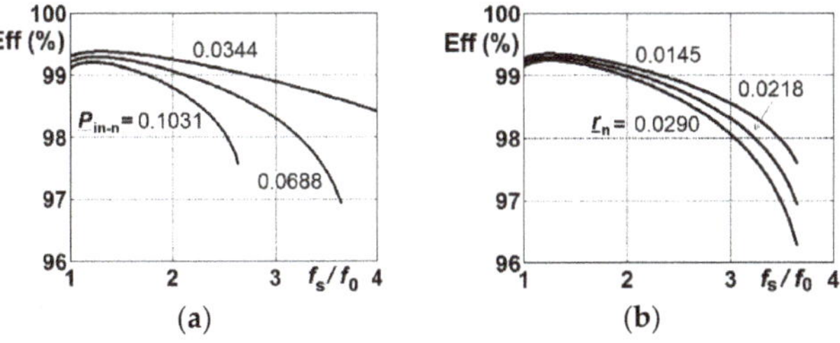

(a)   (b)

Figure 6. Efficiency of SCVD as a function of $f_{Sn} = f_S/f_0$: (**a**) for three values of $P_{in\text{-}n} = P_{in}/(U^2_{in}/\rho)$: 0.0344, 0.0688, and 0.1031 at $r_n = r/\rho = 0.0218$, (**b**) for three values of $r_n = 0.0145$, 0.0218, and 0.0290 at $P_{in\text{-}n} = 0.0688$. $\Delta U_{Dn} = 0.006$. Switching pattern P2.

4.2. The Switching Concept for Maximum Efficiency

The advantages of the application of the GaN switches in the proposed high-frequency SCVD results from the possibility of using the ZVS mode with low switching losses under the operation above the resonant frequency (low C_{OSS} of the switch and short transition time). It can be assumed that the most favorable case of operation (ZVS), from the efficiency point of view, is achieved by turning-off the transistors just before their current reaches zero. The reverse conduction of the transistors should be as short as possible (Figure 7). When the transistors turn off near the zero crossings of the current, the turn-off loss can be neglected. In an SC converter such as the SCVD, with very small inductors, the application of a GaN switch will make it possible to achieve a highly improved efficiency in the ZVS mode keeping both the dead-dime and reverse-time very short.

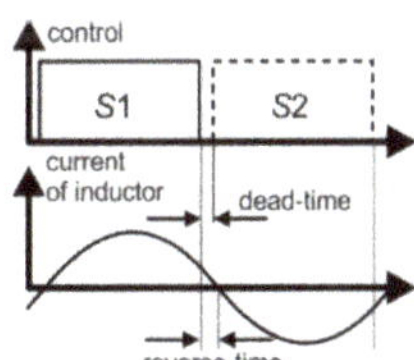

Figure 7. A method for operation with maximum efficiency (Zero Voltage Switching (ZVS), nearly Zero Current Switching (ZCS) (Low-Current Switching—LCS) turn-off and decreased RMS current in comparison to full ZCS case): the use of pattern P2 for high efficiency (pattern P2HE).

5. Mixed Switching Patterns in Applications

To obtain a functional converter with efficient and effective voltage regulation, various switching patterns can be utilized depending on the operational conditions. Furthermore, the proposed switching patterns can be effectively used in more complex systems created on the basis of the SCVD.

5.1. Start-Up of the Converter

During the start-up of the converter, the SCVD is usually overloaded when pattern P1 is used. The maximum power (P_{max}) of this type of voltage multiplier is proportional to the switching frequency and the switched capacitance C [5]. Under pattern P1, the converter can increase the power as far as the switched capacitor is not fully discharged in a switching period. The operation of an SCVD with partial discharge and low voltage ripples under the rated power requires the use of a large-enough switched capacitor (C). In this case, the converter's power is $P_{nom} \ll P_{max}$. When the switches, diodes and the PCB are designed for nominal power P_{nom}, the converter can easily be overloaded. To overcome this issue in conditions of overloading, such as the start-up of the converter, other switching patterns can be used. Figure 8 presents a comparison of waveforms during the start-up with pattern P3 and pattern P1. From these results, it is seen that the overloading of the converter is significantly limited by the appropriate use of the proposed switching pattern P3.

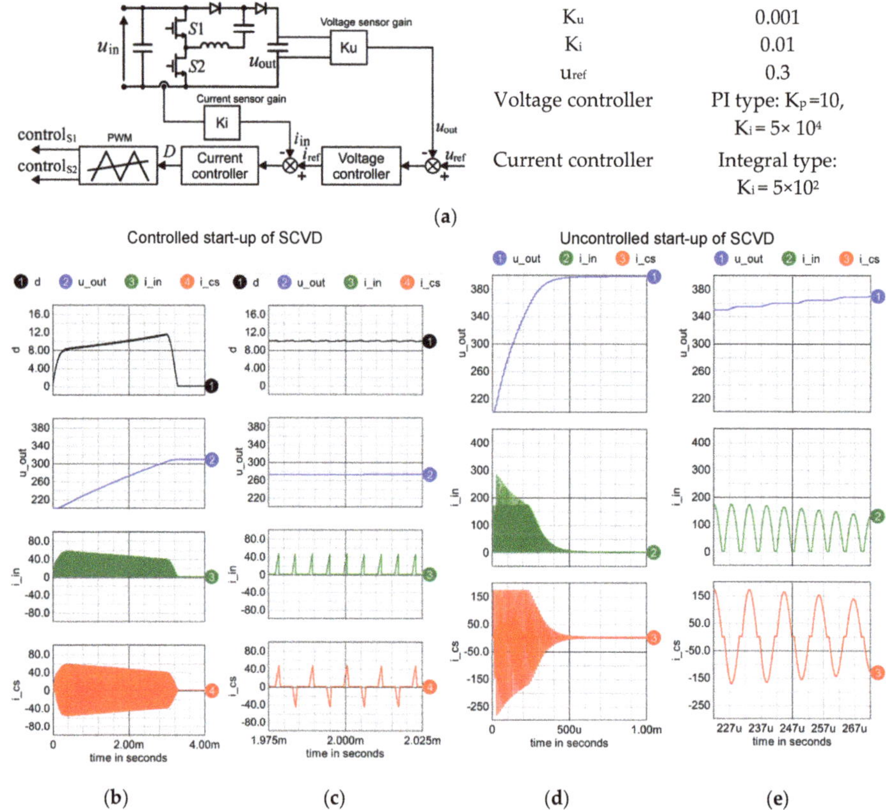

Figure 8. (**a**) Closed-loop voltage control system with overcurrent protection and its parameters (pattern P3 with the duty ratio control $D = T_1/T_S$–Table 1). (**b**,**c**) Controlled start-up of SCVD under pattern P3 in the closed-loop system and uncontrolled start-up of SCVD under pattern P1. Waveforms of duty cycle (symbol d (%)), output voltage u_{out} (V), input current i_{in} (A), current i_{Cs} (A) of capacitor C in the cases: (**b**,**d**) full range, (**c**,**e**) zoom. ICAP/4 simulation results. Circuit parameters as in Table 3.

Pattern P3 can be easily achieved in a classic PWM generator (it requires a constant switching frequency and a variable duty ratio). The implementation of this pattern can be achieved by the use of VCO (Voltage Controlled Oscillator), and pattern P4 could require a special hardware design (e.g., in FPGA technology).

5.2. Bi-Directional Converter

A synchronous SCVD makes it possible to convert energy in both directions, which is required in systems with batteries. When the voltages on both sides are constant, the converter should be able to regulate the voltage gain in both directions. Figure 9 presents the concept of such a synchronous SCVD, the most suitable switching pattern (P4), and its operating states. The operation occurs in the following three states (Figure 9b):

(1)   In the first state, the SC is being charged from the output voltage source. This state is terminated by turning off the switches $S1$ and $S4$;

(2)   State 2—the inductor current goes to zero via $S2$ and $S3$ (reverse conduction);

(3)   In state 3—turning on $S3$ starts an oscillation in a new circuit, and the energy is transferred to the input source. This is advantageous since the oscillation continues until the inductor current reaches zero. Breaking this oscillation by switching off $S3$ would start the current flow to the output and charging the output, which would not be favorable to the efficiency of the conversion.

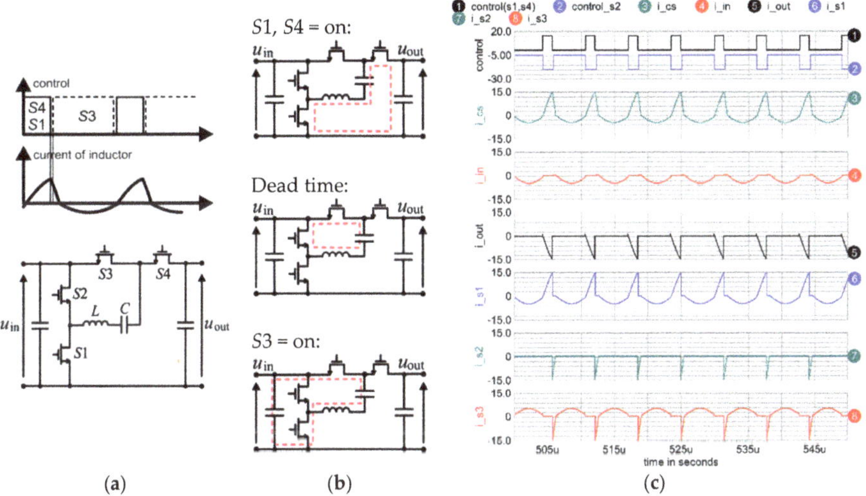

Figure 9. (**a**) Synchronous SCVD and switching pattern P4 with corresponding symbols, (**b**) states in a switching cycle during charging the source by the converter, (**c**) steady-state waveforms during the reverse energy transfer with the use of switching pattern P4 (ICAP/4 simulation results): waveforms of currents (A) and control signals for $S1$, $S3$ and $S4$ (control signal of $S2 = 1$). $P_{in} = 500$ W. Circuit parameters as in Table 3.

Pattern P4 guarantees a proper operation of the synchronous SCVD, which is confirmed by the steady-state waveforms presented in Figure 9. It assures ZVS of switch $S3$, ZCS turn-on of $S1$ and $S4$, voltage regulation and unidirectional currents of the sources. Other switching patterns enable a bi-directional energy transfer with voltage regulation, but with an unrequired current recirculation between SC and the sources.

5.3. A Series-Connected High-Voltage-Gain System

A section containing capacitors allows for designing modular [23] and cascaded [26,27] converters, where such parameters as the output voltage regulation can be improved. In [27], modular converters composed of series-parallel sections are analyzed. It has been proven there that the series-connected voltage doublers (Figure 10) are the most effective voltage multiplier topology taking into consideration the relation of the number of switches to the voltage gain.

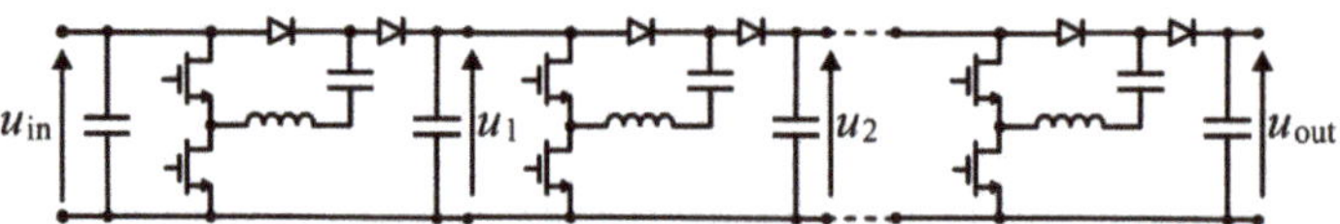

Figure 10. SCVD series in a high-voltage-gain converter.

The SCVD series achieves the voltage gain $G_U = 2^n$, where n is the number of the SCVD converters. In an SCVD series (Figure 10), each internal converter operates in different conditions (voltage and current stress). By suitable use of the proposed switching patterns, the SCVD series can achieve novel unique features such as output voltage regulation and very high efficiency. The most effective mixed switching pattern can depend on such parameters as switching frequency or load. Table 2 presents examples of scenarios for a decision on implementing optimal switching patterns.

Table 2. Examples of scenarios for series SCVD optimal switching.

Scenario	Mixed Switching Patterns
Light-load conditions (low predicted conduction losses, high switching losses)	• The first SCVD, connected directly to the source, regulates the voltage of the converter in the range U_{in}–$2U_{in}$. It operates at the lowest level of voltage with non-significant C_{OSS} losses. It can use pattern P2, and patterns P3 or P4 for a deep voltage decrease with high efficiency. • When the first cell decreases the voltage, the voltages on the switches in all other cells are also decreased, which reduces C_{OSS} losses. It is more beneficial than the voltage regulation by any other cell. • The other cells, except the first, operate with pattern P2HE (Figure 7), which decreases the switching losses significantly.
High-load (predicted conduction losses higher than switching losses in all the cells)	• All the cells should operate with the highest possible voltage to limit conduction losses (pattern P2HE–Figure 7). • The last SCVD, connected directly to the load, regulates the voltage of the converter in the range $(n-1)U_{in}$–nU_{in}, using pattern P2, and patterns P3 or P4. • If a deeper voltage decrease is required, an appropriate middle cell regulates the voltage, e.g., to achieve the range $(n-2)U_{in}$–nU_{in}.

6. Experimental Setup and Test Results

The experimental investigations have been performed in the setup with parameters presented in Table 3 using the equipment listed in Table 4. The efficiency of the GaN-Based DC–DC resonant boost converter was determined using Yokogawa WT1800 power analyzer [35] on the basis of the output to the input power ratio. The range of voltages and currents in the tested converter allows for the use of the internal current and voltage sensors (5 A and 600 V) and achieving an adequate precision of the measurements.

Table 3. Parameters of the laboratory SCVD converter.

Transistors	PGA26E07BA
Diodes	STPSC12065GY-TR
Switched capacitor	220 nF
Inductor	10.4 µH
Input capacitor	4 µF
Output capacitor	4 µF (and 100 µF external bank)
Input voltage	$U_{\text{in}} = 200$ V
Resonant frequency	$f_0 = 105.2$ kHz

Table 4. Parameters of the laboratory test setup.

Purpose	Equipment
PWM signal generator	FPGA-based: DE0-CV Cyclone V Control Board
Measurements	Scope: Tektronix DPO4054 Current probes TCP0030 Voltage probes THDP0200 Power Analyzer Yokogawa WT1800
Supply and load	DC power supply Delta SM300, Rigol DP832, Mixed passive and electronic load LDH400P
IR measurements	FLIR i60

The measurements were intended to verify the basic concepts presented in this paper: the switching strategies and efficiency of the SCVD converter shown in Figure 1. The obtained results confirm the proper operation of the converter under various switching strategies and its high efficiency.

6.1. Switching Pattern P2–Operation with Continuous Capacitor Current Mode

Figure 11 presents examples of selected time waveforms in switching pattern P2. A dead-time of 50 ns has been used; thus, after the turn off of each transistor, the other transistor begins conducting (reverse conduction and, next, forward conduction) with nearly zero turn-on loss and low conduction loss.

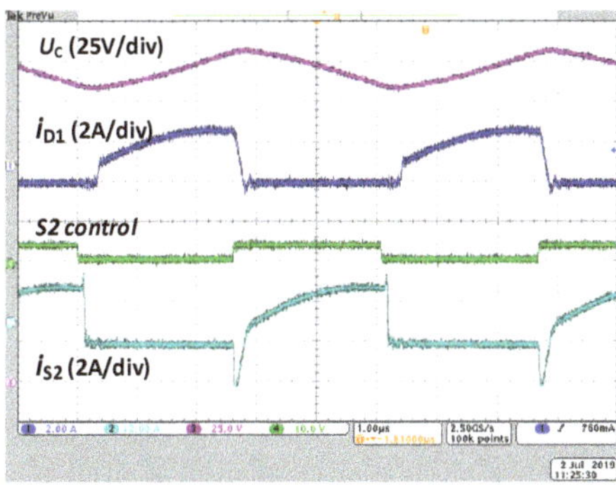

Figure 11. Steady-state operation of SCVD in Continuous Conduction Mode (CCM) and ZVS mode. Waveforms of voltage on switched capacitor C, current of $D1$, control signal of $S2$, and current of transistor $S2$. $U_{\text{in}} = 200$ V, $f_S = 201.6$ kHz, $P_{\text{out}} = 400$ W, $t_M = 50$ ns. Switching pattern P2.

6.2. Comparison of Operation in the ZCS Mode (Pattern P1) and ZVS Mode (Pattern P2)

A comparison of the operation below resonant frequency (ZCS) and above that frequency (ZVS) confirms the concept of operation with high efficiency and output voltage control.

Figure 12 presents the operation of the converter in the ZVS and nearly in the ZCS mode. In these conditions, maximum efficiency is achieved. The ZVS mode is maintained, although the negative current of $S2$ is not as clearly visible as in Figure 11. Before the switch $S2$ is turned on, it conducts a negative current and its voltage is zero. On the other hand, the full ZCS mode is visible in Figure 15a.

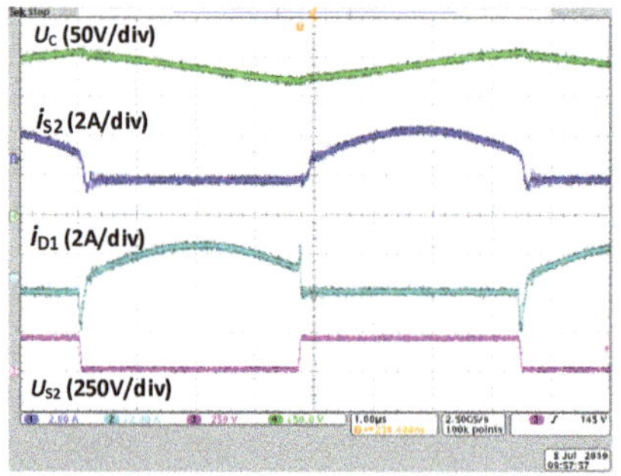

Figure 12. CCM operation of SCVD in steady-state in ZVS and nearly ZCS mode. Waveforms of voltage on switched capacitor C, current of transistor $S2$, current of diode $D1$, voltage on transistor $S2$. $U_{in} = 200$ V, $f_S = 201.6$ kHz, $P_{out} = 400$ W. Switching pattern P2.

Figure 13 presents graphs of voltage gain end efficiency versus switching frequency f_S. The efficiency chart (Figure 13b) clearly shows a substantial increase in efficiency when the operating mode changes from the switching pattern P1 to P2. The peak efficiency occurs slightly above f_0. A further increase in the switching frequency leads to a decrease in output voltage (voltage gain regulation) and results in a decrease in efficiency.

Figure 14a–d present the charts of efficiency of the SCVD versus output power for various values of the switching frequency under pattern P2. Figure 14e depicts the relation between the efficiency and the gain, obtained from the data in Figure 13. The highest achieved efficiency is 99.228% at $f_S = 134.4$ kHz and $P_{out} = 396.23$ W.

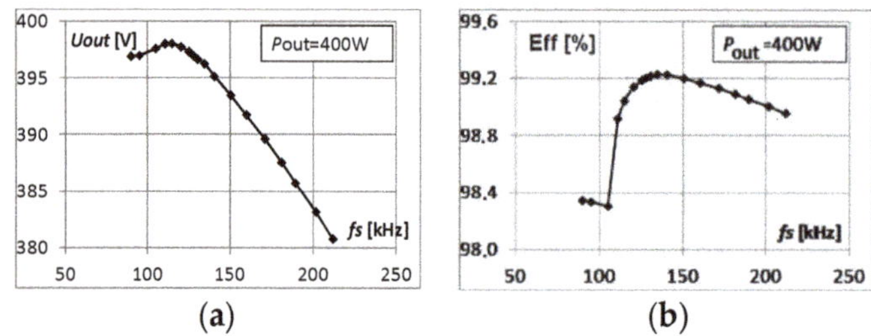

Figure 13. (a) Output voltage U_{out} and (b) efficiency of SCVD as a function of switching frequency f_S. $U_{in} = 200$ V, $P_{out} = 400$ W, $t_M = 50$ ns. Switching pattern P1 (for $f_S < f_0$) and P2 (for $f_S > f_0$).

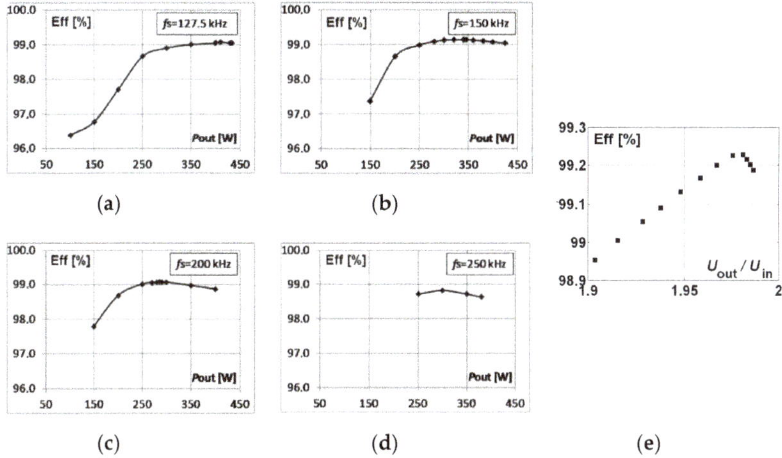

Figure 14. Efficiency of SCVD in CCM mode as a function of: (**a–d**) output power P_{out}, (**e**) U_{out}/U_{in} at P_{out} = 400 W. U_{in} = 200 V, t_M = 50 ns. Switching pattern P2.

Further analysis of the ZVS concept in the SCVD, achieved by the introduction of pattern P2, is presented by comparing the waveforms and thermograms of the converter operating in accordance with patterns P2 and P1. The comparison is presented in Figure 15 for P_{out} = 400 W. The IR photos confirm considerably lower losses and heat generation in the transistors when the converter operates using pattern P2 [see Figure 15b,d] versus the case of the ZCS switching with pattern P1 [see Figure 15a,c].

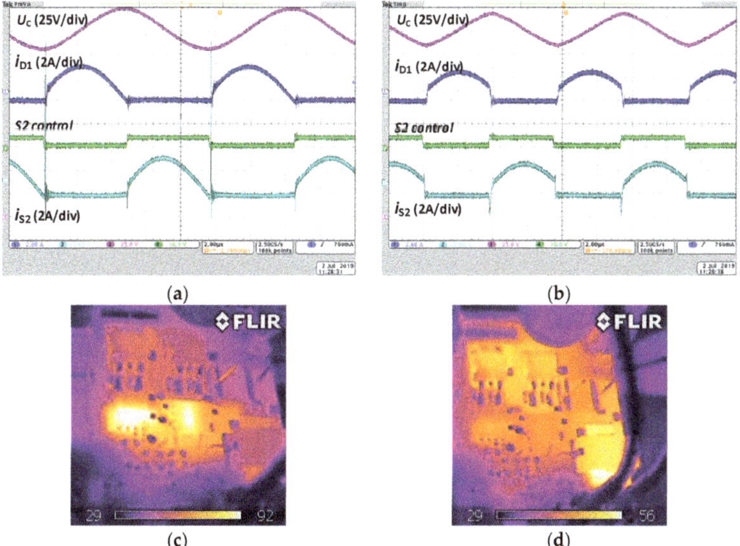

Figure 15. Operation of SCVD in ZCS mode (pattern P1) at f_S = 105 kHz with η = 98.30%, and in ZVS mode (pattern P2) at f_S = 134.4 kHz with η = 99.228%. Steady-state waveforms of voltage on switched capacitor C, current of diode $D1$, control signal of transistor $S2$, current of $S2$; IR photos of converter. U_{in} = 200 V, P_{out} = 400 W, t_M = 50 ns. Results (**a,c**)—switching pattern P1, results (**b,d**)—switching pattern P2.

Figure 16 confirms the proper operation of the converter in switching pattern P3. The inverter operates in discontinuous conduction mode (DCM). Output voltage regulation is possible at a constant switching frequency (Table 1).

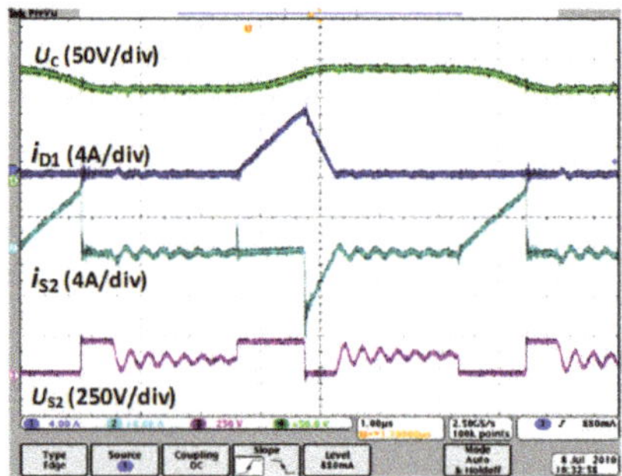

Figure 16. Steady-state operation of SCVD in DCM. Waveforms of voltage on switched capacitor *C*, current of *D1*, current of transistor *S2*, voltage on transistor *S2*. U_{in} = 200 V, f_S = 134.4 kHz, $D = T_1/T_S$ = 15.6%, t_M = 50 ns. Switching pattern P3.

Table 5 shows a comparison of the maximum efficiencies of selected converters presented in recently published papers. The power at that the maximum efficiency was registered, and the type of switches that were used are also listed. (RSCC–resonant switched-capacitor converter, RTBSCC–resonant two-switch boosting switched-capacitor converter, IBC–intermediate bus converter, MRSCC–multilevel resonant switched-capacitor converter). The efficiency of the converter presented in this paper is one of the highest that are reported in the recent bibliography.

Table 5. Comparison of the maximum efficiencies of selected converters presented in recently published papers.

Proposed Solution	Doubler RSCC [3]	RSCC [4]	RTBSCC [14]	High Freq. IBC [33]	MRSCC [31]	Ref. [23]	Ref. [24]
η = 99.228%	η = 99.82%	η = 96%	η = 98.3%	η = 96.7%	η = 98.5%	η = 99.5%	η = 94.6%
P = 400 W	P = 1500 W	P = 10 W	P = 23 W	P = 240 W	P = 5 kW	P = 3 kW	P = 140 W
GaN	GaN	MOSFET	MOSFET	GaN	SiC MOSFET	MOSFET	MOSFET

6.3. Output Voltage Regulation by the Switching Pattern P3

The most suitable method for the output voltage regulation in an SCVD is pattern P3, where the capacitor current is discontinuous (Figure 16). Very good effectiveness of the regulation is confirmed in Figure 17 which shows the output voltage versus duty cycle *D* defined as the ratio of the turn-on time T_1 of a transistor to the switching period T_S. A wide range of the output voltage is achievable with an acceptable efficiency deterioration at low values of the duty cycle.

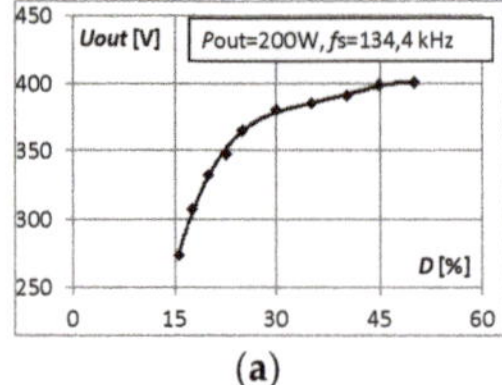

(a)

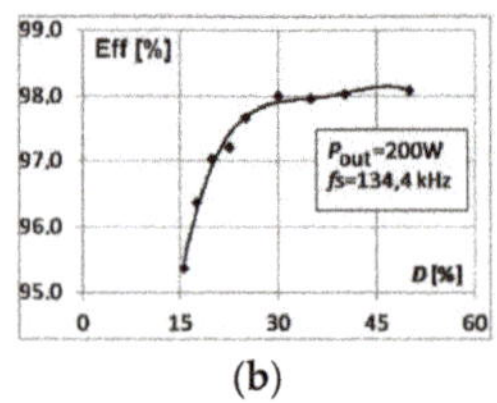

(b)

Figure 17. (**a**) Output voltage U_{out} of SCVD and (**b**) its efficiency in DCM mode as a function of duty cycle $D = T_1/T_S$. U_{in} = 200 V, f_s = 134.4 kHz, P_{out} = 200 W, t_M = 50 ns. Switching pattern P3.

7. Conclusions

In this paper, the concepts of control for a resonant switched-capacitor voltage doubler are presented. They allow the use of the SCVD converter as a fully functional DC–DC converter with output voltage regulation and very high efficiency.

A classic SCVM operates in the ZCS mode, in which switching power losses associated with C_{OSS} of the transistors are significant. The application of GaN switches makes it possible to operate with a high frequency while maintaining high efficiency. However, the efficiency can be significantly improved by the proposed switching patterns of the converter, where the reverse conduction occurs to achieve zero-voltage turn-on of the transistors. The maximum efficiency that was measured in the demonstrated setup exceeds 99.2%. The heat generation in the transistors is reduced significantly as well. In the switching pattern dedicated to maximum efficiency, an output voltage adjustment is possible. In another switching pattern (P3) proposed in this paper, the SCVD converter achieves a very high output voltage regulation range. The developed model of losses matches the experimental results and can be used in the design process. In addition, the results can be applied in other topologies of the SC converters.

The switching pattern P2 allows for a significant improvement in the efficiency of the SCVD by C_{OSS} loss reduction. To accomplish that, very fast switching is required because the cycle sequence: turn-off/dead-time/turn-on should occur on the falling slope of the resonant current, taking only a small part of the oscillation time. The SCVD is a switched-capacitor converter with low resonant inductance, therefore the oscillation time is very short. Three types of switches (superjunction MOSFET, GaN, and SiC) as a prospective adequate solution for high-efficiency operation of the SCVD. Taking into consideration the features of GaN switches, the experimental tests have been performed with the use of GaN GIT switches. The switching pattern P3 allows for easy implementation for overcurrent limitation of the start-up of the converter, and pattern P4 can be used in the bi-directional operation of the synchronous SCVD. Various patterns can be also combined in the cascaded high-voltage-gain system composed of several SCVDs.

Author Contributions: Conceptualization, R.S.; methodology, R.S., Z.W., A.M., A.S. and A.P.; software, R.S., Z.W., A.P., and A.S.; validation, R.S., and Z.W.; formal analysis, R.S., Z.W.; investigation, R.S., Z.W., A.M., A.S. and A.P.; resources, R.S.; data curation, R.S., Z.W., A.M., and A.S.; writing—original draft preparation, R.S., and Z.W.; writing—review and editing; visualization, R.S. and Z.W.; supervision, R.S. and Z.W.; project administration, R.S. All authors have read and agreed to the published version of the manuscript.

Funding: This research received no external funding.

Conflicts of Interest: The authors declare no conflict of interest.

References

1. Ioinovici, A. Switched-capacitor power electronics circuit. *IEEE Circuits Syst. Mag.* **2001**, *1*, 37–42. [CrossRef]
2. Sano, K.; Fujita, H. Performance of a high-efficiency switched-capacitor-based resonant converter with phase-shift control. *IEEE Trans. Power Electron.* **2011**, *26*, 344–354. [CrossRef]
3. Vasić, M.; Serrano, D.; Toral, V.; Alou, P.; Oliver, J.A.; Cobos, J.A. Ultraefficient Voltage Doubler Based on a GaN Resonant Switched-Capacitor Converter. *IEEE J. Emerg. Sel. Top. Power Electron.* **2019**, *7*, 622–635. [CrossRef]
4. Cervera, A.; Evzelman, M.; Mordechai Peretz, M.; Ben-Yaakov, S. A high-efficiency resonant switched capacitor converter with continuous conversion ratio. *IEEE Trans. Power Electron.* **2015**, *30*, 1373–1382. [CrossRef]
5. Waradzyn, Z.; Stala, R.; Mondzik, A.; Penczek, A.; Skala, A.; Pirog, S. Efficiency analysis of MOSFET-based air-choke resonant DC–DC step-up switched-capacitor voltage multipliers. *IEEE Trans. Ind. Electron.* **2017**, *64*, 8728–8738. [CrossRef]

6.   Mondzik, A.; Waradzyn, Z.; Stala, R.; Penczek, A. High efficiency switched capacitor voltage doubler with planar core-based resonant choke. In Proceedings of the 2016 10th International Conference on Compatibility, Power Electronics and Power Engineering (CPE-POWERENG), Bydgoszcz, Poland, 29 June–1 July 2016; pp. 402–409. [CrossRef]

7.   Stala, R.; Waradzyn, Z.; Penczek, A.; Mondzik, A.; Skala, A. A Switched-Capacitor DC-DC Converter with Variable Number of Voltage Gains and Fault-Tolerant Operation. *IEEE Trans. Ind. Electron.* **2019**, *66*, 3435–3445. [CrossRef]

8.   Evzelman, M.; Ben-Yaakov, S. Average-current-based conduction losses model of switched capacitor converters. *IEEE Trans. Power Electron.* **2013**, *28*, 3341–3352. [CrossRef]

9.   Ye, Y.; Cheng, K.W.E. A family of single-stage switched-capacitor–inductor PWM converters. *IEEE Trans. Power Electron.* **2013**, *28*, 5196–5205. [CrossRef]

10.   Wu, G.; Ruan, X.; Ye, Z. Nonisolated high step-up DC-DC converters adopting switched-capacitor cell. *IEEE Trans. Ind. Electron.* **2015**, *62*, 383–393. [CrossRef]

11.   Tang, Y.; Wang, T.; Fu, D. Multicell switched-inductor/switched-capacitor combined active-network converters. *IEEE Trans. Power Electron.* **2015**, *30*, 2063–2072. [CrossRef]

12.   Rosas-Caro, J.C.; Ramirez, J.M.; Peng, F.Z.; Valderrabano, A. A DC-DC multilevel boost converter. *IET Power Electron.* **2010**, *3*, 129–137. [CrossRef]

13.   Li, S.; Xiangli, K.; Zheng, Y.; Smedley, K.M. Analysis and design of the ladder resonant switched-capacitor converters for regulated output voltage applications. *IEEE Trans. Ind. Electron.* **2017**, *64*, 7769–7779. [CrossRef]

14.   Li, S.; Zheng, Y.; Wu, B.; Smedley, K.M. A Family of Resonant Two-Switch Boosting Switched-Capacitor Converter With ZVS Operation and a Wide Line Regulation Range. *IEEE Trans. Power Electron.* **2018**, *33*, 448–459. [CrossRef]

15.   Alam, M.K.; Khan, F.H. A high-efficiency modular switched-capacitor converter with continuously variable conversion ratio. In Proceedings of the IEEE 13th Workshop on Control and Modeling for Power Electronics (COMPEL), Kyoto, Japan, 10–13 June 2012; pp. 1–5. [CrossRef]

16.   Ye, Y.; Cheng, K.W.E.; Liu, J.; Ding, K. A step-up switched-capacitor multilevel inverter with self-voltage balancing. *IEEE Trans. Ind. Electron.* **2014**, *61*, 6672–6680. [CrossRef]

17.   Shen, M.; Peng, F.Z.; Tolbert, L.M. Multilevel DC–DC Power Conversion System with Multiple DC Sources. *IEEE Trans. Power Electron.* **2008**, *23*, 420–426. [CrossRef]

18.   Gunasekaran, D.; Qin, L.; Karki, U.; Li, Y.; Peng, F.Z. A Variable (n/m)X Switched Capacitor DC–DC Converter. *IEEE Trans. Power Electron.* **2017**, *32*, 6219–6235. [CrossRef]

19.   Gunasekaran, D.; Qin, L.; Karki, U.; Li, Y.; Peng, F.Z. Multi-level capacitor clamped DC-DC multiplier/divider with variable and fractional voltage gain—An (n/m)X DC-DC converter. In Proceedings of the 2016 IEEE Applied Power Electronics Conference and Exposition (APEC), Long Beach, CA, USA, 20–24 March 2016; pp. 2525–2532. [CrossRef]

20.   Gunasekaran, D.; Han, G.; Peng, F.Z. Control methods to achieve soft-transition of gains for a variable (n/m)X converter. In Proceedings of the 2017 IEEE Applied Power Electronics Conference and Exposition (APEC), Tampa, FL, USA, 26–30 March 2017; pp. 3365–3372. [CrossRef]

21.   Khan, F.H.; Tolbert, L.M. Multiple-Load–Source Integration in a Multilevel Modular Capacitor-Clamped DC–DC Converter Featuring Fault Tolerant Capability. *IEEE Trans. Power Electron.* **2009**, *24*, 14–24. [CrossRef]

22.   Bento, F.; Cardoso, A.J.M. Fault tolerant DC-DC converters in DC microgrids. In Proceedings of the 2017 IEEE Second International Conference on DC Microgrids (ICDCM), Nuremburg, Germany, 27–29 June 2017; pp. 484–490. [CrossRef]

23.   Setiadi, H.; Fujita, H. Reduction of Switching Power Losses for ZVS Operation in Switched-Capacitor-Based Resonant Converters. *IEEE Trans. Power Electron.* **2021**, *36*, 1104–1115. [CrossRef]

24.   Xie, W.; Brown, B.Y.; Smedley, K. Multilevel Step-down Resonant Switched-Capacitor Converters with Full-range Regulation. *IEEE Trans. Ind. Electron.* **2020**. [CrossRef]

25.   Beck, Y.; Singer, S.; Martinez-Salamero, L. Modular realization of capacitive converters based on general transposed series–parallel and derived topologies. *IEEE Trans. Ind. Electron.* **2014**, *61*, 1622–1631. [CrossRef]

26.   Zamiri, E.; Vosoughi, N.; Hosseini, S.H.; Barzegarkhoo, R.; Sabahi, M. A new cascaded switched-capacitor multilevel inverter based on improved series-parallel conversion with less number of components. *IEEE Trans. Ind. Electron.* **2016**, *63*, 3582–3594. [CrossRef]

27. Stala, R.; Piróg, S. DC–DC boost converter with high voltage gain and a low number of switches in multisection switched capacitor topology. *Arch. Electr. Eng.* **2018**, *67*, 617–627.

28. Dalla Vecchia, M.; Ravyts, S.; Van den Broeck, G.; Driesen, J. Gallium-Nitride Semiconductor Technology and Its Practical Design Challenges in Power Electronics Applications: An Overview. *Energies* **2019**, *12*, 2663. [CrossRef]

29. Neumayr, D.; Guacci, M.; Bortis, D.; Kolar, J.W. New calorimetric power transistor soft-switching loss measurement based on accurate temperature rise monitoring. In Proceedings of the 29th International Symposium on Power Semiconductor Devices and IC's (ISPSD), Sapporo, Japan, 28 May–1 June 2017; pp. 447–450. [CrossRef]

30. Li, K.; Evans, P.; Johnson, M. SiC/GaN power semiconductor devices: A theoretical comparison and experimental evaluation under different switching conditions. *IET Electr. Syst. Transp.* **2018**, *8*, 3–11. [CrossRef]

31. Kawa, A.; Stala, R. SiC-Based Bidirectional Multilevel High-Voltage Gain Switched-Capacitor Resonant Converter with Improved Efficiency. *Energies* **2020**, *13*, 2445. [CrossRef]

32. Zulauf, G.D.; Roig-Guitart, J.; Plummer, J.D.; Rivas-Davila, J.M. Measurements for Superjunction MOSFETs: Limitations and Opportunities. *IEEE Trans. Electron Devices* **2019**, *66*, 578–584. [CrossRef]

33. Reusch, D.; Strydom, J. Evaluation of gallium nitride transistors in high frequency resonant and soft-switching DC-DC converters. *IEEE Trans. Power Electron.* **2015**, *30*, 5151–5158. [CrossRef]

34. Ishida, H.; Kajitani, R.; Kinoshita, Y.; Umeda, H.; Ujita, S.; Ogawa, M.; Tanaka, K.; Morita, T.; Tamura, S.; Ishida, M.; et al. GaN-based semiconductor devices for future power switching systems. In Proceedings of the 2016 IEEE International Electron Devices Meeting (IEDM), San Francisco, CA, USA, 3–7 December 2016; pp. 20.4.1–20.4.4. [CrossRef]

35. *WT1800 Precision Power Analyzer User's Manuals, Features Guide, IM WT1801-01E, Yokogawa Test & Measurement Corporation*, 4th ed.; October 2017. Available online: https://www.yokogawa.com/pdf/provide/E/GW/IM/0000024842/0/IMWT1801-02EN.pdf (accessed on 26 November 2020).

Publisher's Note: MDPI stays neutral with regard to jurisdictional claims in published maps and institutional affiliations.

Article

SiC-Based Bidirectional Multilevel High-Voltage Gain Switched-Capacitor Resonant Converter with Improved Efficiency

Adam Kawa * and Robert Stala

Department of Power Electronics and Energy Control Systems, Faculty of Electrical Engineering, Automatics, Computer Science and Biomedical Engineering, AGH University of Science and Technology, 30-059 Krakow, Poland; stala@agh.edu.pl
* Correspondence: adamkawa@agh.edu.pl

Received: 2 April 2020; Accepted: 9 May 2020; Published: 13 May 2020

Abstract: This paper presents the research results of the bidirectional multilevel resonant switched capacitor converter (MRSCC). The converter can achieve a high voltage ratio in multilevel topology, which limits the voltage stress on switches and is able to operate with high power efficiency. The converter can be applied as an interconnector between DC voltage systems used for various applications. This paper presents a method that significantly improves the efficiency of the MSRCC through topology modification. Furthermore, the feasibility of the converter was demonstrated with the use of SiC and Si MOSFET switches, together with suitable passive components. It was demonstrated that the proposed modification of the topology makes the converter very efficient in SiC-based ones and can significantly improve the efficiency of Si MOSFET converters. The series of test results of the SiC-based converter is a novel aspect presented in this paper and shows promising achievements of efficiency. The results were obtained from the laboratory setup of 5 kW and 0.5/2 kV MRSCC. To demonstrate the bidirectional operation of the converter, a back-to-back setup (0.5/2/0.5 kV) was used. It also demonstrates that such a high-voltage gain converter can be accurately tested with the use of laboratory equipment with a typical voltage range.

Keywords: bidirectional converter; multilevel converter; resonant converter; SiC MOSFET; high-voltage converter; switched capacitor converter

---

1. Introduction

The multilevel topology of the converter is favorable for high-voltage circuits due to the reduction of voltage stress on switches. Aside from typical multilevel converters, such as the flying capacitor, neutral point clamped, and cascade bridges, various concepts of switched capacitor-based converters have been proposed recently. The switched capacitor (SC) technique is suitable for high-voltage gain converters and can be effectively used in power converters, which was proven in [1,2]. On the basis of the SC technique, such multilevel topologies include MRSCC (multilevel mesonant switched capacitor converter) [3–5], modular capacitor clamped [6,7], resonant Ladder [8], MMCCC (multilevel modular capacitor clamped converter), 6X [9], converter with coupled inductors in various levels [10], and converters presented in [11–16], where the topologies include a similar concept to MRSCC.

Another important quality of a power electronic converter is its bidirectional operation capability. It is required in a vast range of applications, which incorporates battery management. The analyzed MRSCC is made up of a basic SC structure, which makes it possible to transfer energy in both directions, similar to the converters presented in [3–5,13–17].

The four-level MRSCC converter analyzed in this study (Figure 1a) is a very favorable solution when compared to well-established topologies. It comprises features of a bidirectional and multilevel topology. In MRSCC, voltage stress on the switches is equal to the voltage of a single cell; therefore, the benefit of switch voltage stress reduction is achieved. In comparison to the other concepts of SC converters, such as SCVMs (switched capacitor voltage multipliers) presented in [18,19], MRSCC is a multilevel concept that can be more favorable for high-voltage applications. Voltage stresses on some switches of SCVMs can reach the output voltage (the highest value). Due to the high voltage gain, low stress on switches, and bidirectional conversion ability, MRSCC can be a beneficial solution for the interconnection between DC voltage systems. Depending on applications, numerous DC voltage systems are often used nowadays [20], and the required DC voltage reaches 1500 V [17]. The microgrid connection is another prospective application of such a converter as well. Furthermore, in MRSCC, the energy is transferred via capacitors, which makes it possible to reduce the weight of the converter in comparison to solutions based on inductive components. In some applications, a low weight can be an important feature due to the requirements related to the assembling system components.

The basic concepts of three-level and four-level MRSCCs are presented in [3], in which a detailed analysis of losses and selection of inductance can be found. In [4], the four-level SC converter with loads connected asymmetrically was analyzed. In [5], the first harmonic approximation analysis and basic experimental results of four-level 1.2 kV MRSCC based on SiC are presented. In [13], as well as [14], the basic structure of a multilevel SC converter is used in cascaded systems. However, the problems of topology modification and efficiency improvement in ZVS conditions were not investigated.

Recently, a lot of research has focused on the elimination or limitation of C_{oss} losses in different types of converters. However, very few of them present a solution for switched capacitor converters and especially multilevel ones. In [15,17], the C_{oss} losses were identified as an important factor in the power efficiency limitation of the voltage doubler based on the MRSCC concept. The solution for C_{oss} elimination is reported in [15] for a MOSFET converter as well as in [16] for GaN and in [17] for SiC. Therefore, the methods for achieving ZVS (zero-voltage switching) in such a converter are demonstrated in those papers, but all of them are similar and limited to two-level (voltage doubler) MRSCC. The methods rely on the application of the special control pattern to switches in higher and lower voltage levels, for additional energy delivery to the resonant inductor before dead time intervals. The energy stored in the resonant inductor is then utilized for the C_{oss} voltage transition in dead time intervals. As reported in [16], this technique is problematic in light load operations, due to significant switching frequency increases. Special control with cycle skipping is required for light load operation, which itself brings some power loss and is problematic due to parasitic oscillations [21]. To mitigate this challenge, the authors in [16,17,21] proposed to move the resonant inductor to the DC side of the converter. It eliminates most disadvantages; however, it still cannot be applied to converters with a number of levels higher than 2, since in higher levels of the converter, the resonant operation could not be achieved. In this paper, the solution with an additional small inductor further referenced as the commutation supporting inductor (L_{SC}) and small (commutation) capacitors connected across resonant branches is presented. This solution is free from the abovementioned limitations. It operates well at any load level, including an idle state, and can be applied for a converter with any number of voltage levels; however, it requires additional components.

C_{oss} losses have an important impact on the efficiency of a high-frequency converter. A model of this type of loss in an SiC-based converter is analyzed in detail in [22]. In the case of SC converters, which contain large numbers of switches, such as SCVM [18,19], and do not operate in ZVS mode, C_{oss} losses can become significant. The method for ZVS operation in SC converters with resonant inductors is an important research subject. It can be introduced to other SC topologies than MRSCC and the results of ZVS operation in an SiC-based converter presented in this paper can have even more general importance. In [23–28], SC converters with ZVS operation are presented. The presented approach for the ZVS problem solution is similar there and the converters use resonant inductors, which makes it possible to apply a phase shift switching or an operation above the resonant frequency for a charge reduction of the switch before it is turned on. The C_{oss} voltage transition method and ZVS operation is accomplished in this paper for four-level MRSCC and can be extended to a higher number of voltage levels, which is the main difference from previously presented solutions. Therefore, this paper confirms the merit of the MRSCC topology with introduced improvements. Furthermore, numerous new experimental results of Si and SiC MOSFETs based on bidirectional four-level MRSCC, such as the efficiency, voltage ratio measurements, and resonant circuit operation, are presented and compared in this paper. In essence, the main goal of this paper was to verify the concept of the topology modification towards efficiency improvement of four-level SiC-based and Si-based MRSCC converters with the major contributions of this paper being as follows:

- A presentation of a modification of the MRSCC topology to achieve ZVS with no restrictions in terms of the number of voltage levels. A significant increase of the efficiency and significant decrease of idle power losses are achieved in comparison to the base topology by (almost) elimination of the C_{oss} losses.
- A comparison the SiC-based and Si-based MRSCC performance based on experimental results. It is demonstrated that the proposed method makes it possible to achieve a performance of the Si MOSFET design close to the outstanding SiC-based one.
- The demonstration of MRSCC operation in a boost mode as well as in buck mode. It is accomplished in a test setup with two cascaded converters with a common high-voltage DC link (2 kV in the tested case). However, the load and the output voltage sensors operate on a low voltage (<1 kV), which is appropriate for precise voltage registration and efficiency measurement, and MRSCC is a bi-directional converter and such concept of a test setup configuration is justified. A similar approach can be found in [15], which presents the regenerative test setup, composed of two resonant SC converters for efficiency. A single DC-DC converter with a high voltage ratio can be used in photovoltaic systems [17] with a common DC link or DC grid on the output. The application of such converters as an interconnector between DC grids is their prospective application as well, especially as SiC-based solutions.

The paper is organized as follows. In Section 2, the basics of operation of the MRSCC and simulation results are presented. In Section 3, the problem of C_{oss} loss is addressed and a modification of base topology is proposed. Section 4 includes a description of the back-to-back experimental setup and experimental results of MRSCC operation.

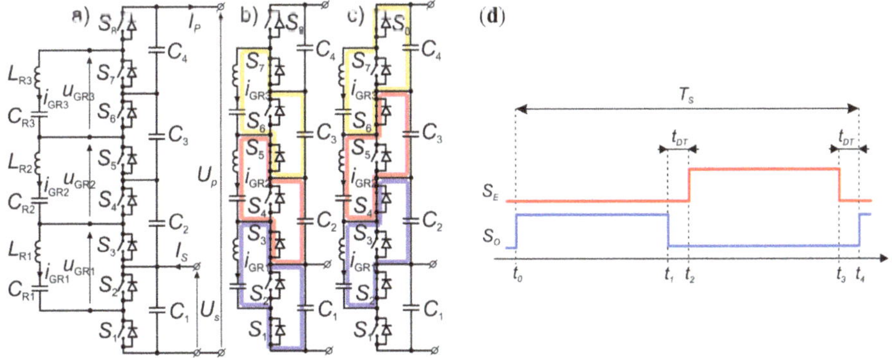

Figure 1. The bidirectional four-level switched capacitor resonant DC-DC converter (**a**) the concept, (**b,c**) the modes of operation, (**d**) the control signals of the switches; S_E—control signal of the even switches; S_O—control signal of the even switches.

2. Principle of Operation of the Multilevel Resonant Switched Capacitor Converter (MRSCC)

Figure 1 presents the topology of the MRSCC. The principle of operation assumes charging and discharging of switched capacitors (C_R) in resonant circuits and energy exchange between DC capacitors C_1–C_4. Each voltage level contains two transistors in HB (half bridge) configuration, with appropriate driving circuits. All the resonant branches $C_{R1}L_{R1}$-$C_{R3}L_{R3}$ are tuned to the same resonant frequency. The control of the converter, in the general concept, requires alternating the switching of odd and even switches with a 50% duty ratio with the resonant frequency of the branches $C_{R1}L_{R1}$-$C_{R3}L_{R3}$. As a result, each resonant branch ($C_{Rk}L_{Rk}$) is switched between two adjacent voltage level capacitors C_k and $C_{(k+1)}$ (Figure 1b,c). The control signals are presented in Figure 1d.

Voltages of the capacitors C_1-C_4 of MRSCC are nearly equalized. As a result, the voltage gain of the idealized converter in the boost-type interpretation is as follows:

$$k_{UTE} = \frac{U_p}{U_s} = n, \tag{1}$$

where U_p-higher side voltage value, U_s-lower side voltage value, and n- number of levels.

During the t_0-t_1 and t_2-t_3 time intervals, the current oscillations occur in resonant branches. During the dead time intervals, the oscillations are stopped, and in the idealized and tuned circuit, the currents of resonant branches are zero during dead time. The current peak stresses of the resonant branches are given by the following relationship:

$$\begin{cases} I_{GR(k)DTmax} = \frac{\pi}{G_{DT}}(n-k)I_P \\ \quad I_S = nI_P \end{cases} \tag{2}$$

where $k = 1, 2, 3, 4, \ldots, (n-1)$; $G_{DT} = 1 - \frac{2t_{DT}}{T_s} \leq 1$; and I_p, I_s-current value of lower and higher voltage side.

The current stresses of the switches are given by:

$$I_{S(2k)RMS}(t) = I_{S(2k-1)RMS}(t) = \frac{\pi}{2}\frac{1}{\sqrt{G_{DT}}}I_p, \tag{3}$$

$$I_{S2RMS}(t) = I_{S1RMS}(t) = \frac{\pi}{2}\frac{1}{\sqrt{G_{DT}}}(n-1)I_p, \tag{4}$$

where $k = 2, 3, 4, \ldots, n$, $\varphi = \varphi_{k-1} = \varphi_k$.

From Equations (3) and (4), it follows that the current stress of switches S1 and S2 is greater than others for $n > 2$. The higher the number of n, the bigger the difference between the current stresses of switches observed in the converter, which is an important conclusion from the switches' selection standpoint. The dead time increases the current stresses of the resonant branches and switches. It corresponds with the G_{DT} coefficient variation in Equations (2)–(4). The example simulation waveforms of the 5 kW and 500 V/2 kV MRSCC are presented in Figures 2 and 3. The results were carried out from two different models and software tools. In the waveforms presented in Figure 2, during the dead time intervals, the oscillations are halted and the current of the resonant branches equals zero. In this simulation, which was performed in MATLAB/SIMULINK software, all parasitic capacitances of the switches were omitted. In a real circuit (Figure 7a), or precise behavioral simulation model designed for PSpice (Figure 3), the distortions can be observed in the dead time intervals. The distortions are caused by an interaction of resonant branches and C_{oss} capacitances of the switches as can be anticipated by comparing the simulation results. Therefore, the distortions in dead time intervals are natural for this topology and are not caused by an improper design of the real converter. For PSpice simulation, the precise models of power transistors provided by the manufactures were used. The simulations parameters are consistent with the experimental setup (Tables 1 and 2) described in Section 4.

Table 1. Parameters for SiC and Si designs for the base configuration.

Number of Voltage Levels	n	4
Switching frequency	f_s	285 kHz
Lower side nominal voltage	V_S	500 V
Higher side nominal voltage	V_P	2000V
Design max power	P_N	5 kW
Capacitance of levels capacitors	$C_4 = C_3 = C_2 = C_1 = 4.7\ \mu F$ $C_{1a} = C_{1b} = = 0.47\ \mu F$	
Resonant branches	$L_{R3} = 3.0\ \mu H$ $L_{R2} = 2.0\ \mu H$ $L_{R1} = 0.9\ \mu H$ $C_{R3} = 100\ nF$ $C_{R2} = 147\ nF$ $C_{R1} = 320\ nF$	
Commutation branches	$C_{C3} = 22\ nF$ $C_{C2} = 22\ nF$ $C_{C1} = 47\ nF$ $R_{C3} = 1\ \Omega$ $R_{C2} = 1\ \Omega$ $R_{C1} = 0.5\ \Omega$	
Switching supporting inductor	$L_{SC} = 54\ \mu H$ ETD29	

Table 2. The measured idle power of a single HB for different transistors used for the SiC and Si setup.

No	Part			Measurement		
	Type	V_{DS} (V)	R_{DSon} (mΩ)	U_{DS} (V)	f (kHz)	ΔP_{CO} (W)
1,2	SCT3030AL	650	30			42
3–8	C3M012090D	900	120	500	285	15
1,2	SiHG33N65EF	650	95			93
3–8	SiHG21N65EF	650	150			60

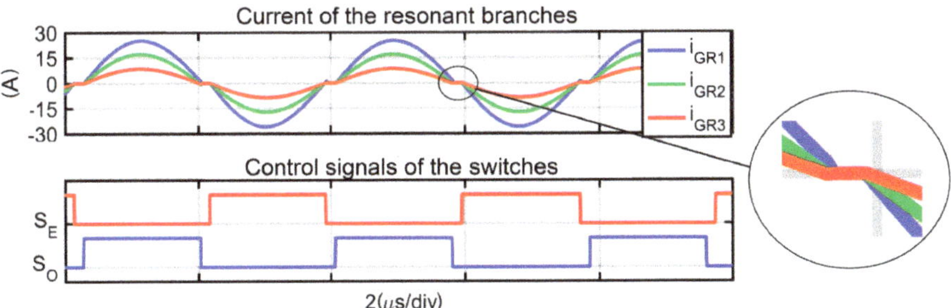

Figure 2. Simulation result of the 5 kW 500V/2 kV MRSCC (Multilevel Resonant Switched Capacitor Converter) in the base configuration in MATLAB/Simulink without C_{oss}.

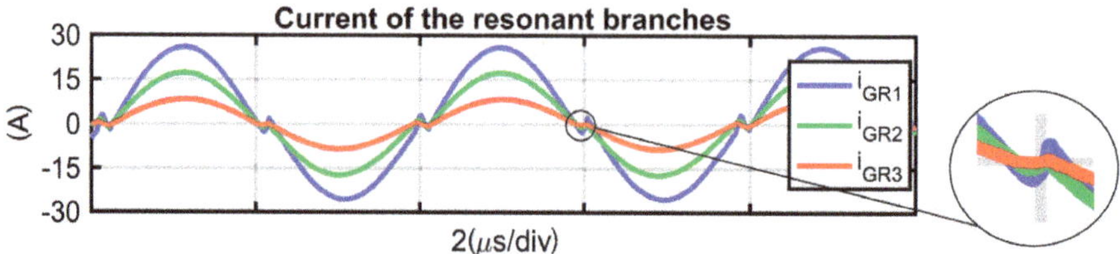

Figure 3. Simulation result of the 5 kW 500 V/2 kV MRSCC in the base configuration in PSpice (plotted in MATLAB) with C_{oss} and other parasitics.

The transition between the boost and buck mode of operation of MRSCC does not require any special control. It depends on the relationships between voltages U_p/U_S. The resonant legs should be tuned to achieve the best possible performance of the converter, and only this case is analyzed in the paper. In [5], the analysis for the general case can be found. For a cost reduction, selected switches can be replaced by diodes, but only unidirectional operation is possible in this case. For example, as presented in the literature [8], a diode-based ladder resonant switched capacitor converter operates in boost mode only.

3. C_{oss} Losses and Its Reduction by Application of the Commutation Supporting Inductor

The converter operates in ZCS (zero current switch) conditions, but switching losses still occur due to charging and discharging of the output capacitances of the switches. The converter consists of a relatively large number of switches; thus, the switching losses may seriously deteriorate its efficiency, especially under the light load operation. C_{oss} losses are also the subject of the research presented in [15,16], where it is eliminated by the special control of the switches. However, a solution for a converter composed of the number of voltage levels above 2 is not presented.

C_{oss} losses are associated with the energy dispersion from C_{oss} by the switch that is turned on as well as the power losses during C_{oss} charging caused by the current flow and commutation in the corresponding switch. In an MOSFET switch, the capacitance C_{oss} is strongly nonlinear and the charge transferred to C_{oss} is (from the DC source U_{HB}, during the lower switch turn on in a half-bridge operating with ZCS):

$$Q_{oss}(U_{HB}) = \int_0^{U_{HB}} C_{oss\_low}(u_{DS\_low})\, du_{DS\_low}. \tag{5}$$

Typically, both switches in HB are the same type and have the same parameters. Therefore, the high side switch will result in the same charge transfer from a DC source. As a result, the charge Q_{oss} is transferred twice in the switching period. Thus, the average value of the C_{oss} losses in a half-bridge of the converter is:

$$\Delta P_{HB\_loss} = 2f_s U_{HB} Q_{oss}(U_{HB}) = 2f_s U_{HB} \int_0^{U_{HB}} C_{oss\_low}(u_{DS\_low})\, du_{DS\_low}. \tag{6}$$

If C_{oss} capacitances are charged to the certain voltage value in the dead time by the external circuit, the start value in the integral in Equation (6) is a nonzero value. Therefore, the losses (Equation (6)) are lower in such a case, and can even be zero (in idealized analysis). The detailed analysis of C_{oss}-related energies and losses in HB can found in [29,30]. The MRSCC converter consists of n number of HBs composed with two various types of switches, resulting from their current stresses described by Equations (3) and (4). The total C_{oss} power loss of the converter is the sum of the C_{oss} losses of all HBs.

For the analysis and optimization of losses in MOSFET power transistors in MRSCC, the following FOM (figure of merit) [31] can be used:

$$F_M = r_{DSon} Q_{oss}, \tag{7}$$

where r_{DSon}-turn on resistance of MOSFET, Q_{oss}-charge calculated as in (5) with $U_{HB} = U_S$.

Based on Equations (3) and (4), the conduction loss in power transistors of the lower HB (S1, S2) and all the higher HBs (S3-S8) can be the following:

$$\Delta P_{Rds\_L} = I_P^2 \frac{\pi^2}{2} r_{DSon\_L}(n-1)^2, \tag{8}$$

$$\Delta P_{Rds\_H} = I_P^2 \frac{\pi^2}{2} r_{DSon\_H}(n-1), \tag{9}$$

where r_{DSon_L}-on-state resistance of the S1-S2 switches, r_{DSon_H}-on-state resistance of the S3–S8 switches.

Utilizing Equations (7) and (6), the C_{oss} loss in power transistors of the lower HB (S1, S2) ΔP_{Coss_L} and all the higher HBs (S3–S8) ΔP_{Coss_H} can be found:

$$\Delta P_{Coss\_L} = 2 f_s F_{ML} U_S \frac{1}{r_{DSon\_L}}, \tag{10}$$

$$\Delta P_{Coss\_H} = 2 f_s F_{MH} U_S (n-1) \frac{1}{r_{DSon\_H}}, \tag{11}$$

where r_{DSon_L}—on resistance of the S1–S2, r_{DSon_H}—on resistance of the S3–S8, f_s—switching frequency, F_{ML}—figure of merit (7) for S1 and S2, F_{MH}—figure of merit (7) for S3–S8.

The total power loss for a given group of switches is the sum of the conduction loss and C_{oss}-related loss. The selection of transistors with higher on-state resistance leads to a higher conduction loss. However, the C_{oss} loss is lower in such a case, if the figure of merit (Equation (7)) is considered constant. Using the above Equations (8)–(11), it is possible to find the optimal value of r_{DSon_L} and r_{DSon_L} for which the total power loss of a transistor is minimal for a given figure of merit. Obviously, the selection of transistors is a discrete problem since only several types of devices with specific parameters are manufactured. The presented solution is continuous, but in spite of this fact, it can be used for the selection of the transistors nearest to the calculated optimal parameters. The optimal on-state resistances of transistors are given by the following relationship:

$$r_{DSon\_L\_opt} = \frac{2n U_S \sqrt{U_N f_s F_{ML}}}{\pi P_S (n-1)}, \tag{12}$$

$$r_{DSon\_H\_opt} = \frac{2n U_S \sqrt{U_N f_s F_{MH}}}{\pi P_S}, \tag{13}$$

where $r_{DSon\_L\_opt}$—optimal on resistance of the S1–S2, $r_{DSon\_H\_opt}$—optimal on resistance of the S3–S8, f_s—switching frequency, F_{ML}—figure of merit (7) for S1–S2, F_{MH}—figure of merit (7) for S3–S8, P_S—power of lower voltage side, for which value the minimum losses should occur.

The above Equations (8)–(13) were evaluated for the real setup parameters, which are presented in Section 4, especially in Tables 1 and 2. The figure of merit was calculated based on Equations (5) and (7) and the C_{oss} curves from the data sheets for the two transistors selected in the initial design step: SCT3030AL as S1, S2; $F_{ML} = 4$ nVs and C3M012090D as S3–S8; $F_{MH} = 5.9$ nVs. The results are presented in Figure 4.

As can be observed in Figure 4, the initially selected transistors are not optimal. However, there is no significant improvement possible since it is about 20% of the total power loss in the switches. A significant improvement of efficiency may be achieved by the introduction of further modifications as described in Section 4, which nearly eliminates the total C_{oss} power loss. In case of the initially selected transistors, the C_{oss} loss accounts for nearly 87% of the total loss in the switches, and in the case of hypothetical optimal switches, 52%. After the introduced modification of the topology, only conduction losses are relevant. The conduction losses are lower for the initially selected transistors (SCT3030AL, C3M012090D) than in the case of any hypothetical optimal ones. Therefore, the initially selected transistors were accepted as the final selection. After the modification, the optimization problem must be defined in another way since there is no longer a tradeoff needed between C_{oss} and the on-state resistance.

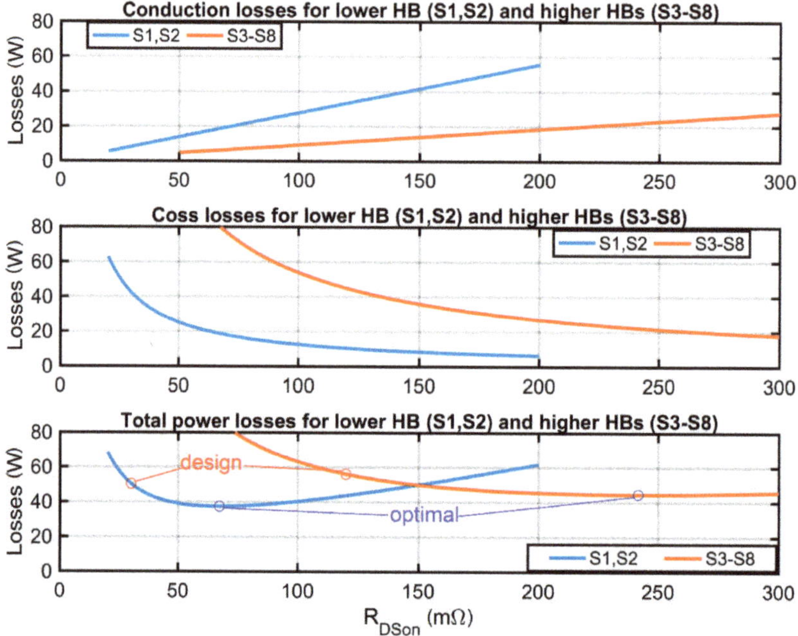

Figure 4. Results of the optimization of losses in power switches in terms of r_{DSon} selection.

3.1. Concept of MRSCC with Commutation Inductor

To reduce the idle power losses, the topology modification is proposed based on the application of the additional commutation inductor L_{SC}, which brings a significant reduction to the switching losses. The proposed concept and commutation sequence from odd to even switches is presented in Figure 5a–c. The inductor L_{SC} is connected between the output of the lowest HB and the output of the voltage divider created in the lowest voltage level. Due to the fact that all HBs are switched with a 50% duty ratio, the waveform of the current in L_{SC} is symmetrically triangular with the positive and negative peaks that occur during the commutation. This current charges and discharges the output capacitances of all switches during the commutation, causing ZVS operation.

When all the HBs are turned into *DT* mode (Figure 5b), the output capacitance of all HBs is charged by the current of inductor L_{SC} (HB1 directly, and the rest via C_{Ck} capacitors). The transition is finished after a certain time, which depends on the peak current of the L_{SC} and the values of the output capacitances of the switches. To finish the switching sequence, all the even switches are turned on (Figure 5c). Because the output capacitances of the switches are charged and discharged by an inductor, their switching losses are significantly reduced (in theory it is lossless).

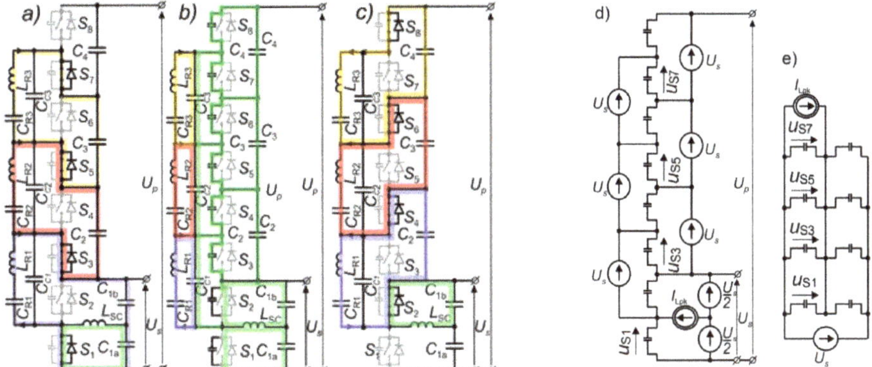

Figure 5. (**a–c**) The sequence of modes for commutation from odd to even switches in MRSCC with the commutation capacitors C_{Cn} and the supporting inductor L_{SC}; (**d**) Simplified equivalent circuit for charging the C_{oss} by L_{CS} current during dead time (mode b); (**e**) further simplification.

3.2. Application of Communication Capacitors in MRSCC Branches

The resonant branches contain series inductors and therefore, for very short commutation processes, they can be considered as the current sources. The values of those current sources can be assumed equal to the instantaneous current of the resonant branches at the beginning of the commutation. In the precisely tuned circuit, those values are close to zero, which means that the resonant branches have almost no effect on the commutation process. For this reason, the commutation capacitors (relatively small) are applied to the circuit, in parallel to the resonant branches. The commutation capacitors C_{Ck} can be considered as the voltage sources with the value of U_S (as well as the all C_k capacitors). They allow the output capacitances of higher-level HBs to be charged and discharged by the current of the L_{SC} inductor flow. The inductor L_{SC} can be considered as the current source with a positive (odd to even switches commutation) or negative (even to odd switches commutation) peak value. The equivalent circuit representing such conditions is shown in Figure 5d. The circuit can be further simplified, to the circuit where all C_{oss} are charged in parallel and supplied by one U_S voltage source, as shown in Figure 5e. The application of the commutation capacitors has an additional effect. In the base circuit, during the dead time, the current oscillations in the resonant branches are stopped, thus the resonant frequency of the converter is lower than the resonant frequency of the branches. The commutation capacitors clamp the resonant branches during the dead time, which allows the oscillation to continue. As a result, the resonant frequency of the converter is equal to the resonant frequency of the resonant branches. Furthermore, an additional benefit is the resonant current is a smooth sinusoidal (Figure 6b), while in the base circuit, some distortion (fast oscillation) can be observed in dead time intervals (Figures 4 and 6a). Such distortions reduce the balancing capability of the resonant branches and increase the series-equivalent output resistance of the converter, which will be presented in Section 4. The application of the L_{SC} inductor and C_{Ck} capacitors has no direct influence on the voltage gain of the converter. However, it improves the power efficiency and eliminates distortions of the oscillations, resulting in a reduction of the series-equivalent resistance of the converter.

Idle mode losses depend on the peak value of the current of L_{SC}, which will be demonstrated in Section 4 together with the research results related to the impact of the I_{LSCpk} on the switch voltage during commutation. The overall efficiency of the MRSCC is also affected by the remaining switching losses and conduction losses, which are out of the scope of this paper.

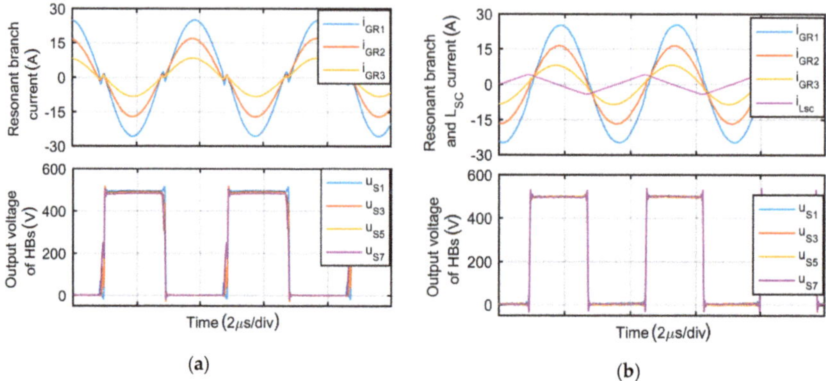

Figure 6. Waveforms of HB (Half Bridge) voltages and resonant branches currents for step-up steady-state operation of: (**a**) base MRSCC at 258 kHz, (**b**) modified MRSCC at 285 kHz. Experimental results of SiC-based converter 5 kW load (oscilloscope data exported and to MATLAB and plotted).

3.3. Selection of L_{SC} and C_{Ck} Values

Firstly, the required peak current of the L_{SC} should be determined. According to Figure 5e, the total charge Q_{ossT} that must be provided by the L_{SC} inductor during the dead time equals the sum of the $Q_{oss}(U_s)$ determined for every transistor. The charge must be transferred during the dead time interval t_{DT}, thus the required peak current of L_{SC} is given as:

$$I_{LSCpk} = \frac{1}{t_{DT}} \sum_{m=1}^{2n} Q_{oss(m)}(U_S), \tag{14}$$

where $Q_{oss(m)}(U_S)$—charge determined based on Equation (6) for the m-th transistor for voltage U_S.

The required inductance L_{SC} could be approximated as:

$$L_{SC} = \frac{U_S}{8 f_s I_{LSCpk}}. \tag{15}$$

The values of the C_{Ck} capacitors should be large enough so that the voltage across does not change significantly in the dead time interval, when the I_{LSC} flows through them, charging the C_{oss} of the transistors. The values of the C_{Ck} capacitors are given by:

$$C_{Ck} = \frac{1}{\Delta U_{CC(k)}} \sum_{m=2k+1}^{2n} Q_{oss(m)}(U_S), \tag{16}$$

where $\Delta U_{CC(k)}$—change of voltage across k-th capacitor in the dead time interval.

The selection of the values of the capacitor according to Equation (16) is therefore dependent on the permitted change of the voltage across in the dead time interval ΔU_{CCk} and C_{oss} of the used transistors. Too high ΔU_{CCk} values (too low capacitance of C_{Ck} capacitors) cause inefficient charging and discharging of C_{oss} of higher-level transistors and stimulation of resonance branches by voltage glitches during the dead time intervals. Too low ΔU_{CCk} values (too high capacitances) are unfavorable not only in terms of size and cost but also due to the increased participation of these capacitors in energy transport between level voltages, which cause unwanted inrush currents. From the conducted experimental research, it follows that the values of ΔU_{CCk} in the range of few volts (for SiC-based devices between 10 and 100 nF) bring suspected favorable effects. From Equation (16), it follows that the higher the index of the capacitor, the lower the capacitance demanded. To dampen very fast oscillation caused by hard switching of C_{Ck} and parasitic inductances, resistors R_{Ck} should be added in series with C_{Ck}. Typically, values in the range 0.1–1 Ω are optimal.

4. Experimental Results

4.1. The laboraTory Setup and Operation of MRSCC

The laboratory setup was designed for performing tests and measurements of a 2 kV four-level MRSCC converter in a comparative manner. To mitigate the problem of high-voltage measurements, the special laboratory setup was designed. Because the topology is bidirectional, twin converters can be connected back-to-back, creating a cascade with the common high-voltage link. The input and output voltage value of the cascade was 500 V, and this allowed for utilization of the low-voltage laboratory equipment.

Figure 7 presents the schematic of the experimental setup. Tables 1 and 2 contain the crucial parameters of the experimental system. Figure 8a presents the picture of the experimental system. The setup consists of two identical converters designed as a PCB module, which contains all the power and auxiliary components. Every voltage level was equipped with two channel gate driver ICs (UCC21520, Texas Instruments, Dallas, USA). The driver incorporates DT logic and a timer, thus only one control signal was sufficient for each voltage level. The fiber wires were used for signal delivery because of the great isolation that they provide. The controller was designed with an FPGA device and provided control for both converters, with eight control signals in total. The control algorithm included simple pattern generation in open loop mode. The resonant branches were designed to achieve a unified voltage ripple across all the resonant capacitors (about 100 $V_{pk\text{-}pk}$), and the same resonant frequency. The resonant branches were composed of the inductors based on ferrite gapped toroid- and FKP1 (WIMA, Mannheim, Germany)-type capacitors.

Figure 6 presents the waveforms of the resonant branch current and the output voltages of every single HB (which in fact are the voltages of the switches with an odd index). The waveforms in Figure 6a were obtained in MRSCC in the base configuration while the waveforms in Figure 6b were measured in the modified MRSCC.

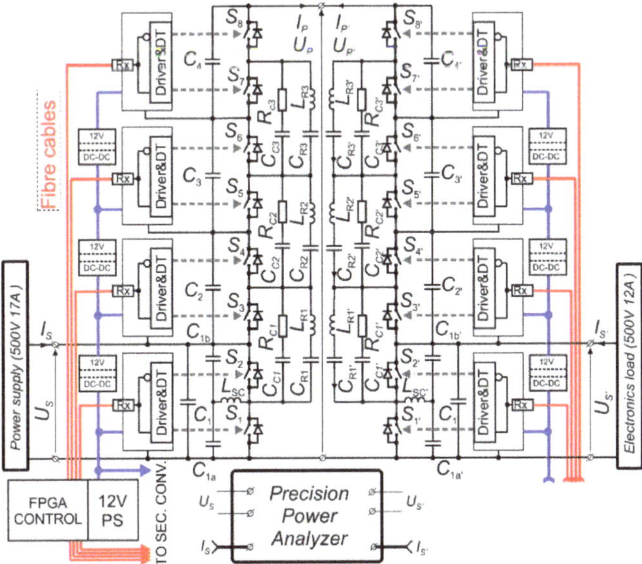

Figure 7. Schematic of the MRSCC laboratory back-to-back setup with modification applied ($L_{SC} + C_{Ck}R_{Ck}$).

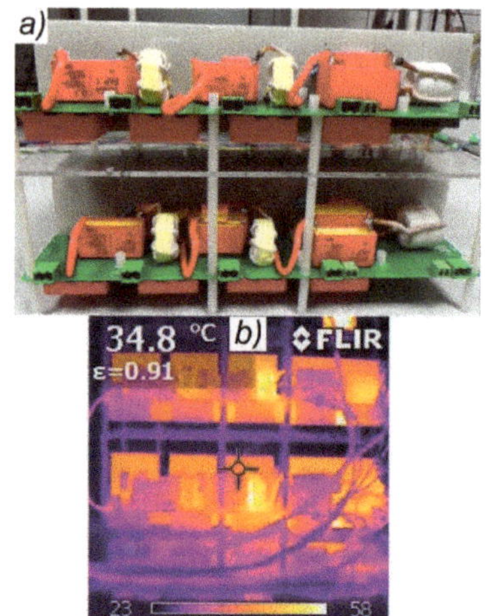

Figure 8. (**a**) Picture of the MRSCC laboratory back-to-back setup (**b**) IR (Infrared) picture under full load (5 kW)—both SiC design.

4.2. Efficiency Measurement Results and Discussion

Figure 9 presents the experimental results related to the efficiency of the MRSCC converter. The measurements presented in Figure 9a,b are related to the single converter but those in Figure 9c are related to the whole cascade. The measurement was performed for different configurations of the converter. The power efficiency of the converter for the base configuration was relatively low (Figure 9c-curves no. 4). The influence of the high idle power losses was significant because a typical peak of the efficiency in the chart efficiency versus power was not observed. Despite this, the efficiency increased with the power load. In the first modification, the $C_{Ck}R_{Ck}$ branches were added but with no inductor L_{SC} (Figure 9c-curves no. 1). The power efficiency was even worse for the low load because the frequency (285 kHz) was higher due to the lack of oscillation breaks in DT intervals. Therefore, the C_{oss} losses were proportionally higher in this case. However, the distortions in dead time intervals were eliminated, which improved of the voltage efficiency. After the application of the commutation supporting inductor L_{SC}, outstanding results were obtained (Figure 9c-curves no. 2). Figure 9a,b presents the detailed results related to the improvement of the efficiency of the converter by a reduction of the C_{oss} losses. As described in Section 3, it was achieved by the application of the supporting inductor (L_{SC}) and operation at the appropriate peak current of L_{SC} during commutations. Figure 9a presents the waveforms of the rising slope of the low side transistor in a half bridge for a different peak value of the L_{SC} current, and Figure 9b presents the respective idle power measurements for the SiC-based design. The results were performed on the experimental setup with SiC switches operating with 285 kHz. The current of L_{SC} should be low but sufficient to effectively perform the transition. A current of L_{SC} that is too high causes hard switching and increases of the losses (Figure 9c). It is remarkable that the idle mode power losses were reduced from approximately 100W in the case when $I_{LSCpk} = 0A$ to nearly 3 W for $I_{LSCpk} = 6A$.

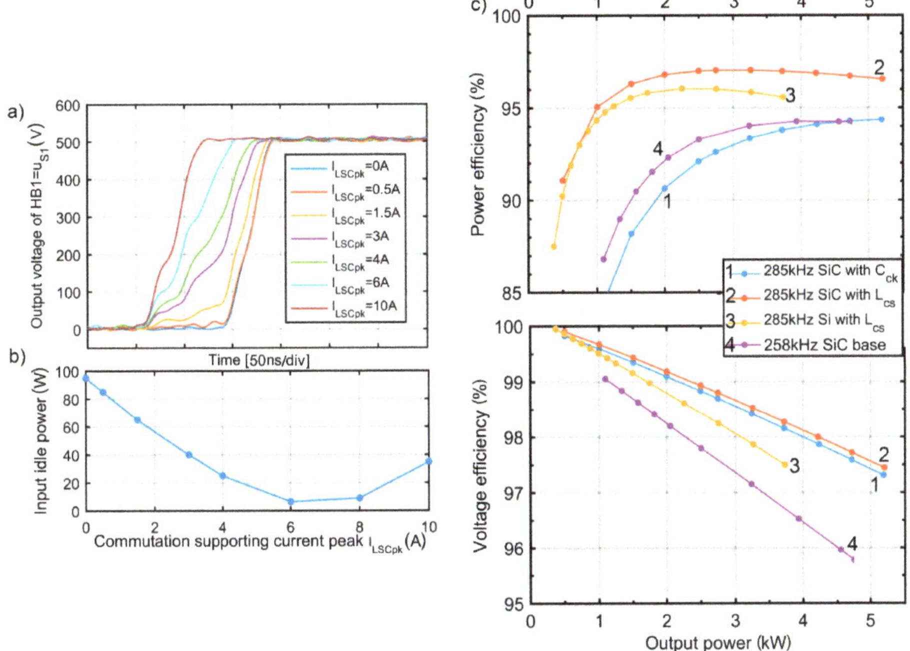

Figure 9. Experimental results: (**a**) waveforms of voltage rising slope HB1. (**b**) idle power losses of the single MRSCC converter-SiC design 285 kHz, (**c**) Power and voltage efficiency characteristics vs. output power for different configurations of the cascaded setup. Yokogawa WT1800 Precision Power Analyzer.

For further references, the power and voltage efficiency curves are presented in Figure 9c-curves no. 3. The parameters of the Si design were insignificantly worse only when compared to outstanding SiC. Higher R_{DSon} losses for Si MOSFETs (Table 1) limited the voltage efficiency and maximum load of the converter. The Si-based design was not able to operate with 285 kHz without the L_{SC} supporting inductor due to huge commutation losses (in theory, 270 W per single converter based on Table 2 and simple calculation). This extreme case shows how effective the proposed modification is. In Table 2, the power consumed by a single non-loaded HB is listed (for different transistors, given frequency, and common DC link voltage). As can be noticed, the power losses are significant, especially for Si devices, even when the types with definitely higher R_{DSon} are compared. The results of the case of the Si-based design L_{SC} inductor and f_s = 285 kHz are presented in Figure 9c-curve no. 3.

A thermography picture presented in Figure 8b shows that the heating of both the converters was nearly the same. It means that the efficiency of a single converter can be properly estimated on the basis of the efficiency results presented in Figure 9c taking into account half of the total losses in the system. Therefore, the peak efficiency of the SiC-based MRSCC with the L_{SC} inductor was approximately equal to 98.5%.

5. Conclusions

In this paper, the research results under the operation of a modified SiC and Si-based MRSCC converters were presented. The MRSCC converter is a relatively novel topology and its improvement was proposed in this paper. The solution assumes the application of a commutation supporting inductor to reduce the switching losses associated with Coss in a converter made up of any number of levels.

The majority of research was performed in the 5 kW laboratory setup, which demonstrates the feasibility of boost and buck operation of the MRSCC. The conversion between the levels of 500 and 2 kV at a switching frequency 285 kHz, with the use of switches with VDS = 900 and VDS = 650 V, was demonstrated. Both the converters were tested simultaneously in the system with a common high voltage DC link. The input and output of the system remained on a low voltage level, which made it possible to perform high-precision efficiency measurements. Furthermore, it is a good example of a low-cost laboratory test setup for high voltage ratio converters.

The major goal of the research focused on a verification of the topology improvement in a four-level bi-directional MRSCC with a 0.5/2 kV voltage conversion ratio with the use of SiC and Si switches. It was accomplished with very promising results and the following conclusions:

- The solution with a commutation supporting inductor in MRSCC is feasible and brings a significant increase of the power efficiency.
- A near-total elimination of the C_{oss} power losses was observed in MRSCC with the commutation supporting inductor and the suitable switching applied.
- An increase of the voltage efficiency was observed as well (lowering of the output-equivalent series resistance). The voltage gain was more stable vs. load and the difference between the theoretical and practical voltage gain was lower than in the case without the proposed improvements.
- The method is very efficient in MRSCC with Si MOSFET as well. Through the application of the commutation supporting inductor, the performance measured in the case of the Si MOSFET MRSCC increased to the level comparable with the converter based on outstanding SiC switches.
- In the SiC-based MRSCC, a high power efficiency of 98.5% was measured.

The efficiency versus power characteristic showed an insignificant decline when the power increased, which is also beneficial. The voltage drop versus power was not significant in the demonstrated design cases of MRSCC. The best solution of 2.5% of the voltage decrease in the 5 kW range was observed. The results of the heat distribution in the converter showed that it can be regular. Overheating of particular cells was not observed.

Author Contributions: Conceptualization, A.K.; methodology, A.K.; software, A.K.; validation, A.K. and R.S.; formal analysis, A.K.; investigation, A.K.; resources, A.K. and R.S.; data curation, A.K.; writing—original draft preparation, A.K. and R.S.; writing—review and editing, A.K., and R.S.; visualization, A.K.; supervision, R.S.; project administration, A.K.; funding acquisition, A.K. and R.S. All authors have read and agreed to the published version of the manuscript.

Funding: This research was funded by Polish National Science Center, grant number 2016/21/N/ST7/02355.

Conflicts of Interest: The authors declare no conflict of interest.

Nomenclature

DT	Dead Time
FOM	Figure Of Merit
HB	Half Bridge
MRSCC	Multi-level Resonant Switched Capacitor Converter
SC	Switched Capacitor
SCVM	Switched-Capacitor Voltage Multipliers
ZVS	Zero Voltage Switch
ZCS	Zero Current Switch
$C_{oss_{low}}\left(u_{DS_{low}}\right)$	Output capacitance characteristic of a MOSFET
F_M	FOM for MOSFET
$I_{GR(k)DTmax}$	Peak current of k-th resonant branch

k_{UTE}	Ideal Voltage Gain
ΔP_{Rds_L}, ΔP_{Rds_H}	Conduction losses for lower HB (S_1, S_2) and for all higher ones (S_3–S_8)
ΔP_{Coss_L}, ΔP_{Coss_H}	losses for lower HB (S_1, S_2) and for all higher ones (S_3–S_8)
P_S	Power of lower voltage side
r_{DSon_L}, r_{DSon_H}	On state resistance for lower switches (S_1, S_2) and for all higher ones (S_3–S_8)
U_p, I_p	Voltage, Current Higher Side
U_S, I_S	Voltage, Current Lower Side
U_{HB}	DC link voltage of a Half Bridge

References

1. Ioinovici, A. Switched-capacitor power electronics circuits. *IEEE Circuits Syst. Mag.* **2001**, *1*, 37–42. [CrossRef]
2. Yeung, Y.P.B.; Cheng, K.W.E.; Ho, S.L.; Law, K.K.; Sutanto, D. Unified analysis of switched-capacitor resonant converters. *IEEE Trans. Ind. Electron.* **2004**, *51*, 864–873. [CrossRef]
3. Gitau, M.; Kala-Konga, C.L. Compact energy efficient switched-capacitor multilevel DC-DC converters for interfacing DC-buses with common ground. In Proceedings of the 2011 IEEE International Symposium on Industrial Electronics, Gdansk, Poland, 27–30 June 2011.
4. Gitau, M. High efficiency multilevel switched-capacitor DC-DC converters for interfacing DC-buses with separate ground. In Proceedings of the IECON 2011—37th Annual Conference on IEEE Industrial Electronics Society, Melbourne, Australia, 7–10 November 2011.
5. Kawa, A.; Stala, R. Bidirectional multilevel switched-capacitor resonant converter based on SiC MOSFET switches. In Proceedings of the 2017 19th European Conference on Power Electronics and Applications (EPE'17 ECCE Europe), Warsaw, Poland, 11–14 September 2017.
6. Cao, D.; Jiang, S.; Peng, F.Z. Optimal design of a multilevel modular capacitor-clamped DC–DC converter. *IEEE Trans. Power Electron.* **2013**, *28*, 3816–3826. [CrossRef]
7. Khan, F.H.; Tolbert, L.M. A Multilevel Modular Capacitor-Clamped DC–DC Converter. *IEEE Trans. Ind. Appl.* **2007**, *43*, 1628–1638. [CrossRef]
8. Li, S.; Xiangli, K.; Zheng, Y.; Smedley, K.M. Analysis and Design of the Ladder Resonant Switched-Capacitor Converters for Regulated Output Voltage Applications. *IEEE Trans. Ind. Electron.* **2017**, *64*, 7769–7779. [CrossRef]
9. Qian, W.; Cao, D.; Cintron-Rivera, J.G.; Gebben, M.; Wey, D.; Peng, F.Z. A Switched-Capacitor DC–DC Converter With High Voltage Gain and Reduced Component Rating and Count. *IEEE Trans. Ind. Electron.* **2012**, *48*, 1397–1406. [CrossRef]
10. Ye, Y.; Cheng, K.W.E.; Chen, S. A High Step-up PWM DC-DC Converter With Coupled-Inductor and Resonant Switched-Capacitor. *IEEE Trans. Power Electron.* **2017**, *32*, 7739–7749. [CrossRef]
11. Andersen, R.L.; Lazzarin, T.B.; Barbi, I. A 1-kW Step-Up/Step-Down Switched-Capacitor AC–AC Converter. *IEEE Trans. Power Electron.* **2013**, *28*, 3329–3340. [CrossRef]
12. Lazzarin, R.T.B.; Andersen, R.L.; Martins, G.B.; Barbi, I. A 600-W Switched-Capacitor AC–AC Converter for 220 V/110 V and 110 V/220 V Applications. *IEEE Trans. Power Electron.* **2012**, *27*, 4821–4826. [CrossRef]
13. Priyadarshi, A.; Kar, P.K.; Karanki, S.B. An inductor-less bidirectional DC-DC converter topology for high voltage gain applications. In Proceedings of the TENCON 2017—2017 IEEE Region 10 Conference, Penang, Malaysia, 5–8 Novemeber 2017.
14. Xiong, S.; Tan, S. Cascaded High-Voltage-Gain Bidirectional Switched-Capacitor DC–DC Converters for Distributed Energy Resources Applications. *IEEE Trans. Power Electron.* **2017**, *32*, 1220–1231. [CrossRef]
15. Sano, K.; Fujita, H. Performance of a High-Efficiency Switched-Capacitor-Based Resonant Converter With Phase-Shift Control. *IEEE Trans. Power Electron.* **2011**, *26*, 344–354. [CrossRef]
16. Vasić, M.; Serrano, D.; Toral, V.; Alou, P.; Oliver, J.A.; Cobos, J.A. Ultraefficient Voltage Doubler Based on a GaN Resonant Switched-Capacitor Converter. *IEEE J. Emerg. Sel. Top. Power Electron.* **2019**, *7*, 622–635. [CrossRef]
17. Stevanović, B.; Serrano, D.; Vasić, M.; Alou, P.; Oliver, J.A.; Cobos, J.A. Highly Efficient, Full ZVS, Hybrid, Multilevel DC/DC Topology for Two-Stage Grid-Connected 1500-V PV System With Employed 900-V SiC Devices. *IEEE J. Emerg. Sel. Top. Power Electron.* **2019**, *7*, 811–832. [CrossRef]

18. Waradzyn, Z.; Stala, R.; Mondzik, A.; Penczek, A.; Skala, A.; Pirog, S. Efficiency Analysis of MOSFET-Based Air-Choke Resonant DC–DC Step-Up Switched-Capacitor Voltage Multipliers. *IEEE Trans. Ind. Electron.* **2017**, *64*, 8728–8738. [CrossRef]

19. Kawa, A.; Stala, R.; Mondzik, A.; Pirog, S.; Penczek, A. High-Power Thyristor-Based DC–DC Switched-Capacitor Voltage Multipliers: Basic Concept and Novel Derived Topology With Reduced Number of Switches. *IEEE Trans. Power Electron.* **2016**, *31*, 6797–6813. [CrossRef]

20. Kumar, D.; Zare, F.; Ghosh, A. DC Microgrid Technology: System Architectures, AC Grid Interfaces, Grounding Schemes, Power Quality, Communication Networks, Applications, and Standardizations Aspects. *IEEE Access* **2017**, *5*, 12230–12256. [CrossRef]

21. Vasic, M.; Serrano, D.; Alou, P.; Oliver, J.A.; Grbovic, P.; Cobos, J.A. Comparative analysis of two compact and highly efficient resonant switched capacitor converters. In Proceedings of the 2018 IEEE Applied Power Electronics Conference and Exposition (APEC), San Antonio, TX, USA, 4–8 March 2018.

22. Li, X. A SiC Power MOSFET Loss Model Suitable for High-Frequency Applications. *IEEE Trans. Ind. Electron.* **2017**, *64*, 8268–8276. [CrossRef]

23. Ding, Y.; He, L.; Liu, Z. Bi-directional bridge modular switched-capacitor-based DC-DC converter with phase-shift control. In Proceedings of the 2016 IEEE Energy Conversion Congress and Exposition (ECCE), Milwaukee, WI, USA, 18–22 September 2016.

24. Li, S.; Zheng, Y.; Wu, B.; Smedley, K.M. A Family of Resonant Two-Switch Boosting Switched-Capacitor Converter With ZVS Operation and a Wide Line Regulation Range. *IEEE Trans. Power Electron.* **2018**, *33*, 448–459. [CrossRef]

25. Li, Y.; Ni, Z.; Cao, D. Optimal Operation of Multilevel Modular Resonant Switched-Capacitor Converter. In Proceedings of the 2018 IEEE 6th Workshop on Wide Bandgap Power Devices and Applications (WiPDA), Atlanta, GA, USA, 15 June 2018.

26. Xie, H.; Li, R. A Novel Switched-Capacitor Converter With High Voltage Gain. *IEEE Access* **2019**, *7*, 107831–107844. [CrossRef]

27. Lei, H.; Hao, R.; You, X.; Li, F. Nonisolated High Step-Up Soft-Switching DC–DC Converter With Interleaving and Dickson Switched-Capacitor Techniques. *IEEE J. Emerg. Sel. Top. Power Electron.* **2019**. [CrossRef]

28. He, L.; Zheng, Z. High step-up DC–DC converter with switched-capacitor and its zero-voltage switching realization. *IET Power Electron.* **2017**, *10*, 630–636. [CrossRef]

29. Oeder, C.; Barwig, M.; Duerbaum, T. Estimation of switching losses in resonant converters based on datasheet information. In Proceedings of the 2016 18th European Conference on Power Electronics and Applications (EPE'16 ECCE Europe), Karlsruhe, Germany, 6–8 September 2016.

30. Elferich, R. General ZVS half bridge model regarding nonlinear capacitances and application to LLC design. In Proceedings of the 2012 IEEE Energy Conversion Congress and Exposition (ECCE), Raleigh, NC, USA, 15–20 September 2012.

31. Hopkins, A.; McNeill, N.; Anthony, P.; Mellor, P. Figure of merit for selecting super-junction MOSFETs in high efficiency voltage source converters. In Proceedings of the 2015 IEEE Energy Conversion Congress and Exposition (ECCE), Montreal, QC, USA, 6 January 2015.

Article

DC-DC High-Voltage-Gain Converters with Low Count of Switches and Common Ground

Robert Stala, Zbigniew Waradzyn * and Szymon Folmer

Department of Power Electronics and Energy Control Systems, Faculty of Electrical Engineering, Automatics, Computer Science and Biomedical Engineering, AGH University of Science and Technology, al. Mickiewicza 30, 30-059 Krakow, Poland; stala@agh.edu.pl (R.S.); folmer@agh.edu.pl (S.F.)
* Correspondence: waradzyn@agh.edu.pl; Tel.: +48-12-617-2811

Received: 1 September 2020; Accepted: 26 October 2020; Published: 29 October 2020

Abstract: This paper presents a new concept and research results of DC-DC high-voltage-gain, high-frequency step-up resonant converters. The proposed topologies are optimized towards minimizing the number of switches and improvements in efficiency. Another relevant advantage of such type of converters is that they have a common input and output negative point. The proposed converters are based on the resonant switched-capacitor voltage multiplier circuit, and that is why they are compared with a classic converter from this family. The included results show the operating principle, possible switching methods with the consideration of their impact on the voltage gain level, as well as the voltage and current ripples. The operating concepts and analytical calculations are confirmed by simulation and experimental results.

Keywords: DC-DC converter; resonant converter; high-voltage-gain converter; switched-capacitor converter; inductiveless converter

---

1. Introduction

Switched capacitor (SC) circuits can be effectively used in power electronic converters [1]. The significant advantages of SC-based DC-DC power converters are high-voltage-gain, low volume, and quasi inductiveless design. To achieve oscillating currents, low-volume inductors can be used in those converters. They can be designed as air-chokes, or even be based on parasitic inductances of the circuits, resulting in a decrease in the weight of the converter. The design without ferrite chokes allows for the use of the converter in high ambient temperature and/or with a low-volume heat sink.

SC DC-DC converters represent one of the classes of non-isolated step-up converters [2–4]. Nowadays, there are a significant number of applications where isolated DC-DC step-up converters are required [3], due to technical reasons and safety requirements. However, various kinds of non-isolated converters are extensively developed as well. One of the prospective applications for non-isolated DC-DC step-up converters proposed in the literature [5–10] are photovoltaic (PV) systems. High step-up DC-DC converters are often required in grid-connected PV systems to transfer the energy from a low-voltage PV source to the grid [5,6]. In transformerless PV systems [7,8], as well as in microinverters [9], dual-stage DC-AC converters are one of the investigated solutions.

The SC step-up DC-DC converter could be a competitive solution to the switch-mode boost converter. An example of such an idea is presented in Reference [10]. The non-isolated step-up converter can be used not only for a single stage supply, but as a part of a system composed of series-connected converters as well. In such systems, isolation can be implemented in another stage of conversion, e.g., by using a series resonant converter [6,11].

High-voltage-gain in SC-based DC-DC converters can be achieved by applying a suitable topology concept. In References [12–14], an SC voltage multiplier (SCVM) has been presented. It is a series-parallel converter in that a high-voltage-gain can be obtained, as it is proportional to the number of switching cells. The advantage of an SCVM is its modular topology; however, the number of required transistors is relatively high. Series-parallel SC converters have also been presented in recent publications [15,16]. In Reference [15], a converter with regulated voltage gain has been discussed. This device utilizes three switches, which means that the voltage gain can reach three. Reference [16] has presented a very effective method that allows the switch count in high-gain series-parallel converters to be decreased. However, the converter presented in Reference [16] does not have a common input and output negative point, and the output voltage is asymmetrically divided. In Reference [17], a converter that combines Dickson-based and ladder SC converter concepts has been presented. In the proposed topology, high-voltage-gain is achieved with limited voltage and current stresses on the switches. The Dickson-based SC concept has also been used in the converter presented in Reference [18] that is composed of an SC part and an interleaved boost converter. The converter achieves a very high-voltage-gain with the output voltage regulation and soft switching operation, using four switches and seven diodes. In Reference [19], high-voltage-gain is achieved in a converter with switched-capacitor and switched-inductor networks. A concept of a family of converters composed of a boost stage and switched-capacitor-inductor cells has been presented in Reference [20]. This increases the voltage gain of the converter significantly with favorable voltage stress levels, efficiency, and component count. References [21–24] have demonstrated high-voltage-gain multilevel converters based on typical multilevel converter concepts. When we take into consideration the number of the utilized components and the reached voltage gain, the multilevel SC converters can be more beneficial in comparison to the SCVMs. The converter described in Reference [21] is based on a modified classic multilevel SC topology; however, it is composed of a significant number of switches. In Reference [22], an improvement in the operation of the multilevel resonant SC converter (MRSCC) has been proposed. The MRSCC makes it possible to operate with high-voltage-gain and limited voltage stress on the switches with the ability of bi-directional energy transfer. In Reference [23], a multilevel structure has been achieved in the converter with two switches and circuits composed of diodes and capacitors. The converter can operate with zero voltage switching (ZVS) and voltage regulation. In Reference [24], a DC-DC bidirectional SC converter has been presented that improves the total device power ratings in comparison to the multilevel modular capacitor clamped converter (MMCCC) and well-established flying-capacitor converters.

One of the major issues of the SC converters is a large number of switches used in the topology. This problem can be solved by the concepts of cascaded or series systems composed of SC units [25,26] or by new concepts of topologies [16,27,28]. In the concept for the switch count reduction presented in Reference [26], a high-power converter has been analyzed in a multi-section topology. The converter is composed of the typical SCVM sections separated by LC filters. According to this concept, a significant reduction in the number of switches has been achieved. However, an increased number of passive components are utilized as LC filters between the sections in the multi-section converters [26]. The problem of the switch count reduction in an SCVM converter has been analyzed in Reference [29], where the charging of the switched capacitors is controlled by a single switch. For high-voltage-gain, the system is significantly simplified. The design of such a cost-effective converter should assume a much higher current stress of the switch that controls the charging of the switched capacitors.

The converters proposed in this paper are optimized towards a low count of transistors (and they are called Low Count of Transistors Switched Capacitor Voltage Multipliers—LCSCVMs). The basic concept of the topology and operation assumes that every second cell has no transistors whatsoever, but the utilization of all the switched capacitors remains possible, and the effect of voltage gain is comparable to that of the multipliers (SCVMs) presented in Reference [13,14]. Furthermore, the optimized concept is introduced into the cost-effective topology presented in Reference [29], which gives a new relevant converter. Taking into consideration the count of switches, SC converters, such as the SCVM [20], may not be in competition with the LLC converters or other established topologies. However, the concepts proposed in this paper demonstrate a development of the SC topologies towards a significant decrease in the number of switches. One of the converters presented in this paper requires only three switches, which is below the number of transistors used in a full-bridge LLC converter. Other advantages of the SC converters, such as: high gain, high power density and low weight (no transformer or bulky choke), fast dynamic response [3], ability for operation in high temperature (no ferrites), and simple control, can make them an alternative solution for existing topologies intended for high-voltage-gain non-isolated DC-DC conversion. SC-based topologies can be suitable for the miniaturization of converters that can be applied in emerging power electronics applications, such as wearable technology.

For the operational parameters of an SC converter, the switching strategy applied for a given topology can be essential, which has been demonstrated in Reference [14]. For the optimization purposes analyzed in this paper, various switching strategies are proposed for the new topologies. This makes it possible to determine the advantages of the presented topologies, also taking into consideration a variety of qualities, other than the count of switches.

The proposed converters are nearly pure switched-capacitor circuits, where a vast majority of energy is transferred via capacitors rather than inductors. The resonant inductors are used to achieve oscillatory currents. The inductors can be designed as air chokes, which reduces the weight of the converters and allows them to work in higher temperatures. However, another trend in the development of very high-voltage-gain converters can be observed in the literature. The concept presented in Reference [30] is based on coupled inductor (CI) converters that achieve good parameters such as voltage ratio, efficiency, low number of switches, or low voltage stress on switches. Notwithstanding, such converters use chokes and, therefore, differ from the presented SC-based concept regarding admissible ambient temperature of operation, weight, and volume. The design comparison can be analyzed in particular case studies.

In this paper, the qualities introduced by the new topologies will be compared with those of a classic SCVM and of other converters discussed in recently published papers.

The paper is organized as follows. Section 2 demonstrates two proposed topologies of the SC converters and presents the principles of their operation. For both converters, switching strategies are analyzed. The discussion is supported by the results of computer simulations of their operation in five cases of switching strategies. Moreover, with the use of the simulation results, a number of parameters of the converters operating under various switching strategies are compared as well. Section 3 contains efficiency models of the proposed converters that demonstrate their efficiency as a function of their parameters. Section 4 presents the laboratory setup and the experimental verification of its operation, including the efficiency of the converter. All the research results are concluded in Section 5.

2. Operating Principle of the Converters

The operating principle of the converters in Figure 1 is similar to that of other SC multipliers, and is based on the charging and discharging of the switched capacitors in consecutive stages (time intervals). However, various switching strategies can be proposed for the new converters, which creates differences in their parameters. In the SCVM, as well as in the case of the converters proposed in this paper, the switched capacitors are recharging in resonant circuits composed of a switched capacitor and a low-volume resonant inductor. This creates ZCS (zero current switching) operating conditions, and limits the current flow between the capacitors and the voltage source connected in parallel.

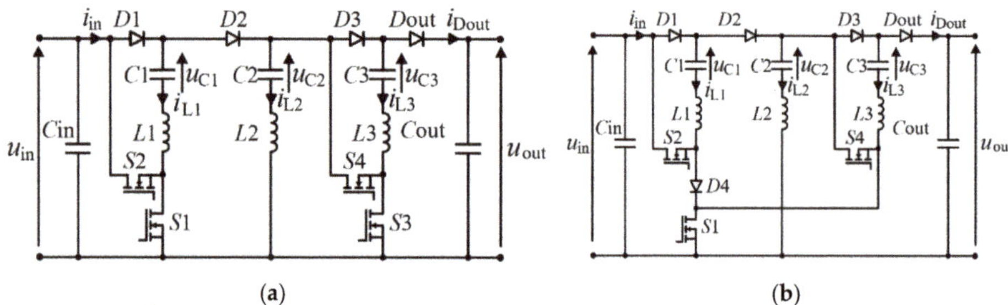

Figure 1. Proposed new resonant converters with low count of switches and common input/output negative point: (**a**) LCSCVMa, (**b**) LCSCVMb.

The main difference between the topology of the proposed converters and the classic SCVMs is that in the former case, an LC circuit that is not a part of a traditional cell is used, usually consisting of a diode and two transistors [13]. This circuit is charged using the energy of the input source and the electric charge of the switched capacitor that is the nearest to the input source. Then, the middle capacitor is discharged to the output capacitor or to the switched capacitor nearer to the output. Its function is to increase the output voltage and the amount of converted power, simultaneously maintaining the same value of the input voltage and the same cell number as in the case of a typical SCVM.

The LCSCVMa (Figure 1a) offers a larger number of strategies than the LCSCVMb (Figure 1b), due to the possibility of independent control of switches S_1 and S_3. The basic switching strategies can be composed of 2, 3, or 5 stages.

2.1. Switching Strategy Concepts for the LCSCVMa

Table 1 presents three switching strategies for the LCSCVMa, and Figures 2–5 depict the corresponding simulation waveforms.

Table 1. Switching strategy concepts of the LCSCVMa. States of switches S_1–S_4.

The Concept for Switching Strategy of LCSCVMa	Description—Stages of Charge Transfer in the Converter
Strategy C1	1. Simultaneous charging of all the switched capacitors 2. Discharging of the capacitor that is the nearest to the source (C_1) to the internal branch (C_2) 3. Charging C_1, and discharging C_2 and the next SC capacitor (C_3) to the output 4. Discharging C_1 to the internal branch (as in 2) 5. Discharging C_2 and C_3 to the output
Strategy C2	1. Simultaneous charging of all the switched capacitors 2. Discharging C_1 to the internal branch (C_2) 3. Discharging C_2 and C_3 to the output
Strategy C3	1. Simultaneous charging of all the switched capacitors (C_1 and C_3) 2. Simultaneous discharging of all the switched capacitors and charging the internal branch capacitor (C_2)

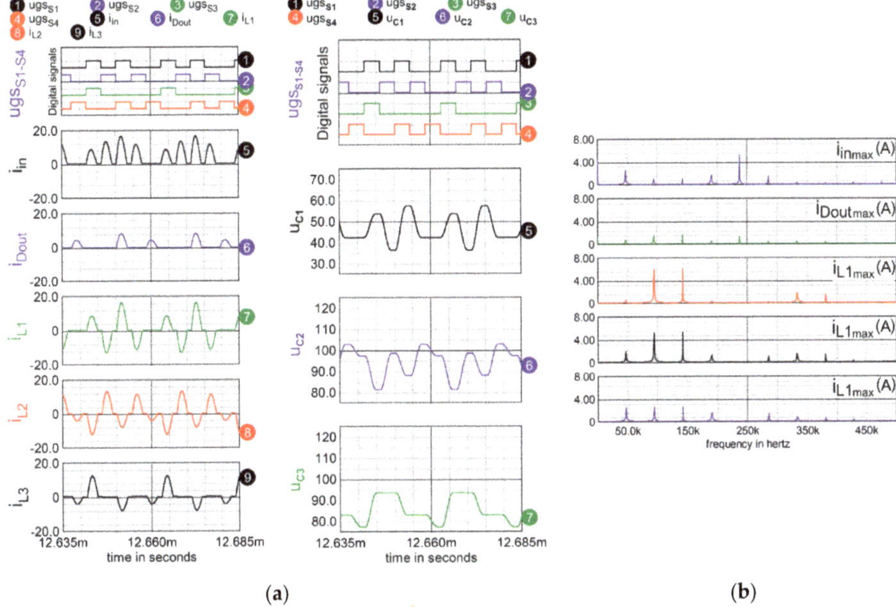

Figure 2. Steady-state operation of the LCSCVMa converter under switching strategy C1: (**a**) waveforms of the gate to source signals of transistors (presented with level shift), input current (in amperes), and voltages (in volts) on capacitors C_1, C_2, and C_3. (**b**) Spectrum of the input current, and currents of switched capacitors and output capacitor. The results were obtained with the use of ICAP/4 simulation software.

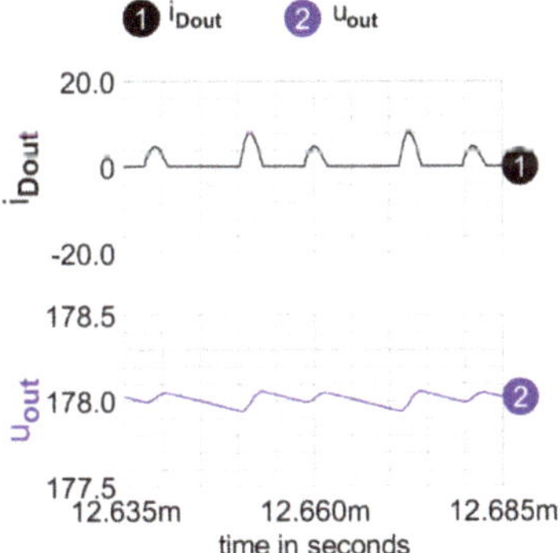

Figure 3. Steady-state output current and voltage waveforms of the LCSCVMb converter under switching strategy C1 (4 A/div and 100 mV/div). The results were obtained with the use of ICAP/4 simulation software.

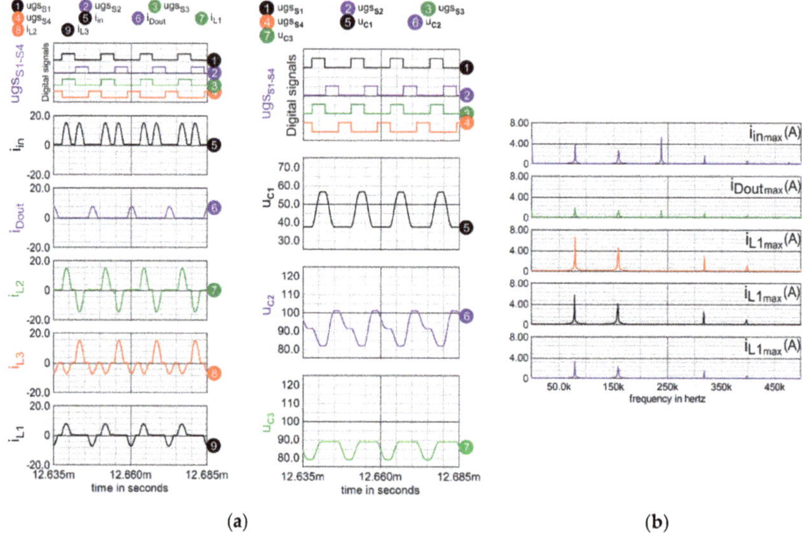

(a)

(b)

Figure 4. Steady-state operation of the LCSCVMa converter under switching strategy C2: (**a**) waveforms of the gate to source signals of transistors (presented with level shift), input current (in amperes), and voltages (in volts) on capacitors C_1, C_2, and C_3. (**b**) Spectrum of the input current, and currents of switched capacitors and output capacitor. The results were obtained with the use of ICAP/4 simulation software.

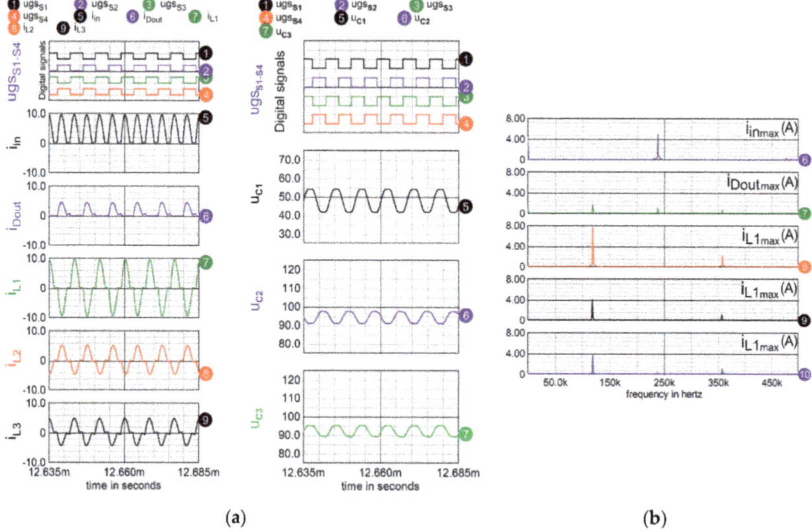

(a)

(b)

Figure 5. Steady-state operation of the LCSCVMa converter under switching strategy C3: (**a**) waveforms of the gate to source signals of transistors (presented with level shift), input current (in amperes), and voltages (in volts) on capacitors C_1, C_2, and C_3. (**b**) Spectrum of the input current, and currents of switched capacitors and output capacitor. The results were obtained with the use of ICAP/4 simulation software.

To characterize the switching strategies, Table 1 contains the idealized control logic waveforms of the transistors (signals S_1 to S_4), as well as the description of the particular operation stages. Dead times have been neglected in Table 1, but they have been taken into account in the simulations and experiments. Capacitor C_2 (Figure 1) is not referred to as a switched capacitor. The maximum switching frequency (strategy C3) is defined as:

$$f_{S_{max}} = \frac{1}{2T_{pulse}} \tag{1}$$

where T_{pulse} is the sum of the duration time $T_0/2$ of a single current pulse of any transistor and the dead time t_d (any period of time denoted as 1–5 in Table 1),

$$T_0 = \frac{1}{f_0} = 2\pi\sqrt{LC} \tag{2}$$

and $L = L_1 = L_2 = L_3$, $C = C_1 = C_2 = C_3$ (Figure 1).

All the simulation results were obtained for the following parameters: $U_{in} = 50$ V, $L_n = 620$ nH, $C_n = 1.47$ µF, $f_0 = 166.7$ kHz, $T_{pulse} = 4.2$ µs ($f_{Smax} = 119$ kHz), $C_{out} = 100$ µF, $P_{out} = 200$ W ($n = 1, 2, 3$). A resistance of 100 mΩ has been inserted into each branch as an equivalent to parasitic resistances. The time period T_{pulse}, as well as the duty cycle of the switching signals of the transistors, remain constant in each switching strategy. The selection of the switching frequency depends on the power of the converter, achievable resonant inductance, and switching losses [13]. This parameter, as well as the others, can be fixed in the following steps. In the ZCS mode, the SC converters' transistors do not operate in the ZVS mode, and during their turn-ons, the output charge is shorted (C_{oss} losses). The limit of C_{oss} losses determines the switching frequency of the transistors taking into consideration their type and voltage stresses. The oscillation frequency should be nearly equal to the switching frequency to minimize conduction losses [13]. This frequency depends on the product of L_nC_n, and allows to select C_n for a known value of L_n. The maximum power of the converter depends on capacitance C_n and the switching frequency [13], and it should be higher than or equal to the rated power for the selected parameters. The simulation results presented in this section have been obtained with the use of ICAP/4 simulation package based on the IsSpice4 simulator.

2.1.1. Simulation Results of the Switching Strategy C1

Figure 2 presents steady-state simulation waveforms of the LCSCVMa controlled according to strategy C1. From all the results, it can be seen that the switched capacitors are recharged by oscillatory currents and each stage of the switching is longer than the half-period of the oscillations.

The entire switching cycle is composed of five stages (Table 1). According to the principle of operation, turning on switches S_1 and S_3 involves the charging of the switched capacitors C_1 and C_3. Capacitor C_3 is being charged from capacitor C_2 of the internal branch whose voltage is going down in this stage. The diode D_2 remains turned off, as $u_{C2} > u_{C1}$ and $u_{C2} > u_{in}$ (Figure 2). In the next stage, switch S_2 is turned on, and capacitor C_2 is being charged from the source u_{in} and capacitor C_1 connected in series with it. The charging of the output capacitor, from capacitors C_2 and C_3 connected in series, occurs in the next stage when the switch S_4 is turned on. At the same time, capacitor C_1 is being charged from the source. In the next two stages, capacitor C_2 and capacitor C_{out} are being charged, consecutively.

The advantage of this switching strategy is reducing the number of the performed switching operations, which leads to switching losses limitation. In three of five stages of the switching period, only one switch is affected.

The input current has various values in each switching state, which is a drawback of this strategy. Therefore, a low-frequency component $f_S = f_{ac\text{-}in} = f_{Smax}/2.5$ appears in current i_{in}, as well as in all other currents and voltages in the circuit. Using this kind of switching requires using a large input filter and a large output capacitor. From the standpoint of the components' volume and input current filtering, this strategy is not favorable.

The output voltage used for the voltage gain calculation in relation (3) has been measured as the average value of the waveform presented in Figure 3 together with the output current. Further results, given in Equations (4)–(7), were obtained in the same manner.

In this strategy, the measured average value of the output voltage of the converter equals $U_{out} = 178$ V. For the input voltage of the converter $U_{in} = 50$ V (maintained by the voltage source in simulations), the voltage gain of the converter under switching strategy C1 equals:

$$G_{UC1} = \frac{U_{out}}{U_{in}} = \frac{178.0}{50.0} = 3.56 \tag{3}$$

2.1.2. Simulation Results of the Switching Strategy C2

Figure 4 presents simulation waveforms in the LCSCVMa controlled according to strategy C2. In this strategy, each switching period consists of three stages. The first two switching stages correspond to those in strategy C1. In the third stage, only transistor S_3 is on. The last two stages of strategy C1 do not occur here, and capacitor C_2 is charged and discharged only once in a switching period. The number of the switching operations is lower in comparison to that in strategy C1. The spectrum of currents and voltages shows more favorable qualities in strategy C2 versus C1, as the 50 kHz components are not present (the lowest frequency is 75 kHz).

In this strategy, the measured average value of the output voltage of the converter equals $U_{out} = 177$ V. For $U_{in} = 50$ V, the voltage ratio is

$$G_{UC2} = \frac{U_{out}}{U_{in}} = \frac{177.0}{50.0} = 3.54 \tag{4}$$

2.1.3. Simulation Results of the Switching Strategy C3

Figure 5 presents simulation waveforms in the LCSCVMa controlled according to strategy C3. In this strategy, there are only two stages. In the first stage, the charging of the switched capacitors takes place (switches S_1 and S_3 are turned on). During the second stage, the output capacitor and C_2 are being charged (with switches S_2 and S_4 turned on).

In this strategy, each switch operates with a much higher frequency than in the case of strategies C1 and C2. This brings an improvement in the spectrum of the currents and voltages, as the lowest frequency is 120 kHz. It is favorable from the passive components volume optimization standpoint.

In strategy C3, the measured average value of the output voltage of the converter equals $U_{out} = 185$ V, and for $U_{in} = 50$ V, the voltage ratio is

$$G_{UC3} = \frac{U_{out}}{U_{in}} = \frac{185.0}{50.0} = 3.7 \tag{5}$$

2.2. Switching Strategy Concepts for the LCSCVMb

The LCSCVMb converter is simpler than the LCSCVMa, and contains three switches only. There is only one stage of charging the switched capacitors, realized by the switch S_1, and two possible stages of discharging them, controlled by switches S_2 and S_3. This creates two switching strategies for this converter, which are presented in Table 2. Figures 6 and 7 depict simulation waveforms of the LCSCVMb controlled according to these strategies.

Table 2. Switching strategy concepts of the LCSCVMb. States of switches S_1, S_2, and S_4.

The Concept for Switching Strategy of LCSCVMb		Description—Stages of Charge Transfer in the Converter
Strategy C4		Similarly to strategy C2 for the LCSCVMa, strategy C4 gives the following characteristic in the LCSCVMb: 1. Simultaneous charging of all the switched capacitors 2. Discharging C_1 to the internal branch (C_2) 3. Discharging C_2 and C_3 to the output
Strategy C5		Similarly to strategy C3 for the LCSCVMa, strategy C5 gives the following characteristic in the LCSCVMb: 1. Simultaneous charging of all the switched capacitors 2. Simultaneous discharging of all the switched capacitors and charging the internal branch (C_2)

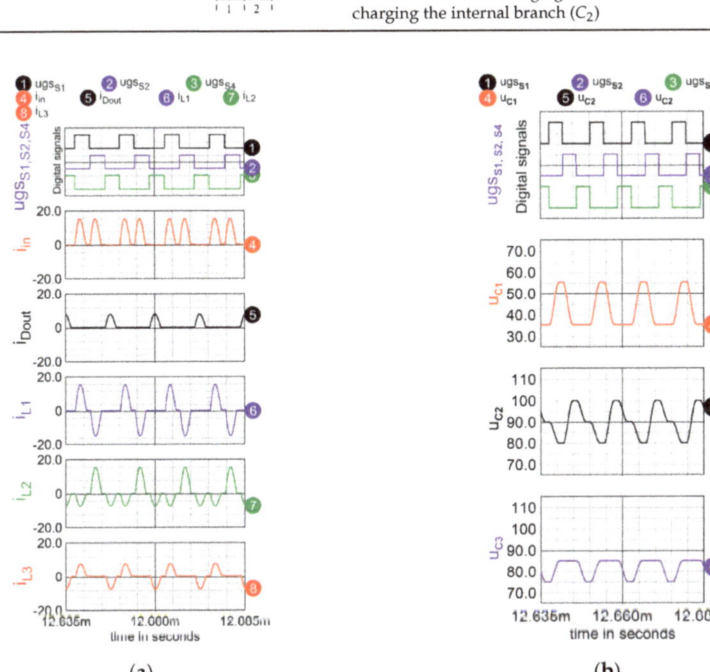

Figure 6. Steady-state operation of the LCSCVMb converter under switching strategy C4: (**a**) Waveforms of the input current, inductor currents, and the current of the output diode (in amperes). (**b**) Voltages (in volts) on capacitors C_1, C_2, and C_3. The results were obtained with the use of ICAP/4 simulation software.

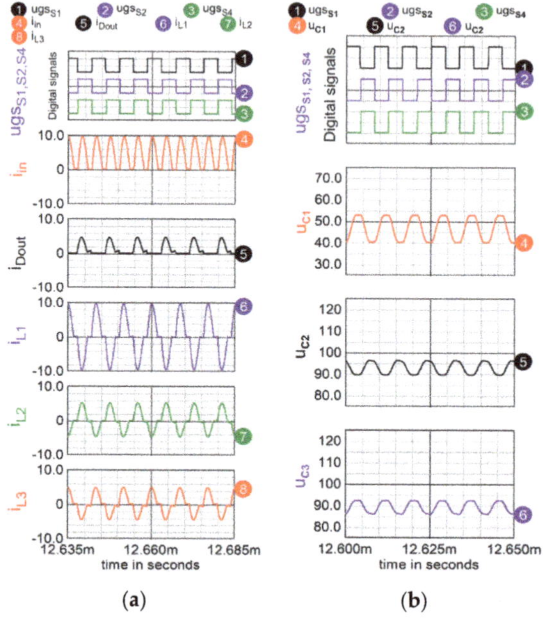

(a)   (b)

Figure 7. Steady-state operation of the LCSCVMb converter under switching strategy C5: (**a**) Waveforms of the input current, inductors currents, and the current of the output diode (in amperes). (**b**) Voltages (in volts) on capacitors C_1, C_2, and C_3. The results were obtained with the use of ICAP/4 simulation software.

2.2.1. Simulation Results of the Switching Strategy C4

Figure 6 presents simulation waveforms of the LCSCVMb controlled according to strategy C4. The current and voltage waveforms in strategy C4 are nearly identical with those in strategy C2.

In this strategy, the measured average value of the output voltage of the converter equals $U_{out} = 172.1$ V, which yields (for $U_{in} = 50$ V):

$$G_{UC4} = \frac{U_{out}}{U_{in}} = \frac{172.1}{50.0} = 3.44 \tag{6}$$

2.2.2. Simulation Results of the Switching Strategy C5

Figure 7 presents simulation waveforms for the LCSCVMb controlled according to strategy C5. The current and voltage waveforms of the strategy C5 are nearly identical with those in strategy C3.

In this strategy, the measured average value of the output voltage of the converter equals $U_{out} = 181.4$ V. For $U_{in} = 50$ V, the voltage ratio is:

$$G_{UC5} = \frac{U_{out}}{U_{in}} = \frac{181.4}{50.0} = 3.63 \tag{7}$$

2.3. Comparison among the Topologies and Switching Strategies

In Section 2, a significant number of waveforms are presented for the particular strategies. The differences in the waveforms of the currents and voltages are clear, but to compare the concepts of the converters and the switching strategies, the following parameters will be taken into consideration and presented in charts:

- Number of components,
- Voltage gain,
- The lowest frequency in the input current (f_{ac_in}),
- The lowest frequency in the output current (f_{ac_out}),
- Voltage pulsation on capacitors ($U_{C1p\text{-}p}$, $U_{C2p\text{-}p}$, $U_{C3p\text{-}p}$),
- rms values of inductor currents (I_{L1_rms}, I_{L2_rms}, I_{L3_rms}),
- Maximum values of inductor currents (I_{L1_max}, I_{L2_max}, I_{L3_max}),
- Symmetry of inductor currents (Sym_i_L).

The data are presented in Table 3, where the parameters of the SCVM (on the basis of Reference [14] for an appropriate strategy) are included as well.

Table 3. Major parameters comparison among the parameters of LCSCVMa and LCSCVMb converters in the tests of 200 W operation.

Parameter	LCSCVMa Strategy			LCSCVMb Strategy		SCVM
	C1	C2	C3	C4	C5	
No. of switches	4	4	4	3	3	6
No. of diodes	4	4	4	5	5	4
U_{out}, V	178.0	177.0	185.0	172.1	181.4	191.2
T_S, µs	21.0	12.6	8.4	12.6	8.4	8.4
f_{ac_in}, kHz	47.6	79.4	238.1	79.4	238.1	238.1
f_{ac_out}, kHz	47.6	79.4	119.0	79.4	119.0	119.0
$U_{C1p\text{-}p}$, V	21.02	19.41	12.4	19.8	12.6	5.98
$U_{C2p\text{-}p}$, V	21.61	19.41	6.54	19.8	6.65	5.98
$U_{C3p\text{-}p}$, V	16.08	9.7	6.18	9.92	6.29	5.98
I_{L1_rms}, A	6.75	7.25	5.65	7.41	5.75	2.73
I_{L2_rms}, A	6.22	6.28	2.92	6.42	2.97	2.73
I_{L3_rms}, A	4.11	3.63	2.75	3.71	2.80	2.73
I_{L1_max}, A	16.1	14.8	9.46	15.1	9.60	4.57
I_{L2_max}, A	13.2	14.8	5.01	15.2	5.09	4.57
I_{L3_max}, A	12.3	7.43	4.74	7.56	4.80	4.57
Symmetry of current i_{L1}	no	yes	yes	yes	yes	yes
Symmetry of current i_{L2}	no	no	yes	no	yes	yes
Symmetry of current i_{L3}	no	yes	yes	yes	yes	yes

Figures 8–10 present a comparison between the values of parameters of the discussed converters, and the corresponding parameter of the SCVM.

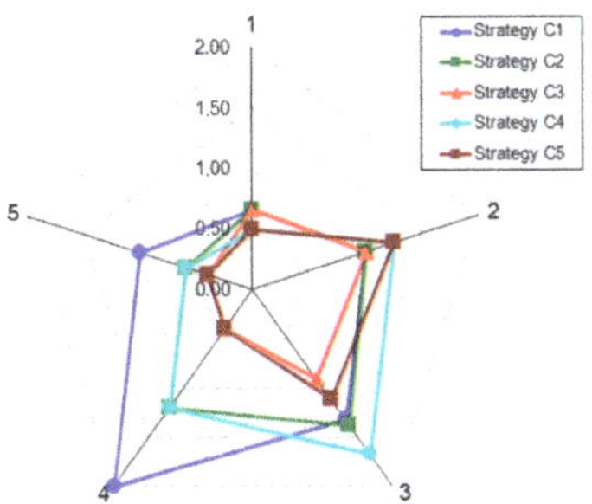

Figure 8. Comparison of converters' parameters under strategies C1–C5: Ratios of number of switches (axis 1) and number of diodes (axis 2) to those in SCVM (on the basis of data in Table 3). Quantity proportional to undesired output voltage decrease: $0.06\cdot(200 - U_{out})$ (axis 3). Ratios of the lowest frequencies in the input and output current: $0.4\cdot f_{ac\_in}$ SCVM/f_{ac_in} (axis 4), $0.4\cdot f_{ac\_out}$ SCVM/f_{ac_out} (axis 5).

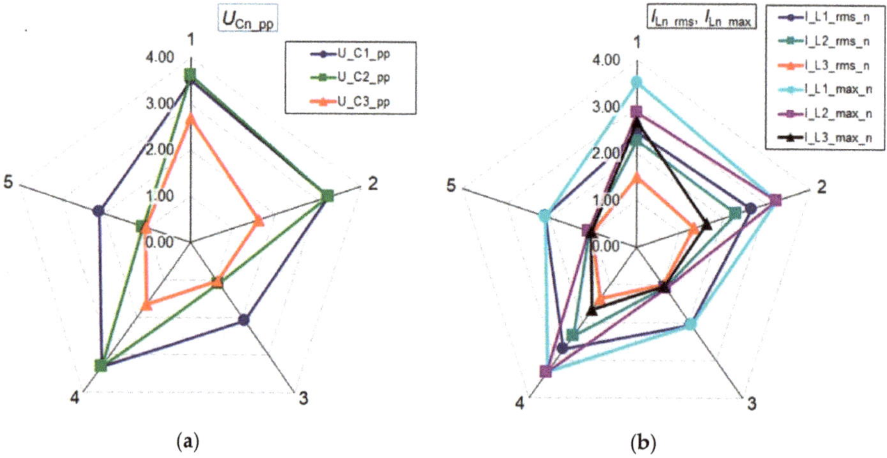

Figure 9. Ratios of the following parameters of strategies C1–C5 (axes 1–5): (**a**) Peak-to-peak voltages across capacitors C_1–C_3, (**b**) rms and maximum values of currents in inductances L_1–L_3. The results are based on the data in Table 3.

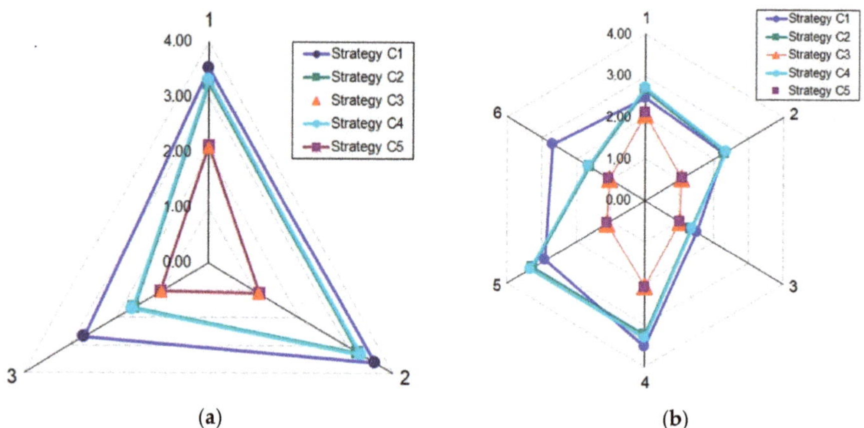

Figure 10. Ratios of the following parameters (strategies C1–C5) to the corresponding parameter of SCVM: (**a**) Peak-to-peak voltages across capacitors C_1–C_3, (**b**) rms (axes 1–3) and maximum values of currents in inductances L_1–L_3 (axes 4–6). The results are based on the data in Table 3.

In Figure 8, the coefficients 0.06 and 0.4 are used respectively, to better visualize the undesired output voltage decrease in regard to the theoretical value of 200 V, and the lowest frequencies in the input and output current of the discussed converters compared to those in the SCVM. In each case, a lower value on the graph is better.

From the chart presented in Figure 9a, it follows that the lowest peak-to-peak (p-p) voltages, in all the strategies, are equal the voltage across capacitor C_3. Moreover, the strategies C3 and C5 show the lowest p-p voltages for all the internal capacitors (C_1–C_3). Figure 9b demonstrates that the currents of inductor L_3 are the lowest, and the strategies with the lowest inductor currents are C3 and C5.

The same qualities are visible in charts presented in Figure 10, which clearly demonstrate that the parameters of strategy C4 are nearly the same as those of strategy C2. The same refers to strategies C5 and C3.

The LCSCVMa and LCSCVMb converters can be further extended to units of higher voltage gain, similarly as in the case of the converters presented in References [13,22,25,26,29]. Taking into consideration the number of switches and diodes, as well as the frequency of the input current, both the proposed converters are very attractive for high-voltage-gain (Table 4). It should be noticed that the converter extension is very effective in the case of the LCSCVMb concept. For voltage gain $G_U = 8$, it requires only four switches, which is an excellent result in comparison to other pure SC converters. Other parameters such as voltage stresses on the switches can be found in the literature.

Table 4. Comparison of the number of switches and diodes, and the lowest frequency of the input current in selected topologies versus the voltage gain. Ref. = Reference.

Parameter		Toplogy							
	Gain	LCSCVMa	LCSCVMb	Ref. [13]	Ref. [22]	Ref. [16]	Ref. [25]	Ref. [26]	Ref. [27]
No. of switches (and diodes)	4	4 (4)	3 (5)	6 (4)	8 (0)	-	8 (0)	4 (4)	4 (6)
	7	-	-	12 (7)	14 (0)	7 (5)	-	-	7 (12)
	8	6 (6)	4 (7)	14 (8)	16 (0)	-	12 (0)	6 (6)	8 (14)
$f_{\text{iin\_min}}/f_{\text{Smax}}$ for all gains(f_{Smax}—in (1))	1	1	0.5	1	1/4	1	1	0.5	

3. Efficiency Model of the LCSCVM Converters

The analysis below concerns the LCSCVMa operating under the strategy C3 (Table 1) and LCSCVMb operating under the strategy C5 (Table 2). In both cases, there are two stages of operation. Figure 11 depicts the current paths in the LCSCVMa. In the LCSCVMb, the switch S_1 conducts the sum of currents i_{L1} and i_{L3} in the stage 1, whereas the current paths in the stage 2 are the same as in the LCSCVMa.

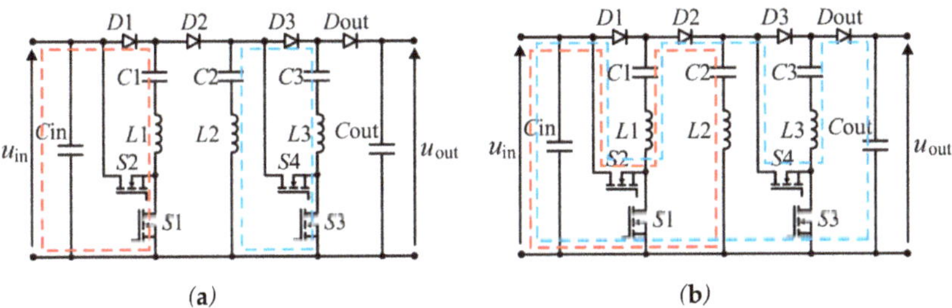

Figure 11. Current paths in the LCSCVMa: (**a**) in the stage 1 and (**b**) in the stage 2.

Assuming ideal power electronic switches, and a constant value of the input (U_{in}) and the output (U_{out}) voltage, as well as neglecting parasitic resistances and voltage drops across the power electronic devices, the currents in the stage 1 (Figure 11a) can be described as follows:

$$i_{L1}(t) = i_{C1}(t) = \frac{U_{\text{in}} - U_{C11}}{\rho} \sin \omega_0 t = I_{1m} \sin \omega_0 t \tag{8}$$

$$i_{L3}(t) = i_{C3}(t) = \frac{U_{\text{in}} - U_{C11}}{2\rho} \sin \omega_0 t = \frac{I_{1m}}{2} \sin \omega_0 t \tag{9}$$

$$i_{L2}(t) = i_{C2}(t) = -i_{L3}(t) \tag{10}$$

With the characteristic impedance and the angular resonant frequency given by

$$\rho = \sqrt{L/C}, \; \omega_0 = 1/\sqrt{LC} \tag{11}$$

where U_{C11} is the initial voltage across capacitor C_1, and I_{1m} and $I_{2m}/2$ are the current amplitudes. Equation (8) presents the current of a typical series LC circuit supplied from a voltage source, and Equation (9) was obtained also taking into account the initial values of the capacitor voltages.

The values of the passive components depend on the assumed nominal power (P_{nom}), switching frequency (f_S), and the volume of the resonant inductor. The values of time T_{pulse} (1), and finally T_0 (2) and ω_0 (11), are assumed taking into account the limit of the switching losses in the converter. The capacitance of the switched capacitors is determined by the charge required to be transferred in a single switching pulse. The maximum power of the SCVM-type converter is achieved when the switched capacitors are fully discharged in a switching cycle (and then charged to the voltage equal to $2U_{in}$). This determines the minimum capacitance, which in the SCVM composed of n switching cells is defined as follows:

$$C_{min} = 2nf_S U_{in}{}^2 / P_{nom}. \tag{12}$$

In a quasi inductiveless SCVM-type converter, the value of resonant inductance (L) is very small (L can be designed as a PCB air choke). Therefore, to achieve the assumed switching frequency, the capacitance of the switched capacitors can be selected considerably bigger than C_{min} (as in the case of the experimental setup presented in this paper). In the stage 2 (Figure 11b), the currents of capacitors C_1–C_3 and inductances L_1–L_3 have the same values (Equations (8)–(10)) as in the stage 1, but with the opposite signs. The voltages across the capacitors C_1, C_2, and C_3 in the stage 1 are given by

$$u_{C1}(t) = (U_{in} - U_{C11})(1 - \cos \omega_0 t) + U_{C11} \tag{13}$$

$$u_{C2}(t) = -\frac{(U_{in} - U_{C11})}{2}(1 - \cos \omega_0 t) + U_{C21} \tag{14}$$

$$u_{C3}(t) = \frac{(U_{in} - U_{C11})}{2}(1 - \cos \omega_0 t) + U_{C31} \tag{15}$$

where U_{C21} and U_{C31} are the initial voltages across capacitors C_2 and C_3, respectively.

In the stage 2 (Figure 11b), the expressions for voltages have similar forms with appropriate signs and initial values.

Based on the formulas mentioned above, all the voltage initial values and the output voltage can be computed as a function of U_{C11}. For example, we obtain

$$U_{out} = 5U_{in} - U_{C11} \tag{16}$$

U_{C11} can be calculated taking into account (8) and the following relation

$$I_{in-av} = I_{L1av} = \frac{2}{\pi} I_{1m} f_{Sn} = \frac{P_{in}}{U_{in}} \tag{17}$$

$$U_{C11} = U_{in} - \frac{\pi \rho P_{in}}{2 f_{Sn} U_{in}} \tag{18}$$

where

$$f_{Sn} = f_S / f_0 \tag{19}$$

From Equations (16) and (18), we have

$$U_{out} = 4U_{in} + \frac{\pi \rho P_{in}}{2 f_{Sn} U_{in}} \tag{20}$$

In practical converters, there are voltage drops across the circuit elements like the diodes and the transistors, which result in a variation of the output voltage with power and frequency.

The efficiency of an SCVM-type converter is determined by the resistances of its components, voltage drops on the diodes and transistors, the input voltage, power, and by the relation between the switching period T_S and period T_0 (2), which can be expressed by f_{Sn} (19). Therefore, it is necessary to calculate the average and rms values of the currents. It is assumed that transistors S_1 and S_3 are IGBTs, and S_2 and S_4 are MOSFETs.

$$I_{D1av} = I_{D2av} = \frac{I_{L1av}}{2} = \frac{1}{\pi} I_{1m} f_{Sn} = \frac{P_{in}}{2U_{in}}, \ I_{D3av} = I_{Dout-av} = \frac{I_{L2av}}{2} = \frac{I_{L3av}}{2} = \frac{I_{L1av}}{4} = \frac{P_{in}}{4U_{in}} \tag{21}$$

$$I_{S2} = \frac{1}{2} I_{1m} \sqrt{f_{Sn}} = \frac{\pi P_{in}}{4U_{in}\sqrt{f_{Sn}}}, \ I_{S4} = \frac{1}{4} I_{1m} \sqrt{f_{Sn}} = \frac{\pi P_{in}}{8U_{in}\sqrt{f_{Sn}}} \tag{22}$$

For the LSCVMa, we have:

$$I_{S1av} = I_{D1av} = \frac{P_{in}}{2U_{in}}, \ I_{S3av} = I_{D3av} = \frac{P_{in}}{4U_{in}} \tag{23}$$

Conduction losses, ΔP_c, in both converters are

$$\Delta P_c = \sum_k r_k I_{Sk}^2 + \sum_l \Delta U_{Dl} I_{Dlav} + \sum_m \Delta U_{Sm} I_{Smav} + \sum_n r_T I_n^2 \tag{24}$$

where r_k denotes the total resistance of the branch with MOSFET transistor S_k ($k = 2, 4$), including the resistance of the transistor. ΔU_{Dl} is the voltage drop across diode D_l, ΔU_{Sm} is the voltage drop across IGBT transistor S_m, r_T is the resistance of each circuit with an IGBT transistor, and I_n is its rms current.
It is assumed that the voltage drops across the devices remain constant in the conducting state. We assume that all the resistances and voltage drops are the same, i.e.

$$r_2 = r_4 = r, \ \Delta U_{S1} = \Delta U_{S2} = \Delta U_S, \ \Delta U_{D1} = \Delta U_{D2} = \Delta U_{D3} = \Delta U_{D4} = \Delta U_{Dout} = \Delta U_D \tag{25}$$

The efficiency of the LSCVMa converter can be calculated as follows. The resistive losses in the circuits containing IGBTs are:

$$\Delta P_{c2} = r_T I_{L11}^2 + 2r_T I_{L31}^2 = \frac{3\pi^2 P_{in}^2 \, r_T}{32U_{in}^2 f_{Sn}} \tag{26}$$

Taking (21)–(26) into account, the conduction losses can be presented as

$$\Delta P_c = \frac{5\pi^2 P_{in}^2 \, r}{64U_{in}^2 f_{Sn}} + \frac{3\pi^2 P_{in}^2 \, r_T}{32U_{in}^2 f_{Sn}} + \frac{3P_{in}}{2U_{in}}\left(\Delta U_D + \frac{1}{2}\Delta U_S\right) \tag{27}$$

The turn-off switching loss is zero, due to the ZCS switching. However, there is a turn-on switching loss, associated with charging and discharging the transistors' output capacitances. The total switching power loss, ΔP_{sw}, is

$$\Delta P_{sw} = \Delta W_{sw} f_S = \Delta P_{sw0} f_{Sn} \tag{28}$$

where ΔW_{sw} is the energy lost at turn-on in the transistor's resistances in a single switching cycle, and $\Delta P_{sw0} = \Delta W_{sw} \cdot f_0$ is power loss at resonant frequency. A way of calculating these losses is presented in Reference [31].
The efficiency is (Equations (27) and (28))

$$\eta = 1 - \frac{\Delta P_c}{P_{in}} - \frac{\Delta P_{sw}}{P_{in}} = 1 - \frac{5\pi^2 P_{in}^2 r}{64U_{in}^2 f_{Sn}} - \frac{3\pi^2 P_{in}^2 r_T}{32U_{in}^2 f_{Sn}} - \frac{3P_{in}}{2U_{in}}\left(\Delta U_D + \frac{1}{2}\Delta U_S\right) - \frac{\Delta W_{sw} f_S}{P_{in}} \tag{29}$$

Introducing normalized quantities:

$$r_n = \frac{r}{U_{in}^2/P_{in}}, \quad r_{Tn} = \frac{r_T}{U_{in}^2/P_{in}}, \quad \Delta U_{Dn} = \frac{\Delta U_D}{U_{in}}, \quad \Delta U_{Sn} = \frac{\Delta U_S}{U_{in}}, \quad \Delta P_{sw0n} = \frac{\Delta P_{sw0}}{P_{in}} \tag{30}$$

We can simplify the efficiency formula to the form

$$\eta = 1 - \frac{5\pi^2 r_n}{64 f_{Sn}} - \frac{3\pi^2 r_{Tn}}{32 f_{Sn}} - \frac{3}{2}\left(\Delta U_{Dn} + \frac{1}{2}\Delta U_{Sn}\right) - \Delta P_{sw0n} f_{Sn} \tag{31}$$

The efficiency of the LSCVMb can be calculated with the use of the following components:

$$I_{S1av} = I_{D1av} + I_{D3av} = \frac{3P_{in}}{4U_{in}}, \quad I_{D4av} = I_{D1av} = \frac{P_{in}}{2U_{in}} \tag{32}$$

where D_4 is the LSCVMb additional diode (Figure 1b).

Conduction losses, ΔP_c, of LSCVMb are as follows:

$$\Delta P_c = \frac{5\pi^2 P_{in}^2 r}{64 U_{in}^2 f_{Sn}} + \frac{3\pi^2 P_{in}^2 r_T}{32 U_{in}^2 f_{Sn}} + \frac{P_{in}}{U_{in}}\left(2\Delta U_D + \frac{3}{4}\Delta U_S\right) \tag{33}$$

and the efficiency of the LSCVMb is

$$\eta = 1 - \frac{5\pi^2 r_n}{64 f_{Sn}} - \frac{3\pi^2 r_{Tn}}{32 f_{Sn}} - \left(2\Delta U_{Dn} + \frac{3}{4}\Delta U_{Sn}\right) - \Delta P_{sw0n} f_{Sn} \tag{34}$$

It can be seen from (30), (31), and (34) that the impact of the voltage drops across the diodes on the efficiency depends only on the ratio of these voltage drops to the supply voltage. The impact of the losses in the resistances is more complex. They increase with rising resistances and rising power, and decrease with rising input voltage and frequency f_S. Switching losses are proportional to switching frequency f_S.

The relationship between the efficiency and normalized frequency $f_{Sn} = f_S/f_0$ for three values of r_n (30): 0.016, 0.0304, and 0.040, ΔU_{Dn} (30) = 0.008 for the LCSCVMa and the LCSCVMb is shown in Figure 12. The value of $r_n = 0.0304$ corresponds to, e.g., $U_{in} = 50$ V, $P_{in} = 200$ W, $L = 500$ nH, $C = 1.5$ µF, $r = 380$ mΩ, and ΔU_{Dn} is equal to 0.008 for, e.g., $\Delta U_D = 0.40$ V and $U_{in} = 50$ V. The value of relative switching losses P_{sw0n} (30) = 0.0101 (Figure 12b) is valid, e.g., for ΔW_{sw} (28) = 11 µJ, $f_0 = 183.8$ kHz, and $P_{in} = 200$ W. The efficiency of the LCSCVMb is slightly lower. In both cases, it increases with increasing normalized frequency, f_{Sn}, and strongly depends on the circuit parasitic resistances. Therefore, it is important to minimize them, and use transistors with low values of $R_{DS(on)}$ and $V_{CE(on)}$.

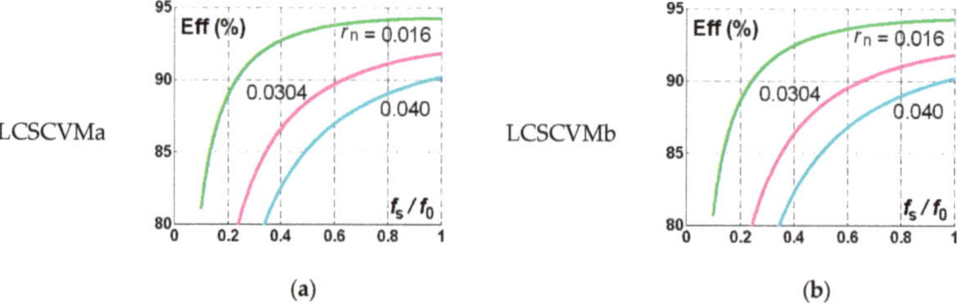

Figure 12. Theoretical charts of efficiency vs. $f_{Sn} = f_S/f_0$ for three values of r_n: 0.016, 0.0304, and 0.040, and $\Delta U_{Dn} = 0.008$: (**a**) LCSCVMa at switching losses $\Delta P_{sw0n} = 0.0138$, (**b**) LCSCVMb at switching losses $\Delta P_{sw0n} = 0.0101$.

The efficiency can be computed in a similar way for the other switching strategies. However, the calculations will be more complex in the case of the strategies with more than 2 stages.

4. Experimental Verification

This chapter presents the experimental results of the LCSCVMb converter operation. All the tests were carried out under switching strategy C5. The experimental verification confirms the proper operation of the converter, according to its concept. The measured voltage gain was on the expected level, and all the relevant waveforms were consistent with the simulation results as well.

4.1. Experimental Setup

All the parameters of the converter used during the experimental research, as well as a photograph of the investigated converter, are collected in Table 5. The parameters of the experimental setup correspond to the simulation model, and the major difference can be found in the inductance of the planar PCB choke. The switching frequency in the experimental measurements has been adjusted to the oscillation period of the switched capacitor currents and differs from the value selected for the simulation tests. An IGBT switch was selected as S_1 in the LSCVMb, as this switch conducts the total charging current. This current can be significant, especially when the converter contains a larger number of the switching cells. In order to generate appropriate control signals, an FPGA evaluation board (INTEL DE0) was utilized. The basic clock frequency of this device was set at 200 MHz, and the time resolution of the generated signals was 5 ns. The test setup is an example design of the converter prepared for the purpose of research, to verify its concepts and feasibility. The tests were conducted with 50 V at the input; however, the voltage range as well as power and the design concept can be rescaled to the parameters of a target application. Moreover, it is important that the prospective applications of the non-isolated DC-DC converter should comply with safety standards.

Table 5. The most important parameters of the laboratory converter.

Parameter	Value	The Laboratory Setup
Input voltage	50 V	
Output load	200 W	
Switching frequency	133 kHz	
Resonant capacitors	1.5 µF (KEMET R76 series)	
Resonant inductances	Planar chokes: L = 500 nH, R_{ESR} = 18 mΩ @ 100 kHz	
Transistors	IKB15N65EH5 (V_{DS} = 650 V, V_{CE} = 1.65 V) as S_1 IPB50R140CP (V_{DS} = 550 V, R_{DSon} = 0.14 Ω) as S_2 and S_4	
Diodes	STTH30L06G (I_F = 30 A, V_F = 1.0 V, V_{RRM} = 600 V)	
PCB	2 layers, 35 µm	
Laboratory equipment	Digital scope: Tektronix MDO3104, current probes: Tektronix TCP0030 150 MHz (input current measurement), Rogowsky coil (switch current measurements) voltage probes: Tektronix THDP0200 200 MHz, Tektronix P5205 100 MHz, power analyzer: Yokogawa WT 1801	

4.2. Test Results

Figure 13a,b presents the waveforms of the switching signals with the input and output current. They confirm that the converter operates correctly according to strategy C5. From the waveforms presented in Figure 13c, it follows that the converter boosts the input voltage. The measured voltage ratio is 3.65. Figure 13d,e presents the input current waveform and the voltages across the resonant capacitors. From the waveforms presented in Figure 13d, the average voltage across the capacitors can be seen. To demonstrate more clearly the magnitude of the oscillation around the average voltage value of each resonant capacitor, the voltage traces in AC coupling mode were recorded as well (Figure 13e). Figure 13f presents voltage stresses across the switches. From these results, it follows that the voltage stresses on switches are significantly below the output voltage of the converter, which is very favorable from the switching losses standpoint.

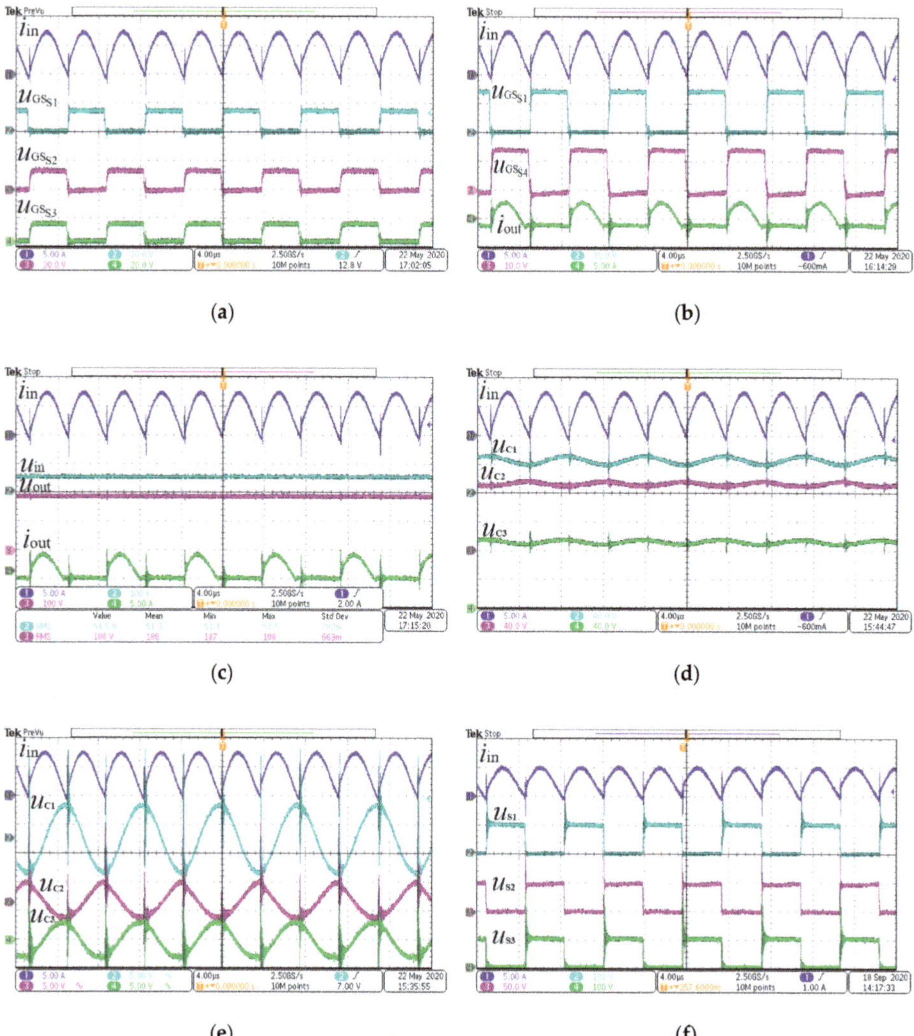

Figure 13. A set of recorded waveforms during experimental tests: (**a**) Switching signals of transistors and the input current, (**b**) input and output current of the converter on the background of switching signals, (**c**) input and output waveforms of the converter (current and voltage traces), (**d**) converter input current and voltages across resonant capacitors recorded in DC coupling mode, (**e**) converter input current and voltages across resonant capacitors recorded in AC coupling mode, and (**f**) voltage stresses across the switches on the background of converter input current. Switching strategy C5.

During the experimental research, the basic operation concept of the investigated converter has been checked. Furthermore, the working correctness of the examined device under different output loads was verified. The tests were carried out for three output load values: 62, 146, and 290 W, focusing especially on the transistor currents and voltages. Figure 14 present the results of the conducted tests for different output load conditions. From the results, it follows that the converter operates properly in low and medium load conditions.

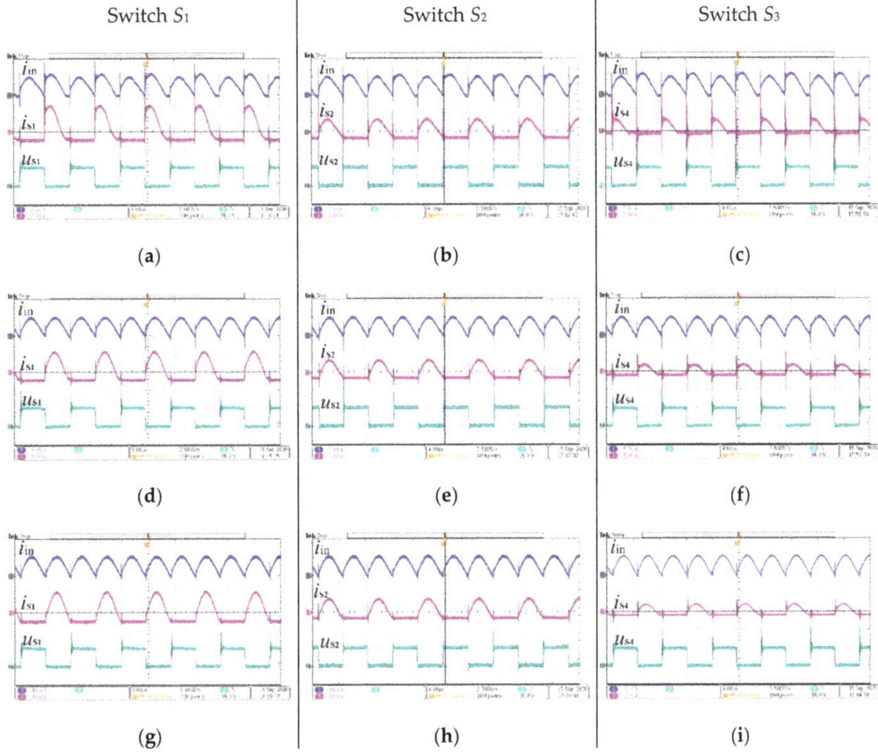

(a)   (b)   (c)

(d)   (e)   (f)

(g)   (h)   (i)

Figure 14. Waveforms of the input current as well as the currents and voltages across switches, during experimental test proceeded with different values of converter output power: (**a–c**) P_{out} = 62 W, 2A/div, 100V/div, (**d–f**) P_{out} = 146 W, 5A/div, 100V/div, (**g–i**) P_{out} = 290 W, 10A/div, 100V/div. Switching strategy C5.

Figure 15 presents the results of the spectral analysis calculated for the input and output currents. The calculations have been carried out with the use of MATLAB software, based on the recorded experimental data. The data was collected by the digital oscilloscope (Tektronix MDO3104) with the sampling rate of 1 MS/s.

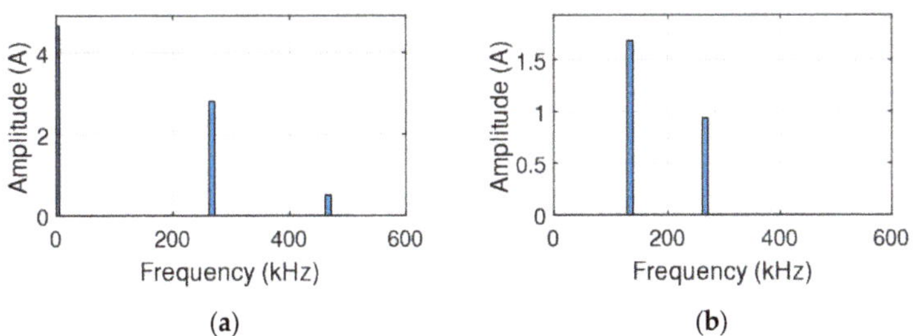

(a)   (b)

Figure 15. Results of spectral analysis for: (**a**) The input current and (**b**) the output current.

The experimental results of the output voltage of the LCSCVMb converter and its efficiency are presented in Figure 16. The efficiency is on an acceptable level. The voltage and efficiency drop versus power is typical for such SC-based converters, and results from their resistive losses. It should be noticed that the presented experimental setup is optimized towards the converter cost reduction. It was designed on a two-layer PCB of 35 μm. To increase the efficiency by reducing the parasitic resistance, a more expensive PCB and switches can be selected.

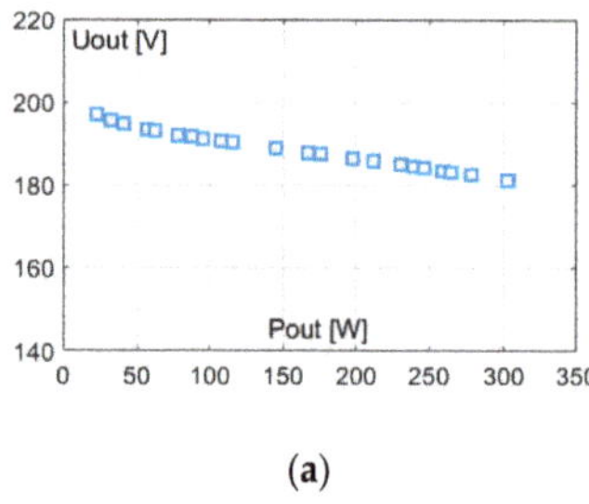

(a)

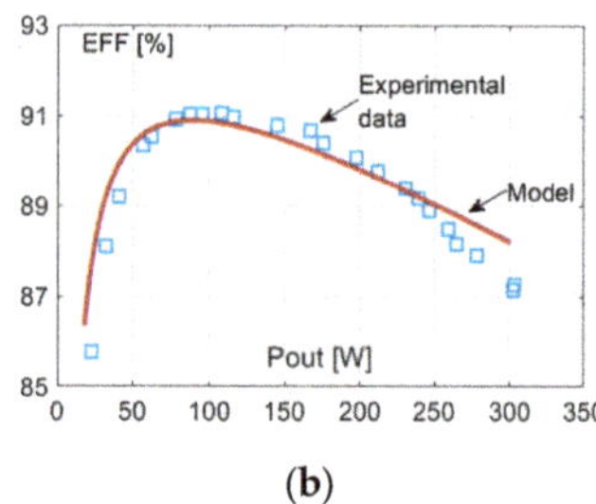

(b)

Figure 16. Experimental and simulation results for LSSCVMb under strategy C5 at U_{in} = 50 V, f_S = 133 kHz: (**a**) Measured output voltage U_{out} vs. P_{out}, (**b**) measured efficiency vs. P_{out} with comparison to theoretical results obtained from (34) for r = 380 mΩ, V_F = 400 mV, $V_{CE(on)}$ = 1 V, W_{sw} = 11 μJ.

5. Conclusions

The presented concepts of the new topologies, as well as the comparison of parameters presented in Table 3, and charts in Figures 8–10, lead to the following conclusions:

- The major idea of the proposed new converters is based on the elimination of the number of switches in a voltage multiplier (SCVM), while maintaining its proper operation. By the modification of an SCVM, the new topology concepts LCSCVMa and LCSCVMb were proposed, with a reduced number of switching cells and redesigned functions of the diodes. Depending on the technology of practical implementation, either of these converters can be more attractive than the other.
- Various switching strategies are possible for the converters, which affect the parameters of operation related to switching losses and the sizing of the passive components of the converter, but also the required input and output filters.
- The converter operates properly with a wide range of output loads.
- From the compared results, it follows that the most effective topology, the LCSCVMb, can operate with nearly the lowest parameters of AC component in the voltages on capacitors, and the highest frequency in the input and output current. This allows for a reduction of the converter volume, especially by optimizing the input and output filters.
- The discussed converters demonstrated an improvement in the SCVM topology, which may result in a prospective cost reduction.

Author Contributions: Conceptualization, R.S.; methodology, R.S. and Z.W.; software, R.S., Z.W. and S.F.; validation, R.S., Z.W. and S.F.; formal analysis, R.S. and Z.W.; investigation, R.S., Z.W. and S.F.; resources, R.S.; data curation, R.S., Z.W. and S.F.; writing—original draft preparation, R.S., Z.W. and S.F.; writing—review and editing, visualization, supervision, R.S., Z.W. and S.F.; project administration, R.S. All authors have read and agreed to the published version of the manuscript.

Funding: This research received no external funding.

Conflicts of Interest: The authors declare no conflict of interest.

References

1.	Ioinovici, A. Switched-capacitor power electronics circuit. *IEEE Circuits Syst. Mag.* **2001**, *1*, 37–42. [CrossRef]

2.   Tofoli, F.L.; de Castro Pereira, D.; de Paula, W.J.; Júnior, D.S.O. Survey on non-isolated high-voltage step-up dc–dc topologies based on the boost converter. *IET Power Electron.* **2015**, *8*, 2044–2057. [CrossRef]

3.   Forouzesh, M.; Siwakoti, Y.P.; Gorji, S.A.; Blaabjerg, F.; Lehman, B. Step-Up DC–DC Converters: A Comprehensive Review of Voltage-Boosting Techniques, Topologies, and Applications. *IEEE Trans. Power Electron.* **2017**, *32*, 9143–9178. [CrossRef]

4.   Wu, G.; Ruan, X.; Ye, Z. Nonisolated High Step-Up DC–DC Converters Adopting Switched-Capacitor Cell. *IEEE Trans. Ind. Electron.* **2015**, *62*, 383–393. [CrossRef]

5.   Li, W.; He, X. Review of Nonisolated High-Step-Up DC/DC Converters in Photovoltaic Grid-Connected Applications. *IEEE Trans. Ind. Electron.* **2011**, *58*, 1239–1250. [CrossRef]

6.   Raghavendra, K.V.G.; Zeb, K.; Muthusamy, A.; Krishna, T.N.V.; Kumar, S.V.S.V.P.; Kim, D.-H.; Kim, M.-S.; Cho, H.-G.; Kim, H.-J. A Comprehensive Review of DC–DC Converter Topologies and Modulation Strategies with Recent Advances in Solar Photovoltaic Systems. *Electronics* **2020**, *9*, 31. [CrossRef]

7.   Liu, W.; Niazi, K.A.K.; Kerekes, T.; Yang, Y. A Review on Transformerless Step-Up Single-Phase Inverters with Different DC-Link Voltage for Photovoltaic Applications. *Energies* **2019**, *12*, 3626. [CrossRef]

8.   Xiao, H. Overview of Transformerless Photovoltaic Grid-Connected Inverters. *IEEE Trans. Power Electron.* **2020**, *36*, 533–548. [CrossRef]

9.   Alluhaybi, K.; Batarseh, I.; Hu, H. Comprehensive Review and Comparison of Single-Phase Grid-Tied Photovoltaic Microinverters. *IEEE J. Emerg. Sel. Top. Power Electron.* **2020**, *8*, 1310–1329. [CrossRef]

10.   Tran, V.-T.; Nguyen, M.-K.; Choi, Y.-O.; Cho, G.-B. Switched-Capacitor-Based High Boost DC-DC Converter. *Energies* **2018**, *11*, 987. [CrossRef]

11.   Lee, J.; Jeong, Y.; Han, B. An Isolated DC/DC Converter Using High-Frequency Unregulated *LLC* Resonant Converter for Fuel Cell Applications. *IEEE Trans. Ind. Electron.* **2011**, *58*, 2926–2934. [CrossRef]

12.   Cao, D.; Peng, F.Z. A family of zero current switching switched-capacitor dc-dc converters. In Proceedings of the 2010 25th Annual IEEE Applied Power Electronics Conference and Exposition (APEC), Palm Springs, CA, USA, 21–25 February 2010; pp. 1365–1372.

13.   Waradzyn, Z.; Stala, R.; Mondzik, A.; Penczek, A.; Skala, A.; Pirog, S. Efficiency analysis of MOSFET-based air-choke resonant DC–DC step-up switched-capacitor voltage multipliers. *IEEE Trans. Ind. Electron.* **2017**, *64*, 8728–8738. [CrossRef]

14.   Penczek, A.; Mondzik, A.; Waradzyn, Z.; Stala, R.; Skała, A.; Piróg, S. Switching strategies of a resonant switched-capacitor voltage multiplier. In Proceedings of the 2017 19th European Conference on Power Electronics and Applications (EPE' 17 ECCE Europe), Warsaw, Poland, 11–14 September 2017; pp. 1–10.

15.   Xie, W.; Li, S.; Zheng, Y.; Smedley, K.; Wang, J.; Ji, Y.; Yu, J. A Step-up Series-parallel Resonant Switched-capacitor Converter with Extended Line Regulation Range. In Proceedings of the 2019 IEEE Applied Power Electronics Conference and Exposition (APEC), Anaheim, CA, USA, 17–21 March 2019; pp. 2150–2154.

16.   Stala, R.; Waradzyn, Z.; Mondzik, A.; Penczek, A.; Skała, A. DC–DC High Step-up Converter with Low Count of Switches Based on Resonant Switched-Capacitor Topology. In Proceedings of the 2019 21st European Conference on Power Electronics and Applications (EPE' 19 ECCE Europe), Genova, Italy, 3–5 September 2019; pp. 1–10.

17.   Xie, H.; Li, R. A Novel Switched-Capacitor Converter with High Voltage Gain. *IEEE Access* **2019**, *7*, 107831–107844. [CrossRef]

18.   Lei, H.; Hao, R.; You, X.; Li, F. Nonisolated High Step-Up Soft-Switching DC–DC Converter with Interleaving and Dickson Switched-Capacitor Techniques. *IEEE J. Emerg. Sel. Top. Power Electron.* **2020**, *8*, 2007–2021. [CrossRef]

19.   He, L.; Zheng, Z. High step-up DC–DC converter with switched-capacitor and its zero-voltage switching realization. *IET Power Electron.* **2017**, *10*, 630–636. [CrossRef]

20.   Chen, M.; Li, K.; Hu, J.; Ioinovici, A. Generation of a Family of Very High DC Gain Power Electronics Circuits Based on Switched-Capacitor-Inductor Cells Starting from a Simple Grap. *IEEE Trans. Circuits Syst.* **2016**, *63*, 2381–2392. [CrossRef]

21.   Kawa, A.; Stala, R. A multilevel switched capacitor DC–DC converter. An analysis of resonant operation conditions. *Power Electron. Drives* **2016**, *1*, 35–53.

22.   Kawa, A.; Stala, R. SiC-Based Bidirectional Multilevel High-Voltage Gain Switched-Capacitor Resonant Converter with Improved Efficiency. *Energies* **2020**, *13*, 2445. [CrossRef]

23. Li, S.; Zheng, Y.; Wu, B.; Smedley, K.M. A Family of Resonant Two-Switch Boosting Switched-Capacitor Converter with ZVS Operation and a Wide Line Regulation Range. *IEEE Trans. Power Electron.* **2018**, *33*, 448–459. [CrossRef]

24. Qian, W.; Cao, D.; Cintron-Rivera, J.G.; Gebben, M.; Wey, D.; Peng, F.Z. A Switched-Capacitor DC–DC Converter with High Voltage Gain and Reduced Component Rating and Count. *IEEE Trans. Ind. Appl.* **2012**, *48*, 1397–1406. [CrossRef]

25. Xiong, S.; Tan, S. Cascaded High-Voltage-Gain Bidirectional Switched-Capacitor DC–DC Converters for Distributed Energy Resources Applications. *IEEE Trans. Power Electron.* **2017**, *32*, 1220–1231. [CrossRef]

26. Stala, R.; Piróg, S. DC–DC boost converter with high voltage gain and a low number of switches in multisection switched capacitor topology. *Arch. Electr. Eng.* **2018**, *67*, 617–627.

27. Kumar, A.; Wang, Y.; Pan, X.; Raghuram, M.; Singh, S.K.; Xiong, X.; Tripathi, A.K. Switched-LC Based High Gain Converter with Lower Component Count. *IEEE Trans. Ind. Appl.* **2020**, *56*, 2816–2827. [CrossRef]

28. Li, S.; Smedley, K.M.; Caldas, D.R.; Martins, Y.W. Hybrid Bidirectional DC–DC Converter with Low Component Counts. *IEEE Trans. Ind. Appl.* **2018**, *54*, 1573–1582. [CrossRef]

29. Waradzyn, Z.; Stala, R.; Skała, A.; Mondzik, A.; Penczek, A. A cost-effective resonant switched-capacitor dc-dc boost converter—Experimental results and feasibility model. *Power Electron. Drives* **2018**, *3*, 75–83. [CrossRef]

30. Wu, G.; Ruan, X.; Ye, Z. High Step-Up DC–DC Converter Based on Switched Capacitor and Coupled Inductor. *IEEE Trans. Ind. Electron.* **2018**, *65*, 5572–5579. [CrossRef]

31. Kazimierczuk, M.K.; Czarkowski, D. *Resonant Power Converters*, 2nd ed.; Wiley: Hoboken, NJ, USA, 2011.

Publisher's Note: MDPI stays neutral with regard to jurisdictional claims in published maps and institutional affiliations.

Article

A New Approach to the PWM Modulation for the Multiphase Matrix Converters Supplying Loads with Open-End Winding

Pawel Szczepankowski [1,*], Natalia Strzelecka [2] and Enrique Romero-Cadaval [3]

[1] Department of Power Electronics and Electrical Machines, Faculty of Electrical and Control Engineering, Gdansk University of Technology, 80-233 Gdansk, Poland

[2] Department of Ship Automation, Faculty of Electrical Engineering, Gdynia Maritime University, 81-225 Gdynia, Poland; n.strzelecka@we.umg.edu.pl

[3] Department of Electrical Electronic and Control Engineering, University of Extremadura, 06006 Badajoz, Spain; eromero@unex.es

* Correspondence: pawel.szczepankowski@pg.edu.pl

Received: 14 December 2020; Accepted: 13 January 2021; Published: 17 January 2021

Abstract: This article presents three variants of the Pulse Width Modulation (PWM) for the Double Square Multiphase type Conventional Matrix Converters (DSM-CMC) supplying loads with the open-end winding. The first variant of PWM offers the ability to obtain zero value of the common-mode voltage at the load's terminals and applies only six switches within the modulation period. The second proposal archives for less Total Harmonic Distortion (THD) of the generated load voltage. The third variant of modulation concerns maximizing the voltage transfer ratio, minimizing the number of switching, and the common-mode voltage cancellation. The discussed modulations are based on the concept of sinusoidal voltage quadrature signals, which can be an effective alternative to the classic space-vector approach. In the proposed approach, the geometrical arrangement of basic vectors needed to synthesize output voltages is built from the less number of vectors, which is equal to the number of the matrix converter's terminals. The PWM duty cycle computation is performed using only a second-order determinant of the voltages coordinate matrix without using trigonometric functions. A new approach to the PWM duty cycles computing and the load voltage synthesis by 5 × 5 and 12 × 12 topologies has been verified using the PSIM simulation software.

Keywords: square-type matrix converters; pulse width modulation; multiphase systems

---

1. Introduction

A fully controlled bidirectional semiconductor switch is an element of the Conventional Matrix Converter (CMC) which offers a direct AC–AC voltages conversion with additional input power factor control functionality. This type of converter, in comparison with the more established Voltage Source Inverter (VSI), has certain individual features that determine the innovation of such a solution [1–3]. It does not contain a bulk dc-link capacitor, thus is far more promising in terms of power density with the inherent four-quadrant operation [4,5]. Compared to traditional variable speed drives with CMC, the multi-phase electric motor drive gives some fundamental advantages. These configurations have grater system redundancy because it can operate during some fault conditions, is characterized by the lower torque ripple and lower per-leg converter rating [6]. Furthermore, the operation of multiphase motors is quieter, allowing for the independent control of two or more series/parallel connected motors [7]. Multiphase

Energies **2021**, *14*, 466; doi:10.3390/en14020466

electric machines can be fed also by the conventional matrix converter with three inputs [8–11]. The matrix topology is also presented as a unit, which control the power flow between the power generator and the electrical grid [12–16]. Multiphase generators have also been used in systems generating electricity such as offshore installations and wind farms [17,18]. Another application of the multiphase matrix converter in straight forward energy conversion is described in [19], where the 6×6 CMC and multi-winding transformer have been used to supply variable reactive power flow to the power system. A modulation approach based on the model of the 5×5 matrix converter with fictitious DC-link was shown in conference paper [20].

Multi-phase electric machines, including three-phase and five-phase variants, can be designed as a machine with open stator winding. Such a solution, especially in combination with a CMC, offers some important features. First, the possibility of direct power supply to both ends of the stator phase winding increases the maximum voltage amplitude within the linear range of the PWM modulator [21]. It should be noted here that the output voltage of the CMC cannot exceed the input voltages envelope. A quite frequency discussed problem in drives controlled by power converters, is the common voltage, which can be understood as a voltage measured between the ground potential and a virtually created star point connected with the stator terminals. The common-mode voltage in terminals of AC drives resulting in bearing currents harmful for the motor drive. The use of certain voltage modulation techniques in a matrix converter allows eliminating this problem [5,22–24]. As indicated in the brief introduction multi-phase drives with CMCs are rather niche applications. However, PWM algorithms are the subject of numerous studies, and almost all of the presented algorithms assume ideal sinusoidal input voltage and are designed for machines with the symmetric construction. The application of these methods without consideration of an input voltage asymmetry or the source harmonics in the calculation results in inaccurate load voltage generation. The approach to voltage synthesis proposed in the article takes this aspect into account and allows for the generation of the appropriate load voltage.

Due to a large number of input and output phases, the complexity of PWM algorithms based on the space-vector approach increases significantly. For a single matrix converter with three inputs and three outputs, the number of switch states is 3^3 (27). In the case of five input, five output converter, it gives 5^5 (3125) and analogically in drive with the open-end stator winding, the number of states is equal to 5^{10}. Therefore, the graphical presentation of voltage vectors corresponding to all switch states, at a given moment of time, becomes very unreadable, which makes it problematic to design and elaborate the dedicated PWM algorithms. The works [3,25] show that the synthesis of output voltages in multi-phase systems can be successfully simplified. These methods can be classified to the direct method of modulation. Compared to the transformation proposed by Clarke, the use of the Hilbert transform leads to the reduction of the number of required vectors. Therefore, for the CMC 5×5 the PWM algorithm can be developed using only 10 vectors. Apart from the modulation methods based on the space-vector approach, a group of methods of direct modulation can be indicated, such as the Venturini solution [1,26,27], while the general Venturini formulas for PWM duty cycles for several multi-phase converters, including square-type matrix converters are presented in [28]. The work in [25] proposes Wachspress formulas, which can be theoretically applied for any number of inputs. The use of either the Venturini solution or the Wachspress functions forces the commutation all switches of the given output cell (shown in the drawing later in the article). This can be explained by the fact that the switch modulating function is continuous, thus results in higher switching losses. All the PWM modulation methods discussed in the article are characterized by a lower number of switching cycles of at most 6 during the modulation period.

The paper is organized as follows. Definitions and principles of the proposed an output voltage synthesis using the DSM-CMC converter are presented in Section 2. This section also presents the method of generating quadrature signals using the Discrete Second-Order Generalized Integrator (DSOGI) structure. Then, the next three sections demonstrate variants of the proposed modulation. Abilities of the input displacement angle control have been also discussed. Results are summarized and discussed in the conclusion section. Due to a huge number of switching elements, the realization of the experimental setup is very expensive. Therefore, the conclusions presented in the article are the result of circuit simulation in PSIM11 software and analytical research only. However, the proposed solution has been verified partially by an experiment and published in [3,25,29].

2. The Principle of an Output Voltage Synthesis in DSM-CMC Converter

The DSM-CMC converter consists of two square-type matrix converters, CMC_P and CMC_N, connected to load terminals as shown in Figure 1, where the simplified diagram of the circuit is depicted. If the number of load phases is equal to n, the total number of bidirectional power electronic switches is equal to $2n^2$. Both converter are connected with the n-phase AC voltage source v_{i1}, v_{i2}, ..., and v_{in}. The voltage of the phase x of the load

$$v_{ox}(t) = v_{Px}(t) - v_{Nx}(t) \tag{1}$$

measured at the load terminals, are synthesized by these converters by using the switch group h_{11}, h_{21}, ..., h_{n1}.

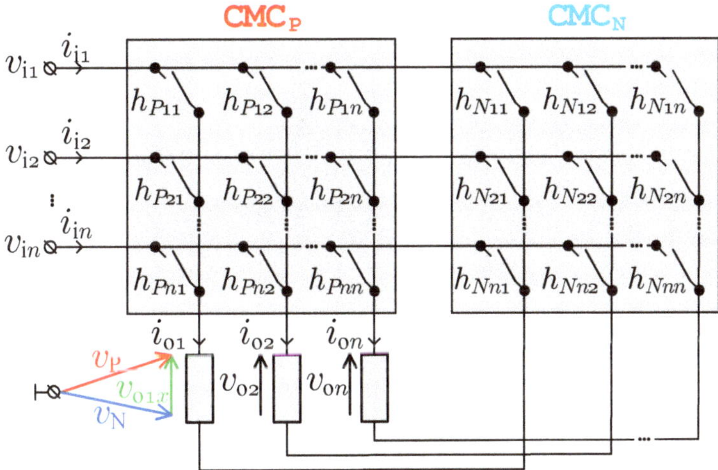

Figure 1. Simplified diagram of the square-type double conventional matrix converter.

As can be seen in Figure 2, switches on both sides of one phase of the load make the single commutation cell with two voltage multiplexers, s_P and s_N, respectively.

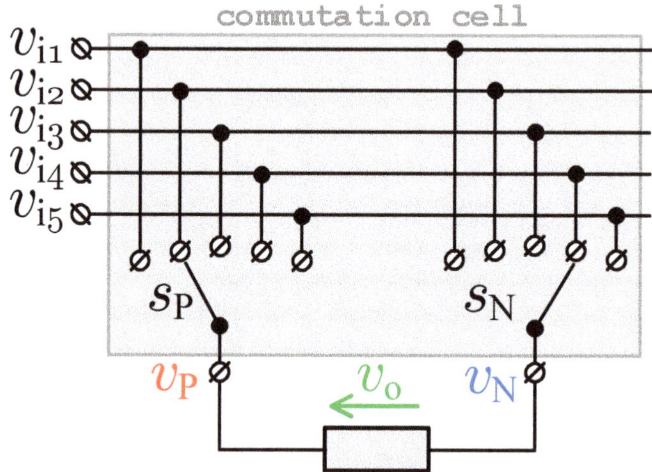

Figure 2. The single commutation cell for DSM-CMC 5×5.

2.1. Case of 5 Phases

The number of multiplexer switches is equal to the number of input voltages, and in this case, takes 5. For a better understanding of the proposal, let the further consideration will be focused on 5×5 topology.

According to the proposed concept of the voltage synthesis described in [3], all input voltages with pulsation ω_i can be represented as a collection of five rotating vectors:

$$\mathbf{v_i} = \begin{bmatrix} v_{i1x} & v_{i1y} \\ v_{i2x} & v_{i2y} \\ v_{i3x} & v_{i3y} \\ v_{i4x} & v_{i4y} \\ v_{i5x} & v_{i5y} \end{bmatrix} \tag{2}$$

with the real

$$\begin{aligned}
v_{i1x} &= V_{i1} \cdot \cos{(\omega_i t)} \\
v_{i2x} &= V_{i2} \cdot \cos{(\omega_i t - 2\pi/5)} \\
v_{i3x} &= V_{i3} \cdot \cos{(\omega_i t - 4\pi/5)} \\
v_{i4x} &= V_{i4} \cdot \cos{(\omega_i t - 6\pi/5)} \\
v_{i5x} &= V_{i5} \cdot \cos{(\omega_i t - 8\pi/5)}
\end{aligned} \tag{3}$$

and the imaginary parts of coordinates

$$\begin{aligned}
v_{i1y} &= V_{i1} \cdot \sin{(\omega_i t)} \\
v_{i2y} &= V_{i2} \cdot \sin{(\omega_i t - 2\pi/5)} \\
v_{i3y} &= V_{i3} \cdot \sin{(\omega_i t - 4\pi/5)} \\
v_{i4y} &= V_{i4} \cdot \sin{(\omega_i t - 6\pi/5)} \\
v_{i5y} &= V_{i5} \cdot \sin{(\omega_i t - 8\pi/5)}
\end{aligned} \tag{4}$$

where V_{i1}, ..., V_{i5} are the amplitudes of these voltages. Due to the analytic signal concept based on the Hilbert transform, for the pure sinusoidal input waveforms, the imaginary coordinates are just quadrature components and an input voltage vectors collection can be presented as shown in Figure 3 as the symmetric system.

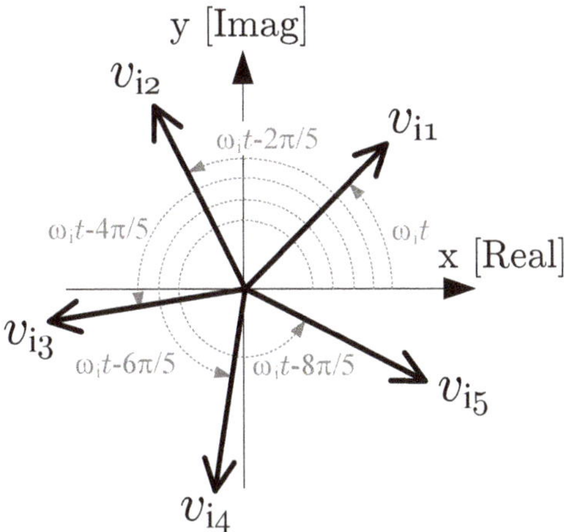

Figure 3. The collection of five the rotating input vectors.

These coordinates can be determined using the Hilbert filter or obtained through FFT/DFT based operation [30–32]. However, the Hilbert filter and algorithms based on DFT, although are quite accurate, are not the simple solution from code developing point of view. Moreover, error signals in the form of DC offsets, glitches, and momentary voltage sags may occur in measurements. Therefore, the input vector coordinates can be calculated in a different manner. A compromise solution, between accuracy and not complicated solution, maybe the use of the Double Second-Order Generalized Integrator with loop feedback extension functioned as Orthogonal Signal Generator (DSOGI-OSG), which in the OSG part prevents unexpected resonance and variables overflow. DSOGI-OSG structure in continuous time-domain is presented in Figure 4.

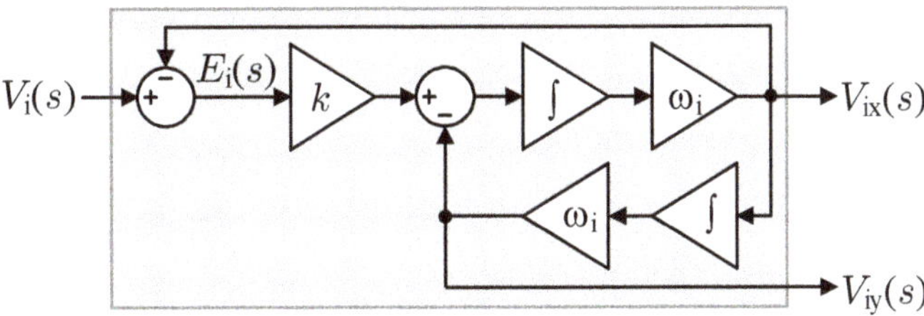

Figure 4. Double Second-Order Generalized Integrator with loop feedback extension functioned as Orthogonal Signal Generator (DSOGI-OSG) structure in continuous time-domain [33–35]: V_i—the input sinusoidal signal, V_{ix}—in-phase component of the input signal, V_{iy}—the quadrature component of the input signal, E_i the error signal, k—the gain block, ω_i—reference pulsation of the input signal, and $\int$ is an integrator block.

The transfer function takes the form of (5) for in-phase output and (6) for orthogonal output, while (7) represents the notch filter equation

$$\frac{V_{ix}(s)}{V_i(s)} = \frac{k \cdot \omega_i \cdot s}{s^2 + k \cdot \omega_i \cdot s + \omega_i{}^2} \tag{5}$$

$$\frac{V_{iy}(s)}{V_i(s)} = \frac{k \cdot \omega_i^2}{s^2 + k \cdot \omega_i \cdot s + \omega_i{}^2} \tag{6}$$

$$\frac{E_i(s)}{V_i(s)} = \frac{s^2 + \omega_i{}^2}{s^2 + k \cdot \omega_i \cdot s + \omega_i{}^2} \tag{7}$$

where the parameter k is a value less than unity (k is taken the value of $1/\sqrt{2}$ here), $E_i(s)$ is the error signal, while ω_i is an input voltage nominal pulsation. If processed signal frequency does not have an exact value, another extension of SOGI structure, called Frequency-Locked Loop (FLL), may be applied [36–38].

The load voltage v_o produced by the single commutation cell, shown in Figure 2, can be analogous represented by two rotating vectors, v_P and v_N, as is depicted in Figure 5. Only the geometrical distance of real (indicated by subscript x) coordinates of these vectors produce the load voltage. While the imaginary coordinate (indicated by subscript y) can generate the reactive power flow at the converter input. In general, there exists some degree of freedom for selecting the instantaneous value of this component because it does not influence on the load currents. The article is focused on the cases, which locus of each output vector is straight a circle. This means that a rotating output voltage vector moves along a circular trajectory and this movement can be clockwise or counterclockwise. Four variants of the PWM modulation scheme are shown in Figure 5.

A vector arrangement in Figure 6a, for the given commutation cell, can be presented as the rotating polygon as illustrated in Figure 6b. The polygon surface is named here as the output voltage synthesis field. All the points, which represent output voltage vectors, have to be located inside the synthesis field. Such a geometric arrangement allows for direct application the Wachspress function for the PWM duty cycles calculation [25,28]. However, the number of switching within the modulation period should be minimal, and for this reason, Venturini and Wachpress solution is not suitable. Decreasing the number of switching can be realized by applying the Nearest Three Vectors (NTV) modulation technique, which relays on the selection of a proper triangle in the synthesis field. Figure 6c shows two selected triangles for the voltages generated by CMC_P and CMC_N converters. Note that both points, representing these voltages, are located in their triangular local synthesis fields. This is the required condition of output voltage synthesizability. The selection of the optimal triangle may consist in finding the appropriate vertex of the synthesis field, which clearly indicates the input vector closest to the output vector. In the case of using NTV technique, this solution is sufficient, because the other two required vectors are adjacent to the selected one. As can be seen, vector v_P is closest to the vertex number 2, while vector v_N is closest to vertex 4. All six required PWM duty cycles can be calculated using the smooth interpolation technique, which is, in the discussed case, nothing more than an appropriate triangle area relation for the NTV modulation [3]. An area of the triangle can be computed using the second-order matrix determinant. Thus, an application of that solution only needs coordinates of the triangle vertices. As mentioned earlier, these coordinates can be computed using the DSOGI-OSG block shown in Figure 4.

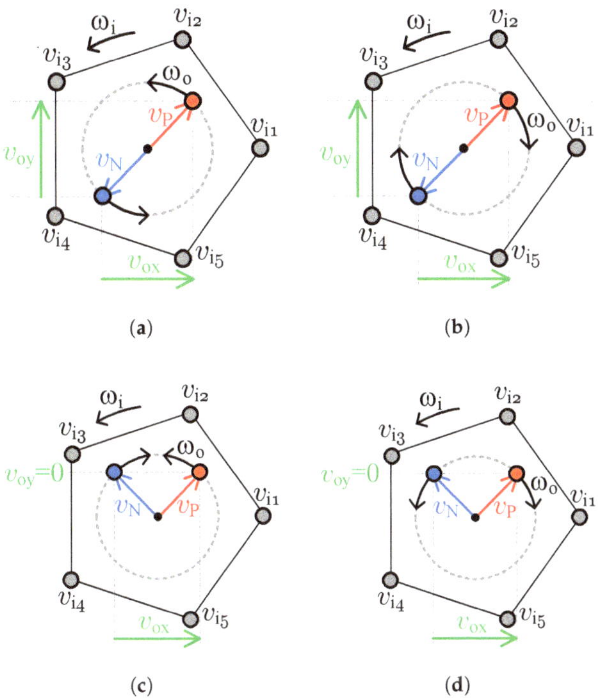

Figure 5. Four variants of output vectors rotation: (**a**) CCV-CCV, (**b**) CV-CV, (**c**) CCV-CV, and (**d**) CV-CCV.

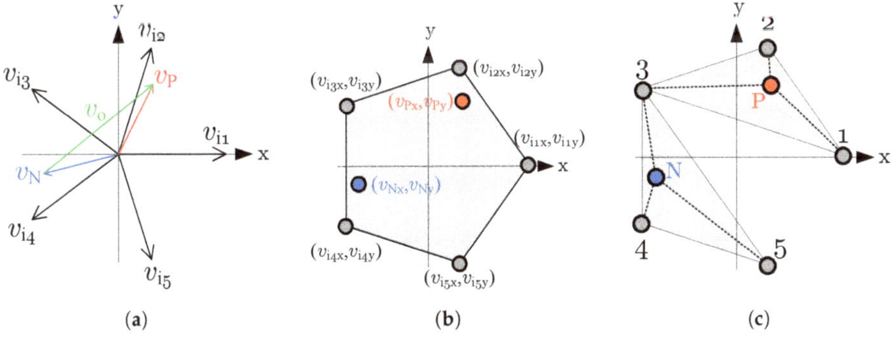

Figure 6. The principle of operation: (**a**) vectors arrangement, (**b**) synthesis field, and (**c**) selected triangles.

2.2. Case of 12 Phases

In the case of more input voltages, for example, when the number of inputs is equal to 12, the choice of the optimal triangle is not so obvious. Now, let us consider the graphical vector arrangements for 12×12 topology expressed as regular polygon shown in Figure 7a.

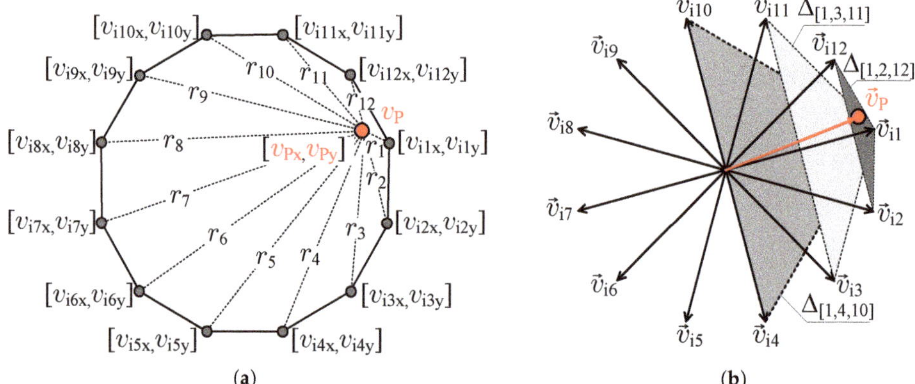

Figure 7. Synthesis field of the 12×12 matrix topology (**a**), the input voltage vectors, and an example reference output voltage $\vec{v}_{o1}$ (**b**).

One of 12 presented input vectors is referred here as the base vector. It means that the distance—defined as $r_1 \cdots r_{12}$ and shown in Figure 7b—between this vector and the reference vector $\vec{v}_{o1}$ is the smallest. There are three triangles with a common upper vertex with coordinates $\{v_{i1x}, v_{i1y}\}$: $\Delta_{[2,1,12]}$, $\Delta_{[3,1,11]}$, and $\Delta_{[4,1,10]}$. A vector $\vec{v}_{i1}$ is the base vector in this case. The given triangle $\Delta_{[p,q,r]}$ satisfies the modulation conditions when the sum

$$\Sigma_{[p,q,r]} = d_p + d_q + d_r \tag{8}$$

where

$$\begin{bmatrix} d_p \\ d_q \\ d_r \end{bmatrix} = \xi \begin{bmatrix} \left| \det \begin{bmatrix} v_{qx} - v_{o1x} & v_{qy} - v_{o1y} \\ v_{rx} - v_{o1x} & v_{ry} - v_{o1y} \end{bmatrix} \right| \\ \left| \det \begin{bmatrix} v_{px} - v_{o1x} & v_{py} - v_{o1y} \\ v_{rx} - v_{o1x} & v_{ry} - v_{o1y} \end{bmatrix} \right| \\ \left| \det \begin{bmatrix} v_{qx} - v_{o1x} & v_{qy} - v_{o1y} \\ v_{px} - v_{o1x} & v_{py} - v_{o1y} \end{bmatrix} \right| \end{bmatrix} \tag{9}$$

and

$$\xi = \left| \det \begin{bmatrix} v_{px} - v_{qx} & v_{py} - v_{qy} \\ v_{rx} - v_{qx} & v_{ry} - v_{qy} \end{bmatrix} \right|^{-1} \tag{10}$$

of PWM duty cycles d_p, d_q, and d_r takes the smallest value, ideally equal unity. When two or more triangles meet this condition, the triangle with the smallest area should be selected for further consideration. In practice, this operation can be performed by using optimized DSP functions like qsort (sorting in required order), vecmin (finding the minimum value within the set), or standard conditional operators.

When the value of transfer voltage ratio $q = V_o/V_i$ of the 12×12 topology, e.g., for CMC_P or CMC_N, is in the range

$$\frac{\cos\left(\frac{\pi}{6}\right)}{\cos\left(\frac{\pi}{12}\right)} \leq q \leq \cos\left(\frac{\pi}{12}\right) \tag{11}$$

a large number of output phases allows generating the output voltage with lower THD, therefore the cost of passive elements can be decreasing. Corresponding simulation results are presented in the further part of the text. PWM duty cycles calculation for CMC_P and CMC_N are explained in two separate subsections. While the two concepts of gating signals generation have been presented in the third subsection.

2.3. PWM Duty Cycles Calculation for CMC_P and Topology 5×5

Referred to the triangle $\Delta_{[1,2,3]}$ in Figure 6, the reference output voltage v_P is synthesized using 3 switches: h_{11}, h_{21}, and h_{31}. Taking into account previous considerations, the following formulas can be proposed for the calculation of PWM duty cycles,

$$d_{1P} = \xi_P \cdot \left| \det \begin{bmatrix} v_{i2x} - v_{Px} & v_{i2y} - v_{Py} \\ v_{i3x} - v_{Px} & v_{i3y} - v_{Py} \end{bmatrix} \right| = \frac{\Delta_{[2,P,3]}}{\Delta_{[1,2,3]}} \tag{12}$$

$$d_{3P} = \xi_P \cdot \left| \det \begin{bmatrix} v_{i1x} - v_{Px} & v_{i1y} - v_{Py} \\ v_{i2x} - v_{Px} & v_{i2y} - v_{Py} \end{bmatrix} \right| = \frac{\Delta_{[2,P,1]}}{\Delta_{[1,2,3]}} \tag{13}$$

$$d_{2P} = 1 - d_{1P} - d_{3P} = \frac{\Delta_{[3,P,1]}}{\Delta_{[1,2,3]}} \tag{14}$$

where det is the determinant of the second-order matrix, and

$$\xi_P = \left| \det \begin{bmatrix} v_{i2x} - v_{i1x} & v_{i2y} - v_{i1y} \\ v_{i3x} - v_{i1x} & v_{i3y} - v_{i1y} \end{bmatrix} \right|^{-1} \tag{15}$$

is the scaling factor, which is equal to the triangle $\Delta_{[1,2,3]}$ surface. Thus, the average value of the CMC_P output voltage can be expressed by the following formula.

$$\bar{v}_P = d_{1P} \cdot v_{i1} + d_{2P} \cdot v_{i2} + d_{3P} \cdot v_{i3} \tag{16}$$

2.4. PWM Duty Cycles Calculation for CMC_N and Topology 5×5

Referred to the triangle $\Delta_{[3,4,5]}$ in Figure 6, the reference output voltage v_N is synthesized by 3 switches: h_{31}, h_{41}, and h_{51}. The corresponded duty cycles can be calculate using the following formulas,

$$d_{3N} = \xi_N \cdot \left| \det \begin{bmatrix} v_{i4x} - v_{Nx} & v_{i4y} - v_{Ny} \\ v_{i5x} - v_{Nx} & v_{i5y} - v_{Ny} \end{bmatrix} \right| = \frac{\Delta_{[4,N,5]}}{\Delta_{[3,4,5]}} \tag{17}$$

$$d_{4N} = \xi_N \cdot \left| \det \begin{bmatrix} v_{i5x} - v_{Nx} & v_{i5y} - v_{Ny} \\ v_{i3x} - v_{Nx} & v_{i3y} - v_{Ny} \end{bmatrix} \right| = \frac{\Delta_{[5,N,3]}}{\Delta_{[3,4,5]}} \tag{18}$$

$$d_{5N} = 1 - d_{3N} - d_{4N} = \frac{\Delta_{[3,N,4]}}{\Delta_{[3,4,5]}} \tag{19}$$

where

$$\xi_N = \left| \det \begin{bmatrix} v_{i4x} - v_{i3x} & v_{i4y} - v_{i3y} \\ v_{i5x} - v_{i3x} & v_{i5y} - v_{i3y} \end{bmatrix} \right|^{-1} \tag{20}$$

is the scaling factor, which is equal to the triangle $\Delta_{[3,4,5]}$ surface. The average value of the CMC_N output voltage can be expressed by the following formula

$$\bar{v}_N = d_{3N} \cdot v_{i3} + d_{4N} \cdot v_{i4} + d_{5N} \cdot v_{i5} \tag{21}$$

2.5. The Concept of Gating Signals Generation

The gate signals can be controlled according to different strategies. Apap et al. [39] compared and presented several PWM signal gating methods. Among them, the cyclic Venturini and Min-Mid-Max (MMM) schemes of modulation are proposed. Two approaches have been applied to the gates signal generation: basic and mentioned MMM scheme.

In the case of the basic solution, the sequences of the switch states always depend on the selected triangle in which the synthesis of the output voltage is realized. Therefore, these sequences can be placed in a lookup table. An overview of the basic sequences is shown in Figure 8, where the value of $\varphi \simeq 1.618$, which is so-called the golden ratio exists in the pentagon.

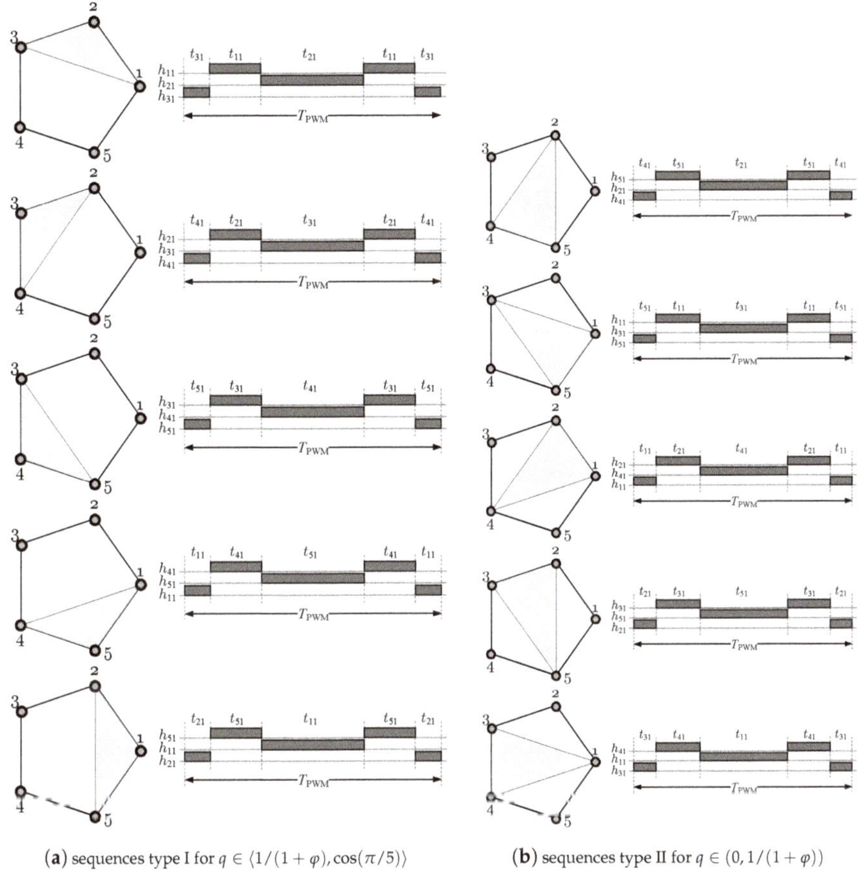

(**a**) sequences type I for $q \in \langle 1/(1+\varphi), \cos(\pi/5)\rangle$

(**b**) sequences type II for $q \in (0, 1/(1+\varphi))$

Figure 8. The switch state sequences in the basic solution of the gating signals for both converter cells CMC_P and CMC_N.

The MMM method is used to improve the quality of the voltage generated by the converter in terms of THD. The switch states sequences for the CMC_P and CMC_N converters are characterized in Figure 9. Note that these sequences correspond to the case illustrated in Figure 6b.

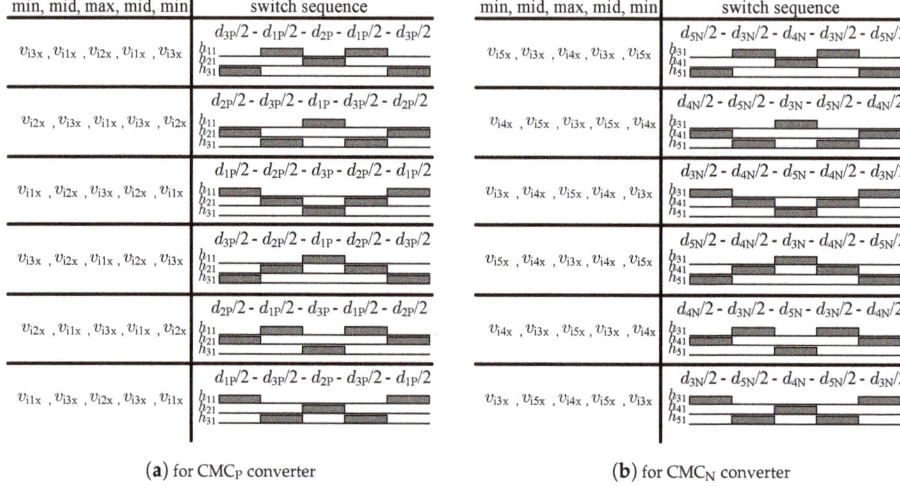

(**a**) for CMC$_P$ converter   (**b**) for CMC$_N$ converter

Figure 9. The MMM type of sequences of the gating signals for both converter cells CMC_P and CMC_N for the case depicted in Figure 6c.

3. The PWM Variant 1—An Output Voltage Synthesis with Zero Value of the Common-Mode Voltage

The common-mode voltage, defined as

$$v_{cm}(t) = (v_{o1}(t) + v_{o2}(t) + v_{o3}(t) + v_{o4}(t) + v_{o5}(t))/5, \tag{22}$$

can lead to the degradation of rolling bearings in electric machines powered by PWM inverters. As indicated in the introduction an open-end windings stator fed by the double matrix converter allows for the PWM modulation without the common-mode voltage generation. The proposed approach to the load voltage synthesis with conjunction with the basic solution of the gating signals control (shown in Figure 8) give the same desired result. Elimination of the common-mode voltage can be performed by all four PWM modulation schemes: CV-CV, CCV-CCV, CV-CCV, and CCV-CV. The use of the first two cases allows obtaining an input displacement angle, which is dependent on the load parameters like in the Venturini methods [1,26].

The referenced k-output voltage vectors of the CCV-CCV modulation scheme can be expressed by following equations.

$$v_{Pxk} = q \cdot \cos(\omega_o \cdot t - ((k-1) \cdot 2\pi/5)) \tag{23}$$

$$v_{Pyk} = -q \cdot \sin(\omega_o \cdot t - ((k-1) \cdot 2\pi/5)) \tag{24}$$

$$v_{Nxk} = -q \cdot \cos(\omega_o \cdot t - ((k-1) \cdot 2\pi/5)) \tag{25}$$

$$v_{Nyk} = q \cdot \sin(\omega_o \cdot t - ((k-1) \cdot 2\pi/5)) \tag{26}$$

Above equation are proposed for the first commutation cell. Equations for the rest of the rotating vectors pairs can be represented by analogous elaboration. The referenced output voltages in CV-CV scheme can be represented by following equations.

$$v_{Pxk} = q \cdot \cos(\omega_o \cdot t - ((k-1) \cdot 2\pi/5)) \tag{27}$$

$$v_{Pyk} = q \cdot \sin(\omega_o \cdot t - ((k-1) \cdot 2\pi/5)) \tag{28}$$

$$v_{Nxk} = -q \cdot \cos(\omega_o \cdot t - ((k-1) \cdot 2\pi/5)) \tag{29}$$

$$v_{Nyk} = -q \cdot \sin(\omega_o \cdot t - ((k-1) \cdot 2\pi/5)) \tag{30}$$

The possibility to change the rotation of the given output vector by changing the sign of the imaginary component allows realizing the PWM modulation, in which the resultant imaginary component of the v_{oy} takes the zero value. If vectors $\vec{v}_P$ and $\vec{v}_N$ rotate in opposite directions, as is typical for last two presented schemes of modulation CV-CCV and CCV-CV, the passive input current component is not generated. Simulation results of the PWM variant 1 for DSM-CMC 5 × 5 and four modulation schemes are shown in Figure 10. Simulation parameters are listed in Table A1, which can be found in an appendix.

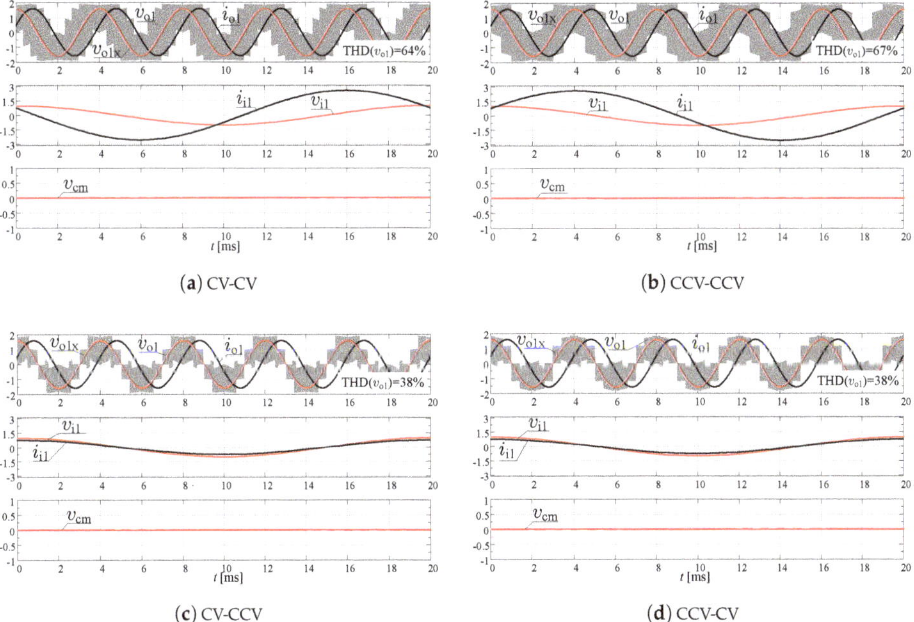

Figure 10. Simulation of the PWM variant 1 for DSM-CMC 5 × 5 and four modulation schemes—RL load case.

Analogous simulation tests have been realized for the proposed converter connected to a 12-phase symmetrical power supply. Figure 11 shows the load voltage generated by DSM-CMC 12 × 12 for an output frequency equal to 10 Hz, while results obtained for 300 Hz are presented in Figure 12.

Simulation parameters for this case are available in Table A2 in Appendix A. The selection of different sets of simulation parameters did not subserve a specific purpose. The simulation tests were carried out with the use of two independent simulation files. However, in the case of the DSM-CMC 12 × 12 simulation, a small calculation step was chosen due to the high modulation frequency. It was set to 100 kHz to get a good PWM resolution at 300 Hz of the fundamental frequency.

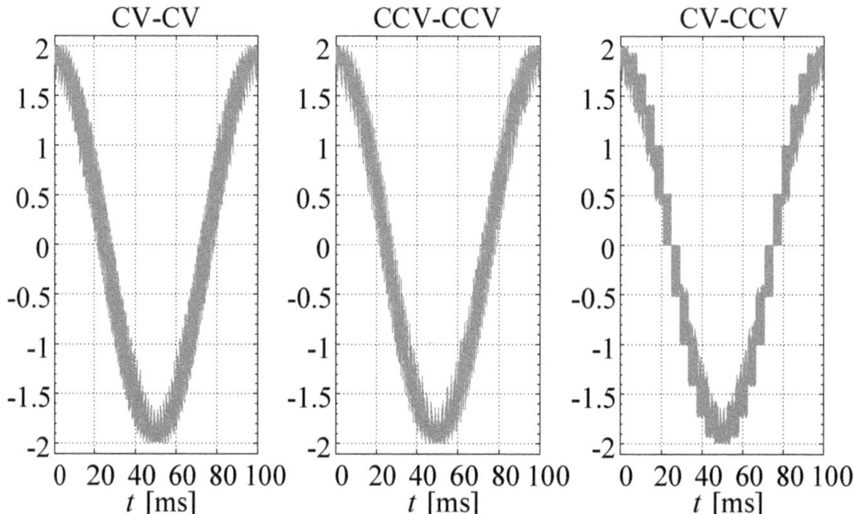

Figure 11. The load voltage v_o for DSM-CMC 12 × 12 converter for the three selected modulation schemes: $f_o = 10$ Hz, $q = 2 \times 0.95$.

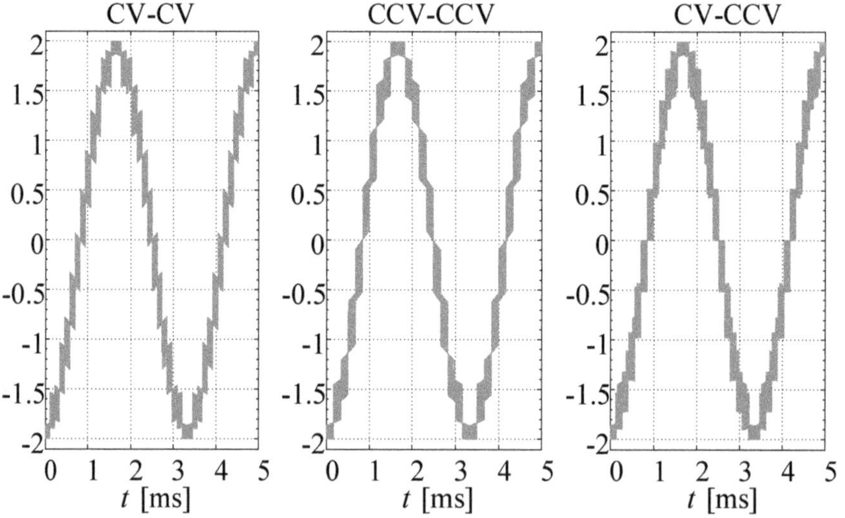

Figure 12. The load voltage v_o for DSM-CMC 12 × 12 converter for the three selected modulation schemes: $f_o = 300$ Hz, $q = 2 \times 0.95$.

Properties of this type of matrix converter, compared with counterpart 5×5, remains the same. In particular, the common-mode voltage is also eliminated by using the basic type of switches state sequences. The voltages shown in these figures are characterized by a low THD, which is about 12%. Comparing to the 5×5 topology, the resulted voltage gain for DSM-CMC is higher and takes optimally the value of 1.93.

4. The PWM Variant 2—An Output Voltage Synthesis with Less Harmonic Distortion

The basic solution of the switch states sequence has been applied in the PWM modulation with eliminating the common-mode voltage. If a lower THD of the load voltage waveform is desired, a more advanced gating signal generation mechanism can be proposed, such as MMM scheme shown in Figure 9. With regard to variant 1, this is the only change. However, the MMM method is more complicated because the input voltage vector collection should be arranged in a specific order $\{min - mid - max - mid - min\}$ within the selected triangular synthesis field. Simulation results are presented in Figure 13. A lower THD of the load voltage v_o is obtained but the common-mode voltage v_{cm} is also generated, as marked in the presented drawings.

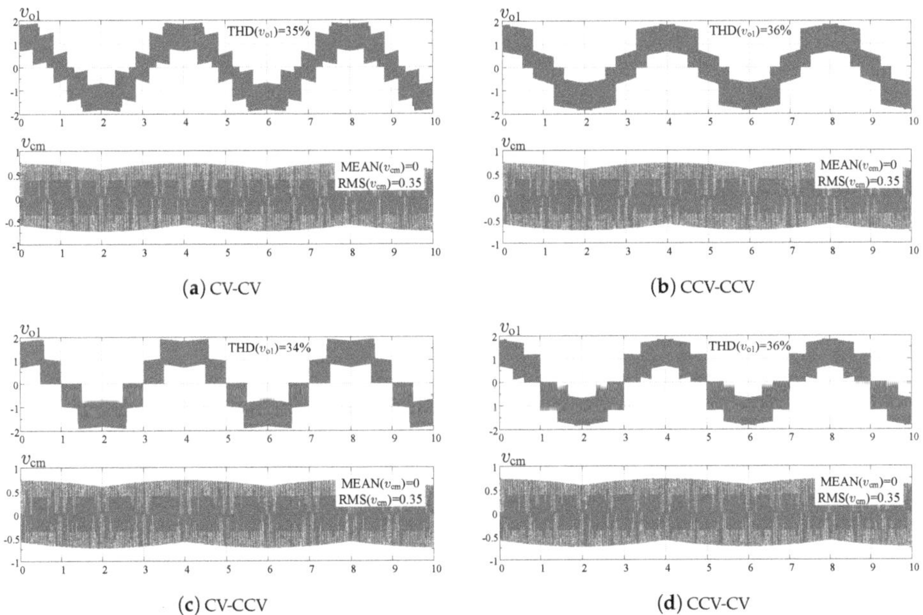

Figure 13. Simulation of the PWM variant 2 for DSM-CMC 5×5 and four modulation schemes: $f_o = 250$ Hz, $q = 2 \times 0.8$.

5. The PWM Variant 3—An Output Voltage Synthesis with Maximum Voltage Transfer Ratio and Minimum Number of Switching

A synthesis field for multi-phase and symmetrical AC voltage sources can be represented by a regular polygon as shown in Figures 6b and 7a. A radius of a circle inscribed of this polygon limits an output voltage amplitude in the linear range of modulation. The maximum voltage transfer ratio for CMC_N

and CMC$_P$, related to the input voltage amplitude and number of inputs equal to n, can be expressed as follows,

$$q_{P\,max} = q_{N\,max} = \cos(\pi/n) \tag{31}$$

Therefore, the maximum load voltage for DSM-CMC 5×5 in *p.u.* is equal to

$$v_{o\,max} = 2 \cdot \cos(\pi/5) = 1.618 \tag{32}$$

The value (32) can be increased by modifying the position of the v_P and v_N vectors. In contrast to the methods described in the previous sections, trajectories of these vectors are not a circle. The locus of each vector is not changing smoothly and contains discontinuities. This type of modulation belongs to the discontinuous group of PWM modulations. Both reference vectors v_P and v_N take exactly five positions, in which they lie on one of five input vectors. An algorithm flowchart for DSM-CMC 5×5 is presented in Figure 14.

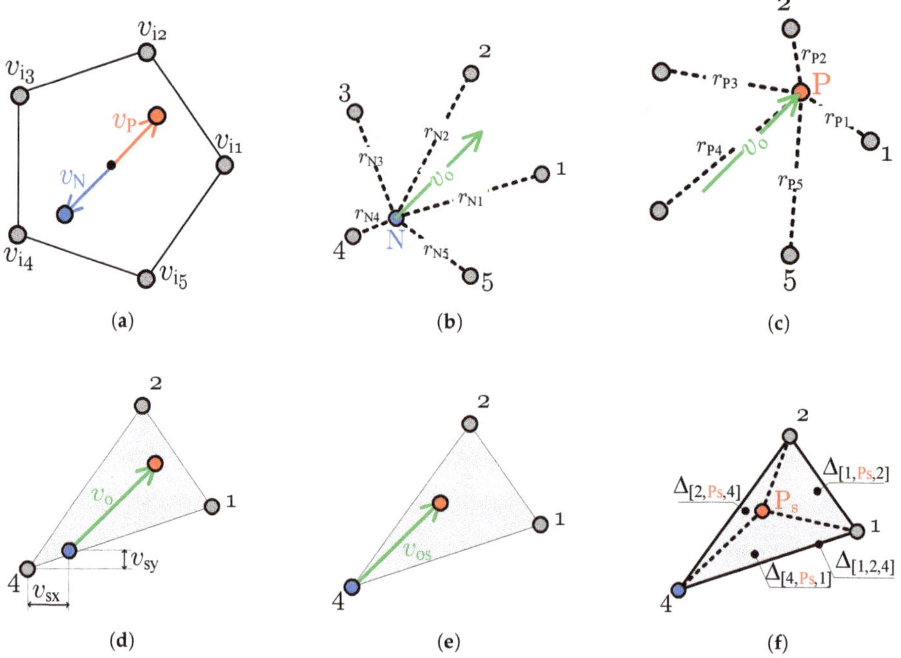

Figure 14. An algorithm flowchart of the variant 3 PWM modulation: (**a**) Step 1: generation of synthesis field and reference voltage vectors v_P and v_N. (**b**) Step 2: calculation of the set of distances between the N point and vertices of the synthesis field. (**c**) Step 3: calculation of the set of distances between the P point and vertices of the synthesis field. (**d**) Step 4: shortest distance selection, setting the origin vertex, and the vector's offset $\{v_{sx}, v_{sy}\}$ calculation. (**e**) Step 5: the reference vector v_o shift resulting in the new coordinates of P and N points. (**f**) Step 6: calculation of four areas of the triangle and PWM duty cycles.

The output voltage synthesis field is generated using the DSOGI blocks at the first step of the proposed algorithm. The reference output voltage vectors coordinates, $\{v_{Nx}, v_{Ny}\}$ and $\{v_{Px}, v_{Py}\}$, are also calculated at this step. The vectors can rotate clockwise (CV-CV scheme) or counterclockwise (CCV-CCV scheme), as shown in Figure 15.

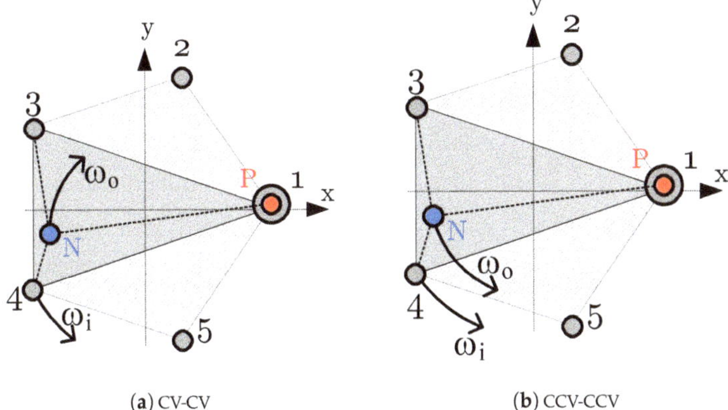

(a) CV-CV   (b) CCV-CCV

Figure 15. An example rotation of the reference output vector.

Based on the analysis of the vector arrangement in Figure 15, it can be written that the maximum length of the voltage vector, in a linear range of modulation, is equal to the following expression.

$$v_{o\,\max\,(variant3)} = 1 + \cos(\pi/5) = 1.809 \tag{33}$$

However, a vector of this length has to be accordingly shifted inside the synthesis field as shown in Figure 14d,e. Therefore, new coordinates of the reference output vector for the given commutation cell can be calculated as follows.

$$v_{osx} = v_{ox} + v_{sx} \tag{34}$$

$$v_{osy} - v_{oy} + v_{sy} \tag{35}$$

Distances between N-point and all the synthesis field vertices are calculated in Step 2. The same procedure is applied for the point P in Step 3. Next, the shortest calculated distance in a N-collection $\{r_{N1}, r_{N2}, r_{N3}, r_{N4}, r_{N5}\}$ is compared with the shortest calculated distance in a P-collection $\{r_{P1}, r_{P2}, r_{P3}, r_{P4}, r_{P5}\}$. Finally, the less value is selected, which correctly indicates the optimal vertex of the synthesis field. The shift coordinates are calculated in Step 4. As can be seen in Figure 14d, vertex number 4 has been chosen. Thus, the PWM duty cycles, for the case illustrated in Figure 14f can be calculated using the following formulas.

$$d_{1Ps} = \Delta_{[2,Ps,4]} / \Delta_{[1,2,4]} \tag{36}$$

$$d_{2Ps} = \Delta_{[4,Ps,1]} / \Delta_{[1,2,4]} \tag{37}$$

$$d_{4Ps} = \Delta_{[1,Ps,2]} / \Delta_{[1,2,4]} \tag{38}$$

and

$$d_{4Ns} \equiv 1 \tag{39}$$

Formula (39) refers to the case where the end of the v_N vector coincides with the v_{i4} vector. It means the permanent connection of the input voltage v_{i4} with one side of the load phase during the PWM modulation period. Figure 15 shows a case, which in the one side of the load is permanently connected to an input voltage v_{i1} during PWM modulation. Sequences of the switch states shown in Figure 16 correspond to the case, which in the s_N switch is connected permanently to the input phase 4. The zero load voltage is generated by using the same switch in both CMC_P and CMC_N matrix converters. Simulation results for maximal voltage transfer ratio (33) are shown in Figure 17. The common-mode voltage v_{cm} is eliminated. An application of the CV-CV and CCV-CCV scheme of modulation resulting in the non-zero input displacement angle.

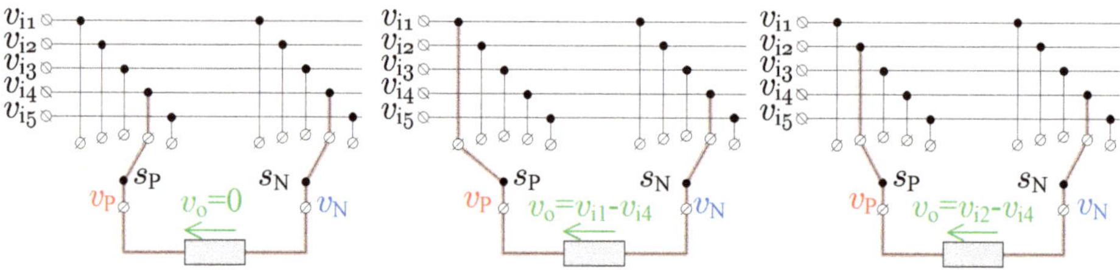

Figure 16. Sequences of the switch states in one commutation cell for the case presented in Figure 14.

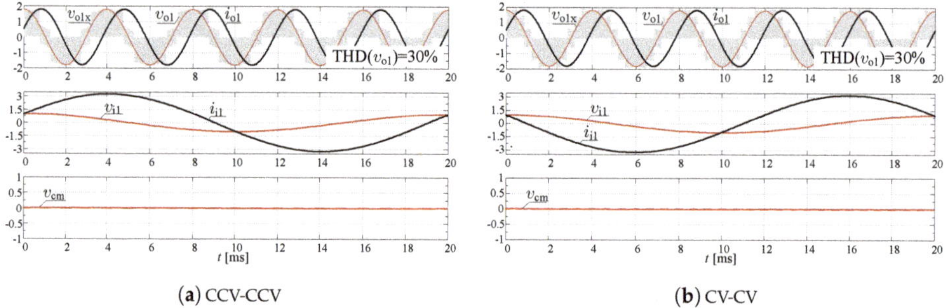

(**a**) CCV-CCV

(**b**) CV-CV

Figure 17. Simulation of the PWM variant 3 for DSM-CMC 5×5 and two modulation schemes: $f_o = 250$ Hz, $q = 1.8$.

Having half the number of switching operations during the PWM modulation period is an advantage of variant 3. In order to obtain the unity power factor at the system input, the sequence types have to be toggled continuously in the order CV-CV, CCV-CCV,..., etc. However, this mode of operation may require to redesign of an input filter. Example simulation results of the PWM variant 3 with toggling mode for DSM-CMC 5×5 have been presented in Figure 18.

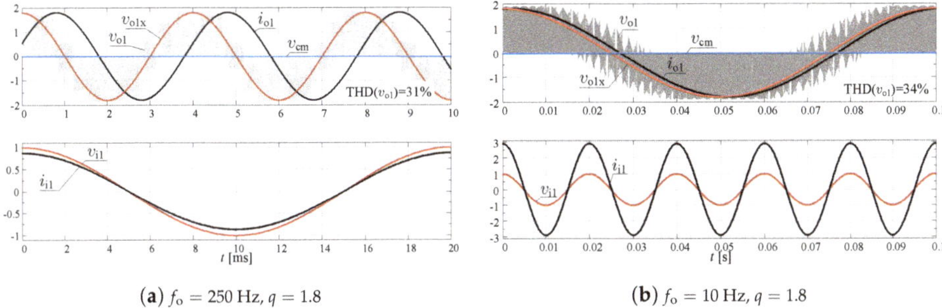

(**a**) $f_o = 250$ Hz, $q = 1.8$ (**b**) $f_o = 10$ Hz, $q = 1.8$

Figure 18. Simulation of the PWM variant 3 with toggling mode for DSM-CMC5 × 5.

6. Summary and Conclusions

This paper presents a new approach to the PWM modulation for the multi-phase matrix converters supplying loads with open-end winding. The proposed approach is an alternative for the methods based on the space-vector modulation. Three variants of PWM modulation were presented. Animations for the first two of them (Figures S1 and S2) are available in Supplementary Materials. The first variant allows for eliminating the common-mode voltage, which is a desired feature from a practical point of view. The second variant based on a specific rearranging switches state sequences can offer quasimultilevel waveforms with low THD. However, a common-mode voltage level can be unaccepted due to influence on the bearings lifetime. Variant 3 of the PWM modulation described in a paper offer over 12% greater voltage transfer gain. Comparison of an input angle value for the proposed variants of PWM modulation is presented in Table 1.

Table 1. Comparison of an input angle value for the proposed variants of PWM modulation.

	Modulation Scheme		
PWM Variant	$\phi_i = \varphi_n$	$\phi_i = -\varphi_o$	$\phi_i = 0$
variant 1	CCV-CCV	CV-CV	CV-CCV or CCV-CV
variant 2	CCV-CCV	CV-CV	CV-CCV or CCV-CV
variant 3	CCV-CCV	CV-CV	toggling

The multi-phase matrix converters, with an equal number of input and outputs, belong to the niche solution. Recently, we can observe an increasing interest in multi-phase systems. Furthermore, the complexity of the modulation algorithms grows up. The described proposal is a research result of analytic signal and an application of the smooth interpolation method in PWM duty cycle computing. Some selected features and properties have been compared with the space-vector method. Table 2 presents such a comparison. The PWM duty cycle computation represented by equations, from (12) to (14), is performed using only a second-order determinant of the voltages coordinate matrix without trigonometry usage. Therefore, the proposed method also naturally extends the applicability of the formulas to unbalanced and distorted AC voltage sources. Moreover, all computation can be realized in the FPGA structure using the simple multipliers and adders. The important contribution of the presented article is a presentation of the novel algorithm, which is much easier than algorithms based on the space-vector approach. This property is essential in multi-phase systems because the number of vectors is very high. The proposed solution uses only vectors that represent the Hilbert analytic signal pair calculated for the input and the reference vectors. The number of vectors needed to realize the output voltage synthesis is equal to only the sum of input and output phases.

Table 2. The comparison of the proposed modulation with the space-vector approach.

	Proposed Modulation	Space Vector Modulation
how the vector map is generated	using the analytic signal concept, which is based on the Hilbert transform	using the Clark transform for multi-phase systems
the difficulty of the vector map generator	comparable with SVPWM, using several methods: triple Clarke, or DSOGI, or DFT	comparable with the proposed, using the Clarke rotation operator
degree of difficulty with more phases	the number of vectors is equal to the number of converter's terminals	the number of vectors is equal to $2^{(2P)}$, where P is the number of the load phase
the common-mode voltage elimination in the multi-phase systems	yes	requires the modification of the modulation using the rotating vectors collection
minimization of the number of switching	possible for variant no. 3, in the general case a sorted and optimized the switch states sequence should be used	the minimization of the number of switching is a natural feature for the space-vector modulation, which is based on the nearest three vectors, however—for that selection can be an additional issue of computation
is it applicable for unbalanced and asymmetrical loads with the open-winding	yes	no applicable, space-vector methods assumed the symmetric loads with the open-winding
the load phase failure	ready for that failure, each load phase is controlled by an individual and independence the cell controller	in the event of the sudden change in the number of load phases, the algorithm (switches' state sequences table) must be thoroughly rebuilt, it is not possible in a real-time system, the modification can only be implemented offline
application of trigonometric functions for PWM duty cycle computing	no (it speeds up the algorithm)	yes

Supplementary Materials: The following are available at http://www.mdpi.com/1996-1073/14/2/466/s1, Supplementary data: Matlab script m-file: energies_1056291_Supplementary_Materials.m, Figure S1: A New Approach to the PWM Modulation for the Multiphase Matrix Converters Supplying Loads with Open-End Winding: two rotating the reference vectors, Figure S2: A New Approach to the PWM Modulation for the Multiphase Matrix Converters Supplying Loads with Open-EndWinding: the input and the output waveforms obtained, and the PWM duty cycles.

Author Contributions: The concept, algorithmization, simulation studies, P.S.; simulation results verification and presubmission editing, E.R.-C. and N.S. All authors have read and agreed to the published version of the manuscript.

Funding: This work was supported by the LINTE^2 Laboratory, Gdansk University of Technology.

Conflicts of Interest: All authors declare no conflict of interest.

Appendix A. Simulation Parameters

Table A1. Simulation parameters for DSM-CMC 5×5.

Parameter	Value
Number of input voltages	5
Number of output voltages	10
Number of an ideal bidirectional switches	50
Source phase voltage amplitude	$V_I = 100$ V
Input frequency	$f_I = 50$ Hz
Output frequencies	$f_o = 10$ Hz, 250 Hz
Voltage gain of CMC_P	$q = 0.8$
Voltage gain of CMC_N	$q = 0.8$
The load parameters	$R_o = 0.5\ \Omega$, $L_o = 1$ mH
An algorithm frequency	$f_s = 10$ kHz
Simulation step	250 ns
Simulation software	PSIM 64-bit Version 11.0.3

Table A2. Simulation parameters for DSM-CMC 12 × 12.

Parameter	Value
Number of input voltages	12
Number of output voltages	24
Number of an ideal bidirectional switches	288
Source phase voltage amplitude	$V_i = 100$ V
Input frequency	$f_i = 50$ Hz
Output frequencies	$f_o = 10$ Hz, 300 Hz
Voltage gain of CMC_P	$q = 0.95$
Voltage gain of CMC_N	$q = 0.95$
The load parameters	$R_o = 10\ \Omega,\ L_o = 0.1$ mH
An algorithm frequency	$f_s = 100$ kHz
Simulation step	100 ns
Simulation software	PSIM 64–bit Version 11.0.3

Simulation research has been performed for symmetric and balanced source and load. Obtained currents and voltages have been presented as *p.u.* values referred to the base voltage $V_{\text{base}} = V_i$ and the base current equal to $I_{\text{base}} = V_i/Z_o$, where Z_o was a load impedance.

References

1. Rodriguez, J.; Rivera, M.; Kolar, J.W.; Wheeler, P.W. A Review of Control and Modulation Methods for Matrix Converters. *IEEE Trans. Ind. Electron.* **2012**, *59*, 58–70. [CrossRef]
2. Friedli, T.; Kolar, J.W. Milestones in Matrix Converter Research. *IEEJ J. Ind. Appl.* **2012**, *1*, 2–14. [CrossRef]
3. Szczepankowski, P.; Wheeler, P.; Bajdecki, T. Application of Analytic Signal and Smooth Interpolation in Pulsewidth Modulation for Conventional Matrix Converters. *IEEE Trans. Ind. Electron.* **2020**, *67*, 10011–10023. [CrossRef]
4. Helle, L.; Larsen, K.B.; Jorgensen, A.H.; Munk-Nielsen, S.; Blaabjerg, F. Evaluation of modulation schemes for three-phase to three-phase matrix converters. *IEEE Trans. Ind. Electron.* **2004**, *51*, 158–171. [CrossRef]
5. Ahmed, S.M.; Abu-Rub, H.; Salam, Z. Common-Mode Voltage Elimination in a Three-to-Five-Phase Dual Matrix Converter Feeding a Five-Phase Open-End Drive Using Space-Vector Modulation Technique. *IEEE Trans. Ind. Electron.* **2015**, *62*, 6051–6063. [CrossRef]
6. Levi, E.; Bojoi, R.; Profumo, F.; Toliyat, H.; Williamson, S. Multi-phase induction motor drives-A technology status review. *IET Elect. Power Appl.* **2007**, *1*, 489–516. [CrossRef]
7. Levi, E. Multi-phase Machines for Variable speed applications. *IEEE Trans. Ind. Electron.* **2008**, *55*, 1893–1909. [CrossRef]
8. Nguyen, T.D.; Lee, H. Development of a Three-to-Five-Phase Indirect Matrix Converter With Carrier-Based PWM Based on Space-Vector Modulation Analysis. *IEEE Trans. Ind. Electron.* **2016**, *63*, 13–24. [CrossRef]
9. Ahmed, S.M.; Salam, Z.; Abu-Rub, H. An Improved Space Vector Modulation for a Three-to-Seven-Phase Matrix Converter With Reduced Number of Switching Vectors. *IEEE Trans. Ind. Electron.* **2015**, *62*, 3327–3337. [CrossRef]
10. Ahmed, S.M.; Iqbal, A.; Abu-Rub, H.; Rodriguez, J.; Rojas, C.; Saleh, M. Simple carrier-based PWM technique for a three-to-nine phase direct AC-AC converter. *IEEE Trans. Ind. Electron.* **2011**, *58*, 5014–5023. [CrossRef]
11. Ahmed, S.M.; Iqbal, A.; Abu-Rub, H. Generalized Duty-Ratio-Based Pulsewidth Modulation Technique for a Three-to-k Phase Matrix Converter. *IEEE Trans. Ind. Electron.* **2011**, *58*, 3925–3937. [CrossRef]
12. El-Khoury, C.N.; Kanaan, H.Y.; Mougharbel, I.; Al-Haddad, K. A review of matrix converters applied to PMSG based wind energy conversion systems. In Proceedings of the IECON 2013—39th Annual Conference of the IEEE Industrial Electronics Society, Vienna, Austria, 10–13 November 2013; pp. 7784–7789. [CrossRef]
13. Liu, X.; Wang, P.; Loh, P.; Blaabjerg, F. A Three-Phase Dual-Input Matrix Converter for Grid Integration of Two AC Type Energy Resources. *IEEE Trans. Ind. Electron.* **2013**, *60*, 20–29. [CrossRef]
14. Pena, R.; Cardenas, R.; Reyes, E.; Clare, J.; Wheeler, P. Control of a Doubly Fed Induction Generator via an Indirect Matrix Converter With Changing DC Voltage. *IEEE Trans. Ind. Electron.* **2011**, *58*, 20–29. [CrossRef]
15. Garces, A.; Molinas, M. A Study of Efficiency in a Reduced Matrix Converter for Offshore Wind Farms. *IEEE Trans. Ind. Electron.* **2012**, *59*, 184–193. [CrossRef]

16. Abdel-Rahim, O.; Funato, H.; Abu-Rub, H.; Ellabban, O. Multiphase Wind Energy generation with direct matrix converter. In Proceedings of the 2014 IEEE International Conference on Industrial Technology (ICIT), Busan, Korea, 26 February–1 March 2014; pp. 519–523. [CrossRef]

17. Yaramasu, V.; Wu, B.; Sen, P.C.; Kouro, S.; Narimani, M. High-power wind energy conversion systems: State-of-the-art and emerging technologies. *Proc. IEEE* **2015**, *103*, 740–788. [CrossRef]

18. Zoric, I.; Jones, M.; Levi, E. Arbitrary Power Sharing Among Three-Phase Winding Sets of Multiphase Machines. *IEEE Trans. Ind. Electron.* **2018**, *65*, 1128–1139. [CrossRef]

19. Sienko, T.; Szczepanik, J.; Martis, C. Reactive Power Transfer via Matrix Converter Controlled by the "One Periodical" Algorithm. *Energies* **2020**, *13*, 665. [CrossRef]

20. Rezaoui, M.M. Study of output voltages of a matrix converter feeding an five AC-induction machine using the strategy calculated modulation PMW. In Proceedings of the 2010 Modern Electric Power Systems, Wroclaw, Poland, 20–22 September 2010; pp. 1–4.

21. Rahman, K.; Iqbal, A.; Al-Hitmi, M.A.; Dordevic, O.; Ahmad, S. Performance Analysis of a Three-to-Five Phase Dual Matrix Converter Based on Space Vector Pulse Width Modulation. *IEEE Access* **2019**, *7*, 12307–12318. [CrossRef]

22. Baranwal, R.; Basu, K.; Mohan, N. Carrier-Based Implementation of SVPWM for Dual Two-Level VSI and Dual Matrix Converter With Zero Common-Mode Voltage. *IEEE Trans. Power Electron.* **2015**, *30*, 1471–1487. [CrossRef]

23. Tewari, S.; Mohan, N. Matrix Converter Based Open-End Winding Drives With Common-Mode Elimination: Topologies, Analysis, and Comparison. *IEEE Trans. Power Electron.* **2018**, *33*, 8578–8595. [CrossRef]

24. Rzasa, J.; Sztajmec, E. Elimination of Common Mode Voltage in the Three-To-Nine-Phase Matrix Converter. *Energies* **2020**, *13*, 631. [CrossRef]

25. Szczepankowski, P.; Bajdecki, T.; Strzelecki, R. Direct Modulation for Conventional Matrix Converters Using Analytical Signals and Barycentric Coordinates. *IEEE Access* **2020**, *8*, 22592–22616. [CrossRef]

26. Malekjamshidi, Z.; Jafari, M.; Zhu, J. Analysis and comparison of direct matrix converters controlled by space vector and Venturini modulations. In Proceedings of the 2015 IEEE 11th International Conference on Power Electronics and Drive Systems, Sydney, Australia, 9–12 June 2015; pp. 635–639. [CrossRef]

27. Dey, A.K.; Mohapatra, G.; Mohapatra, T.K.; Sharma, R. A Modified Venturini PWM scheme for Matrix converters. In Proceedings of the 2019 IEEE International Conference on Sustainable Energy Technologies and Systems (ICSETS), Bhubaneswar, India, 26 February–1 March 2019; pp. 013–018. [CrossRef]

28. Ali, M.; Iqbal, A.; Khan, M.R.; Ayyub, M.; Anees, M.A. Generalized Theory and Analysis of Scalar Modulation Techniques for a $m \times n$ Matrix Converter. *IEEE Trans. Power Electron.* **2017**, *32*, 4864–4877. [CrossRef]

29. Szczepankowski, P.; Nieznanski, J. Application of Barycentric Coordinates in Space Vector PWM Computations. *IEEE Access* **2019**, *7*, 91499–91508. [CrossRef]

30. Reilly, A.; Frazer, G.; Boashash, B. Analytic signal generation-Tips and traps. *IEEE Trans. Signal Process.* **1994**, *42*, 3241–3245. [CrossRef]

31. Marple, L. Computing the discrete-time "analytic" signal via FFT. *IEEE Trans. Signal Process.* **1999**, *47*, 2600–2603. [CrossRef]

32. Asiminoael, L.; Blaabjerg, F.; Hansen, S. Computing the discrete-time "analytic" signal via FFT. *IEEE Ind. Appl. Mag.* **2007**, *13*, 22–33. [CrossRef]

33. Yan, Q.; Zhao, R.; Yuan, X.; Ma, W.; He, J. A DSOGI-FLL-Based Dead-Time Elimination PWM for Three-Phase Power Converters. *IEEE Trans. Power Electron.* **2019**, *34*, 2805–2818. [CrossRef]

34. Xin, Z.; Zhao, R.; Blaabjerg, F.; Zhang, L.; Loh, P.C. An Improved Flux Observer for Field-Oriented Control of Induction Motors Based on Dual Second-Order Generalized Integrator Frequency-Locked Loop. *IEEE J. Emerg. Sel. Top. Power Electron.* **2017**, *5*, 513–525. [CrossRef]

35. Hoffmann, N.; Lohde, R.; Fischer, M.; Fuchs, F.W.; Asiminoaei, L.; Thøgersen, P.B. A review on fundamental grid-voltage detection methods under highly distorted conditions in distributed power-generation networks. In Proceedings of the 2011 IEEE Energy Conversion Congress and Exposition, Phoenix, AZ, USA, 17–22 September 2011; pp. 3045–3052.

36. Rodriguez, P.; Luna, A.; Ciobotaru, M.; Teodorescu, R.; Blaabjerg, F. Advanced Grid Synchronization System for Power Converters under Unbalanced and Distorted Operating Conditions. In Proceedings of the IECON 2006—32nd Annual Conference on IEEE Industrial Electronics, Paris, France, 6–10 November 2006; pp. 5173–5178.
37. Patil, K.R.; Patel, H.H. Modified dual second-order generalised integrator FLL for synchronization of a distributed generator to a weak grid. In Proceedings of the 2016 IEEE 16th International Conference on Environment and Electrical Engineering (EEEIC), Florence, Italy, 7–10 June 2016; pp. 1–5.
38. He, X.; Geng, H.; Yang, G. A Generalized Design Framework of Notch Filter Based Frequency-Locked Loop for Three-Phase Grid Voltage. *IEEE Trans. Ind. Electron.* **2018**, *65*, 7072–7084. [CrossRef]
39. Apap, M.; Clare, J.; Wheeler, P.; Bradley, K. Analysis and comparison of AC-AC matrix converter control strategies. In Proceedings of the IEEE 34th Annual Conference on Power Electronics Specialist, Acapulco, Mexico, 15–19 June 2003; pp. 1287–1292. [CrossRef]

Publisher's Note: MDPI stays neutral with regard to jurisdictional claims in published maps and institutional affiliations.

Article

Application of Grasshopper Optimization Algorithm for Selective Harmonics Elimination in Low-Frequency Voltage Source Inverter

Marcin Steczek *, Włodzimierz Jefimowski and Adam Szeląg

Faculty of Electrical Engineering, Institute of Electrical Power Engineering, Warsaw University of Technology, pl. Politechniki 1, 00-661 Warszawa, Poland; wlodzimierz.jefimowski@ien.pw.edu.pl (W.J.); adam.szelag@ien.pw.edu.pl (A.S.)
* Correspondence: marcin.steczek@ien.pw.edu.pl

Received: 28 October 2020; Accepted: 2 December 2020; Published: 4 December 2020

Abstract: In this paper, an application of the recently developed Grasshopper Optimization Algorithm (GOA) for calculation of switching angles for Selective Harmonic Elimination (SHE) PWM in low-frequency voltage source inverter is proposed. The algorithm is based on insect behavior in the food foraging swarm of grasshoppers. The characteristic feature of GOA is the movement of agents is based on the position of all agents in the swarm. This method represents a higher probability of convergence than Particle Swarm Optimization (PSO) Modifications of GOA have been examined regarding their effect on the algorithm's convergence. The proposed modifications were based on the following techniques: Grey Wolf Optimizer (GWO), Natural Selection (NS), Adaptive Grasshopper Optimization Algorithm (AGOA), and Opposite Based Learning (OBL). The performance of GOA and its modifications were compared with well-known PSO. Areas, where GOA is superior to PSO in terms of probability of convergence, have been shown. The efficiency of the GOA algorithm applied for solving the SHE problem was confirmed by measurements in the laboratory.

Keywords: grasshopper optimization algorithm (GOA); particle swarm optimization (PSO); voltage source inverter (VSI); selective harmonics elimination PWM (SHEPWM)

---

1. Introduction

The Selective Harmonic Elimination (SHE) and Selective Harmonic Mitigation (SHM) [1,2] has been described for the first time in the 1960s in [3] and disseminated by Patel and Hoft [4,5]. Since that time SHE has been introduced in a number of industrial applications where power electronics was proposed [6]. The challenge is progress in the development of techniques for solving SHE/SHM non-linear transcendental equations

Since the early days, iterative techniques such as Newton–Raphson (N–R) [4,5], Gauss–Newton have been employed to solve these equations. The convergence of these methods depends on the initial guess, which is a complex problem and in many cases is not successful. This disadvantage encourages researchers to develop more effective techniques. Thus, Chaison et al. in [7] proposed the method based on the conversion of transcendental equations into an equivalent set of polynomials. The high degree of polynomial requires specialized software to compute it. The combination of Groebner's bases and symmetric polynomials was applied to solve the mentioned polynomials [8]. However, it generates ambiguous solutions which make it less useful. The main disadvantage of iterative techniques is they do not find an optimum solution.

The development of evolutionary algorithms opens new opportunities in the field of solving SHE equations [9]. These algorithms present numerous benefits such as independence from an initial guess, utilization of simple algebra, lower computational costs, formulation of multi-constrained problems.

One of the most popular evolutionary algorithms is Particle Swarm Optimization (PSO) proposed for finding switching angles for PWM VSI inverter to eliminate low order voltage harmonic [10] and to optimize dc-link current harmonics [11]. The application of numerous algorithms are proposed for SHE in the literature: Imperial Colonial Algorithm (ICA) [12], Genetic Algorithm [13], Ant Colony Algorithm [14], bee optimization technique (BA) [15], Bacterial Foraging Algorithm [16], Firefly Algorithm (FA) [17], Shuffled Frog Leaping (SFL) algorithm [18], Backtracking Search Algorithm (BSA) and Differential Search Algorithm (DSA) [9], Whale Optimization Algorithm (WOA) [19]. The use of the Grasshopper Optimization Algorithm (GOA) for SHE has not been studied so far.

The GOA is an algorithm recently developed and introduced by Saremi et al. [20] in 2017. In recent two years, the GOA gained great attention in many research fields due to its high efficiency of solving a different kind of optimization problems. It was tested for constrained and unconstrained test functions with promising results [21]. The GOA was proposed for solving multi-objective optimization problems [22] modified by the application of Opposition-Based Learning (OBL) [23]. Modifications of GOA to improve its performance are has been developed and studied: Adaptive GOA (AGOA), Grey Wolf Optimizer (GWO) and Natural Selection (NS) [24], Gaussian mutation, and Leavy- flight strategy [25].

The efficiency of GOA has been compared with existing evolutionary algorithms utilized for different optimization problems. In [26] GOA adaptation for energy loss reduction and voltage stability factor was proposed and compared with PSO, Gravitational Search Algorithm (GSA), and Artificial Bee Colony (BA) algorithms. In [27], the comparison of GOA with PSO and WOA (Wale Optimization Algorithm) was used to optimize the PI controller parameters in the microgrid. Since the first presentation, GOA has found its implementation in numerous industrial applications such as optimization of the parameters of proton membrane fuel cells (PEMFC) [28], the stability of microgrid applications [29] and energy management [30], medicine [31], the technology of image processing [32], and financial issues [25] as well.

In this paper, the recently developed GOA is applied to eliminate low-order voltage harmonics (5th, 7th, 11th, and 13th) in low-frequency VSI based drive. The hypothesis to prove is that GOA represents a higher probability of convergence than PSO applied for SHE problem with similar computation effort. Results for GOA and modified GOA are compared with PSO. The main criterion of comparison is the probability of convergence. The following modification of GOA are examined: Natural Selection (NS), Adaptive GOA (AGOA), Opposite Based Learning (OBL), and Grey Wolf Optimizer (GWO). Experimental results are presented to validate simulation analysis.

The rapid development of controllers for high and medium power converters provides an opportunity for the application of modulation techniques; a decade ago, they used to be known as difficult to use. This type of modulation is SHE-PWM. When it was invented its applicability was very low and nowadays it competes with the most advanced and popular modulations [33]. Its application is studied for grid connectors [34] and railway vehicles [35] as well. Moreover, the separation of a modulator from the controller brings the possibility of implementation of the SHE-PWM with space vector modulation (SVPWM) [36]. Authors of this paper claim that for railway vehicles the most efficient is hybrid modulation studied in [37] where SHE-PWM and SVPWM are used interchangeably and the choice depends on the operating conditions. This solution is the most reasonable and allows to utilize the advantages of both techniques: dynamics of SVPWM and harmonics control of SHE-PWM.

The attention focused on SHE-PWM stimulates research towards increment of efficiency in the calculation of switching angles. In [38] the comprehensive review of SHE-PWM focused on various aspects, is presented. One of the mentioned aspects is the utilization of optimization-based techniques for solving SHE equations and they were divided into four groups: Genetic Algorithms (GA) Particle Swarm Optimization (PSO) Differential Evolution (DE) and Hybrid. According to the "no free lunch" theorem applied to the bio-inspired optimization algorithms [39], there is the most suitable solver for a specific optimization task. Solving SHE equations developed for voltage source inverter is a task of variable complexity that depends on assumptions like the number of switching angles, modulation index, dead times between switching, switching frequency, and others. Thus, there is a possibility that for different optimization region different algorithm is most suitable. Thus, every recently developed algorithm should be evaluated towards application for solving SHE-PWM equations. In this paper, the authors present the study for the application of the GOA algorithm. The novelty of this study is proof that there is a range of SHE problems where the GOA algorithm gives a higher possibility of convergence with lower computational effort than widely used and appreciated PSO. Results presented in this paper encourage further research to discover the full potential of the GOA algorithm regarding the presented problem by comparing it with a wider representation of bio-inspired algorithms.

The problem undertaken in the study is considered as a single criteria optimization problem. SHE problem could be considered as a multi-objective optimization problem as each harmonic value as a function of optimization variables could be considered as a separate objective function. However, all considered harmonics in the problem of SHE should be eliminated for the same optimization variables, therefore the desirable optimization solution is a utopian solution from the point of view of the multi-optimization approach [40]. Accordingly, the optimization functions have been aggregated to a single optimization variable by means of the dedicated relationship proposed in the article.

2. VSI Model with SHE Control

2.1. Drive's Parameters

The main goal of this work is to study the convergence of GOA applied for the calculation of switching angles for SHE-PWM for a low-frequency VSI drive with an induction motor. Therefore, to prove the validity of results obtained by the examined algorithm, the VSI drive was modeled in MATBAL/SIMULINK and verified in the laboratory. Parameters of utilized induction motor for experiments are presented in Table 1.

Table 1. Parameters of drive's mode.

Symbol	Parameter	Value
P_n	Rated power	2, 5 kW
I_n	Rated phase current	3.9 A
V_n	Rated voltage rms	230/400 V
-	Winding's connection	star
n_n	Rated rotation speed	1465 rpm
L_s	Stator's leakage inductance	0.0108 H
R_s	Stator's resistance	2.8465 Ω
L_r	Rotor's leakage inductance	0.0106 H
R_r	Rotor's resistance	2.7359 Ω
L_m	Core losses inductance	0.27597 H
R_{m_n}	Core losses resistance	1231 Ω

Figure 1a. illustrates the topology of the VSI utilized for the purpose of this work. Figure 1b shows an equivalent circuit of a single phase of the motor applied in the SIMULINK model. Inverter's transistors were controlled by binary switching function ($SF(\omega t)$) formed by switching angles (α) defined as angular "moments" of transistors state change (Figure 2). Switching angles are delivered by solving equations presented in this section.

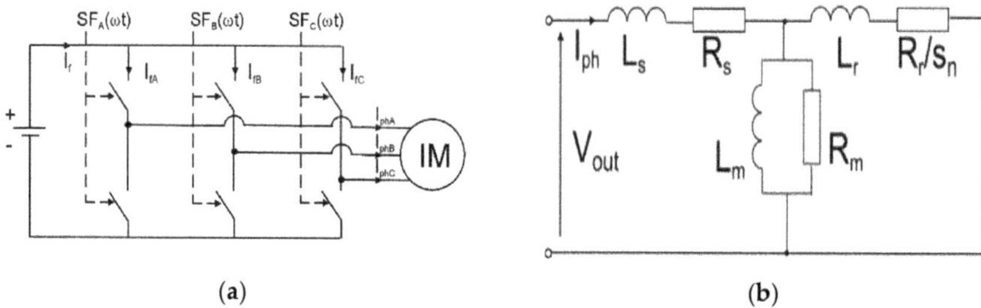

Figure 1. Analytical schema of the considered 3 phase 2-level inverter with load (IM—Model of an induction motor) (**a**) inverter schema (**b**) equivalent circuit of one phase of induction motor (sn—Slip for n-th harmonic).

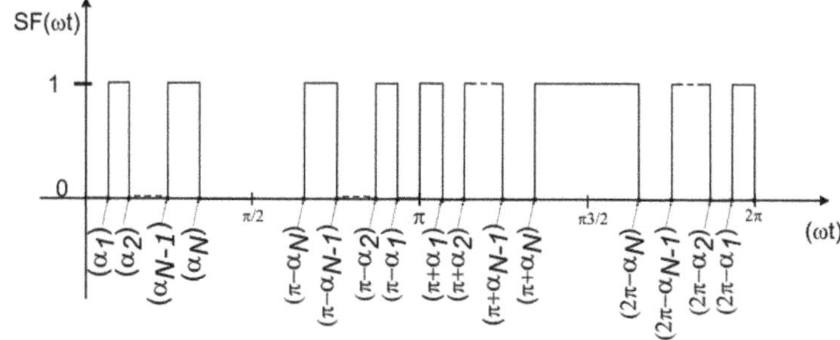

Figure 2. Waveform of switching function SF with quarter-wave symmetry and N switching angles.

2.2. SHE-PWM

In the great number of cases of SHE-PWM application, it is used in electric drives to eliminate low order harmonics, while amplitudes of high order harmonics are reduced by input filters. SHE equations are based on the Fourier series expansion of the inverter output voltage waveform:

$$V(\omega t) = a_0 + \sum_{n=1}^{\infty} [a_n sin(n\omega t) + b_n cos(n\omega t)] \tag{1}$$

where ω is an angular frequency of fundamental component, n is a harmonic order, and an, bn are Fourier coefficients. For quarter-wave symmetry, only coefficient an for the odd n coefficient represents the non-zero value:

$$a_n = \begin{cases} \frac{4U_{DC}}{n\pi}\left[-1-2\sum_{i=1}^{N}(-1)^i cos(n \cdot \alpha_i)\right] & ; \ for \ odd \ n \\ 0 & ; \ for \ even \ n \end{cases} \tag{2}$$

$$b_n = \begin{cases} 0 \ ; \ for \ odd \ n \\ 0 \ ; \ for \ even \ n \end{cases} \tag{3}$$

where U_{DC} is DC-link voltage and n is the number of switching angles per a quarter-period. In this paper, fluctuation and ripplers of DC-link voltage were not of concern.

Assuming the odd quarter-wave symmetry of inverter output voltage, triple harmonics are canceled. The symmetry of the system brings cancelation of even harmonics as well. For $n = 5$

switching angles in quarter-period, 5 non-linear equations can be formulated (4) to satisfy fundamental component (*V1*) and eliminate 5th, 7th, 11th, and 13th harmonics:

$$\begin{cases} \frac{4}{\pi}[-1 + 2\cos(\alpha_1) - 2\cos(\alpha_2) + 2\cos(\alpha_3) - \dots \\ \qquad \dots 2\cos(\alpha_4) + \; 2\cos(\alpha_5)] = M1 \\ \frac{4}{5\pi}[-1 + 2\cos(5\alpha_1) - 2\cos(5\alpha_2) + 2\cos(5\alpha_3) - \dots \\ \qquad \dots 2\cos(5\alpha_4) + \; 2\cos(5\alpha_5)] = 0 \\ \frac{4}{7\pi}[-1 + 2\cos(7\alpha_1) - 2\cos(7\alpha_2) + 2\cos(7\alpha_3) - \dots \\ \qquad \dots 2\cos(7\alpha_4) + \; 2\cos(7\alpha_5)] = 0 \\ \frac{4}{11\pi}[-1 + 2\cos(11\alpha_1) - 2\cos(11\alpha_2) + 2\cos(11\alpha_3) - \dots \\ \qquad \dots 2\cos(11\alpha_4) + \; 2\cos(11\alpha_5)] = 0 \\ \frac{4}{13\pi}[-1 + 2\cos(13\alpha_1) - 2\cos(13\alpha_2) + 2\cos(13\alpha_3) \; - \dots \\ \qquad \dots 2\cos(13\alpha_4) + 2\cos(13\alpha_5)] = 0 \end{cases} \tag{4}$$

where *M1* is for modulation index:

$$V_1 = M1\,\frac{U_{DC}}{2}\,; \quad for\; M1\langle 0,\, \frac{4}{\pi}\rangle \tag{5}$$

In this case the fundamental voltage component (*V1*) is defined as the amplitude of phase voltage of the motor in the star connection of the windings. The main goal of this paper is to adopt GOA for solving Equation (4) to determine switching angles for SHE-PWM end examine its convergence.

3. Formulation of The Optimization Problem

To solve SHE Equation (4) using an optimization algorithm, the fitness function must be formulated. For $n = 5$ switching angles and four harmonics eliminated the fitness function is described by following equation with constraints:

$$Minimize,\; f_{fit}(\alpha_1, \alpha_2, \alpha_3, \alpha_4, \alpha_5)$$
$$= \sigma_1 \cdot \left(V_1 - V_1^*\right)^2 + \sigma_5 \cdot (V_5)^2 + \sigma_7 \cdot (V_7)^2 + \sigma_{11} \cdot (V_{11})^2 + \sigma_{13} \cdot (V_{13})^2 \tag{6}$$

$$subject\; to:\; 0 < \alpha_1 < \alpha_2 < \alpha_3 < \alpha_4 < \alpha_5 < \frac{\pi}{2}$$

where: V_1, V_5, V_7, V_{11}, V_{13} are fundamental component and 5th, 7th, 11th and 13th voltage harmonics (p.u.) respectively, σ_x are penalty weights for the optimization process.

Thus, the aim is to apply an optimization algorithm to minimize fitness function (6) to achieve declared fundamental component (V_1^*) and harmonics elimination. The fundamental component is minimized with the highest weigh (penalty value) that equals $\sigma_1 = 100$. Thus, every 1% of difference between an actual value and the desired one will increase fitness function by 100. Harmonics are minimized with penalty weight $\sigma_5 = \sigma_7 = \sigma_{11} = \sigma_{12} = 10$ One of the essential assumptions of every optimization algorithm is the STOP criterion based on the maximum number of iterations and the minimum value of the fitness function. In this paper, the minimum value of the fitness function assumed to be the success is $f_{fit\_STOP} = 0.0001$. Regarding the necessity of implementation of dead-times in industrial applications (in this paper dt = 5×10^{-6} s), a lower value of the fitness function will not be recognized as higher quality performance. To prove this statement, sample results obtained by GOA with tolerance 1×10^{-3} (Figure 3a) were compared with the results obtained with tolerance 1×10^{-10} (Figure 3b). Figure 3 shows that decreasing the parameter of tolerance does not guarantee better efficiency of eliminated harmonics, only significantly increases computation time. Thus, tolerance 1×10^{-3} was assumed to be sufficient. More results will be presented in Section 5.

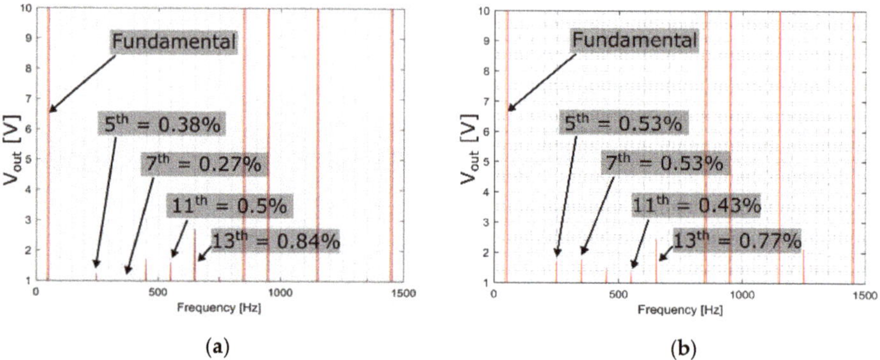

Figure 3. Simulated output voltage spectrum for SHE-SPWM carried out using Grasshopper Optimization Algorithm (GOA) for different tolerance: (**a**) tolerance $< 1 \times 10^{-4}$; M1 = 1.0; x = (0.1234; 0.4242; 0.5199; 1.2197; 1.2791); (**b**) tolerance $< 1 \times 10^{-10}$; M1 = 1.0; x = (0.1225; 0.4259; 0.5206; 1.2186; 1.2783)

4. Grasshopper Optimization Algorithm (GOA)

GOA was developed and introduced by Saremi, Mirjalili, and Lewis in [20]. The first proposed application was for structural optimization to find the optimal shape of a three-bar truss, a 52-bar truss, and a Cantilever beam. In this paper, the authors present the application of GOA to solve SHE equations. The proposed algorithm mathematically models the behavior of grasshopper swarm foraging for food to survive. Their individual behavior and social interactions lead the swarm to the optimal solution. The mathematical model of grasshopper behavior is based on a formula for the position of each grasshopper described by the following equation:

$$X_i = S_i + G_i + A_i \tag{7}$$

where X_i is the position of the i-th grasshopper, S_i is the social interaction between agents in the swarm, G_i is the gravity force acting on the i-th grasshopper, and A_i models the wind effect. However, due to specifics of the problems analyzed in this work effect of gravity and wind were omitted. Thus, only social interactions were taken into account.

$$S_i = \sum_{\substack{j=1 \\ j \neq i}}^{Np} s\left(d_{ij}^*\right) \cdot \vec{d}_{ij} \tag{8}$$

where d_{ij}^* is the normalized distance between the i-th and j-th grasshopper, $s(d_{ij}^*)$ is the function of social forces and $\vec{d}_{ij}$ is an unitary vector from i-th to j-th grasshopper.

If absolute value of the distance between the i-th and j-th agent is formulated as:

$$d_{ij} = |x_i - x_j| \tag{9}$$

thus, the normalized value is defined as

$$d_{ij}^* = 2 + rem\left(d_{ij}, 2\right) \tag{10}$$

where $rem\left(d_{ij}, 2\right)$ is the remainder after division of d_{ij} by 2. Distance normalization allows for the value of the distance to be kept close to the value of 2 what gives the best effect with the *s-function*. Unitary vector $\vec{d}_{ij}$ is defined with the following correlation:

$$\begin{cases} \vec{d}_{ij} = 1; \ for \ x_i - x_j > 0 \\ \vec{d}_{ij} = -1; \ for \ x_i - x_j \leq 0 \end{cases} \tag{11}$$

Thus, $\vec{d}_{ij}$ can be defined by the following formula:

$$\vec{d}_{ij} = \frac{x_i - x_j}{\left|x_i - x_j\right|} \tag{12}$$

A characteristic feature of GOA is a comfort zone shrinking with the iteration number. The Comfort zone is the circle around the best agent. Inside the comfort zone, other agents are being repulsed from the leader and outside the comfort zone, they are being attracted to it. This behavior keeps a balance between exploration and exploitation. The decreasing coefficient c models variation of the comfort zone by changing value typically from 1 to some small number. Regarding the above considerations, Equation (7) for the d-dimensional problem can be expanded as follows:

$$X_i^d = c \left(\sum_{\substack{j=1 \\ j \neq i}}^{Np} c \cdot \frac{ub^d - lb^d}{2} s\left(d_{ij}^{d*}\right) \frac{x_i^d - x_j^d}{\left|x_i^d - x_j^d\right|} \right) + Gbest^d \tag{13}$$

where ub^d is for upper bound of the *d-th* dimension, lb^d is for lower bound of the *d-th* dimension, $Gbest^d$ is for the global best result in the *d-th* dimension, s is for *s—Function* which describes the strength of the interaction between agents and is formulated with the following formula:

$$s\left(d_{ij}^{d*}\right) = F \cdot e^{\left(-\frac{d_{ij}^*}{L}\right)} - e^{\left(-d_{ij}^*\right)} \tag{14}$$

where F and L are coefficients with suggested values 0.5 and 1.5 respectively. Variation of these coefficients will be analyzed in further sections, regarding its influence on the probability of convergence of the algorithm.

Equation (13) reveals the most significant rule of GOA. The position of agents in every iteration is determined with respect to the position of all other agents in the swarm. For instance, in the PSO algorithm position of agents is determined regarding only two vectors: personal best and global best position. That is the reason why GOA requires a lower population to keep the same computational effort. Moreover, an increment of the swarm population may result in lower convergence.

To apply the GOA algorithm for solving SHE Equation (4) the X_i^d must be correlated with the *i-th* vector of switching angles $[\alpha_1 \ \alpha_2 \ \alpha_3 \ \alpha_4 \ \alpha_5]_i$. Thus, in this case, the problem is 5 dimensional. In the following subsections, the modifications of GOA are proposed and tested in section V.

4.1. GOA with GWO Module

Grey Wolf Optimizer is a meta-heuristic algorithm inspired by the behavior of grey wolves and mathematically described by Mirjalili et al. in [41]. The GWO is well studied and can be treated as an independent algorithm. However, its main feature can be implemented in other algorithms. The specifics of the GWO is based on the determination of the three best global solutions called alpha, beta, and gamma. Positions of all particles in the swarm will be updated with respect to the position of the three best global *Ta, Tb, Tc*. Thus, to combine GOA and GWO, formula (13) will be modified to the following form [24]:

$$X_i^d = c \left(\sum_{\substack{j=1 \\ j \neq i}}^{Np} c \cdot \frac{ub^d - lb^d}{2} s(d_{ij}^{d*}) \frac{x_i^d - x_j^d}{\left| x_i^d - x_j^d \right|} \right) + \frac{T_A^d + T_B^d + T_C^d}{3} \tag{15}$$

4.2. GOA with NS Module

The theory of Natural Selection is based on the random elimination of agents from the swarm with a certain probability P with respect to their fitness value. The better result has a higher chance to survive. Thus, NS requires a classification of the agents then the algorithm calculates P for each agent. To adopt NS for the SHE problem the following formula for P for the i-th agent was developed:

$$P_i = P_{min} + \left[(P_{max} - P_{min}) \cdot \left(\frac{f_{fit\_i}}{f_{fit\_swarm}} \right) \right] \tag{16}$$

where P_{min} is the minimum assumed probability of survival assigned for the weakest agent; P_{max} is the maximum assumed probability of survival assigned for the best agent; f_{fit_i} is the fitness of the *i-th* agent; f_{fit_swarm} is the mean value of fitness functions of all agents in the swarm.

The roulette is performed for every single agent regarding its probability of survival. The eliminated agents are replaced by new random solutions.

4.3. Adaptive GOA

Adaptive Grasshopper Optimization Algorithm (AGOA) is based on a dynamic adaptation of the *c* coefficient regarding the Evolutionary Rate (ER) of the swarm of grasshoppers. ER is defined as the ratio between the number of agents whose fitness was improved in the previous iteration to the total number of agents in the swarm *Np*. Thus, *c* for AGOA is defined by the following formula:

$$c(ite) = \left(c_{max} - ite \, \frac{c_{max} - c_{max}}{t_{max}} \right) F_{ER}(ite) \tag{17}$$

where $F_{ER}(ite)$ is the dynamic adjustment function defined by the following correlation:

$$F_{ER}(ite+1) = \begin{cases} \frac{F_{ER}(ite)}{F_0}; & for\ ER < 15\% \\ F_{ER}(ite); & for\ ER \in \langle 15\%; 30\% \rangle \\ F_{ER}(ite) \cdot F_0 ; & for\ ER > 30\% \end{cases} \tag{18}$$

where F_0 is a constant larger than 1. Thus, AGOA presents a dynamic change of comfort zone with a decreasing trend.

4.4. GOA with OBL Module

Opposite Based Learning is a technique of swarm algorithms modification based on the statement that the opposite solution to the developed one might bring better result. Thus, every solution (agent) should be reversed and its fitness should be evaluated. If the fitness of the reversed solution is better than the original, the agent will be replaced. Opposite value $\ddot{X}_i^d$ can be calculated as follows:

$$\ddot{X}_i^d = ub^d + lb^d - X_i^d; \tag{19}$$

$$for\ i = 1, 2, \ldots, Np; d = 1, 2, \ldots, N$$

However, the application of this technique to SHE brings issues regarding the feasibility of calculated solutions. As switching angles are restricted to be sorted according to their feasibility, Equation (19) will reverse their order and make them not feasible. To implement OBL to solve the SHE problem opposite solution $\ddot{X}_i^d$ must be reordered to respect restriction: $0 < \alpha_1 < \alpha_2 < \alpha_3 < \alpha_4 < \alpha_5 < \frac{\pi}{2}$.

5. Simulation Tests and Comparative Study

In this section, a simulation analysis of GOA algorithm performance is presented. Moreover, a comparative study between GOA and modified GOA (NS, AGOA, GWO, and OBL) is carried out. The optimization process has been carried out with the following assumptions:

- STOP criterion of the optimization process is obtained when reaching the assumed maximum number of iterations or the value of the fitness function is below the assumed tolerance 1×10^{-4}
- Every modification module is tested separately. The combination of all modules in one algorithm is not tested. The reason is an increment of computation effort for multi-module algorithm what makes it difficult to compare with single module modifications,
- Swarm population (Np) and maximum number of iterations (max_ite) for comparative study is established regarding similar computational effort (elapsed time of optimization) for compared algorithms.

Figure 4 shows the flow chart of the developed GOA algorithm for SHE with marked modification modules. However, as was mentioned above, only one module is active at the time.

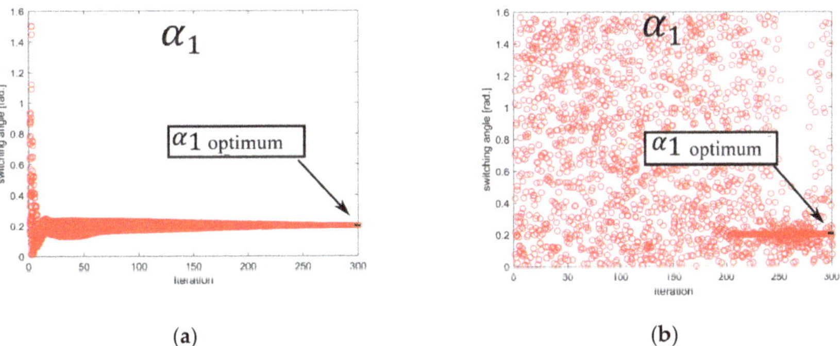

Figure 4. The trajectory of particles in the swarm searching for optimal switching angle α_1 with (a) GOA (b) Particle Swarm Optimization (PSO).

5.1. Comparison between GOA and PSO

The comparative study between GOA and PSO was carried out for $n = 5$ switching angles in a quarter-period, modulation index $M1 = 0.9$, and fundamental frequency $f_f = 50$ Hz what gives 550 Hz switching frequency. In this work 5th, 7th, 11th, and 13th harmonics are eliminated permanently. Figure 5 presents the trajectories of particles during the optimization process with GOA and PSO algorithms, respectively. The GOA algorithm is characterized by a short time of exploration and a long period of exploitation while the PSO algorithm provides a very high intensity of exploration. Figure 6 shows that PSO needs a higher number of iterations to converge comparing with GOA and its modifications.

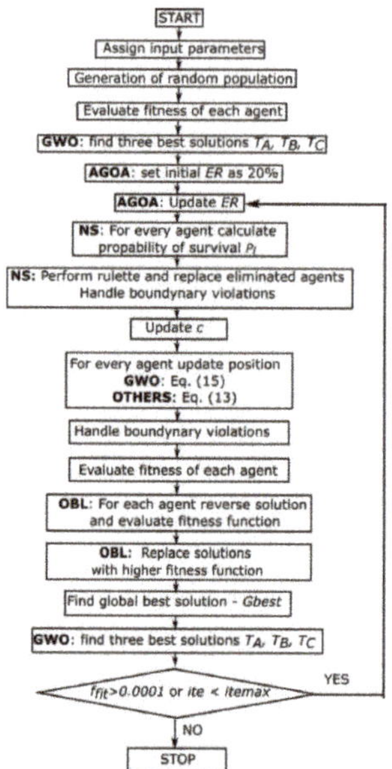

Figure 5. Flow chart of the GOA algorithm with proposed modification.

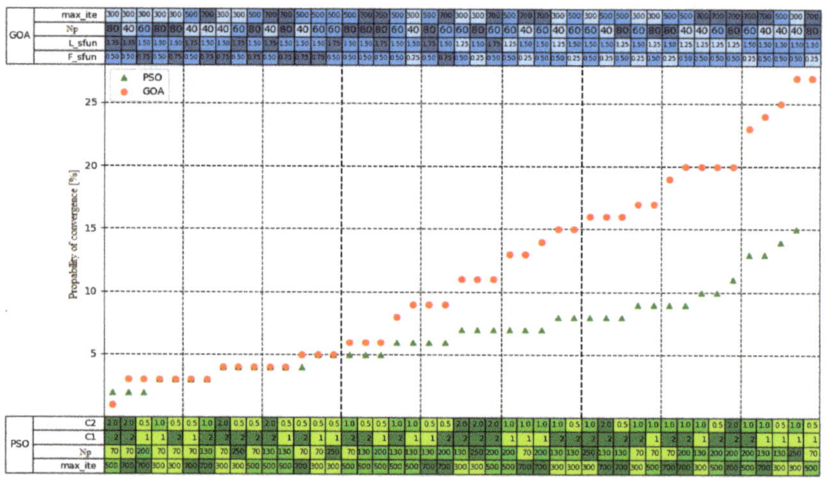

Figure 6. Comparison between PSO and GOA convergence for different values of coefficients.

In the following part of this section, the convergence of GOA and PSO has been compared. The various combinations of parameters were tested. For every set of both GOA and PSO, 100 runs were calculated. For each series of runs, the number of runs that reaches f_{fit} below 1×10^{-4} was recorded as the success. The variation of the following parameters was studied: maximum number of iterations (*max_ite*), size of the population (*Np*), PSO parameters C1 and C2, and GOA parameters used in s-function (L and F). Thus, all results have been presented in Figure 6 where the probability of convergence is compared. The value of population size for PSO and GOA was adjusted to keep similar computation times for both.

The best performance of GOA was recorded for settings: $Np = 40$ *max_ite* $= 300$, $L = 1.5$ and $F = 0.5$. For the aforementioned setting, GOA achieved 27% convergence (time of computation 101 s). The PSO achieved the best performance for: $Np = 250$, $max\_ite = 300$, $C1 = 1$ and $C2 = 0.5$. Thus, GOA presents better performance than PSO regarding higher convergence for lower population size. Results presented in Figure 6 proves that for the SHE problem the GOA presents a significantly higher probability of convergence than the PSO algorithm. The most interesting result was achieved by using the GOA algorithm for population 40 and 300 iterations (the second-highest for this algorithm) where the probability of convergence was 28%. This result proves that the GOA algorithm can be very efficient for low population set up which reduces its computational effort. The highest recorded probability of convergence of PSO during the experiment was 15% (time of computation 260 s). The GOA algorithm gives a better result faster. The computation effort is extremely significant during the application of the optimization task for the task. In this paper, coefficients of the compared algorithm were selected in such a way to keep comparable computation time. The parameters which are affecting computation time are population size Np and the maximum number of iteration max_ite. Results presented in Figure 6 can be divided into 9 *SETS* regarding Np and max_ite. Table 2 presents computation time for the PSO and GOA performed on a computer with processor *Intel Core i7-8557U*. Results from Table 2 are presented in Figure 7. The parameters Np and max_ite were selected to keep similar computation time for GOA and PSO for one SET. The GOA in every iteration calculates the position of every agent (grasshopper) related to every agent in the swarm, PSO calculates the only position of the agent related to the position of the pest particle. That is why GOA needs higher computational effort than PSO for the same Np. Thus, to make it comparable the Np for GOA was reduced in every SET.

Table 2. Sets of coefficients with computation time (100 runs).

No.	PSO			GOA		
	Max_Ite	Np	Time [s]	Max_Ite	Np	Time [s]
SET 1	300	70	99	300	40	101
SET 2	300	130	169	300	60	163
SET 3	300	250	260	300	80	261
SET 4	500	70	165	500	40	152
SET 5	500	130	310	500	60	277
SET 6	500	250	470	500	80	438
SET 7	700	70	245	700	40	264
SET 8	700	130	412	700	60	385
SET 9	700	250	637	700	80	687

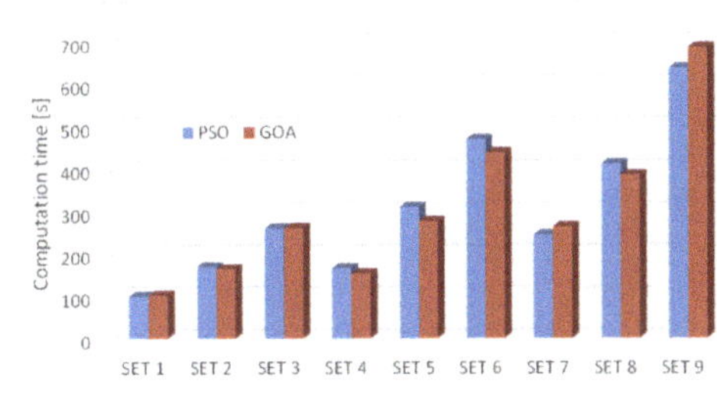

Figure 7. Computation time for different sets of coefficients.

Figure 8 shows the switching angles calculated by PSO and not modified GOA as the function of modulation index M1.

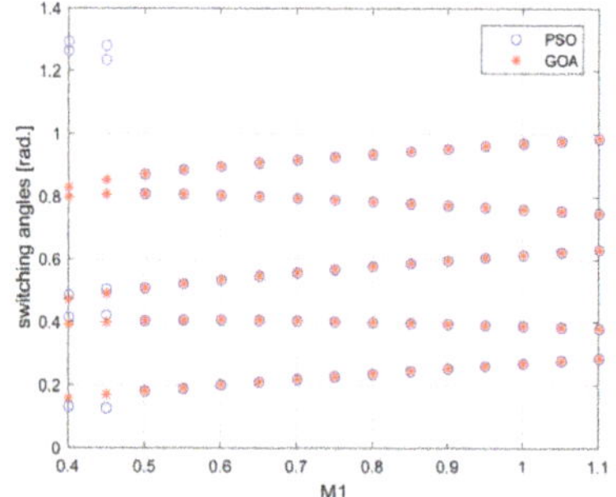

Figure 8. Switching angles as the function of modulation index M1 for SHE-PWM calculated using PSO and GOA.

5.2. Comparative Study between PSO, GOA, and Modified GOA

In this subsection, the convergence of GOA with modifications and PSO algorithms has been carried out. Assumptions from section *V A* and Table 3 are valid in this section. However, the comparison has been conducted for a wider range of modulation index $M1 = 0.4 \div 1.1$.

Table 3. Parameters for all analysis.

Algorithm	Coefficients	Value
GOA	c_{min}	1×10^{-6}
	c_{max}	1
GOA + NS	P_{min}	0.3
	P_{max}	0.95
AGOA	F_0	1.05
PSO	C1	2
	C2	2
	w_{min}	1×10^{-3}
	w_{max}	1

The comparative study reveals that the most efficient with the highest probability of convergence is the GOA algorithm modified by adding the OBL module (Figure 9). The convergence of the studied algorithm was decreasing with an increase of *M1* because higher modulation index requires smaller spaces between switching angles and reduce the feasibility of developed solutions. For *M1* in the range from 0.7 to 0.8 modification based on NS presented very good performance as well. Figure 10 presents the examination of fitness value (f_{fit}) as the function of the iteration number. The conclusion is that GOA-based algorithms present significantly faster convergence than PSO.

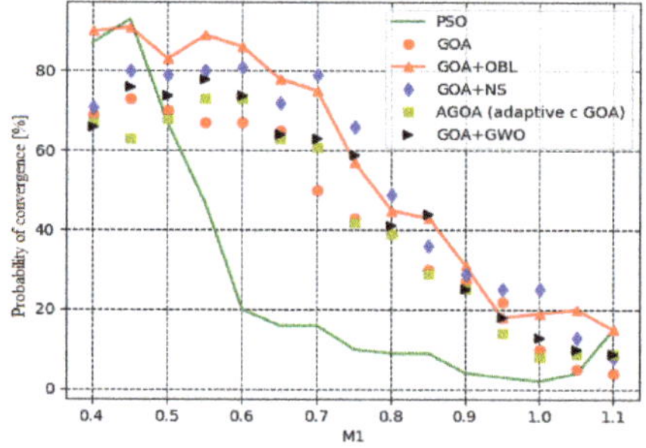

Figure 9. Comparison of the probability of convergence between PSO, GOA, and modified GOA algorithms.

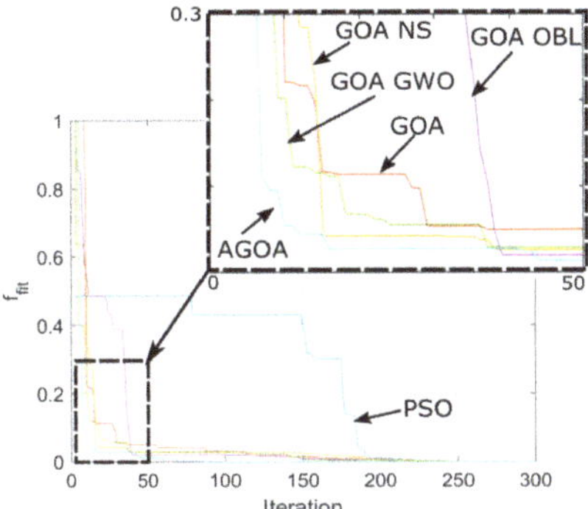

Figure 10. Fitness value versus iteration number for examined algorithms.

6. Experimental Results

The applied GOA algorithm has been experimentally verified. Switching angles calculated by the GOA algorithm were implemented into the laboratory stand (Figure 11) developed according to Figure 1. Parameters of the induction motor applied in the laboratory stand are presented in Table 1. The inverter control by switching angles input was provided by DSpace card 1104. Switching angles were applied as the look-up table for off-line control. Verification was conducted for two operating points with the same fundamental frequency f_f = 50 Hz, and different $M1$ = 0.9 and 1.0, respectively.

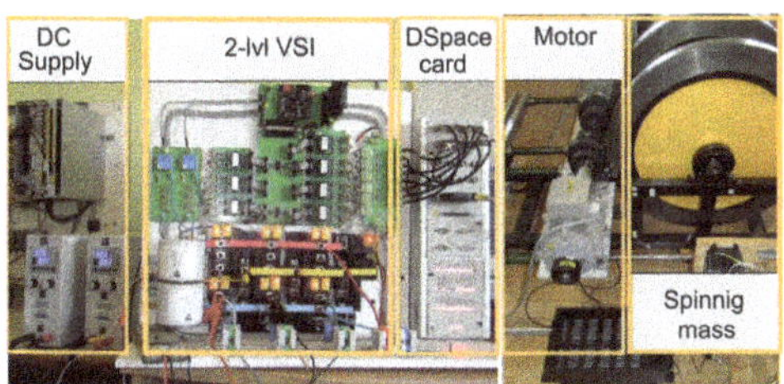

Figure 11. Laboratory stands for experimental tests.

In Figures 12 and 13 it can be noticed that amplitudes of 5th, 7th, 11th, and 13th harmonics in the output voltage have been eliminated and the GOA algorithm has successfully minimized the goal function.

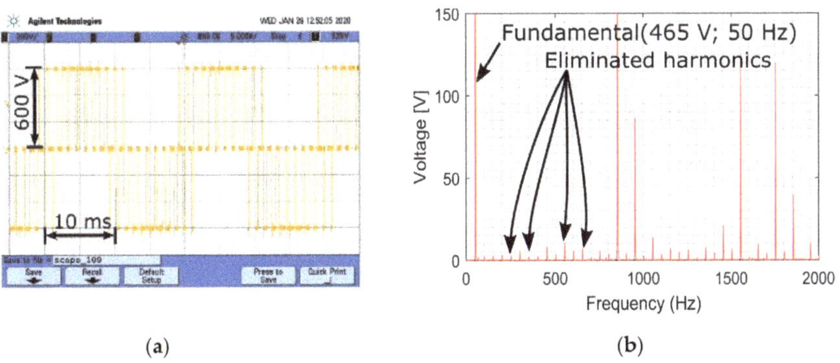

(**a**)   (**b**)

Figure 12. Experimental results for inverter's phase to phase output voltage for M1 = 0.9 (**a**) waveform oscillogram (**b**) spectrum.

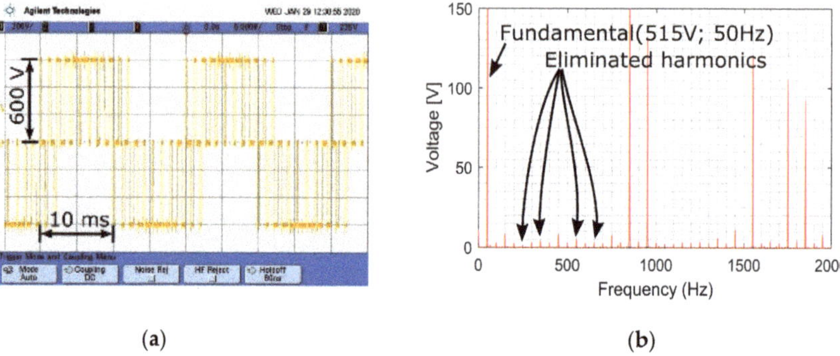

(a)

(b)

Figure 13. Experimental results for inverter's output phase to phase voltage for M1 = 1.0 (**a**) waveform oscillogram (**b**) spectrum.

7. Conclusions

In this paper, the authors investigated a novel application of a recently developed GOA algorithm, for the calculation of switching angles in SHE-PWM inverter modulation. The main goal of this paper was to examine the probability of convergence introduced by GOA applied for solving the SHE problem. Modifications of the GOA algorithm have been implemented and compared with the PSO algorithm. The GOA algorithm is based on the behavior of a swarm of grasshoppers and the most characteristic feature is that the movement of agents depends not only on the position related to the position of the best agent (best global solution) but it depends on the position related to the other agents as well. Thus, the results prove that the GOA algorithm requires a lower population size to converge with computation effort similar to PSO. The most interesting outcome of this study is that the GOA algorithm with OBL elements proves its superiority over the PSO algorithm regarding the probability of convergence for similar computational effort (lower population of particles). The second most efficient combination was the GOA algorithm with NS modification. GOA presents the highest advantage over PSO in the range of modulation index from 0.5 to 1.0. In this range, the convergence of PSO was dramatically reduced (below 5% in the worst case), and meanwhile, the probability of convergence of GOA was between 20 and 80%. The performed measurement experiments proved that the SHE-PWM waveform optimized by the GOA algorithm provided elimination of the chosen harmonics in the inverter's output voltage. In the nearest future, authors will focus their attention on the application of the GOA algorithm for optimization of waveforms generated by multilevel inverters and its applicability in traction drive solutions.

Author Contributions: M.S.: Concept and methodology of the research, development of software and application of algorithms. Formulation of conclusions. W.J.: Laboratory measurements and presentation of results. A.S.: Research supervision and text formatting. All authors have read and agreed to the published version of the manuscript.

Funding: This research received no external funding.

References

1. Lou, H.; Mao, C.; Lu, J.; Wang, D.; Lee, W.J. Pulse width modulation AC/DC converters with line current harmonics minimisation and high power factor using hybrid particle swarm optimisation. *IET Power Electron.* **2009**. [CrossRef]
2. Sharifzadeh, M.; Vahedi, H.; Portillo, R.; Khenar, M.; Sheikholeslami, A.; Franquelo, L.G.; Al-Haddad, K. Hybrid SHM-SHE Pulse-Amplitude Modulation for High-Power Four-Leg Inverter. *IEEE Trans. Ind. Electron.* **2016**, *63*, 7234–7242. [CrossRef]
3. Turnbull, F.G. Selected Harmonic Reduction in Static D-C—A-C Inverters. *IEEE Trans. Commun. Electron.* **1964**, *83*, 374–378. [CrossRef]

4. Patel, H.S.; Hoft, R.G. Generalized Techniques of Harmonic Elimination and Voltage Control in Thyristor Inverters: Part I—Harmonic Elimination. *IEEE Trans. Ind. Appl.* **1973**, *IA-9*, 310–317. [CrossRef]

5. Patel, H.S.; Hoft, R.G. Generalized techniques of harmonic elimination and voltage control in thyristor inverters: Part II–voltage control techniques. *Technology* **1974**, *IA-10*, 666–673. [CrossRef]

6. Sharifzadeh, M.; Vahedi, H.; Portillo, R.; Franquelo, L.G.; Al-Haddad, K. Selective Harmonic Mitigation Based Self-Elimination of Triplen Harmonics for Single-Phase Five-Level Inverters. *IEEE Trans. Power Electron.* **2018**, *34*, 86–96. [CrossRef]

7. Chiasson, J.N.; Member, S.; Tolbert, L.M.; Mckenzie, K.J.; Member, S.; Du, Z. Elimination Problem. *IEEE Trans. Power Electron.* **2004**, *19*, 491–499. [CrossRef]

8. Yang, K.; Zhang, Q.; Yuan, R.; Yu, W.; Wang, J. Harmonic elimination for multilevel converter with Groebner bases and symmetric polynomials. In Proceedings of the 2015 IEEE Energy Conversion Congress and Exposition (ECCE), Montreal, QC, Canada, 20–24 September 2015; Volume 31, pp. 689–694. [CrossRef]

9. Kundu, S.; Burman, A.D.; Giri, S.K.; Mukherjee, S.; Banerjee, S. Comparative study between different optimisation techniques for finding precise switching angle for SHE-PWM of three-phase seven-level cascaded H-bridge inverter. *IET Power Electron.* **2018**, *11*, 600–609. [CrossRef]

10. Ray, R.N.; Chatterjee, D.; Goswami, S.K. An application of PSO technique for harmonic elimination in a PWM inverter. *Appl. Soft Comput.* **2009**, *9*, 1315–1320. [CrossRef]

11. Steczek, M.; Chudzik, P.; Szelag, A. Combination of SHE and SHM—PWM techniques for VSI DC-link current harmonics control in railway applications. *IEEE Trans. Ind. Electron.* **2017**. [CrossRef]

12. Etesami, M.H.; Vilathgamuwa, D.M.; Ghasemi, N.; Jovanovic, D.P. Enhanced Metaheuristic Methods for Selective Harmonic Elimination Technique. *IEEE Trans. Ind. Inform.* **2018**, *14*, 5210–5220. [CrossRef]

13. Dahidah, M.S.A.; Agelidis, V.G. Selective harmonic elimination PWM control for cascaded multilevel voltage source converters: A generalized formula. *IEEE Trans. Power Electron.* **2008**, *23*, 1620–1630. [CrossRef]

14. Sundareswaran, K.; Jayant, K.; Shanavas, T.N. Inverter harmonic elimination through a colony of continuously exploring ants. *IEEE Trans. Ind. Electron.* **2007**, *54*, 2558–2565. [CrossRef]

15. Kavousi, A.; Vahidi, B.; Salehi, R.; Bakhshizadeh, M.K.; Farokhnia, N.; Fathi, S.H. Application of the bee algorithm for selective harmonic elimination strategy in multilevel inverters. *IEEE Trans. Power Electron.* **2012**, *27*, 1689–1696. [CrossRef]

16. Sudhakar Babu, T.; Priya, K.; Maheswaran, D.; Sathish Kumar, K.; Rajasekar, N. Selective voltage harmonic elimination in PWM inverter using bacterial foraging algorithm. *Swarm Evol. Comput.* **2015**, *20*, 74–81. [CrossRef]

17. Gnana Sundari, M.; Rajaram, M.; Balaraman, S. Application of improved firefly algorithm for programmed PWM in multilevel inverter with adjustable DC sources. *Appl. Soft Comput.* **2016**, *41*, 169–179. [CrossRef]

18. Lou, H.; Mao, C.; Wang, D.; Lu, J.; Wang, L. Fundamental modulation strategy with selective harmonic elimination for multilevel inverters. *IET Power Electron.* **2014**, *7*, 2173–2181. [CrossRef]

19. Kar, P.K.; Priyadarshi, A.; Karanki, S.B. Selective harmonics elimination using whale optimisation algorithm for a single-phase-modified source switched multilevel inverter. *IET Power Electron.* **2019**, *12*, 1952–1963. [CrossRef]

20. Saremi, S.; Mirjalili, S.; Lewis, A. Grasshopper Optimisation Algorithm: Theory and application. *Adv. Eng. Softw.* **2017**, *105*, 30–47. [CrossRef]

21. Neve, A.G.; Kakandikar, G.M.; Kulkarni, O. Application of Grasshopper Optimization Algorithm for Constrained and Unconstrained Test Functions. *Int. J. Swarm Intell. Evol. Comput.* **2017**, *6*. [CrossRef]

22. Mirjalili, S.Z.; Mirjalili, S.; Saremi, S.; Faris, H.; Aljarah, I. Grasshopper optimization algorithm for multi-objective optimization problems. *Appl. Intell.* **2018**, *48*, 805–820. [CrossRef]

23. Ewees, A.A.; Abd Elaziz, M.; Houssein, E.H. Improved grasshopper optimization algorithm using opposition-based learning. *Expert Syst. Appl.* **2018**, *112*, 156–172. [CrossRef]

24. Wu, J.; Wang, H.; Li, N.; Yao, P.; Huang, Y.; Su, Z.; Yu, Y. Distributed trajectory optimization for multiple solar-powered UAVs target tracking in urban environment by Adaptive Grasshopper Optimization Algorithm. *Aerosp. Sci. Technol.* **2017**, *70*, 497–510. [CrossRef]

25. Luo, J.; Chen, H.; Xu, Y.; Huang, H.; Zhao, X. An improved grasshopper optimization algorithm with application to financial stress prediction. *Appl. Math. Model.* **2018**, *64*, 654–668. [CrossRef]

26. Sultana, U.; Khairuddin, A.B.; Sultana, B.; Rasheed, N.; Hussain, S.; Riaz, N. Placement and sizing of multiple distributed generation and battery swapping stations using grasshopper optimizer algorithm. *Energy* **2018**, *165*, 408–421. [CrossRef]

27. Jumani, T.A.; Mustafa, M.W.; Rasid, M.M.; Mirjat, N.H.; Leghari, Z.H.; Salman Saeed, M. Optimal voltage and frequency control of an islanded microgrid using grasshopper optimization algorithm. *Energies* **2018**, *11*, 3191. [CrossRef]

28. El-Fergany, A.A. Electrical characterisation of proton exchange membrane fuel cells stack using grasshopper optimizer. *IET Renew. Power Gener.* **2018**, *12*, 9–17. [CrossRef]

29. Barik, A.K.; Das, D.C. Expeditious frequency control of solar photovoltaic/biogas/biodiesel generator based isolated renewable microgrid using grasshopper optimisation algorithm. *IET Renew. Power Gener.* **2018**, *12*, 1659–1667. [CrossRef]

30. Liu, J.; Wang, A.; Qu, Y.; Wang, W. Coordinated Operation of Multi-Integrated Energy System Based on Linear Weighted Sum and Grasshopper Optimization Algorithm. *IEEE Access* **2018**, *6*, 42186–42195. [CrossRef]

31. Tumuluru, P.; Ravi, B. GOA-based DBN: Grasshopper optimization algorithm-based deep belief neural networks for cancer classification. *Int. J. Appl. Eng. Res.* **2017**, *12*, 14218–14231.

32. Liang, H.; Jia, H.; Xing, Z.; Ma, J.B.; Peng, X. Modified Grasshopper Algorithm-Based Multilevel Thresholding for Color Image Segmentation. *IEEE Access* **2019**, *7*, 11258–11295. [CrossRef]

33. Leon, J.I.; Kouro, S.; Franquelo, L.G.; Rodriguez, J.; Wu, B. The Essential Role and the Continuous Evolution of Modulation Techniques for Voltage-Source Inverters in the Past, Present, and Future Power Electronics. *IEEE Trans. Ind. Electron.* **2016**, *63*, 2688–2701. [CrossRef]

34. Napoles, J.; Leon, J.I.; Portillo, R.; Franquelo, L.G.; Aguirre, M.A. Selective harmonic mitigation technique for high-power converters. *IEEE Trans. Ind. Electron.* **2010**, *57*, 2315–2323. [CrossRef]

35. Youssef, M.Z.; Woronowicz, K.; Aditya, K.; Azeez, N.A.; Williamson, S.S. Design and development of an efficient multilevel DC/AC traction inverter for railway transportation electrification. *IEEE Trans. Power Electron.* **2016**, *31*, 3036–3042. [CrossRef]

36. Wang, Y.; Wen, X.; Guo, X.; Zhao, F.; Cong, W. Vector control of induction motor based on selective harmonic elimination PWM in medium voltage high power propulsion system. In Proceedings of the 2011 International Conference on Electric Information and Control Engineering, Wuhan, China, 15–17 April 2011; pp. 6351–6354. [CrossRef]

37. Zhang, Y.; Zhao, Z.; Zhu, J. A hybrid PWM applied to high-power three-level inverter-fed induction-motor drives. *IEEE Trans. Ind. Electron.* **2011**, *58*, 3409–3420. [CrossRef]

38. Dahidah, M.S.A.; Konstantinou, G.; Agelidis, V.G. A Review of Multilevel Selective Harmonic Elimination PWM: Formulations, Solving Algorithms, Implementation and Applications. *IEEE Trans. Power Electron.* **2015**, *30*, 4091–4106. [CrossRef]

39. Orosz, T.; Rassõlkin, A.; Kallaste, A.; Arsénio, P.; Pánek, D.; Kaska, J.; Karban, P. Robust Design Optimization and Emerging Technologies for Electrical Machines: Challenges and Open Problems. *Appl. Sci.* **2020**, *10*, 6653. [CrossRef]

40. Marler, R.T.; Arora, J.S. Survey of multi-objective optimization methods for engineering. *Struct. Multidiscip. Optim.* **2004**, *26*, 369–395. [CrossRef]

41. Mirjalili, S.; Mirjalili, S.M.; Lewis, A. Grey Wolf Optimizer. *Adv. Eng. Softw.* **2014**, *69*, 46–61. [CrossRef]

Publisher's Note: MDPI stays neutral with regard to jurisdictional claims in published maps and institutional affiliations.

Article

Hybrid Modulation for Modular Voltage Source Inverters with Coupled Reactors

Krzysztof Jakub Szwarc [1], Pawel Szczepankowski [1,*], Janusz Nieznański [1], Cezary Swinarski [1], Alexander Usoltsev [2] and Ryszard Strzelecki [1]

[1] LINTE[2] Laboratory, Faculty of Electrical and Control Engineering, Gdańsk University of Technology, 80-216 Gdańsk, Poland; krzysztof.szwarc@pg.edu.pl (K.J.S.); janusz.nieznanski@pg.edu.pl (J.N.); cezary.swinarski@pg.edu.pl (C.S.); ryszard.strzelecki@pg.edu.pl (R.S.)

[2] Faculty od Control Systems and Robotics, ITMO University, 191002 Saint Petersburg, Russia; uaa@ets.ifmo.ru

* Correspondence: pawel.szczepankowski@pg.edu.pl

Received: 3 August 2020; Accepted: 24 August 2020; Published: 27 August 2020

Abstract: This paper proposes and discusses a concept of a hybrid modulation for the control of modular voltage source inverters with coupled reactors. The use of coupled reactors as the integrating elements leads to significant reduction in the size and weight of the circuit. The proposed modulation combines novel coarsely quantized pulse amplitude modulation (CQ-PAM) and innovative space-vector pulse width modulation (SVPWM). The former enjoys very low transistor switching frequency and low harmonic elimination, while the latter ensures high resolution of amplitude control. The SVPWM is based on the use of barycentric coordinates. The feasibility of the proposed solution is verified by simulations and laboratory tests of a 12-pulse modular voltage source inverters with two-level and three-level component inverters.

Keywords: modular; voltage source inverter (VSI); multipulse; 12-pulse; pulse amplitude modulation (PAM); pulse width modulation (PWM); three-level; coupled reactors

---

1. Introduction

The concept of modular voltage source inverters (VSI) with coupled reactors considered in this paper was first reported in [1] (therein dubbed multipulse voltage source converters (VSC) with coupled reactors). The idea has its roots in the well-known multipulse AC–DC converter topologies widely used in a variety of applications, including adjustable speed drives (ASD), high-voltage direct current transmission (HVDC), aircraft power systems, and renewable energy conversion systems [2–5]. The use of multipulse topologies for the DC–AC power conversion has broad coverage in the literature (see, for example, in [6–11]). The topologies proposed in [8–11] are based on the use of a transformer in an integrating circuit, which significantly increases the size of the device and is an expensive solution. If the application does not require galvanic isolation of the DC side from the AC side, the integrated circuit can be based on coupled reactors, as proposed in [1]—a solution which greatly reduces the size and cost of the circuit.

Solutions presented in [1,7–11] are focused on maximizing the achievable magnitude of output voltage, while the control of lower magnitudes is left out. The possible voltage synthesis using all allowed combinations of switch states has not been seriously addressed so far. In this paper, a control of the output voltage of the modular voltage source inverters proposed in [1,12] is considered. These converters contain standard inverter modules, but they are connected by special coupled reactors. The idea of modular VSI with coupled reactors draws on the properties of multipulse diode rectifiers with similar coupled reactors [13–15] and other converters with integrating magnetic circuits [4,16,17]. Coupled reactors were selected as integrating elements because their rated power is below 20% of that of a transformer with similar integrating properties [5,13,14]. As the inverter modules are connected in parallel, these circuits can be used for high-current systems. Where increased operating voltages are required, multilevel inverters can be used as modules.

To clarify the terminology used in the sequel, note that the idea of "pulses" contained in the term "multipulse" has a simple interpretation in the case of diode or thyristor AC–DC converters: it denotes a section of AC input voltage transferred to the DC output. The "number of pulses", denoted M, is used to mean the number of such sections transferred in one fundamental period of the AC voltage. Although for DC–AC inverters with fully controlled power switches this idea of pulses is much less tangible, the number of pulses can still be used for the sake of discussion—as a constructional parameter of the inverter.

Transistor-based modular VSI with coupled reactors can be controlled in a variety of ways. One of the methods mimics the workings of multipulse rectifiers. This method relies on applying a succession of only M inverter voltage vectors in every fundamental period of the synthesized output voltage. This paper demonstrates that by appropriate selection of all available basic voltage vectors, it is possible to achieve coarsely quantized pulse amplitude modulation (CQ-PAM). This control approach, leading to staircase output voltage waveforms, enjoys very low switching frequency of power transistors (equal to the fundamental frequency of the output voltage). A drawback of this modulation method is low amplitude resolution. A workaround might be application of a controlled source of DC voltage, but this option is complex and costly and thus it is left out in this paper.

High-resolution voltage control can be achieved by means of pulse width modulation (PWM). Unlike CQ-PAM, PWM requires relatively high switching frequency (a multiple of the fundamental frequency of output voltage), which can be particularly disadvantageous in high-speed motor drive applications. For example, a two-pole motor operating at 150,000 rpm would call for 5 kHz fundamental frequency of the output voltage, meaning at least several dozens of kilohertz of the transistor switching frequency for PWM controlled inverter. Such high switching frequencies can cause significant mismatches between the transistor on/off commands and the actual turn-on/turn-off timing. What is more, the switching losses and electromagnetic interference resulting from high switching frequency can be prohibitive [18]. The problem of adequate voltage control increases with the increase of motor speed and/or its power. For instance, motor drives in the 1 kW power range are reported to operate at speeds of 500,000 to 1,000,000 rpm [19,20]. Concerning higher power drives, the authors of [21] report on a 60 kW drive operating at 100,000 rpm, while the authors of [22] describe a 300 kW drive operating at 60,000 rpm.

To address the above problems, this paper proposes a hybrid modulation combining CQ-PAM and PWM. This approach is capable of combining advantages of PWM (virtually unlimited resolution of voltage control) and those of CQ-PAM (radically reduced switching frequency). The CQ-PAM is intended for use at steady state, while the PWM ensures smooth passage through the transients. A somewhat similar idea of hybrid modulation was proposed in [23,24], but it relies on a combination of two different PWM methods: space-vector pulse width modulation (SVPWM) is used for smooth transients, while selective harmonic elimination PWM (SHE-PWM) is applied for low switching frequency operation at steady state. The CQ-PAM proposed here is even more effective in the reduction of switching frequency than SHE-PWM and—unlike the latter—can easily be computed in real-time. Both CQ-PAM and PWM can rely on selecting and applying appropriate voltage vectors. At steady state, a succession of only M different vectors per fundamental period is used for M-pulse inverters, with all vectors being applied for time intervals of the same length. At each interval, the voltage vector closest to the reference vector is selected. Concerning the PWM, vector selection and duty cycle computations are carried out using the barycentric coordinates [25]. This approach speeds up the calculations and allows easy inclusion of the DC link voltage fluctuations in the algorithm. The proposed modulation is exposed and discussed using the simplest case of modular VSI, that is, a 12-pulse inverter with two-level component inverter modules. The application of the proposed approach can easily be extended to inverters with higher pulse numbers and/or with multilevel inverter modules [1,12]. Section 2 derives the formula linking the output voltage of modular VSI with coupled reactors with the voltages supplied by component inverter modules and finds the required turns ratio of the coupled reactors. Section 3 presents the proposed modulation method, while Section 4 presents and discusses selected simulation results pertaining to modular VSI using two-level and three-level component inverters and the CQ-PAM. A simulation of the passage between two different steady states using the proposed SVPWM is also demonstrated. The laboratory test results of the same two topologies are presented in Section 5.

2. The 12-Pulse Modular Voltage Source Inverter with Coupled Reactors

Twelve-pulse topology has been known as early as in the 1980s [14,26] and widely used in industry to date [5,27–29]. However, the first attempts to use this topology (with coupled reactors) in the inverter mode of operation were made only several years ago [1,7]. To the best of the authors' knowledge, there have been no other reports on the use of multipulse topologies for DC–AC conversion. This section analyses the output voltage of the considered modular VSI as a result of vector summation of the component inverter voltages and determines the required turns ratio of the coupled reactors. A schematic diagram of the 12-pulse modular VSI with coupled reactors is shown in Figure 1, while Figure 2 establishes vectorial notation of voltages used in the following analysis.

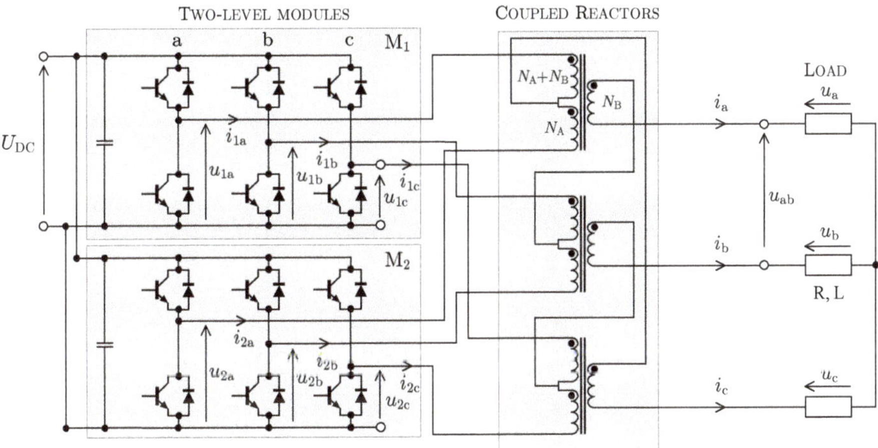

Figure 1. Twelve-pulse modular voltage source inverters (VSI) with coupled reactors.

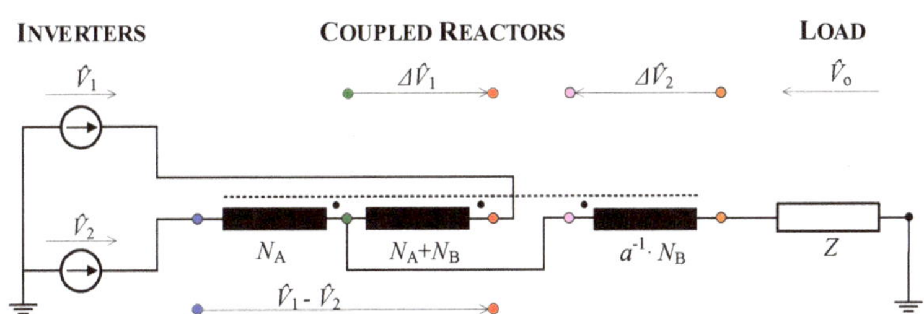

Figure 2. Vectorial equivalent circuit of the 12-pulse modular VSI.

According to the diagram in Figure 2, the output voltage vector of the analyzed modular VSI is given by

$$\hat{V}_o = \hat{V}_1 - \Delta\hat{V}_1 - \Delta\hat{V}_2 \tag{1}$$

where

$$\Delta\hat{V}_1 = \left(\hat{V}_1 - \hat{V}_2\right) \cdot \frac{N_A + N_B}{2 \cdot N_A + N_B} \tag{2}$$

$$\Delta\hat{V}_2 = a^{-1} \cdot \left(\hat{V}_1 - \hat{V}_2\right) \cdot \frac{N_B}{2 \cdot N_A + N_B} \tag{3}$$

$$\hat{V}_1 = U_M \cdot e^{j\frac{\pi}{3} \cdot ent\left[\frac{3}{\pi}\omega t\right]}, \ \hat{V}_2 = U_M \cdot e^{j\frac{\pi}{3} \cdot ent\left[\frac{3}{\pi}(\omega t - \phi)\right]} \tag{4}$$

$$a = e^{j\frac{2 \cdot \pi}{3}} \tag{5}$$

Thus, on the basis of Equations (1)–(3), the output voltage can be described by

$$\hat{V}_o = \hat{V}_1 - \left(\hat{V}_1 - \hat{V}_2\right) \cdot \frac{N_A + N_B}{2 \cdot N_A + N_B} - a^{-1} \cdot \left(\hat{V}_1 - \hat{V}_2\right) \cdot \frac{N_B}{2 \cdot N_A + N_B} \tag{6}$$

After simple algebra, Equation (6) can be rewritten as

$$\hat{V}_o = \hat{V}_1 \cdot \left(\frac{N_A/N_B - a^{-1}}{2 \cdot N_A/N_B + 1}\right) + \hat{V}_2 \cdot \left(\frac{N_A/N_B - a}{2 \cdot N_A/N_B + 1}\right) \tag{7}$$

It is assumed that the control signals of the inverter modules in the considered 12–pulse system are phase-shifted by $\phi = 30°$, meaning the same phase shift between vectors $\hat{V}_1$ and $\hat{V}_2$. As a result, Equation (7) can be converted to

$$\hat{V}_o = \left(\hat{V}_1 + \hat{V}_2\right) \cdot \left[\frac{(N_A/N_B)}{2 \cdot (N_A/N_B) + 1}\right] - \left(\hat{V}_1 \cdot a^{-1} + \hat{V}_2 \cdot a\right) \cdot \left[\frac{1}{2 \cdot (N_A/N_B) + 1}\right] \tag{8}$$

In order to determine the appropriate turns ratios of the reactor coils, assume that the magnitude of voltage across a coil is proportional to its number of turns. Then, from Figure 2 it immediately follows that

$$\frac{\left|\hat{V}_1 - \hat{V}_2\right|}{2 \cdot N_A + N_B} = \frac{\left|\Delta\hat{V}_1\right|}{N_A + N_B} = \frac{\left|\Delta\hat{V}_2\right|}{N_B} \tag{9}$$

The desirable turns ratio is such that ensures symmetric contribution of the component voltages $\hat{V}_1$ and $\hat{V}_2$ to the output voltage $\hat{V}_o$, as illustrated in Figure 3. From the vector diagram in Figure 4, which is an enlargement of the shaded fragment in Figure 3, the following relationship can be found using Thales's theorem,

$$\frac{\left|\hat{V}_1 - \hat{V}_2\right|}{2} \cdot \tan(\lambda) = \left|\Delta\hat{V}_2\right| \cdot \sin\left(60°\right) \tag{10}$$

where $\lambda = \frac{\phi}{2} = 15°$. Now, using this value of λ and substituting Equation (10) to Equation (9), one arrives at

$$\frac{2 \cdot N_A + N_B}{2} \cdot \tan(15°) = N_B \cdot \sin\left(60°\right) \tag{11}$$

whereupon, using basic trigonometric relationships, an explicit formula for the required turns ratio of the reactor coils is found:

$$\frac{N_A}{N_B} = \frac{\sin\left(60° - 15°\right)}{\sin\left(15°\right)} = 2.732 \tag{12}$$

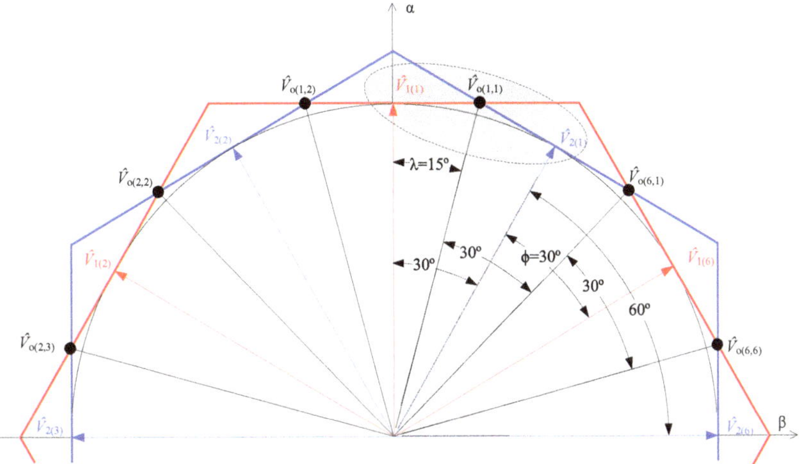

Figure 3. Desirable positions of basic output vectors of component inverters in the 12-pulse modular VSI.

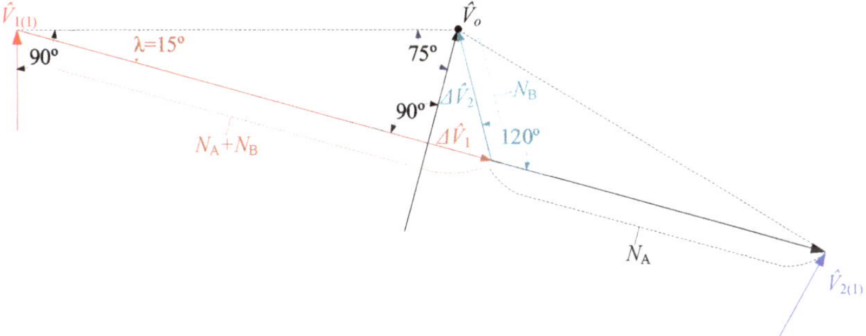

Figure 4. Details of geometric relationship between the output vector $\hat{V}_o$ of the 12-pulse modular VSI and component vectors $\hat{V}_1$ and $\hat{V}_2$.

3. Proposed Hybrid Modulation Method

In one of the proposed operation modes of the considered modular VSI, transistors of inverter modules commute with the fundamental output frequency, which means significant reduction of switching losses and electromagnetic interference (EMI) distortion, and allows high frequencies of output voltages. The problem with this type of control, referred to as the CQ-PAM, is the limited resolution of the output voltage. The proposed solution is the use of PWM as a complementary control method. Both modulation techniques are based on appropriate selection and timing of the available basic vectors, that is, voltage vectors corresponding directly to the on/off states of inverter switches. These vectors can be considered points on a two dimensional $\alpha\beta$ plane. The basic vector diagram for the considered 12-pulse system is shown in Figure 5a. The basic vectors in Figure 5 are obtained in two steps. First, the leg voltages of component inverter modules (marked in Figure 1) are transformed to phase voltages of the modular VSI by

$$\begin{bmatrix} u_a \\ u_b \\ u_c \end{bmatrix} = \begin{bmatrix} u_{1b} & u_{1b} - u_{2b} & u_{1a} - u_{2a} \\ u_{1c} & u_{1c} - u_{2c} & u_{1b} - u_{2b} \\ u_{1a} & u_{1a} - u_{2a} & u_{1c} - u_{2c} \end{bmatrix} \cdot \begin{bmatrix} 1 \\ -k_1 \\ -k_2 \end{bmatrix} \tag{13}$$

where

$$k_1 = \frac{N_A + N_B}{2 \cdot N_A + N_B}, \quad k_2 = \frac{N_B}{2 \cdot N_A + N_B} \tag{14}$$

Then, the $\alpha\beta$ coordinates of the corresponding basic vectors ($\hat{V}_0$) are determined by the Clarke transformation:

$$\begin{bmatrix} v_\alpha \\ v_\beta \end{bmatrix} = \begin{bmatrix} \frac{2}{3} & -\frac{1}{3} & -\frac{1}{3} \\ 0 & \frac{1}{\sqrt{3}} & -\frac{1}{\sqrt{3}} \end{bmatrix} \cdot \begin{bmatrix} u_a \\ u_b \\ u_c \end{bmatrix} \tag{15}$$

In general, the number of different basic vectors for an M-pulse inverter with l-level inverter modules is $l^{\frac{M}{2}}$. Therefore, the considered 12-pulse converter exhibits 64 different basic vectors.

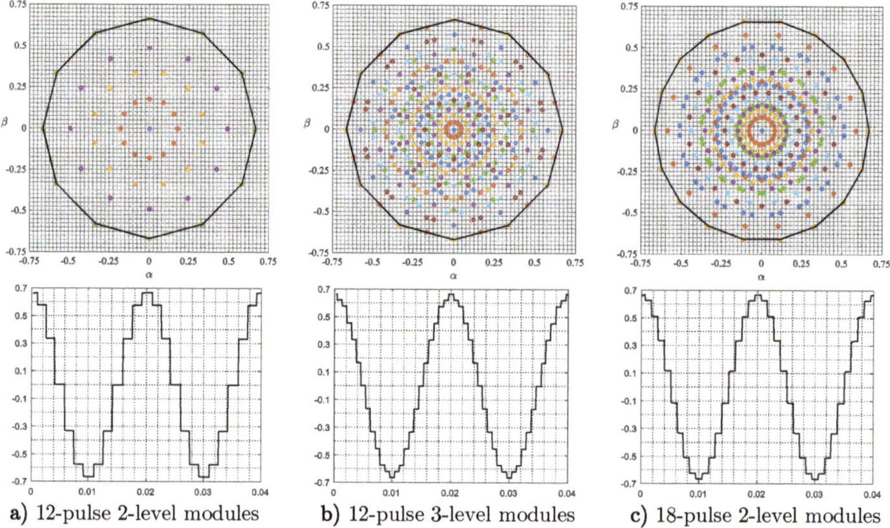

a) 12-pulse 2-level modules   b) 12-pulse 3-level modules   c) 18-pulse 2-level modules

Figure 5. Space-vector diagrams of three modular VSI topologies (top graphs) and example voltage waveforms (bottom graphs) corresponding to the sequences of vectors indicated by connecting lines: (a) 12-pulse 2-level modules; (b) 12-pulse 3-level modules; (c) 18-pulse 2-level modules.

3.1. Coarsely Quantized Pulse Amplitude Modulation (CQ-PAM)

Basically, the CQ-PAM involves applying a succession of only M different basic vectors per fundamental period, with all vectors having the same magnitude and being applied for equal-length time intervals. The selection of basic vectors relies on the criterion of proximity between the reference vector and the basic vectors. As seen in Figure 5a, for $M = 12$ and $l = 2$ only four different non-zero magnitudes are available, meaning very limited resolution. However, as can be noticed in Figure 5b,c, as the number of inverter modules or inverter voltage levels increases, the resolution of the CQ-PAM significantly rises—for an 18-pulse modular VSI it is 16 magnitudes and for a 12-pulse modular VSI with three-level modules it is 24 magnitudes.

For increased M and/or l, some neighboring magnitudes are close to each other, and thus it can be beneficial to use their combinations rather than stick to the same-magnitude principle. This option is illustrated in Figure 5b for the 12-pulse VSI with three-level modules: the sequence of vectors leading to the waveform shown in the lower graph is a succession of 24 (that is, $2 \cdot M$) basic vectors indicated in the upper graph; the magnitude of the shorter vectors is $\cos(\frac{\pi}{M}) = \sim 0.97$ times that of the longer vectors. What is more, some magnitudes can be represented by more than M vectors. For instance, the 12-pulse VSI with three-level modules has seven magnitudes represented by $2 \cdot M = 24$ basic vectors. Again, this fact can be exploited in the CQ-PAM algorithm.

3.2. Space Vector Pulse Width Modulation (SVPWM)

To ensure virtually unlimited amplitude resolution of output voltage control, the PWM can be used whenever necessary or desirable. An economic solution might be the PWM discussed in [17,30] for 12-pulse rectifiers. However, this PWM technique offers only limited voltage control range and does not take advantage of the natural elimination of low harmonics in multipulse systems. Another candidate solution might be selective harmonic elimination PWM (SHE-PWM), which allows to decrease the switching frequency and remove a set of selected harmonics. Such a method was applied for parallel inverters driving a single load separated by line reactors [31]. However, this method requires precomputation of the appropriate PWM patterns, which is hardly possible in real-time [32,33]. This paper proposes an innovative space-vector PWM (SVPWM) based on barycentric coordinates.

In most cases considered in the literature, the output vectors are synthesized using the nearest three vectors (NTV) approach. A similar approach is adopted in this paper. In order to select the nearest basic vectors, the magnitude of reference vector (V_{ref}) is compared with the available magnitudes of inverter basic vectors. The comparisons permit determining two neighboring magnitudes between which the reference vector is located. The closest two basic vectors of either magnitude are then searched for using a simple sorting algorithm and the following vector distance relationship

$$\Delta V = \sqrt{(V_{\text{ref}\alpha} - v_\alpha)^2 + (V_{\text{ref}\beta} - v_\beta)^2} \tag{16}$$

The so obtained closest vectors can be considered vertices of a quadrangle ABCD, as illustrated in Figure 6. The quadrangle can be divided into four triangles. The tasks of the modulation include selecting the most appropriate of the triangles and computing the duty cycles of the three corresponding basic vectors.

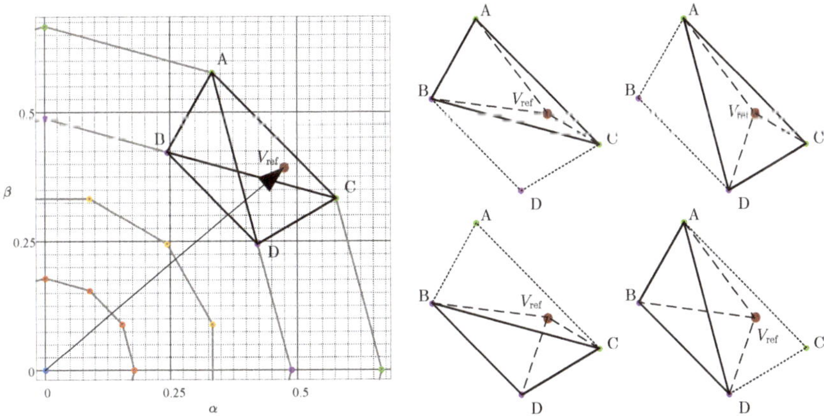

Figure 6. Example reference vector and its representation by means of barycentric coordinates.

Both of the above indicated tasks can be achieved with the aid of barycentric coordinates [25]. Unlike the most popular methods of PWM computations, which are based on trigonometric functions, the use of barycentric coordinates avoids the related inconvenience. Consider the reference vector in Figure 6. It can be considered a point inside triangle ABC or triangle ACD. To fix attention, let us focus on the former triangle. The computation of duty cycles of the corresponding basic vectors is effectively means expressing the position of the reference vector as a linear combination of basic vectors. This is equivalent to expressing the Cartesian coordinates of a point inside a triangle by the barycentric coordinates of that point. Thus, the coordinates of vector (V_{ref}) in Figure 6 can be expressed by the coordinates of points A, B, and C:

$$\begin{bmatrix} V_{ref\alpha} \\ V_{ref\beta} \end{bmatrix} = \begin{bmatrix} A_\alpha & B_\alpha & C_\alpha \\ A_\beta & B_\beta & C_\beta \end{bmatrix} \begin{bmatrix} N_1 \\ N_2 \\ N_3 \end{bmatrix} \tag{17}$$

where N_1, N_2, and N_3 are the barycentric coordinates of V_{ref}, which can be calculated from

$$\begin{bmatrix} N_1 & N_2 & N_3 \end{bmatrix} = \begin{bmatrix} \frac{\triangle_{V_{ref}BC}}{\triangle_{ABC}} & \frac{\triangle_{V_{ref}AC}}{\triangle_{ABC}} & \frac{\triangle_{V_{ref}AB}}{\triangle_{ABC}} \end{bmatrix} \tag{18}$$

with the $\triangle_{ijk}$ symbols representing the areas of the small triangles defined inside triangle ABC by the vertices of the latter and V_{ref}. Stated differently, the barycentric coordinates are equal to the normalized areas of their corresponding small triangles. These areas can be computed direct from the $\alpha\beta$ coordinates of the appropriate basic vectors by

$$\triangle_{ijk} = \frac{1}{2} \cdot \left| \begin{bmatrix} v_{i\alpha} & v_{i\beta} & 1 \\ v_{j\alpha} & v_{j\beta} & 1 \\ v_{k\alpha} & v_{k\beta} & 1 \end{bmatrix} \right| \tag{19}$$

A useful property of the barycentric coordinates is that their sum is equal to unity if they are calculated for a point inside a triangle (e.g., V_{ref} in the triangle ABC or ACD in Figure 6), but is greater if the point lies outside the triangle (e.g., V_{ref} in the triangle ABD or BCD in Figure 6). Thus, a uniform and effective method of finding a triangle or triangles containing a reference vector may be to calculate some candidate barycentric coordinates and then select the smallest (ideally 1).

As can be seen in Figure 6, the reference vector can be located inside two different triangles, and so some additional selection criterion is necessary to make the choice unique. The proposed algorithm selects the triangle for which the distance between the reference vector and the centroid $(C_\triangle)$ is smaller. This criterion was adopted in order to minimize the occurrence of narrow pulses. The coordinates of the centroids are calculated as arithmetic means of the coordinates of vertices:

$$C_\triangle = \begin{bmatrix} C_{\triangle\alpha} \\ C_{\triangle\beta} \end{bmatrix} = \begin{bmatrix} \frac{1}{3}\left(v_{i\alpha} + v_{j\alpha} + v_{k\alpha}\right) \\ \frac{1}{3}\left(v_{i\beta} + v_{j\beta} + v_{k\beta}\right) \end{bmatrix} \tag{20}$$

The distance between V_{ref} and the centroids is determined by Equation (16).

Although the proposed computational approach may seem rather complex for a system with only 64 space vectors, the practical target for the proposed method is modular VSI inverters with higher number of pulses M (notably, 18- and 24-pulse circuits) and using multilevel component inverters [1]. For such systems, the number of basic space vectors increases rapidly with M and the number of inverter levels (see Figure 5), as shown in Table 1.

Table 1. Number of space vectors for M-pulse modular VSI with l-level modules.

$l \backslash M$	12-Pulse	18-Pulse	24-Pulse
2-level	64	512	4096
3-level	729	19,683	531,683
4-level	4096	262,144	16,777,216

It is also important to note that with the increasing number of pulses and levels, the number of available magnitudes of inverter basic vectors also increases rapidly (e.g., 18-pulse and 24-pulse inverters with two-level modules have 16 and 67 output voltage magnitudes respectively and 12-pulse modular VSI with three-level modules has 23 output voltage magnitudes), rendering the CQ-PAM mode a true alternative to the PWM for steady-state operation, with the proposed SVPWM becoming a method for increasing the voltage control resolution, especially during transients.

3.3. Selection of Modulation Method

The selection between CQ-PAM and SVPWM can be based on a variety of criteria, depending on the application (for instance the frequency criterion: SVPWM to be selected for lower fundamental frequencies, e.g., during the start-up, and CQ-PAM for higher fundamental frequencies). For the purpose of this study, the following magnitude proximity criterion is used; CQ-PAM is selected if the reference vector lies inside an annulus A_i defined by two circles—one inscribed in and the other described on a certain M-gon made of vectors of the ith magnitude (V_i); otherwise, SVPWM is chosen.

$$A_i = \left\langle \cos\left(\frac{\pi}{M}\right) \cdot V_i \, , V_i \right\rangle \tag{21}$$

The selection of the modulation method can be based on a decision diagram shown in Figure 7.

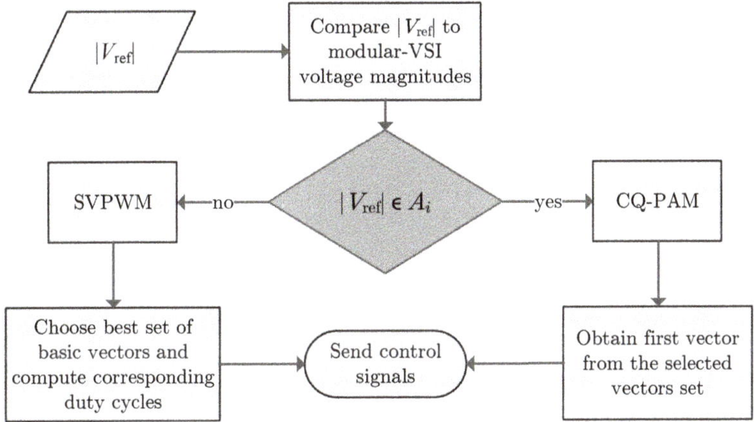

Figure 7. Control algorithm diagram.

4. Simulation Results

The proposed concept of output voltage control for modular VSI with coupled reactors has been verified using the PSIM11 simulation software and the Matlab environment. The simulated topologies included the 12-pulse inverter with two-level modules (Figure 1) and the 12-pulse inverter with three-level modules (Figure 8). The most important circuit and control parameters are given in Table 2.

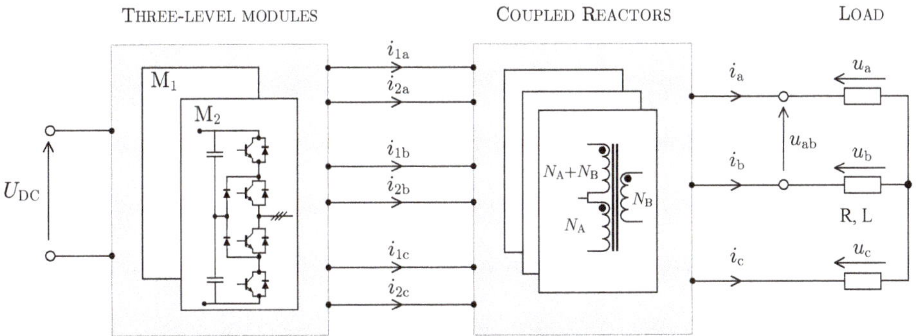

Figure 8. Twelve-pulse modular VSI with coupled reactors and three-level modules.

Table 2. Circuit and modulation parameters used in simulation.

Symbol	Value	Description
U_{DC}	100 V	DC source voltage
R	10 Ω	load resistance
L	0.2 mH	load inductance
f_o	1000 Hz	output fundamental frequency
f_m	30,000 Hz	modulation frequency
N_A	153	number of turns of coils A
N_B	56	number of turns of coils B

The operation of the considered modular VSI with CQ-PAM is illustrated in Figure 9. Characteristic values for output voltages and currents for this control mode are given in Table 3. A comparison of output voltage and current waveforms for two-level and three-level inverter modules is shown in Figure 10. To demonstrate how low the switching frequency is in relation to the output voltage frequency, the above figure also visualizes the leg voltage u_{1a} waveform (scaled down to 25% in Figures 9 and 10). The output voltages shown in Figure 9 are the time waveforms corresponding to 12-pulse space-vector diagrams presented in Figure 5a. Later in this section a passage is illustrated between two steady-state operating points with the aid of SVPWM invoked during the transients (Figure 11).

Table 3. Number of switch commutations per period of the output voltage and the THD (of voltages and currents) for the coarsely quantized pulse amplitude modulation (CQ-PAM) presented in Figure 9.

m_a	Commutations	THD U (%)	THD I (%)
0.179	5	15.58	8.4
0.345	3	15.58	8.4
0.488	3	15.58	8.4
0.67	1	15.58	8.4

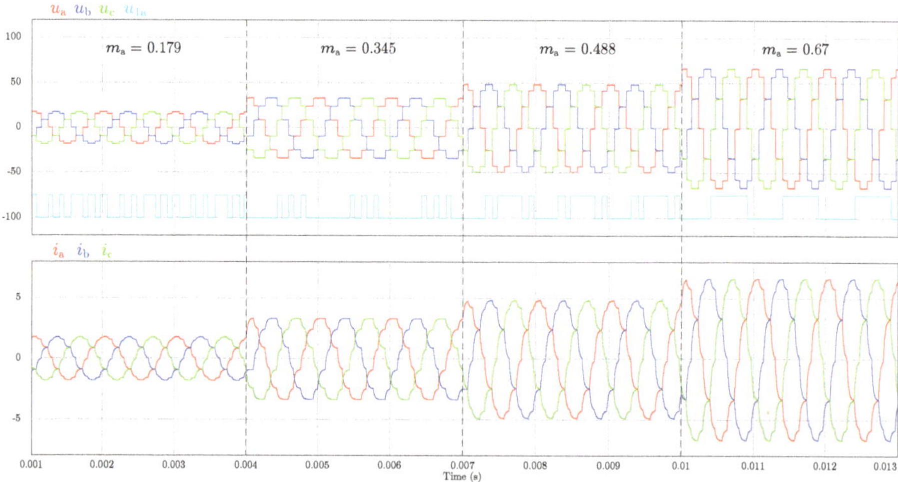

Figure 9. Output voltage and current waveforms of the 12-pulse modular VSI with CQ-PAM control.

Figure 10 illustrates the maximum-magnitude and minimum-magnitude output voltages of 12-pulse modular VSIs with two-level and three-level modules, and the corresponding currents and leg voltages. As can be seen, the use of multilevel modules can increase the number of output voltage steps, so that the total harmonic distortion (THD) of the output voltage decreases (from 15.58% for two-level modules to 10% for three-level modules), and so does the THD of output currents (from 8.4% to 4.4% respectively). Moreover, for multilevel modules the dynamic range and resolution of output voltage magnitudes increases compared to the two-level modules. The minimum voltage for modular VSI with three-level modules is about four times lower than for two-level modules. The number of available non-zero voltage magnitudes also increases—from four in the case of two-level modules to twenty-three for three-level modules.

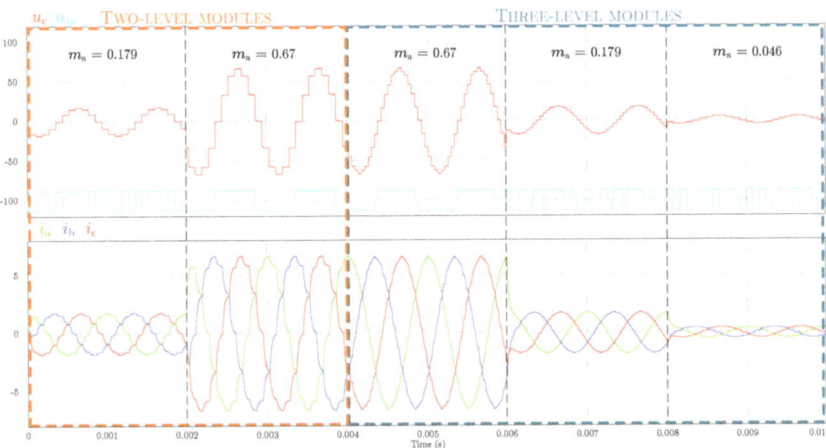

Figure 10. Example voltage and current waveforms of the 12-pulse modular VSI with two-level (**left**) and three-level (**right**) modules.

Figure 11 illustrates combined use of the proposed SVPWM and the CQ-PAM, ensuring smooth passage between different steady states. In this particular example the passage is between steady states at $m_a = 0.345$ and $m_a = 0.488$.

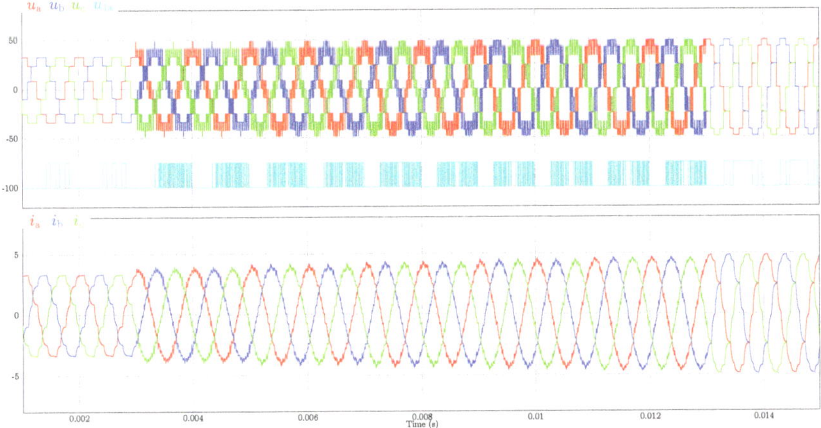

Figure 11. Example output voltage and current waveforms of the 12-pulse modular VSI during the passage between two different steady states.

5. Laboratory Test Results

The laboratory tests have been performed using two prototypes of modular VSI with coupled reactors: one using two-level inverter modules, and the other equipped with three-level modules.

The laboratory setup is presented in Figure 12. The most important circuit and control parameters are the same in both cases and identical to those used in the simulation (cf. Table 2). Measurements were taken by a Tektronix MDO4104B–3 oscilloscope. The control board contained a the two-core Texas Instruments digital signal processor TMS320C6672 and an Intel programmable logic device CYCLONE V. The coupled reactors shown in Figure 12 were designed for 30 kW (10 kW each) and rated frequency of 2.5 kHz. The numbers of turns of reactor coils are given in Table 2.

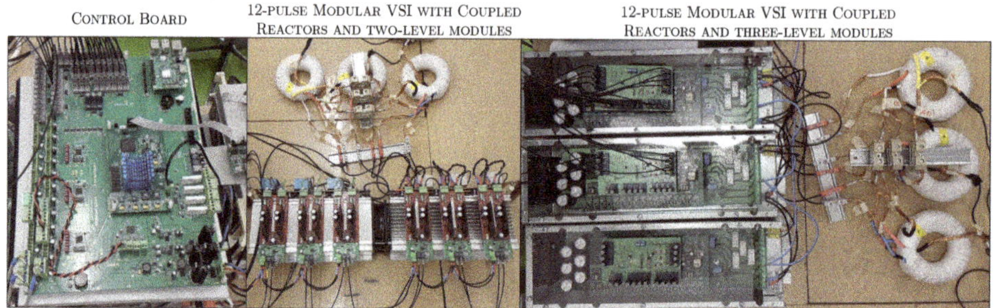

Figure 12. The laboratory set-up.

Figure 13 illustrates output phase voltages and currents, component inverter leg voltages, and the voltage across coil N_B for all output voltage magnitudes and parameters listed in Table 2 (for CQ-PAM controlled VSI with two-level modules). The waveforms correspond to the simulation results shown in Figure 9. The laboratory test results match the results of the simulation, except for the voltage spikes appearing between the steps in the laboratory waveforms. This is a consequence of the use of dead time and the fact that several inverter legs are switched simultaneously (for voltage magnitudes other than the maximum one). However, it can be noted that the amplitudes of these spikes are not significant (smaller than U_{DC}) and do not noticeably affect the current waveforms. The inverter leg voltages indicate the frequency with which the power switches commute (the waveforms confirm the data in Table 3). For the maximum available output voltage the switching frequency is equal to the fundamental output voltage frequency (in this case 1 kHz), and for the smallest CQ-PAM controlled output voltage it is equal to five times the output voltage frequency (5 kHz). One of the most important features distinguishing the coupled reactors from other magnetic integrating elements is their low rated power (related to the power of the overall system). To illustrate this feature, Figure 13 shows the voltage across the N_B coil of the coupling reactor. The voltage is not only significantly lower than U_{DC}, but may even assume values close to zero for some steps (depending on the output voltage magnitude). Therefore, the power transmitted through the reactor is only a fraction of the rated power of the inverter.

The hybrid modulation discussed in Section 3 is based on a combination of the CQ-PAM specific to the considered topologies, and the universal SVPWM—using the proposed computational approach based on barycentric coordinates. Figure 14 illustrates the phase voltages as well as phase-to-phase voltages for both modulation strategies. The voltages in Figure 14 correspond to $m_a = 0.67$ for CQ-PAM, and $m_a = 0.62$ for SVPWM. Figure 15 shows the voltages and currents obtained by SVPWM for two operating points: $m_a = 0.61$ and $m_a = 0.42$. The fundamental frequency was 1000 Hz for the higher output voltage and 600 Hz for the lower.

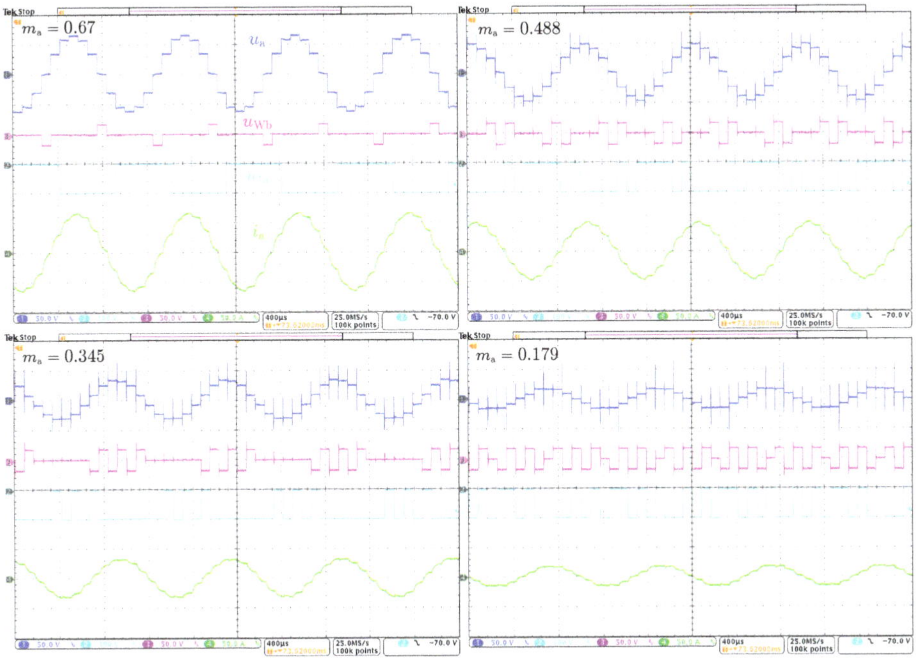

Figure 13. The output phase, leg, and coil N_B voltages and phase currents of the 12-pulse modular VSI with two-level modules for CQ-PAM control.

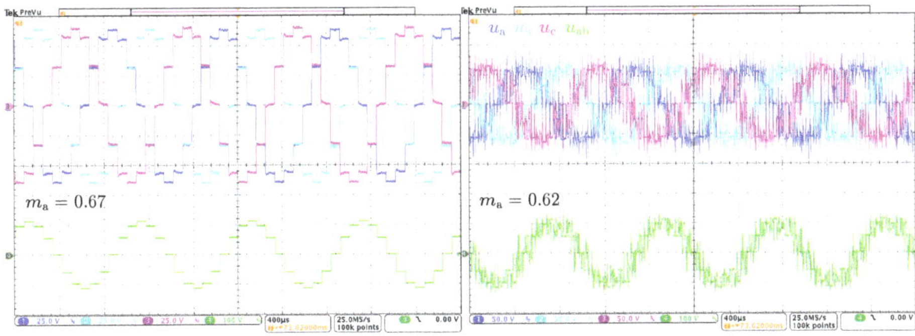

Figure 14. Output voltages for CQ-PAM (**left**) and space-vector pulse width modulation (SVPWM) (**right**) control.

The quality of output voltage and current will improve significantly with the increase in the number of inverter levels, as exemplified in Figures 16 and 17, which provide an initial comparison between the modular VSIs with two-level and three-level inverter modules. What is more, increasing the number of levels significantly increases the amplitude resolution of the CQ-PAM and reduces the voltage stresses of individual transistors. Consequently, the use of multilevel component inverters in the modular VSIs with coupled reactors will contribute to increased attractiveness of the considered topology.

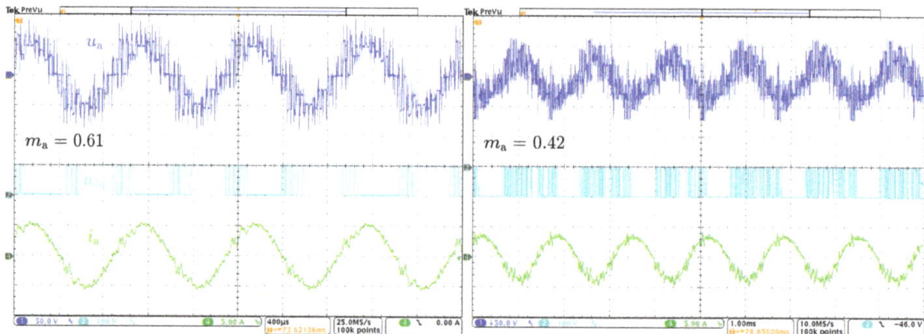

Figure 15. Example output voltages and currents for SVPWM control.

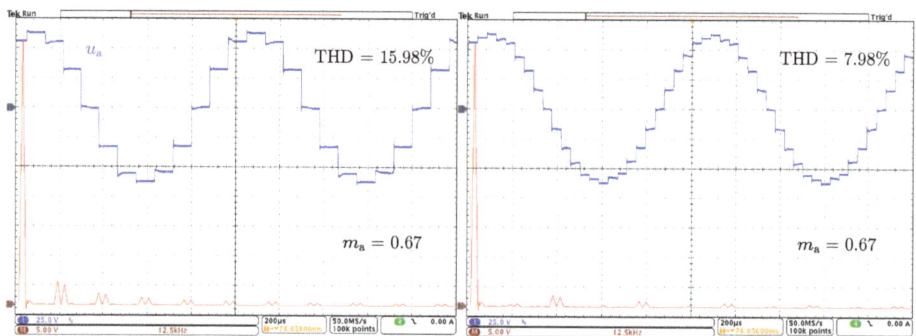

Figure 16. Output voltages and their spectra for the modular VSI with two-level (**left**) and three-level (**right**) modules.

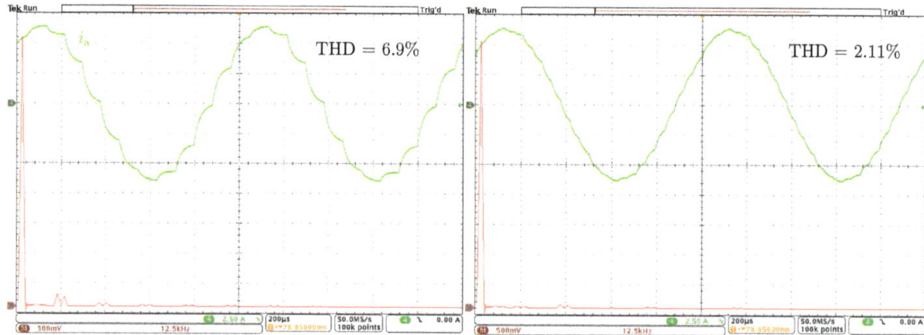

Figure 17. Load currents and their spectra for the modular VSI with two-level (**left**) and three-level (**right**) modules.

The modularity of the considered topology is an important advantage in itself. It allows, inter alia, even distribution of the overall power transferred by the inverter between the reduced-power modules. This feature is illustrated in Figure 18, which shows the phase currents of the component inverters (i_{1a}, i_{2a}) and the resultant load currents (i_a). As can be seen, the component inverter currents have the same amplitude close to half of the load current. The modularity also improves operational reliability of the considered topology, because it can continue working in case of a failure of one component inverter.

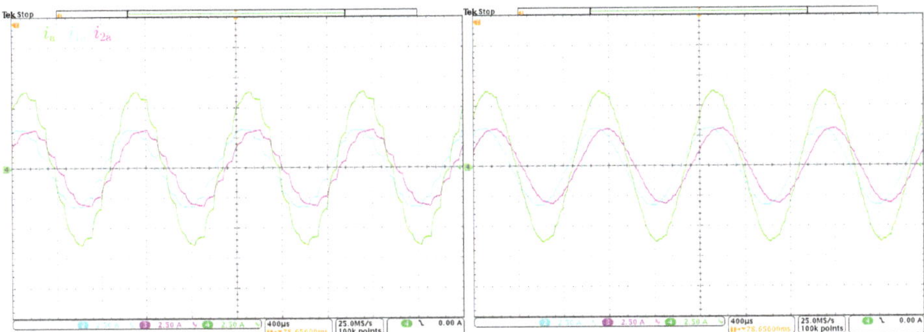

Figure 18. Phase currents of component inverters and the corresponding load currents of the modular VSI with two-level (**left**) and three-level (**right**) component inverters.

6. Conclusions

A hybrid approach to the output voltage control of modular VSI with coupled reactors has been proposed and discussed, including a novel coarsely quantized PAM and space-vector PWM based on the use of barycentric coordinates. Note that the use of these coordinates makes the SVPWM computations feasible and transparent even for such complex space-vector diagrams as those of the considered inverter topologies. The feasibility of the proposed solution has been verified by simulations and laboratory tests of the 12-pulse modular VSI with two-level and three-level component inverters. The use of multilevel inverter modules significantly improves the quality of output voltages and increases the attractiveness of the considered topology and the CQ-PAM, especially for application in high-speed motor drives (research into the latter application is planned for near future). It is also worth noting that the proposed solutions in modulation, although validated for particular inverter topologies, can be equally applicable to other topologies characterized by rich space-vector diagrams, including a variety of modular multipulse inverters.

Author Contributions: Conceptualization, K.J.S. and R.S.; methodology, K.J.S. and R.S.; software, K.J.S. and C.S.; validation, R.S., P.S., and C.S.; formal analysis, P.S. and J.N.; investigation, R.S., K.J.S., and C.S.; resources, K.J.S. and R.S.; data curation, R.S. and P.S.; writing—original draft preparation, K.J.S. and R.S; writing—review and editing, K.J.S., R.S., P.S. J.N., and C.S.; visualization, K.J.S., R.S., and C.S.; supervision, P.S., A.U., and J.N.; funding acquisition, P.S., J.N., and A.U. All authors have read and agreed to the published version of the manuscript.

Funding: This work was supported by the LINTE^2 Laboratory, Gdańsk University of Technology, Grant DS 033784, and the Government of the Russian Federation, Grant 08–08.

References

1. Strzelecki, R.; Sak, T.; Zolov, P.D.; Moradewicz, A.; Grabarek, M. Multi-pulse VSC arrangements with coupled reactors. In Proceedings of the 2016 IEEE 2nd Annual Southern Power Electronics Conference (SPEC), Auckland, New Zealand, 5–8 December 2016; pp. 1–6. [CrossRef]
2. Singh, B.; Bhuvaneswari, G.; Garg, V. Harmonic mitigation using 12-pulse AC-DC converter in vector-controlled induction motor drives. *IEEE Trans. Power Deliv.* **2006**, *21*, 1483–1492. [CrossRef]
3. Choi, S.; Oh, J.; Cho, J. Multi-pulse converters for high voltage and high power applications. In Proceedings of the IPEMC 2000, Third International Power Electronics and Motion Control Conference (IEEE Cat. No.00EX435), Beijing, China, 15–18 August 2000; Volume 3, pp. 1019–1024.
4. Yang, T.; Bozhko, S.; Asher, G. Functional Modeling of Symmetrical Multipulse Autotransformer Rectifier Units for Aerospace Applications. *IEEE Trans. Power Electron.* **2015**, *30*, 4704–4713. [CrossRef]
5. Singh, B.; Gairola, S.; Singh, B.N.; Chandra, A.; Al-Haddad, K. Multipulse AC–DC Converters for Improving Power Quality: A Review. *IEEE Trans. Power Electron.* **2008**, *23*, 260–281. [CrossRef]

6. Walker, L.H. 10-MW GTO converter for battery peaking service. *IEEE Trans. Ind. Appl.* **1990**, *26*, 63–72. [CrossRef]

7. Pietkiewicz, A.; Biner, H. Novel low harmonic three-phase 12-pulse inverter. In Proceedings of the International Symposium on Power Electronics Power Electronics, Electrical Drives, Automation and Motion, Sorrento, Italy, 20–22 June 2012; pp. 837–842. [CrossRef]

8. Singh, B.; Hussain, I. Grid integration of single stage solar PV power generating system using 12-pulse VSC. In Proceedings of the 2014 IEEE 6th India International Conference on Power Electronics (IICPE), Kurukshetra, India, 8–10 December 2014; pp. 1–6.

9. Geethalakshmi, B.; Dananjayan, P. A Combined Multipulse-Multilevel Inverter Suitable For High Power Applications. *Int. J. Comput. Electr. Eng. (IJCEE)* **2010**, *2*, 257–261. [CrossRef]

10. Soto, D.; Green, T.C. A comparison of high-power converter topologies for the implementation of FACTS controllers. *IEEE Trans. Ind. Electron.* **2002**, *49*, 1072–1080. [CrossRef]

11. Chen, Z.; Spooner, E. Voltage source inverters for high-power, variable-voltage DC power sources. *IEE Proc. Gener. Transm. Distrib.* **2001**, *148*, 439–447. [CrossRef]

12. Strzelecki, R.; Sak, T.; Strzelecka, N. Method and System to Converting of Electrical Energy with the Use of Multilevel Inverters Connected in Parallel. Polish Patent PL 228547, 30 April 2018.

13. Oguchi, K.; Maeda, G.; Hoshi, N.; Kubata, T. Coupling rectifier systems with harmonic cancelling reactors. *IEEE Ind. Appl. Mag.* **2001**, *7*, 53–63. [CrossRef]

14. Paice, D.; Society, I.I.A. *Power Electronic Converter Harmonics: Multipulse Methods For Clean Power*; Electrical Engineering/Power and Energy Engineering; IEEE Press: Piscataway, NJ, USA, 1996.

15. Iwaszkiewicz, J.; Mysiak, P. Supply System for Three-Level Inverters Using Multi-Pulse Rectifiers with Coupled Reactors. *Energies* **2019**, *12*. [CrossRef]

16. Shih, D.; Hung, L.; Young, C. The harmonic elimination strategy for a 24-pulse converter with unequal-impedance phase-shift transformers. In Proceedings of the 2013 1st International Future Energy Electronics Conference (IFEEC), Tainan, Taiwan, 3–6 November 2013; pp. 783–788. [CrossRef]

17. Biela, J.; Hassler, D.; Schönberger, J.; Kolar, J.W. Closed-Loop Sinusoidal Input-Current Shaping of 12-Pulse Autotransformer Rectifier Unit With Impressed Output Voltage. *IEEE Trans. Power Electron.* **2011**, *26*, 249–259. [CrossRef]

18. Schwager, L.; Tüysüz, A.; Zwyssig, C.; Kolar, J.W. Modeling and Comparison of Machine and Converter Losses for PWM and PAM in High-Speed Drives. *IEEE Trans. Ind. Appl.* **2014**, *50*, 995–1006. [CrossRef]

19. Ismagilov, F.R.; Uzhegov, N.; Vavilov, V.E.; Bekuzin, V.I.; Ayguzina, V.V. Multidisciplinary Design of Ultra-High-Speed Electrical Machines. *IEEE Trans. Energy Convers.* **2018**, *33*, 1203–1212. [CrossRef]

20. Zwyssig, C.; Duerr, M.; Hassler, D.; Kolar, J.W. An Ultra-High-Speed, 500000 rpm, 1 kW Electrical Drive System. In Proceedings of the 2007 Power Conversion Conference—Nagoya, Nagoya, Japan, 2–5 April 2007; pp. 1577–1583. [CrossRef]

21. Buticchi, G.; Gerada, D.; Alberti, L.; Galea, M.; Wheeler, P.; Bozhko, S.; Peresada, S.; Zhang, H.; Zhang, C.; Gerada, C. Challenges of the Optimization of a High-Speed Induction Machine for Naval Applications. *Energies* **2019**, *12*. [CrossRef]

22. Gerada, D.; Mebarki, A.; Brown, N.L.; Gerada, C.; Cavagnino, A.; Boglietti, A. High-Speed Electrical Machines: Technologies, Trends, and Developments. *IEEE Trans. Ind. Electron.* **2014**, *61*, 2946–2959. [CrossRef]

23. Liang, Y.; Liang, D.; Jia, S.; Chu, S.; He, J. A Full-Speed Range Hybrid PWM Strategy for High-Speed Permanent Magnet Synchronous Machine Considering Mitigation of Current Harmonics. In Proceedings of the 2019 IEEE Energy Conversion Congress and Exposition (ECCE), Baltimore, MD, USA, 29 September–3 November 2019; pp. 1886–1890.

24. Zhang, Y.; Xu, D.; Yan, C.; Zou, S. Hybrid PWM Scheme for the Grid Inverter. *IEEE J. Emerg. Select. Top. Power Electron.* **2015**, *3*, 1151–1159. [CrossRef]

25. Szczepankowski, P.; Nieznański, J. Application of Barycentric Coordinates in Space Vector PWM Computations. *IEEE Access* **2019**, *7*, 91499–91508. [CrossRef]

26. April, G.; Olivier, G. A Novel Type of 12-Pulse Converter. *IEEE Trans. Ind. Appl.* **1985**, *IA-21*, 180–191. [CrossRef]

27. Gong, G.; Drofenik, U.; Kolar, J.W. 12-pulse rectifier for more electric aircraft applications. In Proceedings of the 2003 IEEE International Conference on Industrial Technology, Maribor, Slovenia, 10–12 December 2003; Volume 2, pp. 1096–1101. [CrossRef]
28. Alabduljabbar, A.A.; Gerçek, C.Ö.; ATALIK, T.; Koç, E.; PARLAK, D.; Alsalem, F.S. Design and Implementation of a 12-Pulse TCR-Based SVC for Voltage Regulation. In Proceedings of the 2019 IEEE Jordan International Joint Conference on Electrical Engineering and Information Technology (JEEIT), Amman, Jordan, 9–11 April 2019; pp. 186–193. [CrossRef]
29. Lambert, G.; Costabeber, A.; Wheeler, P.; De Novaes, Y.R. A Unidirectional Insulated AC–DC Converter Based on the Hexverter and Multipulse-Rectifier. *IEEE Trans. Power Electron.* **2020**, *35*, 2363–2371. [CrossRef]
30. Mino, K.; Gong, G.; Kolar, J.W. Novel hybrid 12-pulse boost-type rectifier with controlled output voltage. *IEEE Trans. Aerosp. Electron. Syst.* **2005**, *41*, 1008–1018. [CrossRef]
31. Omara, A.M.; El-Nemr, M.K.; Moschopoulos, G. Parallel operation of voltage source inverters using SHE-PWM based on genetic algorithm. In Proceedings of the 2017 Nineteenth International Middle East Power Systems Conference (MEPCON), Cairo, Egypt, 19–21 December 2017; pp. 1292–1297.
32. Yang, K.; Feng, M.; Wang, Y.; Lan, X.; Wang, J.; Zhu, D.; Yu, W. Real-Time Switching Angle Computation for Selective Harmonic Control. *IEEE Trans. Power Electron.* **2019**, *34*, 8201–8212. [CrossRef]
33. Janabi, A.; Wang, B.; Czarkowski, D. Generalized Chudnovsky Algorithm for Real-Time PWM Selective Harmonic Elimination/Modulation: Two-Level VSI Example. *IEEE Trans. Power Electron.* **2020**, *35*, 5437–5446. [CrossRef]

Article

The Conceptual Research over Low-Switching Modulation Strategy for Matrix Converters with the Coupled Reactors

Pawel Szczepankowski [1,*], Jaroslaw Luszcz [1], Alexander Usoltsev [2], Natalia Strzelecka [3] and Enrique Romero-Cadaval [4]

[1] Department of Power Electronics and Electrical Machines, Faculty of Electrical and Control Engineering, Gdansk University of Technology, 80-233 Gdansk, Poland; jaroslaw.luszcz@pg.edu.pl

[2] Department of Electrical Engineering and Precision Electro-Mechanical Systems, ITMO University, 197101 St. Petersburg, Russia; uaa@ets.ifmo.ru

[3] Department of Ship Automation, Faculty of Electrical Engineering, Gdynia Maritime University, 81-225 Gdynia, Poland; n.strzelecka@we.umg.edu.pl

[4] Department of Electrical Electronic and Control Engineering, University of Extremadura, 06006 Badajoz, Spain; eromero@unex.es

* Correspondence: pawel.szczepankowski@pg.edu.pl

Academic Editors: Tomonobu Senjyu and Ahmed Abu-Siada

Received: 15 December 2020; Accepted: 27 January 2021; Published: 28 January 2021

Abstract: In this paper, different Pulse Width Modulation (PWM) strategies for operating with a low-switching frequency, a topology that combines Conventional Matrix Converters (CMCs), and Coupled Reactors (CRs) are presented and discussed. The principles of the proposed strategies are first discussed by a conceptual analysis and later validated by simulation. The paper shows how the combination of CMCs and CRs could be of special interest for sharing the current among these converters' modules, being possible to scale this solution to be a modular system. Therefore, the use of coupled reactors allows one to implement phase shifters that give the solution the ability to generate a stair-case load voltage with the desired power quality even the matrix converters are operated with a low-switching frequency close to the grid frequency. The papers also address how the volume and weight of the coupled reactors decrease with the growth of the fundamental output frequency, making this solution especially appropriate for high power applications that are supplied at high AC frequencies (for example, in airport terminals, where a supply of 400 Hz is required).

Keywords: matrix converter; pulse width modulation; multipulse voltage converter; nearest voltage modulation; pulse width regulation; low-switching modulation technique; multipulse matrix converter with coupled reactors

---

1. Introduction

The energy conversion in the AC grid realized by power electronics devices always needs efficiency, reliability, and compatibility [1]. The first element is significantly affected by conduction and switching losses of the applied semiconductors. Reliability can lead to the elimination of weak construction elements, which most often fail. Demands for Electromagnetic Intereference (EMI) compatibility have also increased in recent years [2]. Moreover, the ecological aspect of the energy-saving cannot be omitted today [3,4]. The paper proposes the Multipulse Matrix Converter with Coupled Reactors (MMCCR) [5–7] as an alternative solution to the Variable Frequency Drive (VFD) based on the classic AC–DC–AC topology. The use of PWM techniques with a high-switching frequency in these applications can lead to significant dynamic losses but it is required to maintain the good quality of the generated AC voltage.

Reduction of switching frequency of the power electronic devices without a significant decreases of the output voltage quality can be achieved by using multilevel AC–DC–AC topology [8]. This device converts an AC input voltage into the DC voltage and the voltage smoothing bulk electrolytic capacitor in the DC-link circuit is required for this purpose. Capacitor bank stores the energy, which is converted back into AC voltage with the desired frequency using the PWM inverter [9–12]. Another concept of the AC voltage quality improvement uses magnetic elements and most often involves the use of multiphase transformers with an appropriately designed winding configuration [13]. This solution is characterized by low switching frequency but the main disadvantage of such approach related to transformers is cost. Without a doubt, the overall cost is driven by the transformer price, which can be reduced by applying instead the coupled reactors arrangement [14]. The main advantage of using coupled reactors is a significant size reduction, thus for the same load power coupled reactors will be designed for power around five times lower then transformer [15].

The Conventional Matrix Converter (CMC), in comparison with the classic AC–DC–AC frequency converter, has certain individual features that determine the innovation of such a solution [16–19]. This converter is fully bidirectional and operates without a large capacitor, with different frequencies at inputs and outputs of the system. Moreover, a matrix converter allows the power factor regulation [20]. The topology, which contains four matrix converters with coupled reactors, has been already demonstrated in literature [21] but without the inclusion of the amplitude voltage control. The proposed converters arrangement can be used in a turbine generator system equipped with a high-speed synchronous generator with permanent magnets. In such a system, the turbine transmits torque to the shaft of the electric machine directly or through a mechanical transmission, as illustrated in Figure 1.

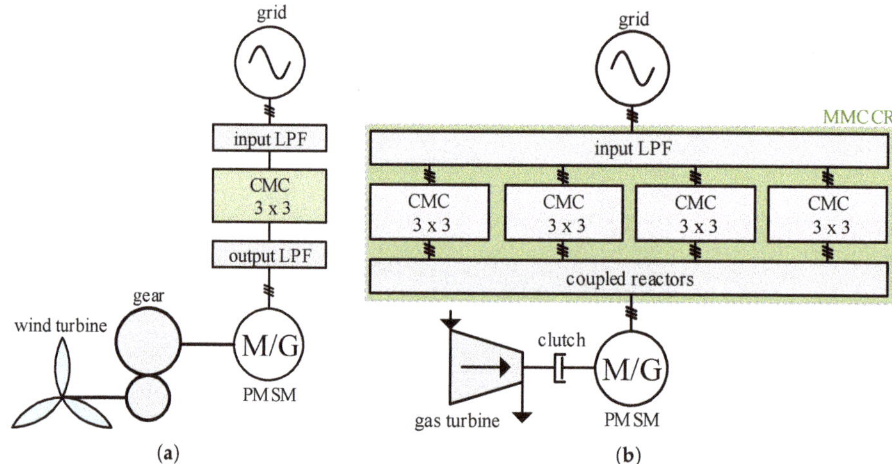

Figure 1. Examples of simplified application diagrams of turbines: (**a**) a wind turbine with a gear, (**b**) a gas turbine with a clutch. Conventional Matrix Converter (CMC) 3 × 3—conventional matrix converter with 3 inputs and 3 outputs, M/G—motor/generator, LPF—low-pass filter.

The synchronous generator produces AC voltage with a frequency dependent on the rotational speed of the turbine [22]. To transfer the obtained energy to the grid, the generated voltage should be converted and synchronized with the three-phase source. This task is performed with the use of power electronic converters, by the matrix converter in particular. Both mechanical transmission using the gear and converter losses determine the efficiency of the system. The smaller the difference between the input and output speed in the mechanical transmission, the smaller the losses [23]. Therefore, a modulation method which decreases the switching number in power converters with a small impact on the quality of voltage and current waveforms are desired. High-speed electrical machines are characterized by greater overall power than machines made for standard speeds, higher frequency of voltage at the terminals, and higher current [24].

Another factor legitimizing the frequency increase is the possibility of eliminating large electrolytic capacitors, which are the fastest deteriorating element in converters. This can be done by using a matrix converter that does not require such energy storage at all. Unfortunately, the use of standard power electronic switches and classic matrix converter topologies is limited by the upper allowable switching frequency. Therefore, the choice of such a solution may be resulting in a significant increase in the complexity of passive filters and a limit the converter dynamic, which is essential in small microgrids with distributed generation elements.

The paper proposes to use four matrix converters operating in parallel due to the modularity of such a solution and the increase in the range of operating currents. Due to new conditions, such as higher voltage frequency, modular nature of the topology and no requirement of galvanic isolation, the use of the coupled reactors circuits is an interesting idea. In addition, the leakage inductance of such reactors can also be used in controlling the power flow between the generator and the grid. The set and arrangement of the base vectors in the alpha–beta plain allow for the implementation of modulation methods with a lower switching frequency compared to Space-Vector Pulse Width Modulation (SVPWM) but with relatively good waveform quality. The switching frequency is equal to the generator frequency in particular. The purpose of conceptual research is shortly presented in the next subsection.

A multiphase transformer for multipulse rectifiers and similar topologies with the coupled reactors are known solutions. However, such an approach is mostly applied to systems that operate with the grid frequency. Considering the price of copper, these solutions are relatively expensive. The dimensions, as well as the price of these components, decrease with increasing nominal frequency. Thus, high-speed electrical machines seem to be the perspective area for the proposed converter topology. The SVPWM modulation allows for linear adjustment of both frequency and amplitude. However, it requires power electronic switches to operate at high frequency. The efficiency of that solution can be improved using the hybrid strategy of modulation. Consider the simplified gas turbine work profile shown in Figure 2.

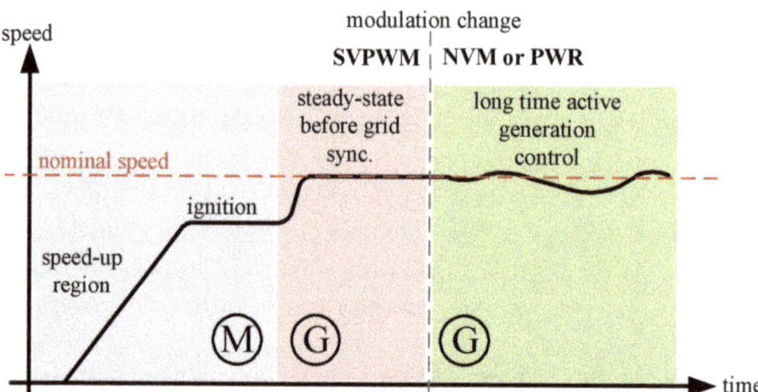

Figure 2. The simplified speed profile of the gas turbine: M—PMSM operates as a motor, G—PMSM operates as a generator, SVPWM—high-frequency modulation method based on the space-vector concept, Nearest Vector Modulation (NVM) and Pulse Width Modulation (PWR)—the low-switching frequency type of PWM modulation.

In the speed-up region, the required energy is supplied from the utility grid through the inverter converter controlled using the SVPWM algorithm. This modulation is used until the synchronisation with the grid. When the speed is stable, the long time active generation control begins, which can be performed using the proposed low-switching modulation approach. The main goal of the authors of the publication was to develop such modulation and preliminary simulation studies.

1.1. The Discrete Projection of Voltage Vectors

The essence of the solution is to achieve a very low operating frequency of power electronic switches. The forming of the output voltage in the proposed group of converters, while maintaining the power switches in the lowest operating frequency, can be formally presented as an effect of discrete projection of the reference vector, as shown in Figure 3.

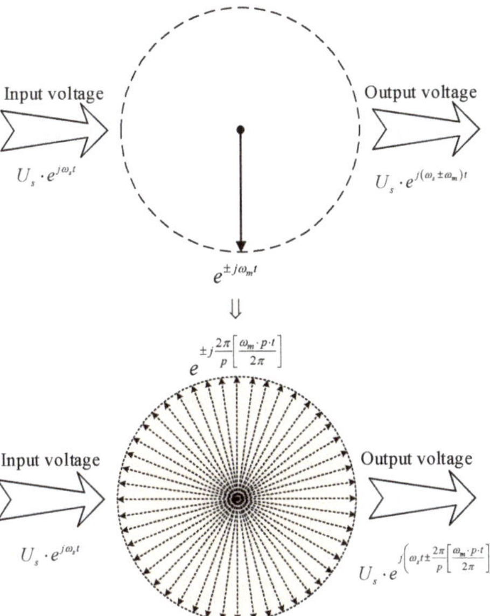

Figure 3. The general discrete reference vector projection concept: U_s—voltage amplitude, ω_s and ω_m pulsations, p a number of discrete voltage vectors [7].

Applying the idea of the projection in solutions composed of conventional 3×3 matrix converters, a multipulse output voltage can be obtained. Figure 4 shows a conventional matrix topology as the 3-pulse system, which does not contain any coupled reactors yet. The need of using the coupled reactors appears in the 6-pulse system. Such a converter is shown in Figure 5.

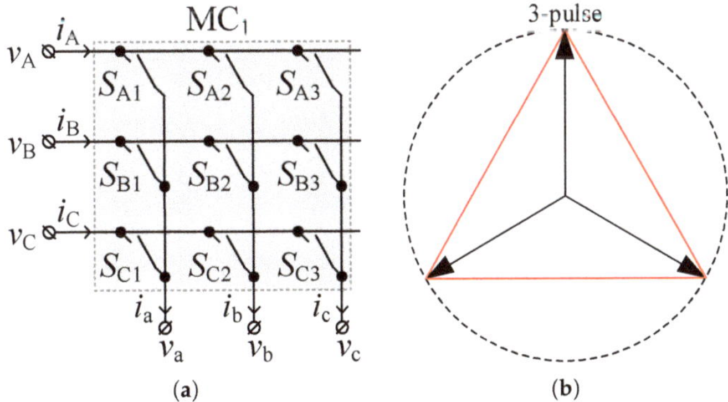

(a) (b)

Figure 4. Conventional matrix converter as a 3-pulse system: (**a**) schematic diagram, (**b**) the voltages discrete projection.

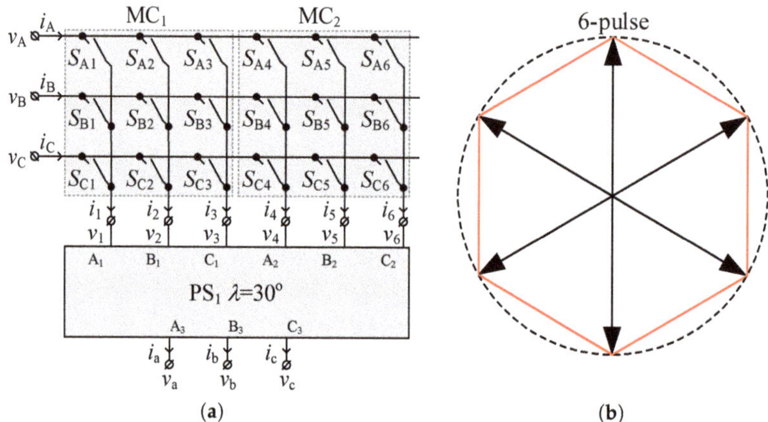

Figure 5. Two conventional matrix converters make the 6-pulse system: (**a**) schematic diagram, (**b**) the voltages discrete projection.

1.2. Coupled Reactors

Both solutions, 6-pulse and 12-pulse, require an appropriately coupled reactors arrangement [25]. The proposed topology, shown in Figure 6a, contains 36 bidirectional switches S_{A1}–S_{C12}, which are elements of four 4 matrix converters CMC$_1$–CMC$_4$ and 3 Phase Shifters (PS) PS$_1$, PS$_2$ and PS$_3$ respectively. The connection diagram of the simulation model of a three-phase type coupled reactor is shown in Figure 7.

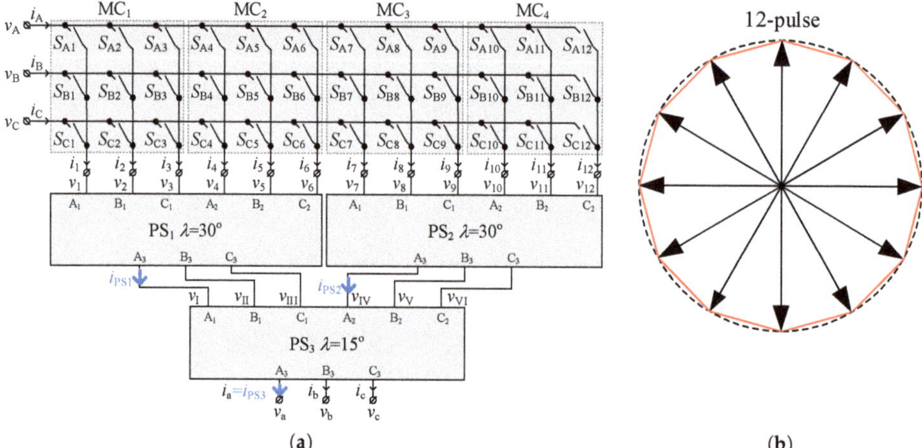

Figure 6. Four conventional matrix converters make the 12-pulse system: (**a**) schematic diagram, (**b**) the voltages discrete projection.

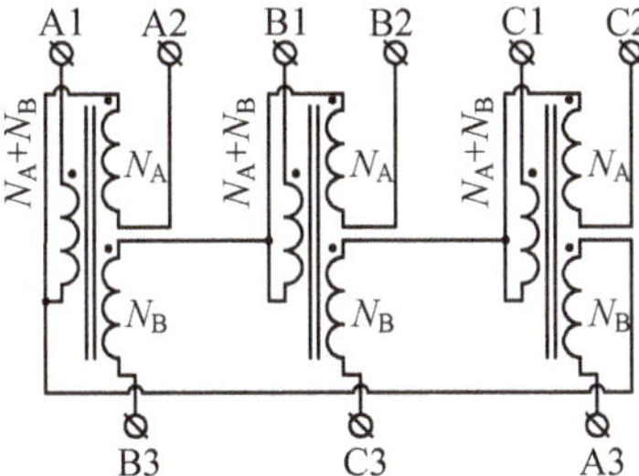

Figure 7. Phase Shifter schematic.

To obtain the desired phase shift angle λ the ratio of turns number N_A to N_B, should meet the following condition

$$n_{AB} = \frac{N_A}{N_B} = \frac{\sin\left(\frac{4\pi}{p_n} - \lambda\right)}{\sin(\lambda)} \tag{1}$$

where

p_n is the number of pulses of a given system and in the discussed example takes 12,
λ is a desired phase shift angle,
N_A and N_B is the ratio of turns number.

An example turns number for two shift angles are presented in Tables 1 and 2. The final number of windings depends on the adopted design parameters, in particular on the power of the system and the reactors' voltage spectrum. This aspect is not covered in this article.

Table 1. Turns number for shift angle equal to 30° (conversion values).

$N_A + N_B$	<100	<300	<1000
N_A	15	56	209
N_B	15	56	209

Table 2. Turns number for shift angle equal to 15° (conversion values).

$N_A + N_B$	<100	<300	<1000
N_A	41	153	571
N_B	15	56	209

The values of the turns number determine not only the shift angle but also certain properties of the presented topology, such as the amount of reactive power circulating between the coupled three-phase reactors and also the value of the maximum amplitudes of the output phase voltage. For simplicity of the rest text, let us assume that magnetic elements in circuits shown in Figure 7 are linear and lossless, and bidirectional power switches are ideal. The further part of the paper concerns the 12-pulse system only. The article is organised as follows. The space of the rotating vectors for 12-pulse MMCCR and the load voltage synthesis basic are presented in Section 2. While the control of output voltage amplitude using the low switching frequency modulation is proposed in the next section. Simulation results are shown and discussed in Section 4.

2. The Space of the Rotating Vectors for 12-Pulse MMCCR

The switch state is 0, if it is switched off, and takes unity if it is switched on. The states of all bidirectional switches can be defined by four switch state matrices $\mathbf{S}_{MC1}$–$\mathbf{S}_{MC4}$ expressed as follows

$$\mathbf{S}_{MC1} = \begin{bmatrix} S_{A1} & S_{B1} & S_{C1} \\ S_{A2} & S_{B2} & S_{C2} \\ S_{A3} & S_{B3} & S_{C3} \end{bmatrix} \tag{2}$$

$$\mathbf{S}_{MC2} = \begin{bmatrix} S_{A4} & S_{B4} & S_{C4} \\ S_{A5} & S_{B5} & S_{C5} \\ S_{A6} & S_{B6} & S_{C6} \end{bmatrix} \tag{3}$$

$$\mathbf{S}_{MC3} = \begin{bmatrix} S_{A7} & S_{B7} & S_{C7} \\ S_{A8} & S_{B8} & S_{C8} \\ S_{A9} & S_{B9} & S_{C9} \end{bmatrix} \tag{4}$$

$$\mathbf{S}_{MC4} = \begin{bmatrix} S_{A10} & S_{B10} & S_{C10} \\ S_{A11} & S_{B11} & S_{C11} \\ S_{A12} & S_{B12} & S_{C12} \end{bmatrix} \tag{5}$$

The modulation techniques presented in this article were developed for a system containing four conventional matrix converters. The proposed approach uses only six switch states among the 27 available. These selected vectors belong to the group of the rotating vectors [16]. Three of them rotate in a clockwise direction (Table 3), while the remaining three are counterclockwise (Table 4). In general, two collections of switch states can be proposed for the modulation. The first collection contains all combinations, which utilise the counterclockwise rotating voltage vectors. These states are described in Appendix A. The second collection, presented in Appendix A, comprises the clockwise rotating vectors. In summary, the total number of switch state combinations for MMCCR is equal to 3^4.

Table 3. Allowed $\mathbf{S}_{MCp}$ switch states.

	States $\mathbf{S}_{MCp}$	
$\mathbf{S}_{Ip}$	$\mathbf{S}_{IIp}$	$\mathbf{S}_{IIIp}$
$\begin{bmatrix} 0 & 0 & 1 \\ 0 & 1 & 0 \\ 1 & 0 & 0 \end{bmatrix}$	$\begin{bmatrix} 0 & 1 & 0 \\ 1 & 0 & 0 \\ 0 & 0 & 1 \end{bmatrix}$	$\begin{bmatrix} 1 & 0 & 0 \\ 0 & 0 & 1 \\ 0 & 1 & 0 \end{bmatrix}$

Table 4. Allowed $\mathbf{S}_{MCn}$ switch states.

	States $\mathbf{S}_{MCn}$	
$\mathbf{S}_{In}$	$\mathbf{S}_{IIn}$	$\mathbf{S}_{IIIn}$
$\begin{bmatrix} 0 & 0 & 1 \\ 1 & 0 & 0 \\ 0 & 1 & 0 \end{bmatrix}$	$\begin{bmatrix} 0 & 1 & 0 \\ 0 & 0 & 1 \\ 1 & 0 & 0 \end{bmatrix}$	$\begin{bmatrix} 1 & 0 & 0 \\ 0 & 1 & 0 \\ 0 & 0 & 1 \end{bmatrix}$

The output voltage values for each converter depicted in Figure 7 can be calculated as follows

$$\begin{bmatrix} v_1 & v_2 & v_3 \end{bmatrix}^T = \mathbf{S}_{MC1} \cdot \begin{bmatrix} v_A & v_B & v_C \end{bmatrix}^T \tag{6}$$

$$\begin{bmatrix} v_4 & v_5 & v_6 \end{bmatrix}^T = \mathbf{S}_{MC2} \cdot \begin{bmatrix} v_A & v_B & v_C \end{bmatrix}^T \tag{7}$$

$$\begin{bmatrix} v_7 & v_8 & v_9 \end{bmatrix}^{\mathrm{T}} = \mathbf{S}_{\mathrm{MC3}} \cdot \begin{bmatrix} v_{\mathrm{A}} & v_{\mathrm{B}} & v_{\mathrm{C}} \end{bmatrix}^{\mathrm{T}} \tag{8}$$

$$\begin{bmatrix} v_{10} & v_{11} & v_{12} \end{bmatrix}^{\mathrm{T}} = \mathbf{S}_{\mathrm{MC4}} \cdot \begin{bmatrix} v_{\mathrm{A}} & v_{\mathrm{B}} & v_{\mathrm{C}} \end{bmatrix}^{\mathrm{T}} \tag{9}$$

According to the shown topology scheme, the matrix converters' output is connected with the PS PS_1 and PS_2 respectively. A simple circuit analysis, shown in Figure 7, leads to the following voltage synthesis matrices

$$\begin{bmatrix} v_{\mathrm{I}} \\ v_{\mathrm{II}} \\ v_{\mathrm{III}} \end{bmatrix} = \begin{bmatrix} v_2 & v_2 - v_5 & v_1 - v_4 \\ v_3 & v_3 - v_6 & v_2 - v_5 \\ v_1 & v_1 - v_4 & v_3 - v_6 \end{bmatrix} \begin{bmatrix} 1 \\ -k_1 \\ -k_2 \end{bmatrix} \tag{10}$$

$$\begin{bmatrix} v_{\mathrm{IV}} \\ v_{\mathrm{V}} \\ v_{\mathrm{VI}} \end{bmatrix} = \begin{bmatrix} v_8 & v_8 - v_{11} & v_7 - v_{10} \\ v_9 & v_9 - v_{12} & v_8 - v_{11} \\ v_7 & v_7 - v_{10} & v_9 - v_{12} \end{bmatrix} \begin{bmatrix} 1 \\ -k_1 \\ -k_2 \end{bmatrix} \tag{11}$$

where the values of coefficients k_1 and k_2 can be calculated using the number of turns listed in Table 1

$$k_1 = \frac{N_{\mathrm{A}(30°)} + N_{\mathrm{B}(30°)}}{2N_{\mathrm{A}(30°)} + N_{\mathrm{B}(30°)}} \quad k_2 = \frac{N_{\mathrm{B}(30°)}}{2N_{\mathrm{A}(30°)} + N_{\mathrm{B}(30°)}} \tag{12}$$

The final output voltage synthesis is realised by the third PS PS_3, according to the equation

$$\begin{bmatrix} v_{\mathrm{a}} \\ v_{\mathrm{b}} \\ v_{\mathrm{c}} \end{bmatrix} = \begin{bmatrix} v_{\mathrm{II}} & v_{\mathrm{II}} - v_{\mathrm{V}} & v_{\mathrm{I}} - v_{\mathrm{IV}} \\ v_{\mathrm{III}} & v_{\mathrm{III}} - v_{\mathrm{VI}} & v_{\mathrm{II}} - v_{\mathrm{V}} \\ v_{\mathrm{I}} & v_{\mathrm{I}} - v_{\mathrm{IV}} & v_{\mathrm{III}} - v_{\mathrm{VI}} \end{bmatrix} \begin{bmatrix} 1 \\ -k_3 \\ -k_4 \end{bmatrix} \tag{13}$$

where, as before, the coefficients k_3 and k_4 values can be calculated using the number of turns listed in Table 2 resulting in the following formula

$$k_3 = \frac{N_{\mathrm{A}(15°)} + N_{\mathrm{B}(15°)}}{2N_{\mathrm{A}(15°)} + N_{\mathrm{B}(15°)}} \quad k_4 = \frac{N_{\mathrm{B}(15°)}}{2N_{\mathrm{A}(15°)} + N_{\mathrm{B}(15°)}} \tag{14}$$

Assuming that the MMCCR is supplied by a three-phase balanced AC voltage source and the load is symmetrical, the space vector α–β coordinates can be obtained using the simplified amplitude invariant Clarke transform

$$\begin{aligned} v_\alpha &= v_{\mathrm{a}} \\ v_\beta &= \frac{v_{\mathrm{b}} - v_{\mathrm{c}}}{\sqrt{3}} \end{aligned} \tag{15}$$

The space-vector diagram for $\mathbf{S}_{\mathrm{MC}p}$ switch state types, and $n_{\mathrm{AB1}} = n_{\mathrm{AB2}} = 209/209$, $n_{\mathrm{AB3}} = 571/209$, is shown in Figure 8. While the space-vector diagram for $\mathbf{S}_{\mathrm{MC}n}$ switch state types is illustrated in Figure 9. The obtained space-vector diagrams are not stationary and rotate with the frequency of the grid voltage.

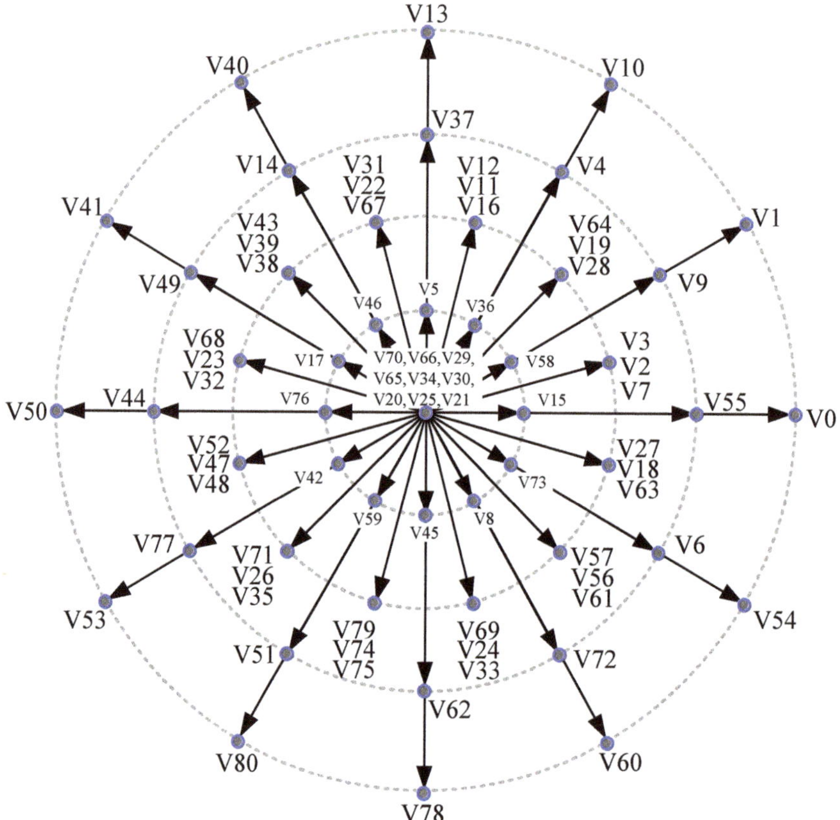

Figure 8. The space-vector diagram for $\mathbf{S}_{\mathrm{MC}p}$ switch states.

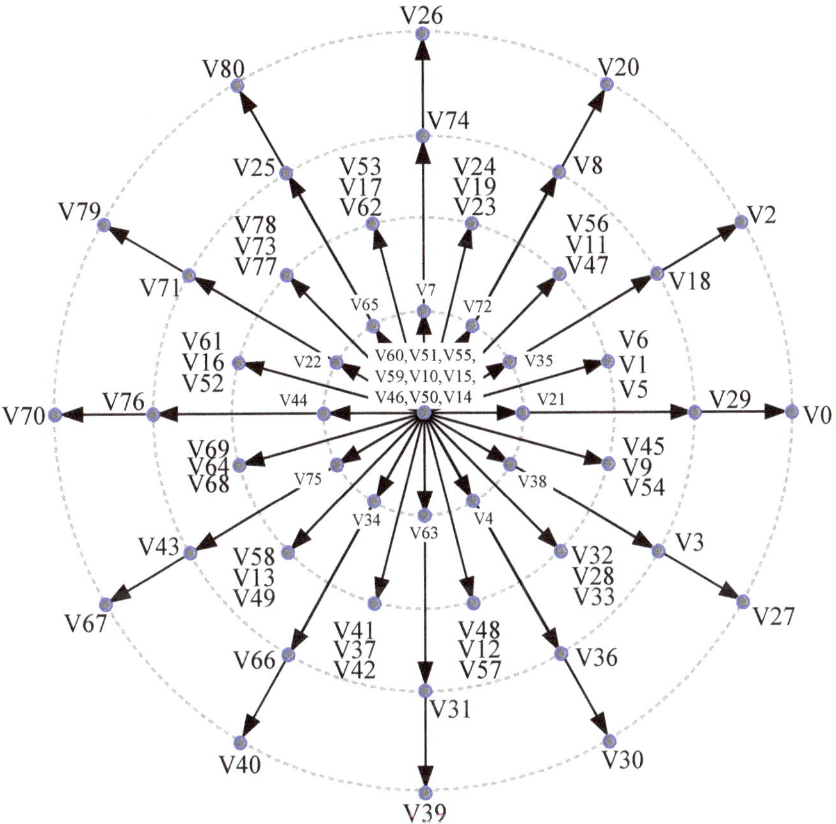

Figure 9. The space-vector diagram for $\mathbf{S}_{\mathrm{MC}n}$ switch states.

3. The Control of Output Voltage Amplitude Using the Low Switching Frequency Modulation

This section proposes two low switching frequency modulation methods—the Pulse Width Regulation PWR and Nearest Vector Modulation NVM. Both methods are successfully verified using PSIM simulation software. The load voltage is represented by one vector, which is rotating on the stationary, orthogonal α–β reference frame. This frame is built from the basic voltage vectors correspond to the switches states. The total number of switch state is equal to the M^N, where M is a number of allowed state combination across the one commutation cell, while N is a number of cells. Thus theoretically the total number of switch states is 3^{12} for the proposed converter topology. Due to the concept of magnetically coupled using the coupled reactors presented in [25] only the rotating vectors are allowed. The stationary and the zero vectors from the conventional matrix converter space-vector frame are not suitable for that kind of the reactors' circuit. There are six rotating vectors allowed for each of the conventional matrix converters. Two collections can be distinguished—the first covers vectors, that rotate clockwise—while the second set represents vectors rotate counterclockwise. Thus, if the number of the conventional matrix converter is equal to 4, as shown in Figure 6 the total number of selected vectors is reduced to 3^4. The load voltage can be synthesised using a variety of switch states sequence. However, due to the requirement to minimise losses during modulation, only the nearest vectors are applied within the modulation period.

3.1. Pulse Width Regulation

The synthesis of voltages in the MMCCR with modular structure can be directed to the high efficiency of the power conversion system also oriented to decreasing the number of switching. However, the operating frequency should be chosen so as to preserve the multipulse nature of the generated phase voltages with the assumption of the amplitude output voltage regulation. Considering the vectors arrangement shown in Figure 10, the vectors V0–V54 belong to the outer circle with radius is equal to 1.0 p.u., while the vectors V55–V6 are located on an inner circle with radius of 0.732 p.u.. The shown reference voltage vector V_{REF} lies exactly between vectors V9 and V1 area, which refers to a sector 2 in propose modulation scheme. To obtain the symmetry effect of the states sequence in the time window corresponding to the modulation period T_{PWR}, the three–step switch states sequence has been proposed. The first three sequences are shown in Table 5. The PWR duty cycles, for each sector, can be calculated as follows

$$d_H = \frac{V_{REF} - V_L}{V_H - V_L}$$
$$d_L = 1 - d_H \tag{16}$$

where $V_H = 1.0$, $V_L = 0.732$, and d_H corresponds to the longer vector. An example waveform of the output voltage is shown in Figure 11.

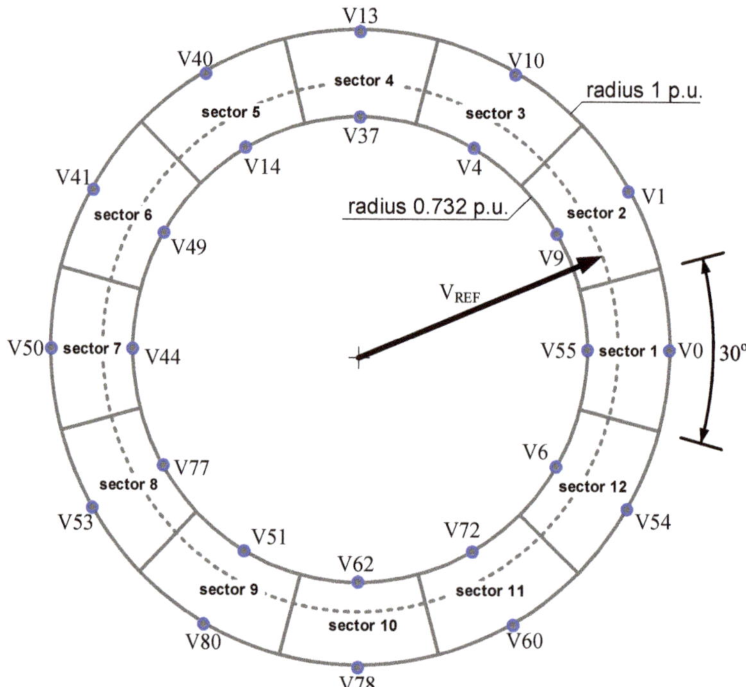

Figure 10. The PWR modulation workspace for $\mathbf{S}_{MCp}$ switch states.

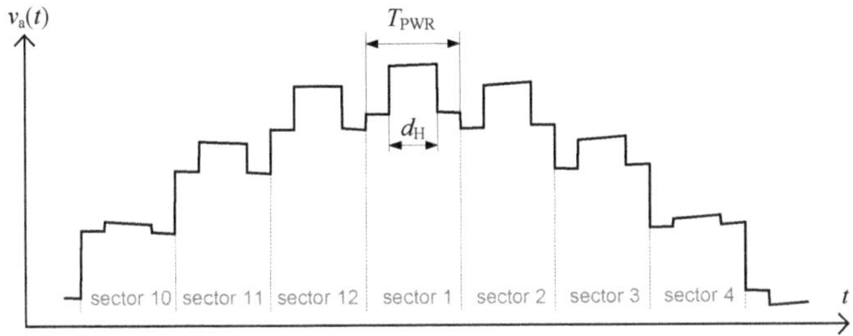

Figure 11. Fragment of phase voltage waveform for the PWR modulation using the S_{MCn} states.

The coordinates of the reference output voltage V_{REF} can be defined according to the selected type of switch state matrices. For selected S_{MCn} the $\alpha\beta$ coordinates of the V_{REF}, can be calculated

$$
\begin{aligned}
v_\alpha^* &= q \cdot \cos\left(\omega_i \cdot (k_f + 2) \cdot t\right) \\
v_\beta^* &= q \cdot \sin\left(\omega_i \cdot (k_f + 2) \cdot t\right)
\end{aligned}
\tag{17}
$$

while for S_{MCp} we obtain

$$
\begin{aligned}
v_\alpha^* &= q \cdot \cos\left(\omega_i \cdot (k_f - 2) \cdot t + \tfrac{2\pi}{3}\right) \\
v_\beta^* &= q \cdot \sin\left(\omega_i \cdot (k_f - 2) \cdot t + \tfrac{2\pi}{3}\right)
\end{aligned}
\tag{18}
$$

where

v_α^*—reference α coordinate,
v_β^*—reference β coordinate,
ω_i—input voltage pulsation,
$k_f = \omega_o / \omega_i$—pulsation ratio, and
$q = V_{REF}/V_i$ is the voltage transfer ratio.

Table 5. Example the switch states sequences in the first 3 sectors.

Type	n_{ABn}	Sector 1	Sector 2	Sector 3
S_{MCp}	571/209	V55–V0–V55	V9–V1–V9	V4–V10–V4
S_{MCn}	571/209	V29–V0–V29	V18–V2–V18	V8–V20–V8

3.2. The Nearest Vector Modulation

The second approach is based on the minimum distance selection criterion, in which a distance is measured between the reference vector and basic vectors belonging to the collection of rotating voltage vectors. Another control concept, also leading to the minimisation of the number of switching, depends on choosing the one space-vector within the modulation period, which is geometrically closest to the reference vector with coordinates v_α^* and v_β^*. Theoretically, the number of required distances depends on the regulation range of the output phase voltage and takes the maximum value equal to 49 (48 active vectors and one zero vector). Since the following minimum value is needed

$$
r_k = \sqrt{\left(v_\alpha^* - v_{\alpha k}\right)^2 + \left(v_\beta^* - v_{\beta k}\right)^2}
\tag{19}
$$

in the decision process, finally another expression can be chosen

$$
g_k = \left(v_\alpha^* - v_{\alpha k}\right) \cdot \left(v_\alpha^* - v_{\alpha k}\right) + \left(v_\beta^* - v_{\beta k}\right) \cdot \left(v_\beta^* - v_{\beta k}\right)
\tag{20}
$$

where

> k—the switch state index, form 1 to 49,
> g_k—the proposed distance function,
> $v_{\alpha k}$—vector α coordinate for k switch state,
> $v_{\beta k}$—vector β coordinate for k switch state.

The new proposed expression contains no square root operation. Further optimisation of the algorithm may consist of taking into account the redundancy of certain switch states. The redundancy, in this case, means that the same voltage vector can be assigned to at least two switch states. In this paper, this aspect is omitted in further discussion. The algorithm calculations should be performed quite frequently to maintain best output voltage quality as possible. In order to counteract the appearance of undesirable effects associated with the so-called a short impulse, attention should be paid to the commutation capabilities of the used bi-directional switch. The commutation process should be appropriately performed according to the dynamic properties of the switch included in the datasheet. This issue is more important in medium and high power application characterised by the large currents values. Overvoltage across the switch can damage it. The proposed solution is able to be discussed for the nominal frequencies (50 Hz or 60 Hz) but is promising for higher frequencies used in gas turbines, ultrasounds, and high-speed drives. Such applications require modern transistors based on GaN or SiC technology. The time of the commutation process is much less compared to the IGBT switch counterparts. That feature has critical importance for NVM modulation, wherein the short impulse problem can appear. This aspect in the algorithm can be limited to the adoption of a specific frequency of the algorithm's call or using the hysteresis mechanisms in the sequence selection process. The quality of the proposed modulation method can be assessed using the error rate defined as follows

$$\varepsilon_{\text{RMS}} = \sqrt{\frac{1}{T_{50\text{Hz}}} \cdot \int_0^{T_{50\text{Hz}}} \left(\frac{v_a^* - v_a}{v_a^*}\right)^2 dt} \tag{21}$$

The results are presented in Figure 12. The shown waveform contains four local optimum q values, which correspond to the optimal multipulse operation.

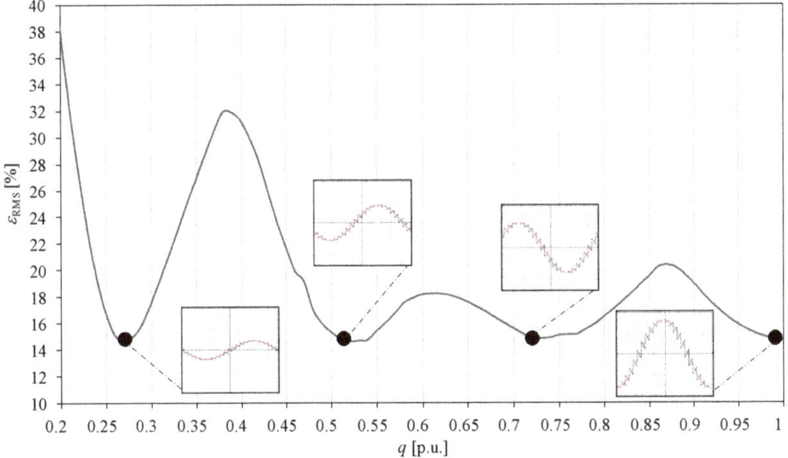

Figure 12. An error rate, defined by (21), for proposed NVM method: $n_{AB1} = n_{AB2} = 209/209$, $n_{AB3} = 571/209$, $k_f = 1/12$.

4. Research

Initial simulation studies focused on selected modulation techniques. The converter MMCCR topology is new; therefore, it was necessary to develop modulation algorithms from scratch. Three types of modulation have been developed:

- SVPWM—this type of modulation belongs to the high-frequency switching technique, is characterised by good accuracy. The nearest three vector modulation has been adopted and developed for a new space-vector diagram presented in the previous section. The formulas of the PWM duty cycles can be found in the paper [26].
- PWR—this type of modulation allows for an amplitude control only in the limited range of modulation index with the number of switching is less than in the SVPWM. However, the precise amplitude control is worst in comparing with the SVPWM.
- NVM—modulation with the minimum number of switching operations, dedicated to working in a steady state. The field of application may be inverters operating with a constant output frequency.

The results are presented in two subsections. Voltage and current waveforms for inverter operation with RL load are presented first. The next subsection is dedicated to a potential application in a system with a high-speed PMSM generator, as mentioned in the introduction.

4.1. PWR and NVM Modulations in the MMCCR Inverter Mode of Operation

Simulation has been performed using DLL block as a DSP platform emulator works with 50 μs step. The markings used in demonstrated figures are: v_a—load phase voltage, v_A—grid phase voltage, $i_{a,b,c}$—load currents, and $i_{A,B,C}$—input currents. Two load models parameters sets are applied, which are listed in Table 6. All waveforms are presented in p.u. unit.

Table 6. 20 kVA/400 V load models simulation parameters.

	R [Ω]	L [H]	Z [Ω]	PF
model-1	7.75	0.0008	8.0	0.97
model-2	4	0.0028	8.0	0.5

Simulation tests for PWR modulation were carried for the model-1, which was characterised by a power factor of 0.97. The proposed PWR modulation has been verified for both types of switches state sequence. Results for $\mathbf{S}_{MCn}$ switches state sequence are shown in Figures 13 and 14, while the waveforms obtained for opposite rotation, $\mathbf{S}_{MCp}$ switch state sequences, are illustrated in Figures 15 and 16.

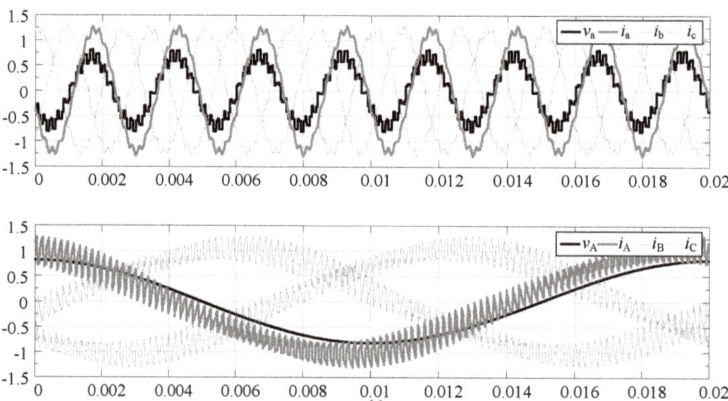

Figure 13. The waveforms of the phase voltages and currents at the output and input of the PWR controlled MMCCR for the $\mathbf{S}_{MCn}$ switch state sequences: load model-1, $n_{AB1} = n_{AB2} = 209/209$, $n_{AB3} = 571/209$, $q = 0.866$, $k_f = 8$.

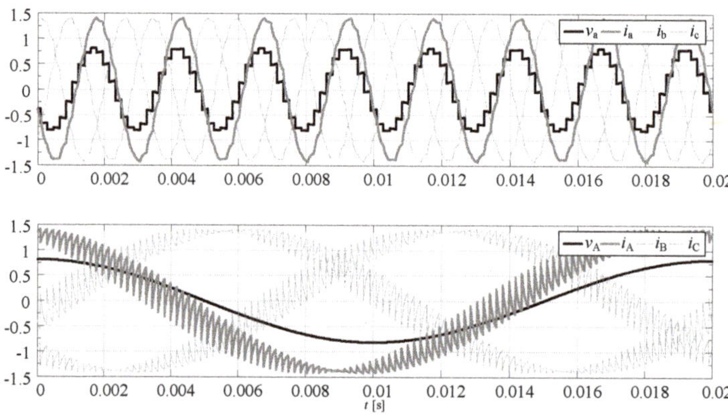

Figure 14. The waveforms of the phase voltages and currents at the output and input of the PWR controlled MMCCR for the $\mathbf{S}_{MCn}$ switch state sequences: load model-1, $q = 1.0$, $k_f = 8$.

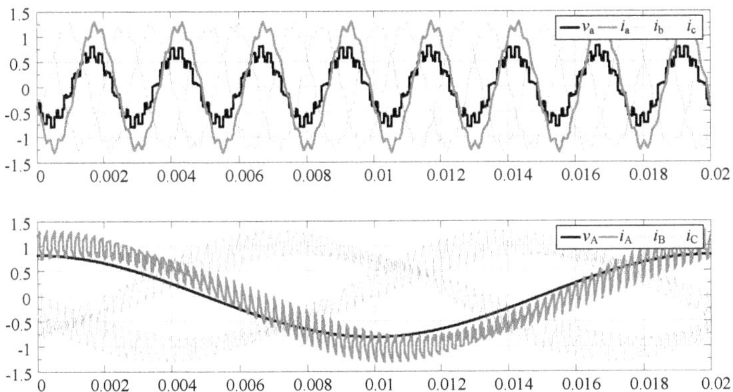

Figure 15. The waveforms of the phase voltages and currents at the output and input of the PWR controlled MMCCR for the $\mathbf{S}_{MCp}$ switch state sequences: load model-1, $q = 0.866$, $k_f = 8$.

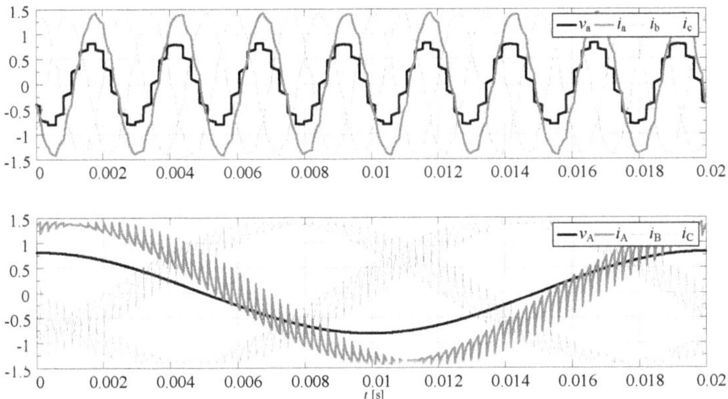

Figure 16. The waveforms of the phase voltages and currents at the output and input of the PWR controlled MMCCR for the $\mathbf{S}_{MCp}$ switch state sequences: load model-1, $q = 1.0$, $k_f = 8$.

According to the proposed concept, for q equal to unity, the output voltage is formed based only on the vectors located on the outer circle. In this case, the system works in 12-pulse mode with the optimal number of switching. Note that in all simulations, a standard input low pass filter has not been used, to demonstrate the unfiltered input current shape.

Comparison of load current Total Harmonic Distortion (THD) and average switching frequency for PWR modulation using the $\mathbf{S}_{MCn}$ and $\mathbf{S}_{MCp}$ switch state sequences for $k_f = 8$ is characterised in Table 7.

Table 7. Comparison of load current THD and average switching frequency for PWR modulation using the $\mathbf{S}_{\text{MC}n}$ switch state sequences and $k_f = 8$.

	$\mathbf{S}_{\text{MC}n}$ Switch State Type						
q	0.732	0.75	0.8	0.85	0.9	0.95	1
THD i_a [%]	3.15	3.35	4.8	5.4	5.2	4.3	3.1
f_{switch} [Hz]	1350	3125	3125	3125	3125	3125	450
	$\mathbf{S}_{\text{MC}p}$ Switch State Type						
q	0.732	0.75	0.8	0.85	0.9	0.95	1
THD i_a [%]	4.15	4.6	6.2	7.0	6.5	5.5	4.1
f_{switch} [Hz]	1050	2450	2450	2450	2450	2450	350

As can be deduced low switching frequency MMCCR operation is achieved for voltage q equal to the radius shown in Figure 10. An important feature of the matrix topology is the ability to the regulation of input angle, defined as displacement between the input current and grid voltage. The range of an angle regulation relies on the load parameters. In a simple way, the desired input angle can be selected using the $\mathbf{S}_{\text{MC}n}$ or $\mathbf{S}_{\text{MC}p}$ switch state sequences. The result of the rapid change of the switch state sequence for model-2 parameters is shown in Figure 17, in which an input angle ϕ is approximately equal to $\pi/3$. Note that the proposed PWR method is elaborated only for a limited range.

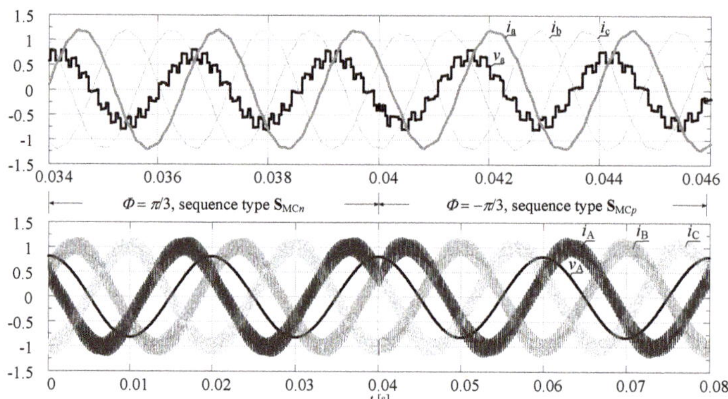

Figure 17. The zoom of the rapid change of switch sequence type for PWR modulation and load model-2, $q = 0.866$, and $k_f = 8$.

The proposed PWR modulation is not an accurate voltage synthesis method. Formula (16) calculates the proportion only in an estimated way, assuming that the voltages are constant in the modulation period.

A simulation has been performed in which the Root Mean Square (RMS) value and THD of the load current have been calculated. Figure 18 shows maximal magnetic flux values referred to the case of the 50 Hz output waveform generation in function of voltage transfer ratio q and output frequency ratio k_f. On the basis of the collected data, it was confirmed that the magnetic flux value is linearly dependent on the output frequency. On the other hand, a smaller flux allows for a smaller design of the coupled reactor circuit. Therefore, the proposed solution will work better in applications with a higher fundamental frequency.

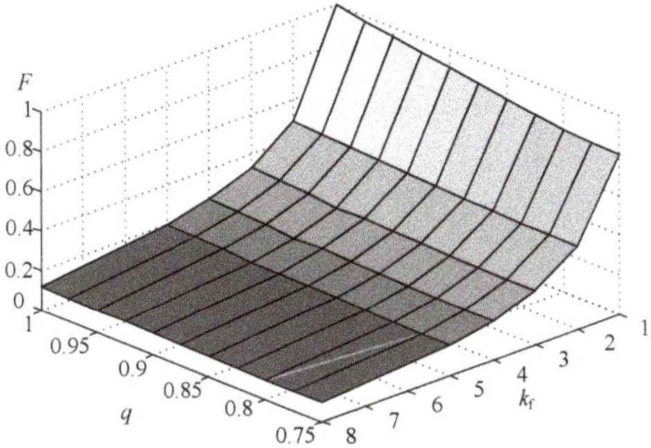

Figure 18. The maximal relative magnetic flux F in PS$_3$ coupled reactor in function of voltage transfer ratio q and output frequency ratio k_f.

Figure 19 presents an example of current and voltage waveforms in the case of 400 Hz voltage generation using the NVM method. The presented fragment of the waveforms refers to the case when the output voltage amplitude is gradually increased. The results were obtained for the load model-1.

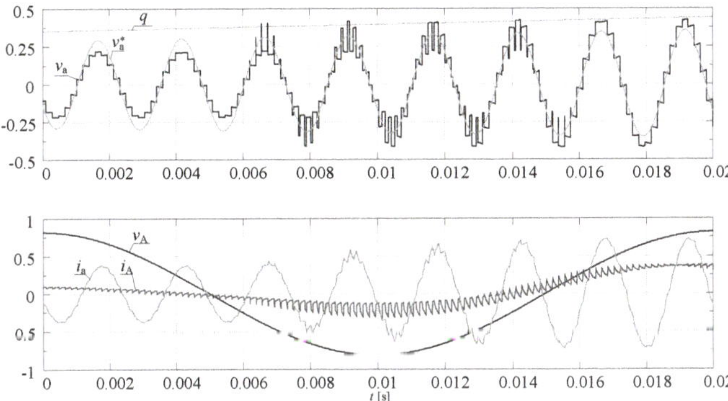

Figure 19. Example output voltage waveforms, for NVM modulation, during gradually increasing the voltage gain q from 0.27 to 0.43 for $k_f = 8$.

Figure 20a shows a comparison of THD modulation methods described in the article. Obviously, curves in the figure are not similar because they represent different approaches to low-frequency modulation. To formulate conclusions one should also consider the results presented in Figure 20b, in which the RMS load currents have been compared to the reference current.

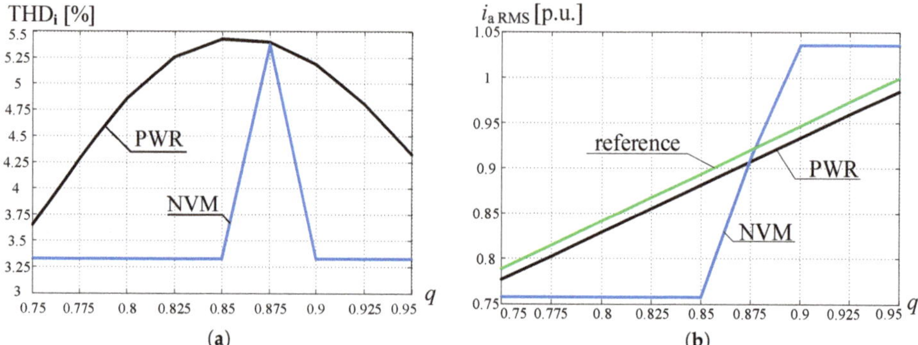

Figure 20. The comparison of the PWR and NVM modulation methods during the inverter operation: (**a**) the load current total harmonic distortion in the applicable range of modulation index, (**b**) the load current RMS value comparison of reference and proposed modulation.

4.2. A Potential Application in a System with a High-Speed PMSM Generator

The proposed MMCCR converter can be applied in a system with a high-speed PMSM generator. Figure 21 shows a simplified circuit for simulation tests, while the proposed control diagram is presented in Figure 22. The purpose of the simulation tests was as follows:

- comparing the converters CMC and MMCCR controlled by SVPWM modulation,
- investigating the possibility of the system control for active NVM modulation with the low-frequency switching within a steady-state operation (shown in Figure 2).

Precise amplitude control is required during generator start-up and synchronisation with the grid. The SVPWM method based on the three nearest vectors selection has been used. At this stage, simulation studies focused on the verification of the possibility of controlling the active component of the generator current, estimating the THD for selected waveforms, and determining the number of switching of power electronic switches.

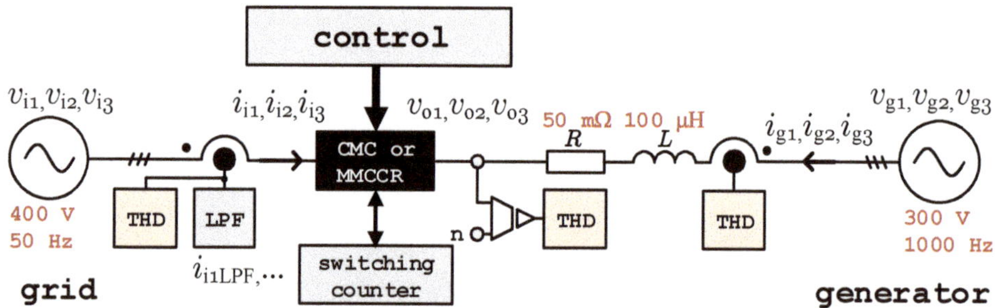

Figure 21. A potential application scheme with a high-speed PMSM generator: CMC—conventional matrix converter, MMCCR—the proposed converter, THD—calculation of the total harmonic distortion block, LPF—the low–pass filter, and "n"—the star point for the phase voltage measurement.

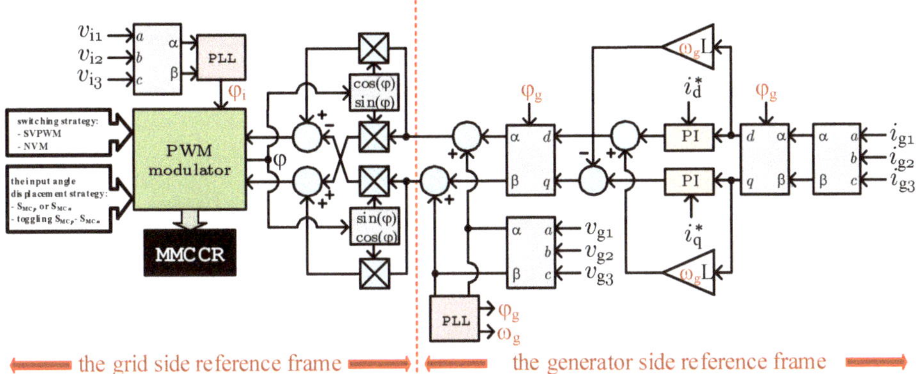

Figure 22. The proposed control scheme: PLL—phase-locked loop, MMCCR—the proposed converter, PI—standard proportional integral controller, φ_i—grid's voltage angle, φ—the synchronisations angle, φ_g—generator's voltage angle, i_d^*—an active reference current for generator, i_q—a reactive reference current for generator, $\omega_g = \omega_m$—for simplicity, mechanical pulsation is equal to electrical pulsation.

Selected results of the comparison CMC and MMCCR converters during SVPWM modulation are shown in Table 8. All presented measurements were carried out for the modulation period equal to 10 µs and the set active current of the generator 200 A. For the MMCCR converter, compared to the CMC topology, the THD factor of the input current and the output voltage is over 2.5 times lower, while maintaining a constant value of the modulation period. In addition, the quality of the generator current is better.

Table 8. Selected results of the comparison CMC and MMCCR converters during SVPWM modulation.

Converter Type	THD(i_{i1})	THD(v_{o1})	THD(i_{g1})	The Relative Number of Switching Operations
CMC	74%	70%	1.6%	100%
MMCCR	28%	26%	0.85%	≃50%

Example waveforms of currents and voltages are shown in Figure 23. The simulation performed for the step change of the reference active generator current, from 0 to 200 A (approximately 75 kW power) in $t = 0.07$ s. As can be seen, the shape of the input current and the voltage generated by the converters have differed. The MMCCR generates a quasi multilevel voltage and the input current shape is near to the sinusoidal waveform. The input current spectrums for both converters are presented in Figure 24.

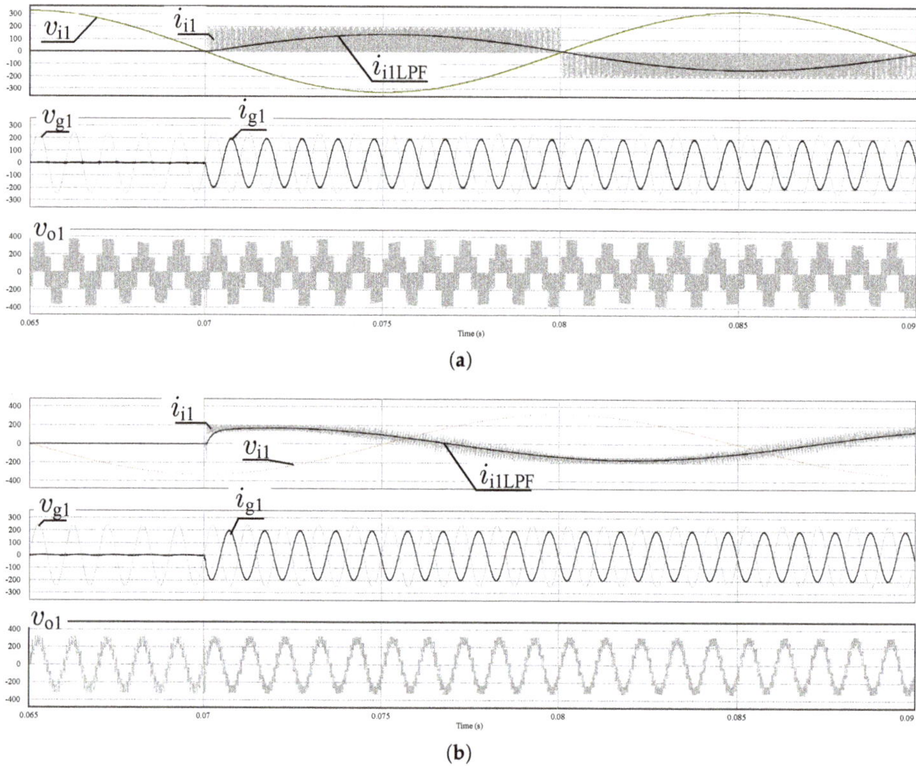

Figure 23. The step change 0–200 A of the reference active generator current in $t = 0.07$ s: (**a**) for CMC converter, (**b**) for MMCCR converter.

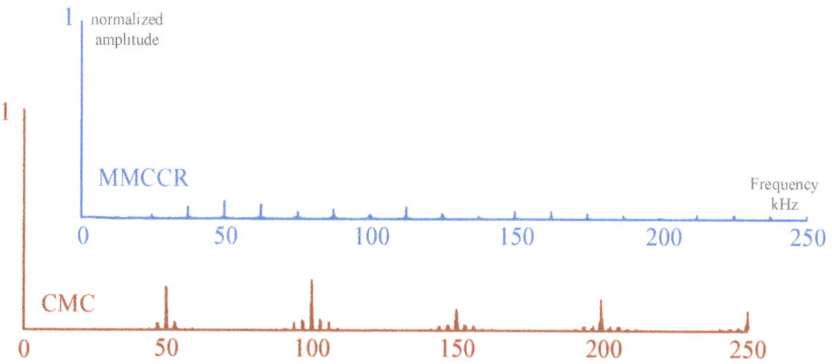

Figure 24. The input current spectrums for both converters.

The proposed MMCCR topology contains four matrix converters and three-phase shifter circuits. As mentioned in an introduction section, the power is equally divided with among these matrix converters. Example PS currents i_{PS1}, i_{PS2}, and i_{PS3} are shown in Figure 25.

In steady-state it is possible to change the control strategy. The SVPWM method can be replaced by NVM modulation, which is characterised by a much lower operating frequency of power electronic switches. However, new PI controllers settings should be selected in this case to keep the staircase character of the generated voltage. An obtained example of converter's voltage and other waveforms, during an active NVM modulation, is shown in Figures 26 and 27.

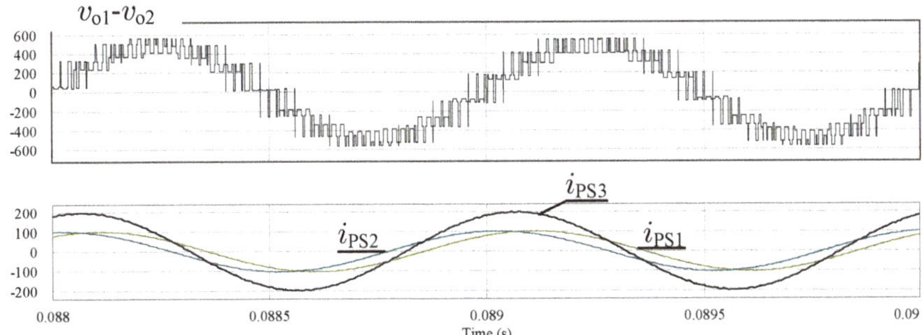

Figure 25. Converter's line–to–line output (correspond to Figure 21) voltage and the current sharing among the PS shown in Figure 6—SVPWM modulation.

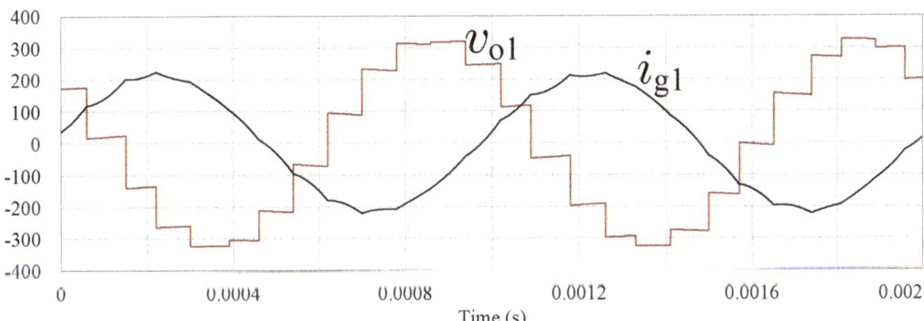

Figure 26. MMCCR converter' output voltage v_{o1} and the generator current i_{g1} for NVM modulation: $\text{THD}(i_{g1}) = 3.5\%$ and $\text{THD}(v_{o1}) = 16\%$.

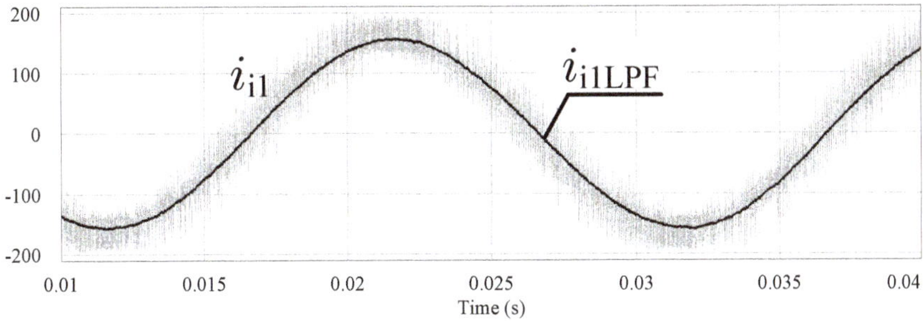

Figure 27. An input current i_{i1} and its filtered waveform $i_{i1\text{LPF}}$ for NVM modulation: $\text{THD}(i_{i1}) = 26\%$.

5. Conclusions

This article studies nature and presents conceptual research and discusses the different Pulse Width Modulation (PWM) strategies for operating, with a low-switching frequency for the proposed topology. It shows how the unconventional combination of CMC modules and CR could improve the quality of energy conversion. The paper also presents how this solution may be specifically appropriate for the high power systems, that are supplied by the high AC frequency sources, such as the high-speed generators or airport terminals' supply of 400 Hz.

The main features of the proposed approach are as follows:

- The use of coupled reactors allows the generation of voltages with staircase character (see Figure 26) with relatively low switching frequency operation, which depends on the value of the reference voltage. However, the staircase-type waveforms of the load voltages are obtained only for maximum voltage gain (for both NVM and PWR modulations) or for the certain values of the NVM modulation index, as shown in Figure 12.
- Increasing the number of CMC devices allows for the achievement of the voltage gain close to unity. For CMC, the voltage gain is no greater than 0.866, while for the proposed topology, the generated maximum amplitude is 11% greater. However, four PS circuits are required. Therefore, and also due to the volumes of coupled reactors, the application of the discussed system is the most rational in the case of high frequencies at the output and/or output of the CMC, e.g., in gas turbines.
- The content of higher harmonics in the voltage, generated using PWR and NVM algorithms, is not linearly dependent on the reference load voltage value, this can make application of the proposed solution problematic in a full range of load voltage amplitude changes. Therefore, the proposed type of amplitude control should be used and optimised preferably in systems with a fixed output frequency. Only the SVPWM method, based on the rotating vectors, can be utilised in a full range of the modulation index.
- In comparison with high-frequency methods such as PWM, both PWR regulation and NVM modulation are not precise methods generating output voltages. Therefore, the proposed solution will not find application even in electric drive with high requirements for dynamics. However, it can be assumed that apart from reducing switching losses, the EMI level will also decrease. Due to these properties, the presented modulation methods can be applied in converters for high-speed generators/motors and also the onboard power supply.

In the case of MMCCR control with NVM modulation, it is possible to achieve a reasonable compromise between low THD value and relatively low frequency of power switches. However, as shown in Figure 20, in case of NVM modulation, it is difficult to control the output current/voltage with reference amplitude changes. Special switchable algorithms are required. Much better control possibilities, albeit at the cost of increasing THD and about 2 times the switching frequency, can be achieved when implementing the PWR algorithm. A further increase in the precision of the generation and control range of the output current/voltage is possible by means of SVPWM modulation for two rotating spatial vectors. This results in higher switching frequency and control calculation problems for CMC systems, in particular for this 12-pulse MMCCR. In order to reduce the importance of these problems and the related requirements on the capabilities of processor controllers, the authors initially propose to use the barycentric coordinate method [19], which unifies and simplifies calculations.

The research of this method in relation to the MMCCR system, including the hybrid method [25], allowing to minimize the frequency of connections in the converter at fixed points will be presented in the next paper. This study article does not in any way pretend to present the full spectrum of problems associated with MMCCR systems. Its main objective was only to present the possibilities and basic properties of equalising typical CMC modules with chokes coupled with overall power of no more than about 30% of the rated power of the whole system. The above objective also includes a general discussion of dedicated control algorithms. Considering the generality and the material's size, only the most important theoretical issues verified by simulation are presented. Experimental research on specific cases will be the subject of further publications in the near future.

Author Contributions: Conceptualisation: P.S. and N.S., investigation and simulation: P.S. and J.L., validation, formal analysis: A.U., final editing: P.S. and E.R.-C. All authors have read and agreed to the published version of the manuscript.

Funding: This work was supported by the LINTE^2 Laboratory, Gdansk University of Technology. It was also supported by ITMO University Saint Petersburg, Russian Federation.

Conflicts of Interest: The authors declare no conflict of interest.

Abbreviations

The following abbreviations are used in this manuscript:

CMC	Conventional Matrix Converter
CMCs	Conventional Matrix Converters
CRs	Coupled Reactors
EMI	Electromagnetic Intereference
MMCCR	Multipulse Matrix Converter with Coupled Reactors
NVM	Nearest Vector Modulation
PS	Phase Shifters
PWM	Pulse Width Modulation
PWR	Pulse Width Modulation
RMS	Root Mean Square
SVPWM	Space-Vector Pulse Width Modulation
THD	Total Harmonic Distortion
VFD	Variable Frequency Drive

Appendix A

Table A1. The S_{MCn} Switch States.

	$S_{A1}\ldots S_{C3}$	$S_{A4}\ldots S_{C6}$	$S_{A7}\ldots S_{C9}$	$S_{A10}\ldots S_{C12}$
V0	001-100-010	001-100-010	001-100-010	001-100-010
V1	001-100-010	001-100-010	001-100-010	010-001-100
V2	001-100-010	001-100-010	001-100-010	100-010-001
V3	001-100-010	001-100-010	010-001-100	001-100-010
V4	001-100-010	001-100-010	010-001-100	010-001-100
V5	001-100-010	001-100-010	010-001-100	100-010-001
V6	001-100-010	001-100-010	100-010-001	001-100-010
V7	001-100-010	001-100-010	100-010-001	010-001-100
V8	001-100-010	001-100-010	100-010-001	100-010-001
V9	001-100-010	010-001-100	001-100-010	001-100-010
V10	001-100-010	010-001-100	001-100-010	010-001-100
V11	001-100-010	010-001-100	001-100-010	100-010-001
V12	001-100-010	010-001-100	010-001-100	001-100-010
V13	001-100-010	010-001-100	010-001-100	010-001-100
V14	001-100-010	010-001-100	010-001-100	100-010-001
V15	001-100-010	010-001-100	100-010-001	001-100-010
V16	001-100-010	010-001-100	100-010-001	010-001-100
V17	001-100-010	010-001-100	100-010-001	100-010-001
V18	001-100-010	100-010-001	001-100-010	001-100-010
V19	001-100-010	100-010-001	001-100-010	010-001-100
V20	001-100-010	100-010-001	001-100-010	100-010-001
V21	001-100-010	100-010-001	010-001-100	001-100-010
V22	001-100-010	100-010-001	010-001-100	010-001-100
V23	001-100-010	100-010-001	010-001-100	100-010-001
V24	001-100-010	100-010-001	100-010-001	001-100-010
V25	001-100-010	100-010-001	100-010-001	010-001-100
V26	001-100-010	100-010-001	100-010-001	100-010-001
V27	010-001-100	001-100-010	001-100-010	001-100-010
V28	010-001-100	001-100-010	001-100-010	010-001-100
V29	010-001-100	001-100-010	001-100-010	100-010-001

Table A1. *Cont.*

	$S_{A1}...S_{C3}$	$S_{A4}...S_{C6}$	$S_{A7}...S_{C9}$	$S_{A10}...S_{C12}$
V30	010-001-100	001-100-010	010-001-100	001-100-010
V31	010-001-100	001-100-010	010-001-100	010-001-100
V32	010-001-100	001-100-010	010-001-100	100-010-001
V33	010-001-100	001-100-010	100-010-001	001-100-010
V34	010-001-100	001-100-010	100-010-001	010-001-100
V35	010-001-100	001-100-010	100-010-001	100-010-001
V36	010-001-100	010-001-100	001-100-010	001-100-010
V37	010-001-100	010-001-100	001-100-010	010-001-100
V38	010-001-100	010-001-100	001-100-010	100-010-001
V39	010-001-100	010-001-100	010-001-100	001-100-010
V40	010-001-100	010-001-100	010-001-100	010-001-100
V41	010-001-100	010-001-100	010-001-100	100-010-001
V42	010-001-100	010-001-100	100-010-001	001-100-010
V43	010-001-100	010-001-100	100-010-001	010-001-100
V44	010-001-100	010-001-100	100-010-001	100-010-001
V45	010-001-100	100-010-001	001-100-010	001-100-010
V46	010-001-100	100-010-001	001-100-010	010-001-100
V47	010-001-100	100-010-001	001-100-010	100-010-001
V48	010-001-100	100-010-001	010-001-100	001-100-010
V49	010-001-100	100-010-001	010-001-100	010-001-100
V50	010-001-100	100-010-001	010-001-100	100-010-001
V51	010-001-100	100-010-001	100-010-001	001-100-010
V52	010-001-100	100-010-001	100-010-001	010-001-100
V53	010-001-100	100-010-001	100-010-001	100-010-001
V54	100-010-001	001-100-010	001-100-010	001-100-010
V55	100-010-001	001-100-010	001-100-010	010-001-100
V56	100-010-001	001-100-010	001-100-010	100-010-001
V57	100-010-001	001-100-010	010-001-100	001-100-010
V58	100-010-001	001-100-010	010-001-100	010-001-100
V59	100-010-001	001-100-010	010-001-100	100-010-001
V60	100-010-001	001-100-010	100-010-001	001-100-010
V61	100-010-001	001-100-010	100-010-001	010-001-100
V62	100-010-001	001-100-010	100-010-001	100-010-001
V63	100-010-001	010-001-100	001-100-010	001-100-010
V64	100-010-001	010-001-100	001-100-010	010-001-100
V65	100-010-001	010-001-100	001-100-010	100-010-001
V66	100-010-001	010-001-100	010-001-100	001-100-010
V67	100-010-001	010-001-100	010-001-100	010-001-100
V68	100-010-001	010-001-100	010-001-100	100-010-001
V69	100-010-001	010-001-100	100-010-001	001-100-010
V70	100-010-001	010-001-100	100-010-001	010-001-100
V71	100-010-001	010-001-100	100-010-001	100-010-001
V72	100-010-001	100-010-001	001-100-010	001-100-010
V73	100-010-001	100-010-001	001-100-010	010-001-100
V74	100-010-001	100-010-001	001-100-010	100-010-001
V75	100-010-001	100-010-001	010-001-100	001-100-010
V76	100-010-001	100-010-001	010-001-100	010-001-100
V77	100-010-001	100-010-001	010-001-100	100-010-001
V78	100-010-001	100-010-001	100-010-001	001-100-010
V79	100-010-001	100-010-001	100-010-001	010-001-100
V80	100-010-001	100-010-001	100-010-001	100-010-001

Table A2. The S_{MCp} Switch States.

	$S_{A1}\ldots S_{C3}$	$S_{A4}\ldots S_{C6}$	$S_{A7}\ldots S_{C9}$	$S_{A10}\ldots S_{C12}$
V0	001-010-100	001-010-100	001-010-100	001-010-100
V1	001-010-100	001-010-100	001-010-100	010-100-001
V2	001-010-100	001-010-100	001-010-100	100-001-010
V3	001-010-100	001-010-100	010-100-001	001-010-100
V4	001-010-100	001-010-100	010-100-001	010-100-001
V5	001-010-100	001-010-100	010-100-001	100-001-010
V6	001-010-100	001-010-100	100-001-010	001-010-100
V7	001-010-100	001-010-100	100-001-010	010-100-001
V8	001-010-100	001-010-100	100-001-010	100-001-010
V9	001-010-100	010-100-001	001-010-100	001-010-100
V10	001-010-100	010-100-001	001-010-100	010-100-001
V11	001-010-100	010-100-001	001-010-100	100-001-010
V12	001-010-100	010-100-001	010-100-001	001-010-100
V13	001-010-100	010-100-001	010-100-001	010-100-001
V14	001-010-100	010-100-001	010-100-001	100-001-010
V15	001-010-100	010-100-001	100-001-010	001-010-100
V16	001-010-100	010-100-001	100-001-010	010-100-001
V17	001-010-100	010-100-001	100-001-010	100-001-010
V18	001-010-100	100-001-010	001-010-100	001-010-100
V19	001-010-100	100-001-010	001-010-100	010-100-001
V20	001-010-100	100-001-010	001-010-100	100-001-010
V21	001-010-100	100-001-010	010-100-001	001-010-100
V22	001-010-100	100-001-010	010-100-001	010-100-001
V23	001-010-100	100-001-010	010-100-001	100-001-010
V24	001-010-100	100-001-010	100-001-010	001-010-100
V25	001-010-100	100-001-010	100-001-010	010-100-001
V26	001-010-100	100-001-010	100-001-010	100-001-010
V27	010-100-001	001-010-100	001-010-100	001-010-100
V28	010-100-001	001-010-100	001-010-100	010-100-001
V29	010-100-001	001-010-100	001-010-100	100-001-010
V30	010-100-001	001-010-100	010-100-001	001-010-100
V31	010-100-001	001-010-100	010-100-001	010-100-001
V32	010-100-001	001-010-100	010-100-001	100-001-010
V33	010-100-001	001-010-100	100-001-010	001-010-100
V34	010-100-001	001-010-100	100-001-010	010-100-001
V35	010-100-001	001-010-100	100-001-010	100-001-010
V36	010-100-001	010-100-001	001-010-100	001-010-100
V37	010-100-001	010-100-001	001-010-100	010-100-001
V38	010-100-001	010-100-001	001-010-100	100-001-010
V39	010-100-001	010-100-001	010-100-001	001-010-100
V40	010-100-001	010-100-001	010-100-001	010-100-001
V41	010-100-001	010-100-001	010-100-001	100-001-010
V42	010-100-001	010-100-001	100-001-010	001-010-100
V43	010-100-001	010-100-001	100-001-010	010-100-001
V44	010-100-001	010-100-001	100-001-010	100-001-010
V45	010-100-001	100-001-010	001-010-100	001-010-100
V46	010-100-001	100-001-010	001-010-100	010-100-001
V47	010-100-001	100-001-010	001-010-100	100-001-010
V48	010-100-001	100-001-010	010-100-001	001-010-100
V49	010-100-001	100-001-010	010-100-001	010-100-001
V50	010-100-001	100-001-010	010-100-001	100-001-010

Table A2. *Cont.*

	$S_{A1}\ldots S_{C3}$	$S_{A4}\ldots S_{C6}$	$S_{A7}\ldots S_{C9}$	$S_{A10}\ldots S_{C12}$
V51	010-100-001	100-001-010	100-001-010	001-010-100
V52	010-100-001	100-001-010	100-001-010	010-100-001
V53	010-100-001	100-001-010	100-001-010	100-001-010
V54	100-001-010	001-010-100	001-010-100	001-010-100
V55	100-001-010	001-010-100	001-010-100	010-100-001
V56	100-001-010	001-010-100	001-010-100	100-001-010
V57	100-001-010	001-010-100	010-100-001	001-010-100
V58	100-001-010	001-010-100	010-100-001	010-100-001
V59	100-001-010	001-010-100	010-100-001	100-001-010
V60	100-001-010	001-010-100	100-001-010	001-010-100
V61	100-001-010	001-010-100	100-001-010	010-100-001
V62	100-001-010	001-010-100	100-001-010	100-001-010
V63	100-001-010	010-100-001	001-010-100	001-010-100
V64	100-001-010	010-100-001	001-010-100	010-100-001
V65	100-001-010	010-100-001	001-010-100	100-001-010
V66	100-001-010	010-100-001	010-100-001	001-010-100
V67	100-001-010	010-100-001	010-100-001	010-100-001
V68	100-001-010	010-100-001	010-100-001	100-001-010
V69	100-001-010	010-100-001	100-001-010	001-010-100
V70	100-001-010	010-100-001	100-001-010	010-100-001
V71	100-001-010	010-100-001	100-001-010	100-001-010
V72	100-001-010	100-001-010	001-010-100	001-010-100
V73	100-001-010	100-001-010	001-010-100	010-100-001
V74	100-001-010	100-001-010	001-010-100	100-001-010
V75	100-001-010	100-001-010	010-100-001	001-010-100
V76	100-001-010	100-001-010	010-100-001	010-100-001
V77	100-001-010	100-001-010	010-100-001	100-001-010
V78	100-001-010	100-001-010	100-001-010	001-010-100
V79	100-001-010	100-001-010	100-001-010	010-100-001
V80	100-001-010	100-001-010	100-001-010	100-001-010

References

1. Blaabjerg, F.; Yang, Y.; Ma, K. Power electronics—Key technology for renewable energy systems—Status and future. In Proceedings of the 2013 3rd International Conference on Electric Power and Energy Conversion Systems, Istanbul, Turkey, 2–4 October 2013; pp. 1–6. [CrossRef]
2. Ahmad, I.; Fandi, G.; Muller, Z.; Tlusty, J. Voltage Quality and Power Factor Improvement in Smart Grids Using Controlled DG Units. *Energies* **2019**, *12*, 3444. [CrossRef]
3. Gentvilaite, R.; Kander, A.; Warde, P. The Role of Energy Quality in Shaping Long-Term Energy Intensity in Europe. *Energies* **2015**, *8*, 133–153. [CrossRef]
4. Momete, D.C. Analysis of the Potential of Clean Energy Deployment in the European Union. *IEEE Access* **2018**, *6*, 54811–54822. [CrossRef]
5. Strzelecki, R.; Sak, T.; Zolov, P.D.; Moradewicz, A.; Grabarek, M. Multi-pulse VSC arrangements with coupled reactors. In Proceedings of the 2016 IEEE 2nd Annual Southern Power Electronics Conference (SPEC), Auckland, New Zealand, 5–8 December 2016; pp. 1–6. [CrossRef]
6. Strzelecki, R.; Sak, T.; Strzelecka, N. Modular multipulse voltage source inverters with integrating coupled reactors. In Proceedings of the 2016 13th Selected Issues of Electrical Engineering and Electronics (WZEE), Rzeszow, Poland, 4–8 May 2016; pp. 1–6. [CrossRef]
7. Strzelecki, R.; Sak, T.; Roslaniec, L.; Dmitrović, Z.P.; Strzelecka, N. Coupled inductors based filter for matrix converters. In Proceedings of the 2016 10th International Conference on Compatibility, Power Electronics and Power Engineering (CPE-POWERENG), Bydgoszcz, Poland, 29 June–1 July 2016; pp. 358–363. [CrossRef]
8. Rodriguez, J.; Lai, J.S.; Peng, F.Z. Multilevel inverters: A survey of topologies, controls, and applications. *IEEE Trans. Ind. Electron.* **2002**, *49*, 724–738. [CrossRef]
9. Zhang, J.; Xu, S.; Din, Z.; Hu, X. Hybrid Multilevel Converters: Topologies, Evolutions and Verifications. *Energies* **2019**, *12*, 615. [CrossRef]
10. Ventosa-Cutillas, A.; Montero-Robina, P.; Umbría, F.; Cuesta, F.; Gordillo, F. Integrated Control and Modulation for Three-Level NPC Rectifiers. *Energies* **2019**, *12*, 1641. [CrossRef]

11. Pires, V.F.; Monteiro, J.; Silva, J.F. Dual 3-Phase Bridge Multilevel Inverters for AC Drives with Voltage Sag Ride-through Capability. *Energies* **2019**, *12*, 2324. [CrossRef]

12. Verdugo, C.; Kouro, S.; Rojas, C.A.; Perez, M.A.; Meynard, T.; Malinowski, M. Five-Level T-type Cascade Converter for Rooftop Grid-Connected Photovoltaic Systems. *Energies* **2019**, *12*, 1743. [CrossRef]

13. Chivite-Zabalza, J.; Rodríguez Vidal, M.A.; Izurza-Moreno, P.; Calvo, G.; Madariaga, D. A Large Power, Low-Switching-Frequency Voltage Source Converter for FACTS Applications with Low Effects on the Transmission Line. *IEEE Trans. Power Electron.* **2012**, *27*, 4868–4879. [CrossRef]

14. Mysiak, P.; Sleszynski, W.; Cichowski, A. Experimental test results of the 150 kVA 18-pulse diode rectifier with series active power filter. In Proceedings of the 2016 10th International Conference on Compatibility, Power Electronics and Power Engineering (CPE-POWERENG), Bydgoszcz, Poland, 29 June–1 July 2016; pp. 380–383. [CrossRef]

15. Oguchi, K.; Maeda, G.; Hoshi, N.; Kubata, T. Coupling rectifier systems with harmonic cancelling reactors. *IEEE Ind. Appl. Mag.* **2001**, *7*, 53–63. [CrossRef]

16. Rodriguez, J.; Rivera, M.; Kolar, J.W.; Wheeler, P.W. A Review of Control and Modulation Methods for Matrix Converters. *IEEE Trans. Ind. Electron.* **2012**, *59*, 58–70. [CrossRef]

17. Helle, L.; Larsen, K.B.; Jorgensen, A.H.; Munk-Nielsen, S.; Blaabjerg, F. Evaluation of modulation schemes for three-phase to three-phase matrix converters. *IEEE Trans. Ind. Electron.* **2004**, *51*, 158–171. [CrossRef]

18. Friedli, T.; Kolar, J.W. Milestones in Matrix Converter Research. *IEEJ J. Ind. Appl.* **2012**, *1*, 2–14. [CrossRef]

19. Szczepankowski, P.; Wheeler, P.; Bajdecki, T. Application of Analytic Signal and Smooth Interpolation in Pulse Width Modulation for Conventional Matrix Converters. *IEEE Trans. Ind. Electron.* **2019**. [CrossRef]

20. Empringham, L.; Kolar, J.W.; Rodriguez, J.; Wheeler, P.W.; Clare, J.C. Technological Issues and Industrial Application of Matrix Converters: A Review. *IEEE Trans. Ind. Electron.* **2013**, *60*, 4260–4271. [CrossRef]

21. Strzelecki, R.; Sak, T.; Strzelecka, N. A System for Power Conversion Using Matrix Converters Connected in Parallel. Polish Patent PL 227934, 15 January 2018.

22. Jafari, S.; Miran Fashandi, S.A.; Nikolaidis, T. Modeling and Control of the Starter Motor and Start-Up Phase for Gas Turbines. *Electronics* **2019**, *8*, 363. [CrossRef]

23. Gülen, S.C. *Gas Turbines for Electric Power Generation*; Cambridge University Press: Cambridge, UK, 2019; [CrossRef]

24. Morimoto, M.; Aiba, K.; Sakurai, T.; Hoshino, A.; Fujiwara, M. Position sensorless starting of super high-speed PM Generator for micro gas turbine. *IEEE Trans. Ind. Electron.* **2006**, *53*, 415–420. [CrossRef]

25. Szwarc, K.J.; Szczepankowski, P.; Nieznański, J.; Swinarski, C.; Usoltsev, A.; Strzelecki, R. Hybrid Modulation for Modular Voltage Source Inverters with Coupled Reactors. *Energies* **2020**, *13*, 4450. [CrossRef]

26. Szczepankowski, P.; Nieznanski, J. Application of Barycentric Coordinates in Space Vector PWM Computations. *IEEE Access* **2019**, *7*, 91499–91508. [CrossRef]

Publisher's Note: MDPI stays neutral with regard to jurisdictional claims in published maps and institutional affiliations.

Article

Unidirectional DC/DC Converter with Voltage Inverter for Fast Charging of Electric Vehicle Batteries

Jerzy Ryszard Szymanski [1], Marta Zurek-Mortka [1], Daniel Wojciechowski [2,*] and Nikolai Poliakov [3]

[1] Faculty of Transport, Electrical Engineering and Computer Science, Kazimierz Pulaski University of Technology and Humanities, 29 Malczewski Str., 26-600 Radom, Poland; j.szymanski@uthrad.pl (J.R.S.); m.zurek-mortka@uthrad.pl (M.Z.-M.)

[2] Faculty of Electrical and Control Engineering, Gdańsk University of Technology, 11/12 Gabriel Narutowicz Str., 80-233 Gdańsk, Poland

[3] Faculty of Control Systems and Robotics, ITMO University, Saint Petersburg 197101, Russia; polyakov.n.a@gmail.com

* Correspondence: daniel.wojciechowski@pg.edu.pl

Received: 12 August 2020; Accepted: 8 September 2020; Published: 14 September 2020

Abstract: The paper proposes the adaptation of the industrial plant's power network to supply electric vehicle (EV) fast-charging converters (above 300 kW) using renewable energy sources (RESs). A 600 V DC microgrid was used to supply energy from RESs for the needs of variable speed motor drives and charging of EV batteries. It has been shown that it is possible to support the supply of drive voltage frequency converters (VFCs) and charging of EV batteries converters with renewable energy from a 600 V DC microgrid, which improves the power quality indicators in the power system. The possibility of implementing the fast EV batteries charging station to the industrial plant's power system in such a way that the system energy demand is not increased has also been shown. The EV battery charging station using the drive converter has been presented, as well as the results of simulation and laboratory tests of the proposed solution.

Keywords: EV battery; electric vehicles; fast battery charging; local transport; DC micro grid; drive voltage frequency converter; big power DC/DC converter

1. Introduction

The fast development of industrial power electronics gives the opportunity to replace traditional solutions not only in electric drives, but also contributes to the construction of scalable high-power modular charging stations for electric vehicles, e.g., 300 kW. The authors propose the use of low-voltage frequency converter modules, that are commonly used in electric drives, in fast-charging stations. This approach will significantly facilitate the construction of a fast vehicle charging station and should significantly reduce the cost of their manufacture.

The literature review shows that there are different methods and topologies for electric vehicle (EV) battery charging. In [1], the authors present simulation models of selected topologies of EV fast-charging systems and their research, where particularly interesting are the converter topologies with regulated rectified voltage using a thyristor half-bridge three-phase rectifier. This solution does not use galvanic separation between the charged battery and the converter. Other models shown use a single-phase inverter and a high-frequency isolation transformer. In [2], the authors focus on reviewing all the useful data available on EV configurations, battery energy sources, electrical machines, charging and optimization techniques, impacts, trends, and possible directions of future developments. In [3] various charging station topologies compared and evaluated based on microgrid support, power density, modularity and other factors. The authors use full-bridge single-phase inverters to supply the individual phases of a three-phase high-frequency transformer that separates the output power stages of the DC/DC converter. In [4], mainly on-board converters for charging EV batteries powered from single- and three-phase grids are presented. Various power level chargers and infrastructure configurations are presented, compared, and evaluated based on amount of power, charging time and location, cost, equipment, and other factors. The paper [5] shows power electronics converters for EV fast charging stations, where a three-branch DC/DC converter is used that uses a half-bridge inverter structure to produce DC charging for EV batteries. The properties of the DC/DC converter model from [5] were described by the authors of this paper on the basis of simulation tests carried out.

The most popular charging method is constant current charging of EV battery to about 80% of its capacity [6,7]. Whereas the reference [8] shows off-board EV fast battery charger based on a dual-stage power converter (AC/DC and DC/DC) sharing the same DC link. Publications presenting the possibility of application of three-phase half-bridge voltage inverters as components for DC/DC converters in the EV charging stations are rarely seen. In particular, these three-phase Pulse Width Modulation (PWM) inverters are used in the induction motor drives [9]. The majority part of literature on the subject describes solutions with single-phase full-bridge PWM inverters [1,10]. Converters for charging EV batteries mainly use single-phase inverters. Three-phase converters are required for high charging powers. The proposals from the literature do not use the three-phase voltage frequency drives voltage frequency converters (VFCs) to obtain the charging voltage of an EV battery.

Particularly, paper [11] presents a comparison of a current-source converter and a voltage-source converter (VSC) for three-phase EV fast battery chargers, where it is possible to control the output voltage of VSCs in a wide range of values but no more than 560 V, which is the maximum instantaneous value of the phase-to-phase voltage.

An important aspect is that the power supply of the EV fast-charging station should come from renewable energy sources (RESs). The concept of powering from RES dedicated for EV fast-charging station is described in [12]. The use of an AC/DC converter to supply a three-phase diode rectifier, generating the charging voltage of an EV battery is presented. The elimination of distorted currents in three-phase networks was achieved by means of a resonant LC filter. The PWM inverter is connected to the rectifier via the differential-mode voltage filter, which additionally allows the voltage regulation on the rectifier. In [13] the methods of supporting DC power supply to drive converters from a PV source is presented. This solution reduces the harmonics of the current in the three-phase AC network. The main idea in [14] is to reduce the number of DC/DC converters in an off-board DC/AC charging station powered from a PV source. The vehicles are equipped with on-board AC/DC converters. The [15] presents calculations of the demand for renewable energy for the needs of EV battery charging, taking into account energy storages.

The combination of many different sources enables more efficient use of the production capacities of systems using RES, increases the reliability and quality of power supplied to consumers and ensures independence from the supply. This integration of different energy sources is ensured by a microgrid. The paper [16] presents the basic assumptions of the idea of connecting various generation units of distributed generation cooperating within the so-called "microgrids" on the example of DC microgrid, properties of renewable energy sources and economic aspects of energy production in the DC microgrid. The use of DC microgrids in the EV battery charging stations is described in detail in [17], as well as battery manufacturing technologies and charging strategies. In [18], the possibility of using a hybrid DC/AC microgrid to power an EV charging station is demonstrated.

It is also important to limit the harmonic content of low orders in the phase current caused by rectifiers located in EV chargers. In [19] the authors analyse the operating principle of charging current in a continuous and discontinuous mode in case of EV charger with three-phase uncontrolled rectifier with a passive method of power factor correction (current total harmonic distortion—THD). The use of a 12-pulse rectifier [20,21] or resonant filters [12] is justified by the low cost and significant reduction of current harmonics for demanding industry applications, typical above 250 kW.

The rest of the introduction contains two main sections. Section 1.1 presents a model of an exemplary Li-ion battery pack for an EVs and its basic parameters: nominal voltage, charging voltage, internal resistance. A proprietary method of determining the equivalent resistance of a cell and the entire set of batteries was proposed. The aim of the paper is to demonstrate the possibility of adapting high-power drive converters to generate a constant voltage with an adjustable value for fast-charging with a constant current of an EV battery set. Section 1.2 presents the described in the literature and already implemented for production of direct voltage–direct current (DC/DC) converters using the half-bridge structure of a three-phase inverter to generate direct current charging an EV battery pack by using appropriate PWM control and branch output chokes of the converter. The authors presented their own models for simulation tests of such converters to show the differences between PWM control of DC/DC converters and PWM control of a drive inverter (DC/AC converter).

Further parts present the possibilities of using DC microgrid with hybrid power supply using RES to supply clean energy, both for drive converters operating as converters to supply AC motors or as converters for charging EV batteries (e.g., internal transport). Simulation tests of a high-power converter model adapted for charging a battery set with DC current were carried out. The main circuit diagrams and the control of the converter models are given precisely to enable the reader to verify the obtained results and to further develop the presented research. The final section of the paper presents the results of laboratory tests in which an industrial low-power drive converter with a rectifier unit was used, where the energy supplied from the rectifier was lost on the power resistor. The aim of this study was to demonstrate that, according to simulation studies, it is possible to automatically maintain a constant current at a given value, regardless of the value of the load resistance. The industrial drive converter automatically adjusted the AC voltage of the inverter to the set rectifier load current value. The applied rectifier load power resistor with a given value replaced the EV battery set. The obtained experimental results confirm the possibility of using high power drive converters for fast-charging of EV batteries. Finally, in the Discussion section, the comparison of sinusoidal and triangular modulation was presented and the use of triangular modulation in the PWM inverter to EV battery charging was proposed for further research.

1.1. Charging of High Capacity, High Power Batteries

For the Li-ion type LFP100AHA battery cell [22] with the following parameters: $V_{bn} = 3.2$ V (from 20% to 100% State Of Charge—SOC) and $Q_{no} = 100$ Ah, it is possible to build a battery pack by connecting cells in series to determine the voltage of battery set and in parallel to increase the capacity of set.

With a series connection of 100 cells, a set of 100 cells multiplied by 3.2 V is obtained, giving the nominal voltage of the battery set equal to $V_{zbn} = 320$ V and a capacity $Q_{zbn1} = 100$ Ah. By connecting eight chains of 100 cells in parallel (100cx8ch), the final battery set 100cx8ch with the parameters 800 Ah/320 V is obtained, which corresponds to the capacity of 256 kWh. An equivalent diagram of an EV battery with a voltage DC/DC charging converter is shown in Figure 1.

According to Figure 1, it was assumed that a DC/DC converter with a minimum power of approx. $P_{(1C)} = 276$ kW should be used to charge the battery set in 1 h with 1 C current or with a power of nearly 828 kW to charge the battery in about 20 min. with 3 C current. C rate is derived from Coulomb's Law. The value of the charging current resulting from the battery capacity specified in Ah, e.g., 100 Ah means 1 C = 100 A. The possibility to charge an EV battery with 3 C current depends on its cooling capability, as the losses during charging increase three times compared to 1 C current. For 1 C current is $P_s = R_{zb} \cdot I_{zb}^2 = 12.5$ m$\Omega \cdot (800$ A$)^2 = 8$ kW and for 3 C current is 3×8 kW $= 24$ kW. During continuous operation and while charging the battery, its temperature may not exceed 65 °C [22]. For customer is better, when the charging this battery current is 3 C (2400 A). The EV battery in this case will be charge in 20 min, but it is associated with a shorter battery life.

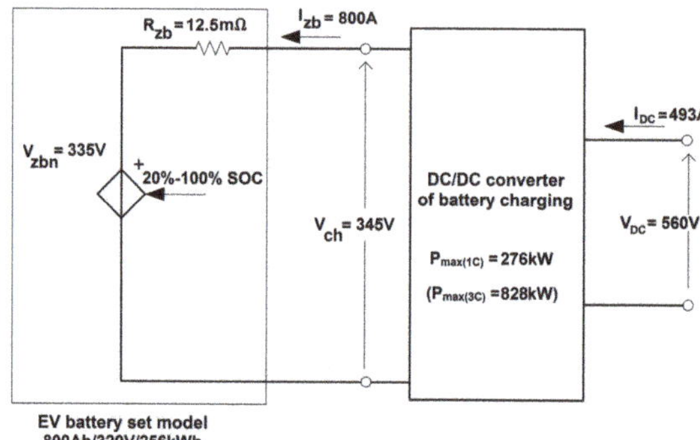

Figure 1. Model of 100cx8ch battery set made of Li-ion cells LPF100AHA-800 Ah/320 V/256 kWh with direct voltage–direct current (DC/DC) converter powered from 600 V DC microgrid [23,24].

Different strategies are used to charge EV batteries, e.g., Constant Current (CC) 3 C (20–80% SOC) and Constant Voltage (CV) 80–100% SOC or charging with 10 C current pulse (20–80% SOC). A compromise should be found between charging time and battery temperature.

The equivalent internal resistance R_{zb} of the battery pack intended to estimate charging power loses, can be determined based on the equivalent resistance of a single cell R_b. Several methods for determining R_b are known [25]. The authors of this paper propose a method based on reading the cell charging voltage value at 1 C ($U_{ch(1C)}$ = 3.45 V) and 3C ($U_{ch(3C)}$ = 3.65 V) charging currents, which is given in the catalogue card [26,27]. Charging voltages for the Li-ion cells of the LFP100AHA type are linearly approximated and presented in Figure 2. To the further simulation studies, an equivalent resistance was used, which reflects the power given to the battery by the converter.

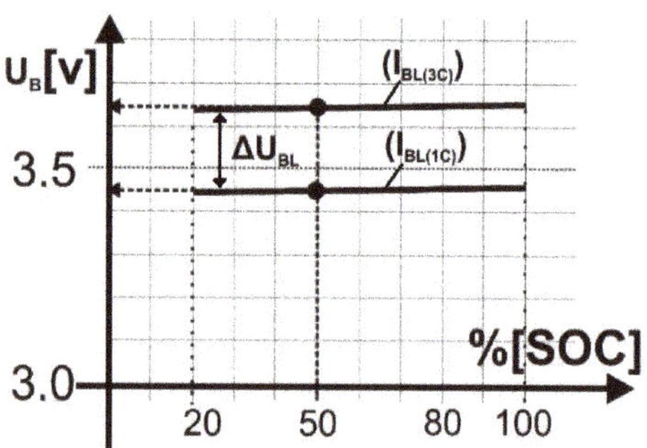

Figure 2. Linear approximated Li-ion cell voltages of the LFP100AHA type for the 20–100% State of Charge (SOC) range when charged at 1C and 3C [22].

Based on the reading of voltages for 50% SOC from Figure 2, the equivalent internal resistance R_b of the single cell can be calculated (1) as follows:

$$R_b = \frac{\triangle U_b}{\triangle I_b} = \frac{U_{b(3C)(50\%SOC)} - U_{b(1C)(50\%SOC)}}{I_{b(3C)(50\%SOC)} - I_{b(1C)(50\%SOC)}} = \frac{0.2V}{200A} = 1 \text{ m}\Omega \tag{1}$$

where:

R_b—equivalent resistance of a single cell,
$\triangle U_b$—difference of cell voltages for 3 C and 1 C charging currents,
$\triangle I_b$—difference of cell charging currents 3 C (300 A) and 1C (100 A), respectively.

Then the equivalent resistance R_{zb} of the 100c8ch battery set is equal to:

$$R_{zb} = \frac{R_b \cdot 100 \ cells}{8 \ chains} = 12.5 \ \mathrm{m\Omega} \tag{2}$$

1.2. Bidirectional DC/DC Converter

DC output power supplies, such as energy storages, uses a bidirectional DC/DC converter for direct connection to DC microgrid. The type of DC/DC converter, shown in Figure 3a, allows the converter current to be split into 3 branches, which are controlled with an 120° interleave method [28]. Figure 3b,c show switching states and waveforms of currents in operating states of boost and buck mode, respectively.

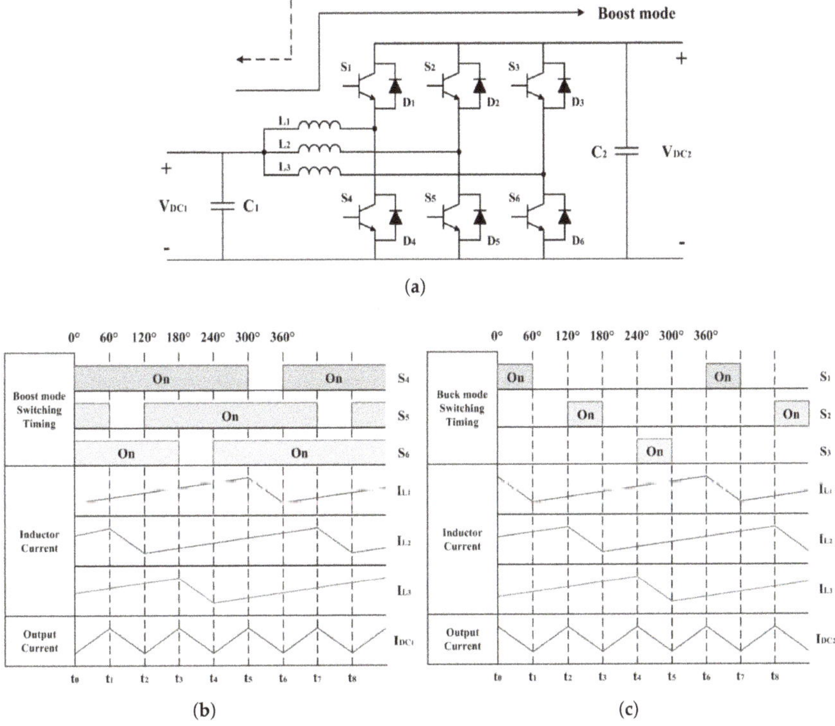

Figure 3. Three branches half-bridge bidirectional DC/DC converter: (**a**) diagram of the converter, (**b**) switching states and currents in boost mode, (**c**) switching states and currents in buck mode [28].

Bidirectional DC/DC converters can be basically divided into isolated [29,30] and non-isolated types. In general, an insulated bidirectional DC/DC converter has the advantage of easily controlling voltage step-up and step-down through a transformer inside the converter, but the transformer used in the insulated type has a large volume. It has a disadvantage that the size and weight of the transformer are increasing with the power of the converter. On the other hand, the non-isolated bidirectional DC/DC converter has the advantage of being relatively simple in structure, has high efficiency, and reduced weight in comparison to the insulated type [31]. In the non-isolated bidirectional DC/DC converter structure, various methods have been proposed to aim for higher efficiency, but among them, the interleaved method reduces the current stress of the power element because the magnitude of the load current (battery storage, DC microgrid) is divided into multiple phases. It has a feature of reducing the power device size and its current rating [29,32].

Tests of the battery (energy storage) charging current from the DC voltage line as a function of the converter control factor D (S1) are shown in Figure 4a,b. They present the models of boost type and buck type of DC/DC converter respectively, which are simulated in ANSYS Simplorer. The inverter of each one is controlled by means of a state graph. Triangular modulation is based on the analysis of waveforms controlling IGBT power transistors and is implemented for each phase separately (TRIANG1 to TRIANG3). A constant value S1 representing the control factor D is given. It is presented on Figure 5a.

Whereas, Figure 5b presents phase currents L1, L2, L3 and voltage waveforms at selected points of the models from Figure 4a. The DC/DC boost converter supplies energy to the 600 V DC voltage line (E1 source-VM1 voltmeter) from an energy storage with an initial voltage of 320 V (E2 source-VM2 voltmeter). To ensure constant charging current AM1 at different values of its voltage, the converter's control factor D should be changed accordingly. The operation of a buck converter is analogous (model from Figure 4b).

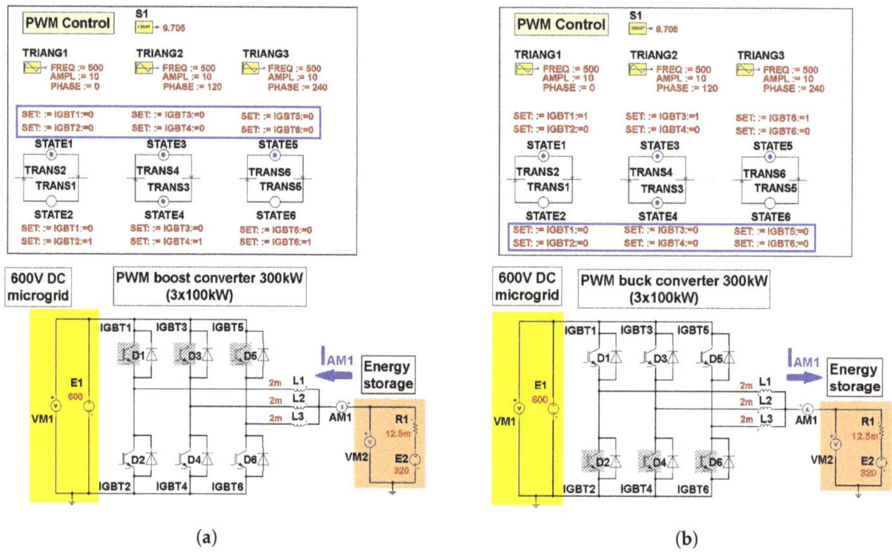

Figure 4. Model of DC/DC converter: (a) boost type—energy is transferred from energy storage to DC microgrid, (b) buck type—energy is transferred from DC microgrid to energy storage.

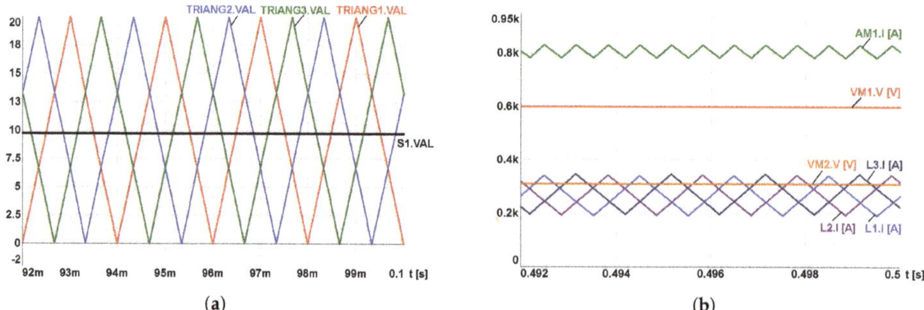

Figure 5. Waveforms in the boost mode (model from Figure 4a): (a) control, (b) currents and voltages.

The authors proposed an alternative solution. In the same power electronic structure as the bidirectional DC/DC converter, it was produced AC voltage instead of DC voltage, because the different control of converter is used. The unidirectional converter used in regulated industrial drives to control the speed and torque of induction motors was investigated. The maximum voltage of the EV battery charging converter constructed in this way is determined by the value of the DC link voltage.

2. Drive Voltage Frequency Converters Used in the EV Charging Stations

In a typical "load sharing" drive application, each VFC normally supplies power to a motor via the AC supply line. If one or more motors are driven in regenerative mode, they deliver power to the common DC bus. Then this power is used by other VFC and in this way the installation is more efficient, because in many situations the brake resistors can be omitted. In this situation the DC voltage (intermediate voltage) can be slightly different in each converter. This is due to minor differences in the rectifiers, different temperature, output power, etc. This small difference in DC voltage makes it necessary to use small line reactors in the AC main supply and fuses in the DC bus. The load sharing DC grid, which connect a few intermediate circuits of VFC, is presented in Figure 6a [33]. A large number of VFCs located in the various places of the local grid enables the location of EV battery charging points as close as possible to places of using the autonomous electric work machines and various types of EVs. Currently, all drives in which the engine speed is controlled, use indirect AC/DC/AC converters.

The I_{limit} battery charging current can be set between the minimum current I_{min} and the maximum current I_{max} of the inverter, Figure 6b. For the maximum frequency of sinusoidal voltage f_0, the condition is $f_c/f_0 = 10$ [21]. Increasing the frequency of the sinusoidal waveform f_0 to a value above 50 Hz ensures a reduction of the alternating component in the rectified DC.

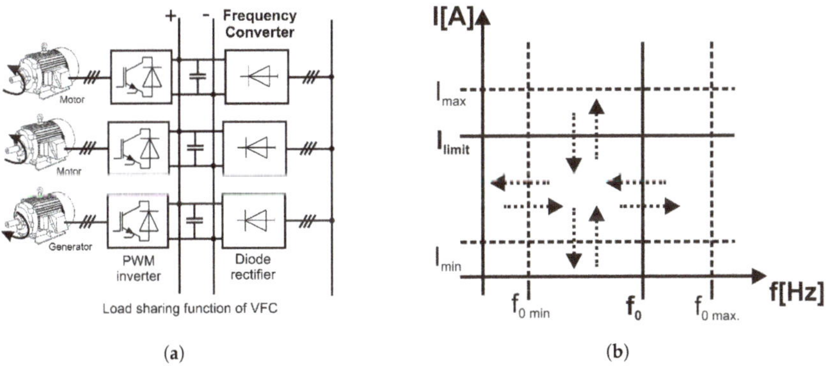

Figure 6. Functions of drive voltage frequency converter (VFC): (**a**) load sharing in drive VFCs, (**b**) current limiter [21,33].

An example of basic drive VFC with scalar control, which is adapted to DC/DC converter, is shown in Figure 7a. The CC battery charging strategy can use the inverter output current limitation function. The control systems of these converters can ensure any voltage characteristic as a function of frequency, typically $u/f^2 = const.$, or $u/f = const.$ It enables to realise a special characteristic like the one shown in Figure 7b, which is used for battery charging. This shape of u/f characteristic limits the frequency changing when the current limiter is active, Figure 6b. The arrows indicate the possibility of setting any frequency and limit current values in the drive FC. When EV charger achieves the charging current I_{min}, it means that the battery is fully charged. The sinusoidal frequency of the inverter output voltage f_0 is limited by the carrier frequency f_c of PWM. Frequency changes of 150 Hz causes the voltage to change between 250 V and 400 V. It is possible to choose a battery charging characteristic from four variable sets of VFC parameters, Figure 7a [21].

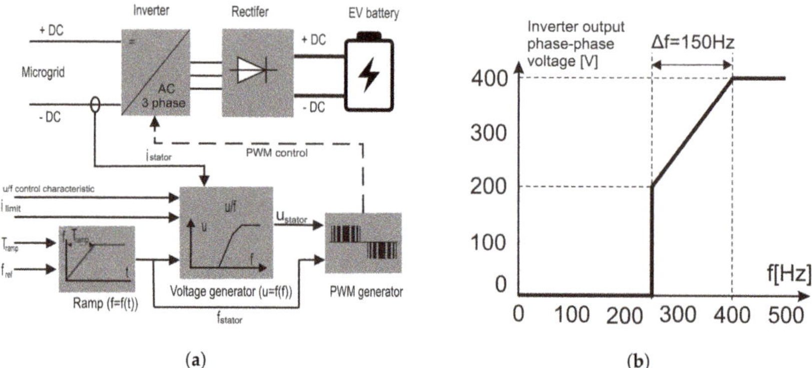

(a)   (b)

Figure 7. Drive VFC implemented as an electric vehicle (EV) fast charger: (**a**) control of industrial Pulse Width Modulation (PWM) drive converter with output rectifier, (**b**) special characteristic of inverter phase-phase voltage.

Frequently the control of a complex voltage vector in the drive VFC is used in the recuperative DC/AC converter that transfers energy from a DC microgrid to a three-phase AC grid. Drive VFC commonly uses the inverter current limiting function to protect motors against overload. This function can be used to set charging current of an EV battery.

3. DC Microgrid

By connecting all DC links of the sources and loads, a DC microgrid is formed, Figure 8. The DC microgrid does not directly connect to the prevalent three-phase AC utility grid, like the AC microgrid, but via a bidirectional DC/AC converter for common integration. The use of this type of solution gives wide possibilities of cooperation of various generating units, such as RES. Functional microgrids focused on EV battery charging can become a basic element of charging infrastructure. The possibility of eliminating many stages of AC/DC and DC/AC conversion, could significantly reduce the cost of network components and power losses, and additionally increase the reliability of network systems. Moreover, the lack of reactive power, absence of harmonics and asymmetry of voltages and currents in the DC system, make the DC microgrid one of the key areas of application that contribute to significant benefits [16,17].

The own concept of hardware integration of the EV fast-charging station with the local LV DC microgrid (3) and the MV industrial power system (4) supplying the drive VFC (14) of the squirrel-cage induction motor (16) is presented in Figure 8. The local DC microgrid is powered from three sources: renewable energy (solar or wind) (1), energy storage (e.g., Li-ion battery) (9) and MV power system (4) via LV grid (5). The hybrid DC power supply is optional, however, the extension of the power supply system with additional RES provides "clean energy" to the production process and relieves the power system. Reducing the so-called carbon footprint in products is now a mandatory requirement due to the increased effort to protect the environment. The EV battery is attached to the output rectifier (11). Implementation of the battery charging strategy, e.g., CC charging in the range of 20–80% SOC is controlled by a PLC controller (17), which for this purpose communicates with the inverter (12) of drive VFC (14).

A parallel resonant filter (15) is connected to the power supply of the drive VFC, which reduces the harmonic current of the input rectifier (13). To limit the effect of capacitive reactive power of the filter, it is switched on, if the VFC load exceeds the set value, e.g., when the drive VFC (14) load current exceeds 50% of the nominal current. The ES (9) accumulates excess energy generated by RES and maintains power supply for devices connected to the local DC microgrid during interruptions in the supply of energy from other sources. The ES is coupled to the DC microgrid via a bidirectional DC/DC converter (8). The task of this converter is to ensure charging of the ES according to the set strategy or supplying energy to the DC microgrid in accordance with the algorithm implemented in the PLC software (17).

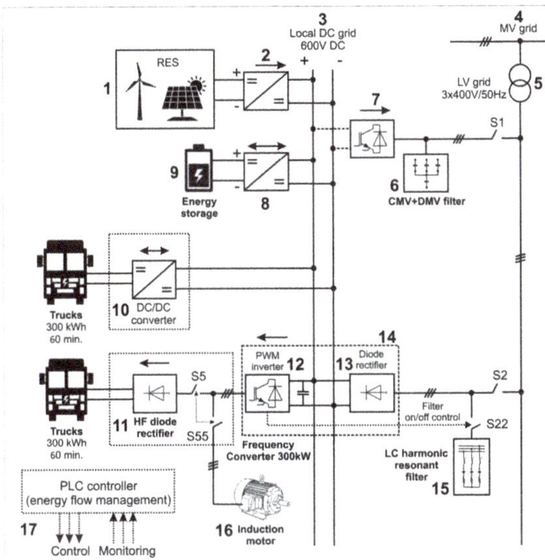

Figure 8. Hybrid 600 V DC microgrid for EV fast charging.

The local RES power plant (1) supplies DC voltage to the microgrid (3) via a unidirectional DC/DC converter (2). The task of this converter is to supply renewable energy to the DC grid and minimize the energy consumed from the power system (4). During a significant deterioration of power quality indicators in the power system or the occurrence of surplus renewable energy, it can be sent to the system via a DC/AC inverter (7), which cooperates with the LCL filter group (6) to limit the content of high-frequency differential mode (DM) and common-mode (CM) voltage produced by the PWM Active Front End (AFE) recuperative inverter (7) [34,35].

The authors carried out research showing the possibility of using drive VFC as inverter components for fast charging of EV batteries, which is depicted in Figure 9.

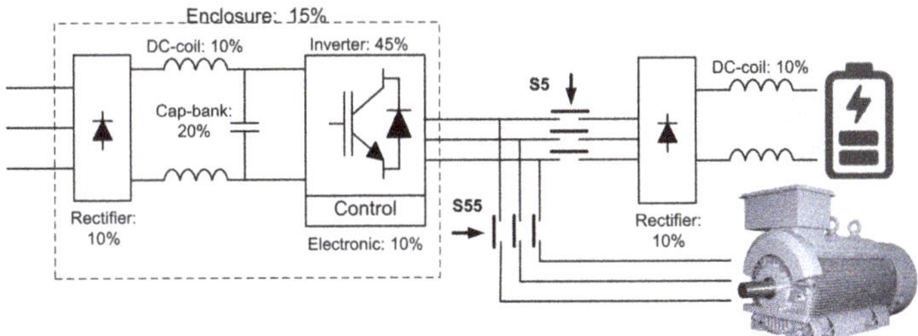

Figure 9. Use of drive VFC to charging EV batteries.

The mathematical model of the bipolar PWM inverter of the drive VFC connected with a six-pulse diode rectifier is presented in Figure 10. The model is written with electric symbols representing ideal energy sources, ideal passive elements and linearized models of controlled and non-controlled power electronics. The electrical differential equations of the DC/AC/DC converter are solved using an ANSYS Simplorer. The obtained results of simulation tests of the presented model confirm the possibility of controlling the direct voltage value of the diode rectifier charging the EV battery. Experimental tests carried out on the laboratory stand confirmed the results of simulation tests. The battery constant voltage obtained here is characterized by stability and negligible ripple. A detailed description of the operation of the EV battery charging rectifier and the PWM inverter is not the subject of this study.

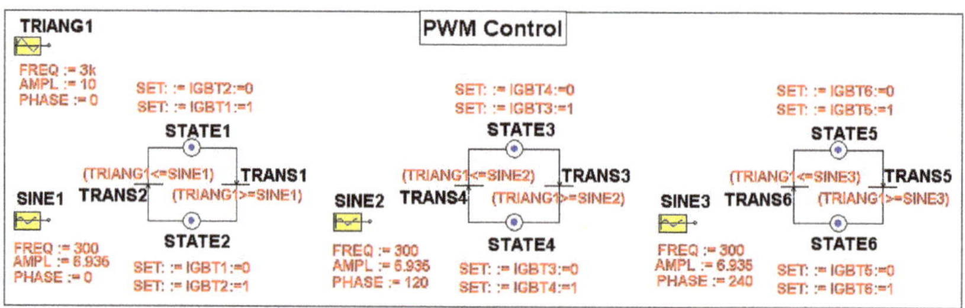

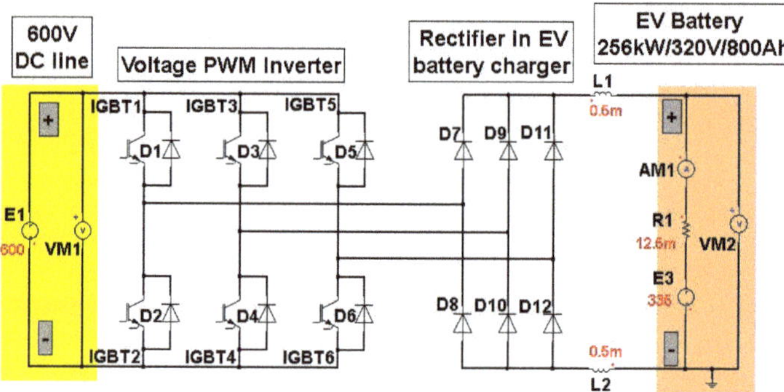

Figure 10. DC/AC/DC converter model with a two-level PWM voltage inverter and a six-pulse diode rectifier.

Drive VFCs with two-level voltage inverters are usually available with a six-pulse diode rectifier or less often with an AFE rectifier [21]. The basic types of drive converters used in LV grid are shown in Figure 9. The digital designations of the VFC components in Figure 9 refer to the hardware configuration of the hybrid EV fast-charging station shown in Figure 8. Proposed hardware combination of a hybrid EV charging station power supply system contains a significant part of the components used in industrial drive systems.

4. Simulation and Experimental Tests of Novel EV Charger

The electrical diagram of the simulation linear circuit model of a three-phase EV charger is depicted in the Figure 10. The EV battery is represented with a resistor R1 and voltage source E3. Using different values of sinusoids frequency and its amplitude in the inverter PWM control (SINE1, SINE2 and SINE3 PWM control-Figure 10), it was possible to test the EV charger properties for different values of the modulation factor M and the frequency of modulating voltages. The frequency of the triangular carrier waveform of PWM modulation was set in the TRIANG1 module and $f_c = 3$ kHz was used. It is a typical carrier frequency for inverters in the industrial high power drives. The frequency of SINE modules is 300 Hz, and it was the maximum output sinusoidal frequency of industry drive FC used in the laboratory stand. The modulation factor M can be changed between 0 and 1.25 to control output inverter voltage.

The use of the PWM modulator model described by the state graph made it easy to control the inverter IGBT transistors in relation to the sinusoidal PWM pattern. It was assumed that all model elements used in electrical circuits had linearized parameters.

To receive the constant charging current 800 A, the control of the modulation factor value M was used. Figure 11a,b show the obtained results of current and voltage waveforms, which are depended on the modulation factor's value (according to the model from Figure 10, M = 0.6935). The EV battery charging current 1 C was used in the test. The value of the charging current results from the technical specification of tested the Li-ion battery [22]. In the case of continuous modulation (the maximum value of modulation factor is M = 1), the EV battery voltage increased and the current exceeded the permissible charging value, which could destroy the EV battery. Therefore, it is important to choose the appropriate value of the modulation factor M, which allows battery charging with constant current for nominal pack voltages at the level to about 500 V [36].

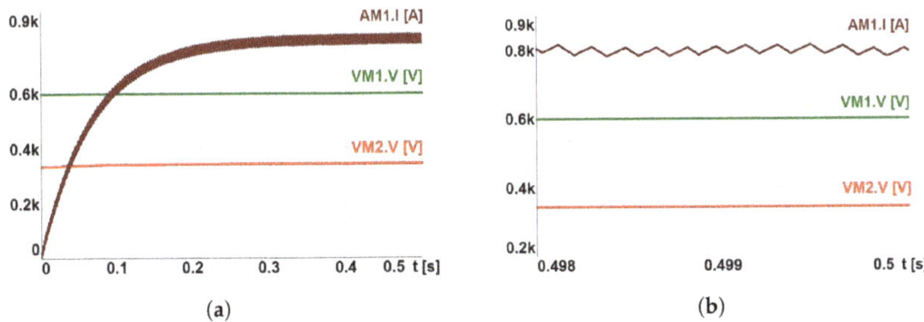

Figure 11. Constant-current battery charging measured in the time interval up to (**a**) 0.5 s, (**b**) 2 ms.

The tests of a diode rectifier powered by an inverter were performed under the following conditions:

1. without chokes of LC filters (3, 4) downstream of the inverter and in the rectifier,
2. with two types of LC filters of inverter DM voltage (3, 4),
3. without DM filter, but with DC chocks in the rectifier.

In case 1, it was not possible to adjust the value of the rectified voltage. In case 2, the impedance of the LC filter chokes caused an unfavourable significant drop in the rectifier supply voltage. Case 3 made it possible to control the rectified voltage in a wide range. Moreover, the elimination of the capacitors on the DC side of the rectifier did not significantly increase the AC component in the rectified voltage.

The specification of laboratory stand depicted in Figure 12a is presented in Table 1.

Table 1. Specification of the laboratory stand from Figure 12a.

No.	Name of Component	Parameters
1	Frequency converter VLT 3004	2.2 kW, 3 × 400 V/50 Hz, I_N = 5 A
2	RFI filter	I_N = 16 A
3	LC filter 1	16 A, 3 × L = 4 mH, 3 × C = 3 μF-Y
4	LC filter 2	16 A, 3 × L = 4 mH, 3 × C = 2 μF-△
5	Capacitors battery for CM voltage suppression	3 × C2 = 1 μF (connected to supply AC phases)
6	Drive frequency converter 5.5 kW	5.5 kW, 3 × 400 V/50 Hz, I_N = 12 A
7	Load resistors for drives VFC in rows 2 and 6	3 × (100 Ω–500 Ω)
8	Passive current harmonic filter for drives VFC (rows 2 and 6)	3 × 400 V/50 Hz, I_N = 10 A

When the negative pole of the load (EV battery) is grounded, the CM voltage is absent in the rectified voltage. Therefore, there is no need to filter the inverter CM voltage by using the capacitors (No. 5, Table 1) attached to the DM filter from one side and AC phase voltages from the second side.

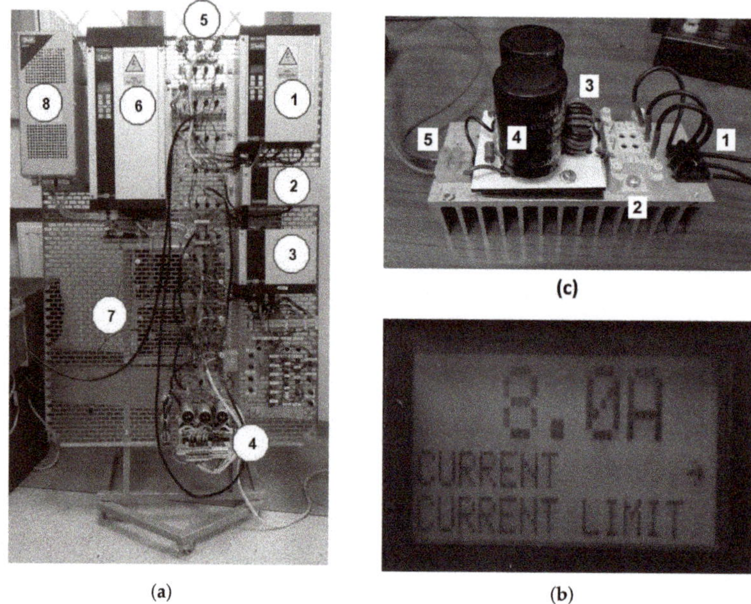

Figure 12. Laboratory stand: (**a**) equipped with two drive VFCs with built-in rectifying units attached to PWM inverters - the detailed specification is in Table 1, (**b**) with three-phase rectifier unit I_n = 15 A built into the drive VFC, (**c**) programmed value of the maximum output current in 5.5 kW drive VFC.

Given by the VFC drive current limitation, the rectifier load current maintained a constant value thanks to automatically lowering the rectifier supply voltage by the inverter control system. Figure 12b shows the programmed value of the maximum output current 8 A of low voltage (3 × 400 V/50 Hz) and small power (5.5 kW) industrial VFC drive (No. 6, Table 1). The maximum value of the rectified current did not exceed 10 A, which resulted from the power balance ($P_{AC} = P_{DC}$).

Figure 12c shows the six-pulse diode rectifier built for drive VFCs of the laboratory stand. Drive VFC outputs (1) were connected to a fast six-diode rectifier (2) with ferrite anti-distortion filter (3) and capacitor bank on the constant voltage side (4) of the rectifier, thus it was possible to charge EV batteries with DC current (5). The DC voltage fluctuations did not depend substantially from the value of the capacitor bank, because the inverter frequency of fundamental voltage harmonic was set to 300 Hz.

The received output DC voltages and DC currents are presented in Figure 13, for the capacitor bank equal to C = 16 μF. By comparing Figure 13a,b, the effective operation of the EV battery charging current stabilizer was visible. The rectifier voltage depends on the value of load resistance. A two-times decrease of the resistance resulted in a decrease of the charging voltage from 500 V to 300 V. In Figure 13a the DC voltage and charging current had a constant value and the output power was about 1 kW. When the rectifier was loaded with the resistance of 30 Ω (Figure 13b), the voltage supplying of the six-pulse diode rectifier automatically decreased. The load current was at the same value of 10 A in accordance with the set point of current limiter in the drive VFC.

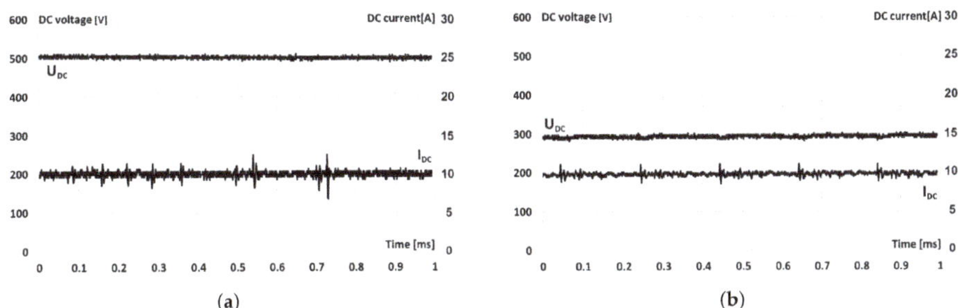

Figure 13. The DC output voltage and current of EV charger when current limiter in drive VFC is active at different value of rectifier load: (**a**) R = 50 Ω—the current limiter of the inverter is inactive, (**b**) R = 30 Ω—the current limiter of the inverter is active I_{limit} = 8.0 A.

The operation of the drive VFC could be programmed to perform expected functions of an EV battery charger. The experimental tests done for low power setup confirmed the correctness of performed simulation tests and the possibility of using the drive VFC as the basic DC/DC converter component of EV fast chargers.

5. Discussion

When charging the EV battery, the voltage inverter does not use freewheeling diodes (they are inactive), because there is a unidirectional energy flow from the DC microgrid to the diode rectifier. Therefore, it should be assumed that the efficiency of charging system will be similar to the efficiency of a drive VFC. The rectifier diodes cause losses similar to those in the inverter freewheeling diodes when supplying an induction motor.

If the drive converter is powered only from the DC microgrid, then it is possible to use one integrated circuit with an inverter and a rectifier to build a DC/DC converter for charging EV batteries.

When building a new converter for battery charging purposes, it is possible to replace sinusoidal modulation, e.g., with triangular modulation. The advantage of using the triangular PWM Figure 14b instead of the sinusoidal PWM (Figure 14a) is the proportional dependence of the value of rectified voltage and modulation factor M. As the amplitude of the triangular of the modulating wave increases linearly (TRIANG11-Figure 14b), there is a directly proportional increase of width modulated pulse. Such proportionality does not occur if the modulating waveform is a sine wave and the modulated waveform is a triangular wave [34]. The spectral analysis of the inverter CM voltage for sinusoidal and triangular modulation shows that there are no significant differences in the CM voltages, in these both kind of PWM modulations as shown in the Figure 14c,d respectively.

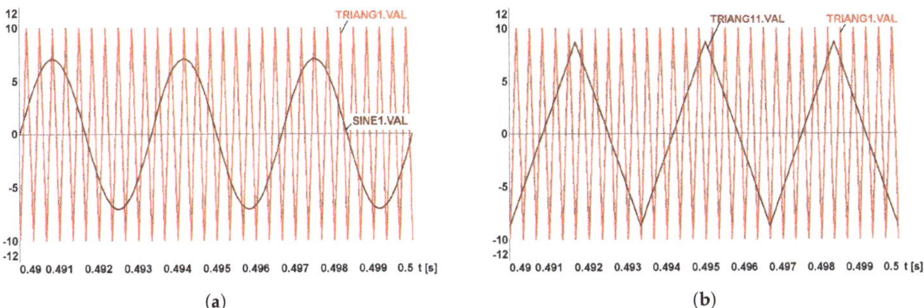

Figure 14. *Cont.*

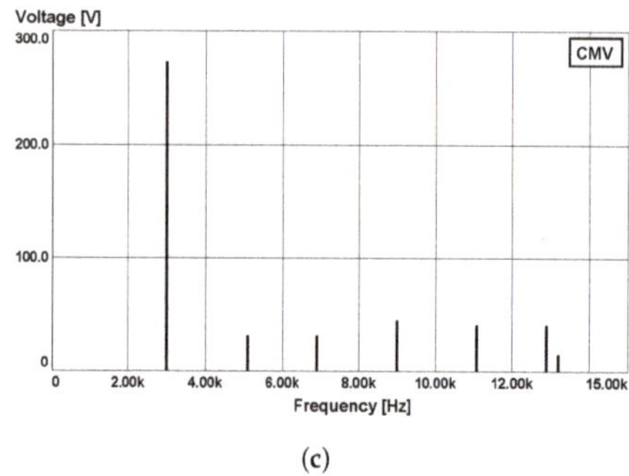
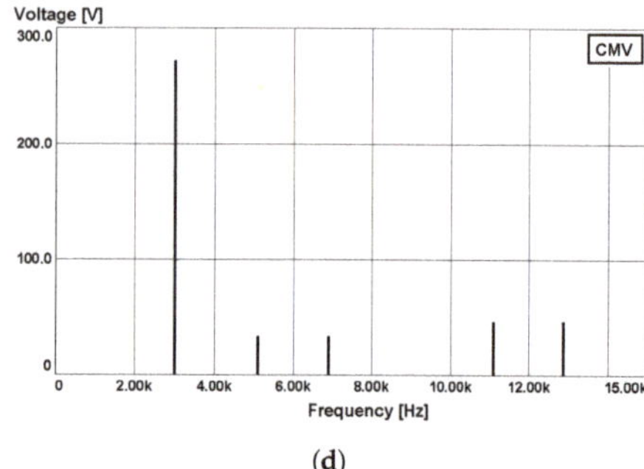

(c)   (d)

Figure 14. Comparison of using different modulations: (**a**) sinusoidal modulation, (**b**) triangular modulation, (**c**) harmonics spectrum in the common-mode (CM) voltage using the sinusoidal modulation, (**d**) harmonics spectrum in the CM voltage using the triangular modulation.

6. Conclusions

The authors proposed a DC 600 V microgrid, which is connected to the intermediate circuits of drive VFC used in the induction motor drives. Thanks to this solution, the efficiency of electric drives has increased, as energy losses on brake resistors for drive converters have been eliminated. The actual efficiency of the converter has not been experimentally tested. The efficiency can be estimated on the basis of the efficiency of the driving frequency converters (VFCs), which reaches values of 98% [21].

In the proposed solution, there are also losses in the rectifier, but the reactive component of the current does not flow via the freewheeling diodes of the inverter while charging the EV battery. The reactive component of the current flows via the freewheeling diodes of the inverter while supplying induction motors. Therefore, the authors conclude that the efficiency of the proposed converter for charging EV batteries will be similar to the efficiency of the drive converter.

The bidirectional converter is an adjustable current source for battery and it was used as an example of fast charging battery converter in this paper. The authors' solution is a converter with adjustable EV battery voltage source. Using of the drive frequency converter as an EV battery charging converter is a novel solution where EV battery is charged via a constant voltage source with regulated value.

The energy supplied by the generator is transferred to the energy storage or other converters connected to the microgrid (load sharing). RES and ES cooperate with the microgrid, which has a hybrid DC converter power supply system for fast charging of EV batteries. Scheduled EV battery charging was used, which is carried out in such a way that when the motor is powered by a frequency converter, it is used to charge an EV battery or a mobile electric work machine. The battery charging converter has been developed through the adaptation of drive VFC, consisting of the attachment of a diode rectifier. The VFC drive is used to set the rectifier DC voltage value. The phase voltage value and frequency are controlled by the PWM drive parameters of the drive VFC inverter.

The use of a microgrid provides the opportunity to integrate a hybrid power supply system for fast EV charging stations in such a way that the battery charging energy does not increase the load of the power system and in addition has an impact on worsening the power quality indicators in the power system.

7. Patents

There are three patent applications resulting from the work presented in this manuscript:

1. Power electronic converter with the conversion of alternating voltage into regulated direct voltage for fast charging of batteries in electric vehicles. Patent application no. P-434784 dated 24 July 2020.

2. Power electronic converter with inverter and rectifier for fast charging of electric vehicle batteries. Patent application no. P-434786 dated 24 July 2020.
3. Power electronic converter with a mobile rectifier set for fast charging of electric vehicle batteries. Patent application no. P-434787 dated 24 July 2020.

Author Contributions: Conceptualization, J.R.S., M.Z.-M., D.W. and N.P.; methodology and software, J.R.S. and M.Z.-M.; validation and formal analysis, D.W. and N.P.; investigation and resources, J.R.S., M.Z.-M.; data curation, D.W.; writing—original draft preparation, writing—review and editing, D.W. and M.Z.-M.; visualization, M.Z.-M.; supervision, J.R.S.; project administration, J.R.S., M.Z.-M., D.W. and N.P.; funding acquisition, D.W., N.P. All authors have read and agreed to the published version of the manuscript.

Funding: This research received no external funding.

Acknowledgments: In the part of the article related to the modelling of the electric circuit, the ANSYS national scientific software license has been used, which was funded by a computational grant obtained by Kazimierz Pulaski University of Technology and Humanities in Radom, Poland. This work was also supported by Gdańsk University of Technology and the Government of the Russian Federation, Grant 08–08.

Conflicts of Interest: The authors declare no conflict of interest.

Abbreviations

The following abbreviations are used in this manuscript:

AFE	Active Front End
CC	Constant Current
CV	Constant Voltage
CM	Common-mode Voltage
DM	Differential-mode Voltage
ES	Energy Storage
EV	Electric Vehicle
PLC	Programmable Logic Controller
PWM	Pulse Width Modulation
RES	Renewable Energy Sources
SOC	State of Charge
THD	Total Harmonic Distortion
VFC	Voltage Frequency Converter
VSC	Voltage Source Converter

References

1. Trivedi, N.; Gujar, N.S.; Sarkar, S.; Pundir, S.P.S. Different fast charging methods and topologies for EV charging. In Proceedings of the IEEMA Engineer Infinite Conference (eTechNxT), New Delhi, India, 13–14 March 2018. [CrossRef]
2. Un-Noor, F.; Padmanaban, S.; Mihet-Popa, L.; Mollah, M.N.; Hossain, E. A Comprehensive Study of Key Electric Vehicle (EV) Components, Technologies, Challenges, Impacts, and Future Direction of Development. *Energies* **2017**, *10*, 1217. [CrossRef]
3. Ahmadi, M.; Mithulananthan, N.; Sharma, R. A review on topologies for fast charging stations for electric vehicles. In Proceedings of the IEEE International Conference on Power System Technology (POWERCON), Wollongong, Australia, 28 September–1 October 2016; pp. 1–6. [CrossRef]
4. Yilmaz, M.; Krein, P.T. Review of Battery Charger Topologies, Charging Power Levels, and Infrastructure for Plug-In Electric and Hybrid Vehicles. *IEEE Trans. Power Electron.* **2013**, *28*, 2151–2169. [CrossRef]
5. Pinto, J.G.; Monteiro, V.; Exposto, B.; Barros, L.A.M.; Sousa, T.J.C.; Monteiro, L.F.C.; Afonso, J.L. Power Electronics Converters for an Electric Vehicle Fast Charging Station with Storage Capability. In *Lecture Notes of the Institute for Computer Sciences, Social Informatics and Telecommunications Engineering, Proceedings of the Green Energy and Networking, Guimarães, Portugal, 21–23 November 2018*; Springer International Publishing: Cham, Switzerland, 2019; pp. 119–130.
6. Collin, R.; Miao, Y.; Yokochi, A.; Enjeti, P.; von Jouanne, A. Advanced Electric Vehicle Fast Charging Technologies. *Energies* **2019**, *12*, 1839. [CrossRef]

7. Dusmez, S.; Cook, A.; Khaligh, A. Comprehensive Analysis of High Quality Power Converters for Level 3 Off-board Chargers. In Proceedings of the IEEE Vehicle Power and Propulsion Conference, Chicago, IL, USA, 6–9 September 2011; pp. 1–10. [CrossRef]

8. Monteiro, V.; Ferreira, J.C.; Meléndez, A.N.; Couto, C.; Afonso, J.L. Experimental Validation of a Novel Architecture Based on a Dual-Stage Converter for Off-Board Fast Battery Chargers of Electric Vehicles. *IEEE Trans. Veh. Technol.* **2018**, *76*, 1000–1011. [CrossRef]

9. Shi, R.; Semsar, S.; Lehn, P.W. Constant Current Fast Charging of Electric Vehicles via a DC Grid Using a Dual-Inverter Drive. *IEEE Trans. Ind. Electron.* **2017**, *64*, 6940–6949. [CrossRef]

10. Yan, X.; Li, J.; Zhang, B.; Jia, Z.; Tian, Y.; Zeng, H.; Lv, Z. Virtual Synchronous Motor Based-Control of a Three-Phase Electric Vehicle Off-Board Charger for Providing Fast-Charging Service. *Appl. Sci.* **2018**, *8*, 856. [CrossRef]

11. Monteiro, V.; Pinto, J.G.; Exposto, B.; Afonso, J.L. Comprehensive Comparison of a Current-Source and a Voltage-Source Converter for Three-Phase EV Fast Battery Chargers. In Proceedings of the 9th International Conference on Compatibility and Power Electronics (CPE), Costa da Caparica, Portugal, 24–26 June 2015. [CrossRef]

12. Szymanski, J.; Zurek-Mortka, M. Harmonic Resonant Filters in Quick Battery Charging Station of Motor Vehicles. In Proceedings of the 23rd International Conference Electronics, Palanga, Lithuania, 17–19 June 2019. [CrossRef]

13. Mansoor, A.; Rajagopalan, S.; Lai, J.S.; Khan, F.H. Photovoltaic Integrated Variable Frequency Drive. U.S. Patent 8,334,616, 18 December 2012.

14. Oulad-Abbou, D.; Doubabi, S.; Rachid, A. Solar Charging Station for Electric Vehicles. In Proceedings of the 3rd International Renewable and Sustainable Energy Conference (IRSEC), Marrakech, Morocco, 10–13 December 2015; pp. 1–5. [CrossRef]

15. Mithulananthan, N.; Hung, D.; Lee, K. *Intelligent Network Integration of Distributed Renewable Generation*; Springer: Berlin/Heidelberg, Germany, 2017. [CrossRef]

16. Paska, J.; Michalski, Ł.; Molik, Ł.; Kocęba, M. Usage of DC microgrids for integration of distributed energy sources. *Energy Market* **2010**, *2*, 118–123.

17. Veneri, O. *Technologies and Applications for Smart Charging of Electric and Plug-In hybrid Vehicles*; Springer: Berlin/Heidelberg, Germany, 2016; pp. 1–307. [CrossRef]

18. Chen, Y.; Wei, W.; Zhang, F.; Liu, C.; Meng, C. Design of PV hybrid DC/AC microgrid for electric vehicle charging station. In Proceedings of the IEEE Transportation Electrification Conference and Expo, Asia-Pacific (ITEC Asia-Pacific), Harbin, China, 7–10 August 2017; pp. 1–6. [CrossRef]

19. Jian, F.; Yong, W.; Le, L.; Shengnan, L. Frequency Domain Harmonic Model of Electric Vehicle Charger Using Three-Phase Uncontrolled Rectifier. In Proceedings of the China International Conference on Electricity Distribution (CICED 2016), Xi'an, China, 10–13 August 2016; pp. 1–5. [CrossRef]

20. Szymanski, J. *Model and Simulation Searching of Modular 12-Pulse Diode Rectifiers for Traction Substation with 6-Pulse Rectifier Used in Industry Frequency Drives Converters*; TTS Technika Transportu Szynowego: Pomorskie, Poland, 2016; Volume 12.

21. Danfoss Technical Note, Design Guide VLT Automation Drive FC 300 90-1200kW. MG34S2020 Rev. 2013-08-1. 2013. Available online: http://files.danfoss.com/download/Drives/MG34S222.pdf (accessed on 4 July 2020)

22. GW, Battery Cell Technical Specification, Winston LFP200AHA Cell. Available online: www.gwl.eu (accessed on 10 June 2020).

23. Marra, F. Electric Vehicles Integration in the Electric Power System with Intermittent Energy Sources—The Charge/Discharge Infrastructure. Ph.D. Thesis, Technical University of Denmark, Kgs. Lyngby, Denmark, March 2013.

24. Anseán, D.; García, V.M.; Viera, J.C.; Garcia, V.M.; Alvarez, J.C.; Blanco, C. Electric Vehicle Li-ion Battery Evaluation based on Internal Resistance Analysis. In Proceedings of the 2014 IEEE Vehicle Power and Propulsion Conference (VPPC), Coimbra, Portugal, 27–30 October 2014; Volume 3, ISSN 2032-6653. [CrossRef]

25. Marra, F.; Yang, G.Y.; Traeholt, C.; Larsen, E.; Rasmussen, C.N.; You, S. Demand Profile Study of Battery Electric Vehicle under Different Charging Options. In Proceedings of the IEEE Power Energy Society General Meeting, San Diego, CA, USA, 22–26 July 2012. [CrossRef]

26. Danfoss Technical Note, VLT High Power Drives to fit your application. Selection Guide. DKDD.PB.404.A4.22. 2016. Available online: http://files.danfoss.com/download (accessed on 4 July 2020).
27. Fuji Electric, Data Sheet of Power Circuit: 7MBR100VX120-50. Available online: http://www.fujielectric.com (accessed on 2 July 2020).
28. Han, J.; Doo-Ung, K.; Yun-Sik, O.; Gi-Hyeon, G.; Chul-Ho, N.; Tack-Hyun, J.; Chul-Hwan, K. Modeling of Bi-directional DC/DC Converter for Connecting DC Distribution System using EMTP. *Trans. Korean Inst. Electr. Eng.* **2014**, *63*, 615–621. [CrossRef]
29. Betten, J.; Kollman, R. Interleaving DC/DC Converters Boost Efficiency and Voltage-Though somewhat more complex than single-phase designs, interleaved-boost converters run cooler, occupy less space, and can cost less. *EDN* **2005**, *50*, 77–88.
30. Nair, A.C.; Fernandes, B. A Solid State Transformer based Fast Charging Station for all Categories of Electric Vehicles. *IEEE Proc.* **2018**. [CrossRef]
31. Schupbach, R.M.; Balda, J.C. Comparing DC-DC converters for power management in hybrid electric vehicles. In Proceedings of the Electric Machines and Drives Conference IEMDC, Madison, WI, USA, 1–4 June 2003; Volume 3, pp. 1369–1374. [CrossRef]
32. Zhang, J.; Lai, J.S.; Kim, R.Y.; Yu, W. High-Power Density Design of a Soft-Switching High-Power Bidirectional DC–DC Converter. *IEEE Trans. Power Electron.* **2007**, *22*, 1145–1153. ISSN: 1941-0107. [CrossRef]
33. Danfoss Technical Note, Load Sharing. MI.50.N2.02, 2002. Available online: http://files.danfoss.com (accessed on 4 July 2020).
34. Szymanski, J.; Zurek-Mortka, M.; Sadhu, P.K.; Goswami, A. Mitigation Methods of Ground Leakage Current Caused by Common-Mode in Voltage Frequency Drives. *Energy Syst. Drives Autom.* **2020**, *664*, 1–10, ISSN: 1876-1100. [CrossRef]
35. Schneider Electric, Transformation from Six-Pulse to Low Harmonic, Three-Level, Active Rectification Technologies. 2017. Available online: https://download.schneider-electric.com (accessed on 4 July 2020).
36. Pistoia, G.; Liaw, B. *Behaviour of Lithium-Ion Batteries in Electric Vehicles. Battery Health, Performance, Safety, and Cost*; Springer International Publishing: Cham, Switzerland, 2018; ISSN: 1865-3529. [CrossRef]

Article

Suppression of Supply Current Harmonics of 18-Pulse Diode Rectifier by Series Active Power Filter with LC Coupling

Wojciech Sleszynski [1], Artur Cichowski [1] and Piotr Mysiak [2,*]

[1]   Faculty of Electrical and Control Engineering, Gdansk University of Technology, 11/12 Narutowicza St., 80-233 Gdansk, Poland; wojciech.sleszynski@pg.edu.pl (W.S.); artur.cichowski@pg.edu.pl (A.C.)

[2]   Faculty of Electrical Engineering, Gdynia Maritime University, 81-87 Morska St., 81-225 Gdynia, Poland

*   Correspondence: p.mysiak@we.umg.edu.pl; Tel.: +48-58-347-2954

Received: 27 October 2020; Accepted: 17 November 2020; Published: 19 November 2020

Abstract: The reported research aims at improving the quality of three-phase rectifier supply currents. An effective method consists of adding properly formed booster voltages to the fundamental supply voltages using a series active filter. In the proposed solution, the booster voltages are generated by three single-phase systems consisting of inverters, LC filters, and single-phase transformers. The application of LC couplings ensures low emission of disturbances, but may provoke compensator stability problems. The article presents the current control system for a series active filter designed to suppress the dominant harmonics in the supply currents of an 18-pulse rectifier, without interference into fundamental current components. A proportional control is proposed in combination with integral terms implemented in the orthogonal coordinate systems, which synchronically rotate with frequencies equal to those of the harmonic components to be eliminated. The use of complex gains in integral terms allows a simple phase correction of the output signals. A description is given of the method to determine controller parameters based on the mathematical model of the control object. Sample results of experimental tests performed in steady-state and transient conditions are included to illustrate the quality of performance of the series active filter as compared to the results recorded for the rectifier alone, and for the rectifier with additional line reactor. The applied control method of active filter significantly reduces harmonic distortion of the grid current, which is particularly advantageous at nonideal supply voltage and low loads.

Keywords: series active power filters; multipulse converters; power conditioning; coupled reactors

---

1. Introduction

Diode rectifiers are frequently used in industry, because of their low cost, high reliability, and low-level emission of disturbances. Unfortunately, the simplest rectifier solutions usually draw a highly distorted current from the electrical grid. However, after many years of deploying them in the industry, effective methods have been developed to improve the quality of the input current. One of these methods is the use of multipulse rectifiers [1–4], whose supply line current has a multistep shape which is characterized by a lower content of higher harmonics. If the galvanic separation is not required and there is no need to adjust the voltage between the supply line and the load, then the multipulse diode rectifiers with coupled reactors are a good solution. The main advantage, in comparison to multipulse converters with transformers, is the much smaller limiting power of the required electromagnetic elements, resulting in the smaller dimensions and weight of the entire rectifying device [5,6]. The downside is that when the supply voltage is unsymmetrical with higher harmonics it results in distortions of the supply current [5]. In such cases, it is advisable to use an additional smaller rated Series Active Power Filter (S-APF), which further improves the power quality [7].

Several control methods of the S-APF connected to the input of a multipulse rectifier are presented in the literature [8–12]. The first developed control algorithms were implemented using analogue control techniques [8,9]. The fundamental harmonic was removed from the supply currents, and the remaining signal was properly amplified and added to the rectifier supply voltages, thus reducing current distortion. However, in the case of digital control, due to unavoidable delays in the S-APF control system, this method did not bring satisfactory results [10]. That is why a DFT (discrete Fourier transform) based control algorithm [10] has been proposed, in which only dominating harmonics of the supply current are extracted and suppressed. Unfortunately, the authors did not use a switching-ripple filter in the S-APF system and provided only general guidelines for the selection of the regulator parameters.

Digital control of S-APF as a current source based on the hysteresis controller was proposed in [11] by the authors of this article. In [12] instead of current controllers with a large bandwidth, proportional-integral controllers in multiple reference frames were used for selective line current harmonic suppression. Both solutions ensure very good quality of rectifier supply currents, but insufficiently suppress current and voltage ripples caused by transistor switching. This article presents the research results of the S-APF system additionally equipped with the LC ripple filter, and the current controller with a simple structure, which enables additional phase correction of the outputs of integral terms.

2. Converter System Characteristic

A simplified schematic diagram of the analyzed ac/dc supply system is shown in Figure 1.

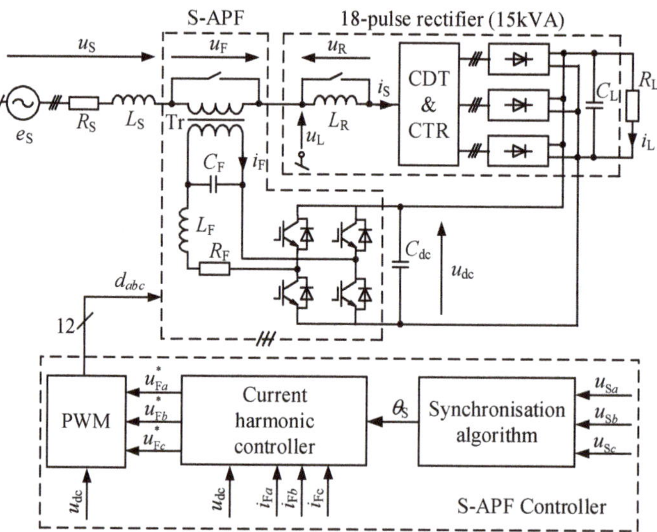

Figure 1. Schematic diagram of the proposed ac/dc supply system based on 18-pulse rectifier and series active power filter with LC output stage.

Three-phase supply is modeled by the voltage source e_S, resistance R_S, and inductance L_S, which also represents the leakage inductance of a rectifier's magnetic circuits. The system is composed of two separate modules: an 18-pulse rectifier, and a series active filter. Three system configurations are possible: (1) only 18-pulse rectifier, (2) rectifier with additional series reactor, and (3) rectifier with the series active filter. The main element of the system is the 18-pulse rectifier based on current dividing transformer (CDT) for preliminary current division, and the set of coupled three-phase reactors (CTR) [5]. The above magnetic elements compose three 3-phase voltage systems, shifted by 20° in relation to each other. Six-pulse rectifiers with a shared output capacitor are connected to CTR outputs. The 18-pulse rectifier enables reduction of undesired higher harmonics from the supply network currents, mainly of the order of 5, 7, 11, and 13.

The series active filter consists of three single-phase circuits, each composed of a voltage source inverter (VSI) with IGBTs transistors. The dc circuits of these inverters are connected to the output of the 18-pulse rectifier. The ac sides of the inverters are series-connected to the supply voltage via LC filters and step-up transformers (Tr) with voltage ratio 1:12. During system start-up and operation only with the rectifier, the S-APF is bypassed by a contactor.

The parameters of the converter system are given in Table 1. The supply resistance and inductance were measured using the loop impedance meter. The series injection transformer (Tr) was selected through simulation research, based on the results presented in [8,13]. The transformer is represented by its classical circuit model, excluding a magnetizing branch. The transformer parameters listed in Table 1 were determined on the basis of a short-circuit test.

Table 1. Parameters of the proposed ac/dc supply system.

Symbol	Description	Value
E_S	Phase voltage of the supply (50 Hz)	230 V
L_{Sp}	Supply inductance referred to the primary side of the transformer	7.2 mH
R_{Sp}	Supply resistance referred to the primary side of the transformer	57.6 Ω
P_{REC}	Nominal output power of the 18-pulse rectifier	15 kW
C_L	Rectifier output capacitance	10 mF
S_T	Nominal power of the series injection transformer (Tr)	800 VA
U_{Tp}	Nominal primary voltage of the transformer Tr	300 V
I_{Tp}	Nominal primary current of the transformer Tr	2.9 A
N_T	Turns-ratio of the series injection transformer Tr	12
L_T	Leakage inductance of the windings of the transformer referred to the primary side	3.46 mH
R_T	Resistance of the windings of the transformer referred to the primary side	3.7 Ω
L_{TS}	Equivalent inductance, sum of L_T and L_{Sp}	10.66 mH
R_{iS}	Equivalent resistance, sum of R_T and R_{Sp}	61.3 Ω
L_F	Inductance of the switching ripple filter inductor	20 mH
R_F	Resistance of the switching ripple filter inductor	0.5 Ω
C_F	Capacitance of the switching ripple filter	0.56 μF
T_d	Delay introduced by control system and VSI	75 μs
f_s	Sampling and PWM switching frequency	20 kHz

The inductance (L_F) and capacitance (C_F) of the switching ripple filter were selected assuming the maximum ripple of the inductor current and the capacitor voltage. Figure 2a shows a simplified circuit diagram of the VSI and its output filter to define the signals, the waveforms of which are presented on Figure 2b. It shows the branch voltages u_1, u_2 and the output voltage u_o of the VSI, as well as the capacitor voltage u_C and its averaged value over the sampling period $u_{C,avg}$. In addition to the voltages, the figure also shows, in an idealized way, the VSI output current i_o and its averaged value over the sampling period $i_{o,avg}$. In the case of unipolar modulation with the double update mode the largest ripples of inductor current occur when the duty cycle $m = u_o/u_{dc}$ is equal to 0.5. The simplified waveforms in Figure 2b apply to this case.

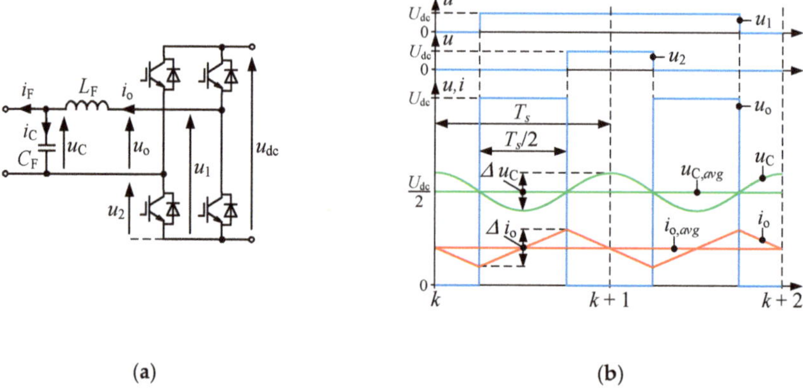

(a)   (b)

Figure 2. Simplified circuit diagram of the VSI and switching ripple filter (**a**) and waveforms of the characteristic signals in the case of maximum ripples of capacitor voltage and inductor current (**b**).

Taking into account the time interval of length $T_s/2$ (Figure 2b), in which the current i_o increases from the minimum value to the maximum, the peak-to-peak value of the current ripple Δi_o can be calculated from the formula [14]:

$$\Delta i_o = \frac{U_{dc}}{4L_F f_s},\tag{1}$$

where U_{dc} is the maximum *dc* link voltage equal to 500 V.

Lower current ripple reduces inductor high frequency losses and for this reason a choke with a relatively high inductance $L_F = 20$ mH was selected. The maximum current ripple is at 11% of the peak nominal input current of the 18-pulse rectifier current, referred to the primary side of the transformer Tr.

Considering the time interval $T_s/2$, in which the capacitor voltage u_C varies from the minimum to the maximum value (Figure 2b), the maximum capacitor voltage ripple Δu_C can be estimated by the equation [14]:

$$\Delta u_C = \frac{\Delta i_o}{8C_F f_s} = \frac{U_{dc}}{32L_F C_F f_s^2},\tag{2}$$

It was assumed that the capacitor voltage ripple should not exceed 1% of the peak output voltage of the VSI, which is equal to U_{dc} in the worst case. Finally, the value of $C_F = 560$ nF was selected, for which maximum Δu_C equals 3.5 V (0.7% of the U_{dc}).

The control algorithm was implemented in the microprocessor controller based on the digital signal processor TMS320C6713 and the programmable system FPGA Cyclone IV. To execute the control algorithm, measurements were performed of the supply network phase voltages (u_{Sa}, u_{Sb}, u_{Sc}), transformer phase currents (i_{Fa}, i_{Fb}, i_{Fc}) on the inverter side, and the rectifier output voltage u_{dc}.

3. The Structure of Multiple Reference Frame Current Controller

The task of the series active filter is to improve the quality of the supply current by suppressing higher harmonics and compensating the asymmetry of the fundamental components. A simplified schematic diagram of the S-APF control system is shown in the lower part of Figure 1. Three functional blocks are singled out in this part: pulse width modulator (PWM), synchronization algorithm [12,15], and the current controller, which will be described in detail further in the article. It was developed on the basis of the research results reported in [12,16–18].

In order to implement the control system, the three-phase quantities x_a, x_b, and x_c were converted using the space vector defined as:

$$\underline{x}_{\alpha\beta} = x_{\alpha\beta} + jx_{\alpha\beta} = \frac{2}{3}\left(x_a + x_b e^{j2\pi/3} + x_c e^{-j2\pi/3}\right). \tag{3}$$

In steady state, the current space vector $\underline{i}_{F\alpha\beta}$ can be approximated by a complex Fourier series given by the formula:

$$\underline{i}_{F\alpha\beta} = \sum_{m=-\infty}^{\infty} \underline{I}_{F\alpha\beta m} e^{jm\omega_1 t} \cong \sum_{m\in M} \underline{I}_{F\alpha\beta m} e^{jm\theta_1}. \tag{4}$$

where $\underline{I}_{F\alpha\beta m} = |I_{F\alpha\beta m}| e^{j\varphi_m}$ is the complex-valued amplitude of m-th current harmonic belonging to the set of M dominant harmonics to be suppressed; ω_1 is the frequency of the fundamental component, and θ_1 is its instantaneous phase.

The 18-pulse rectifier draws the distorted current from the supply source. The dominating frequencies in this current are of the order of $m = -17, 19, -35, 37, \ldots$ while the harmonics of the space vector of supply network voltage are usually of the order of $m = -1, -5, 7, -11, 13, \ldots$, where the component $m = -1$ represents the asymmetry of fundamental components of phase voltages. Suppressing the harmonics of the above orders is the basic task of the proposed current controller. Additional S-APF functions, requiring the adjustment of the fundamental harmonic of the rectifier supply voltage, such as the power factor compensation or stabilization of the rectifier's output voltage are not implemented in considered system. They require the use of a series transformer with a higher rated power and an appropriate voltage ratio and also increase the S-APF power losses [13].

The block diagram of the current controller is shown in Figure 3. To limit the S-APF power losses related to the first harmonic, the proposed current controller should not affect the fundamental component of the current taken from the supply network. Consequently, this harmonic was removed from the space current vector $\underline{i}_{F\alpha\beta}$ using the fundamental component filter [12] based on recursive discrete Fourier transform. The signal $\underline{e}_{F\alpha\beta}$ created in the above way is the control error, assuming that the reference value for all compensated current harmonics is zero.

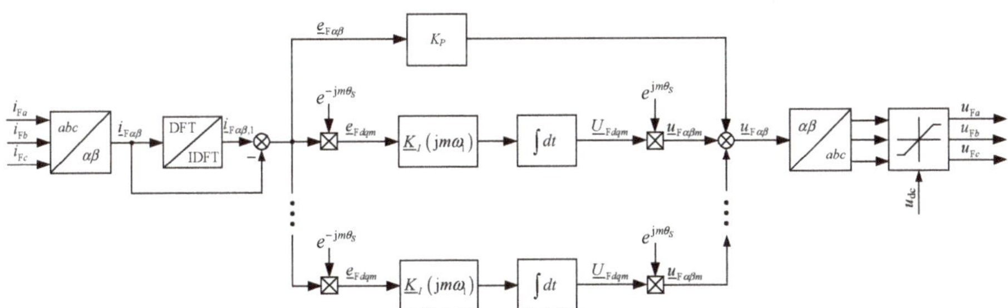

Figure 3. Simplified block diagram of the current harmonic controller.

The error signal is given to the input of the proportional term of the current controller, and to the inputs of the integral terms implemented in the synchronous coordinate systems dq_m, the number of which is equal to the number of compensated current harmonics. The error signal is converted to m synchronous coordinate systems by multiple Park transformations defined by the formula:

$$\underline{e}_{Fdqm} = \underline{e}_{F\alpha\beta} e^{-jm\theta_S}, \tag{5}$$

where $\underline{e}_{Fdqm}$ is the error signal converted to the coordinate system dq_m rotating with frequency $m\omega_S$, ω_S is the estimated frequency of fundamental component, and $\theta_S = \omega_S t$ is the estimated instantaneous phase. After transformation, the control error harmonic of the order m becomes a constant component in dq_m

frame and is amplified by the integral term of current controller which corresponds to this harmonic. Simultaneously, the remaining harmonic components in the output signal $\underline{U}_{Fdqm}$ are suppressed:

$$\underline{U}_{Fdqm}[k] = \sum_{n=k_0}^{k} \underline{K}_i(jm\omega_1)\underline{e}_{Fdqm}[n], \tag{6}$$

where k_0 is the time of control algorithm activation, and $\underline{K}_i(jm\omega_1)$ is the complex-value gain of the frequency-dependent integral term of the controller.

In the proposed controller implementation, the integral gain is a complex number which also determines relevant phase shift of the output signal of this block after its reconversion to the coordinate system $\alpha\beta$ using the inverse Park transform:

$$\underline{u}_{F\alpha\beta m} = \underline{U}_{Fdqm}e^{-jm\theta_s}. \tag{7}$$

The sum of output signals from particular integral terms and from the proportional part is the controller output signal in the coordinate system $\alpha\beta$.

$$\underline{u}_{F\alpha\beta} = K_p\underline{e}_{F\alpha\beta} + \sum_{m\in M} \underline{u}_{F\alpha\beta m}. \tag{8}$$

when it is too large, the modulus of the output voltage space vector $\underline{u}_{F\alpha\beta}$ is limited to the voltage in the *dc* circuit. The calculated output signal from the current controller is passed to the input of pulse width modulator.

4. Selection of Controller Settings

When selecting controller settings, the magnetizing branch in the transformer model was omitted. Then, the transfer function of the circuit coupling the inverter with the supply network takes the form:

$$G_o(s) = \frac{1}{s^3 L_F L_{TS} C_F + s^2 C_F (L_F R_{TS} + L_{TS} R_F) + s(L_F + L_{TS} + C_F R_F R_{TS}) + R_F + R_{TS}} e^{-sT_d}. \tag{9}$$

Selection of controller settings started with determining the gain of the proportional part of the controller. This setting was calculated based on the assumed gain margin for the open-loop system working only with the proportional controller. For the assumed gain margin of 10 dB, the calculated proportional coefficient was equal to $K_p = 44$.

To select the integral gains, the current control system was treated as a multiloop scheme, the inner feedback loop of which consists only of a proportional term [18]. Then, the integral gains were selected in such a way as to compensate the remaining errors for the frequencies of dominant harmonics. The error which should be compensated by the integral terms is given by the following transfer function:

$$G_e(s) = \frac{E_{F\alpha\beta}(s)}{U_I(s)} = \frac{G_o(s)}{1 + K_p G_o(s)} = \frac{1}{K_p}G_{cp}(s), \tag{10}$$

where $E_{F\alpha\beta}(s)$ is the Laplace transform of the control system error, $U_I(s)$ is the Laplace transform of the output signal of the integral part (corresponding to the sum of output signals from individual integral terms), and $G_{cp}(s)$ is the transfer function of the closed-loop system when only the proportional controller is used:

$$G_{cp}(s) = \frac{K_p G_o(s)}{1 + K_p G_o(s)}. \tag{11}$$

Figure 4 shows the Bode plots of the open and closed-loop control system with only the proportional term. The proportional controller with fixed setting is not able to suppress harmonics effectively. The task of the integral terms is to increase controller gain for selected frequencies. Considering the formula (8), the transfer function of the integral term of the proposed controller in the stationary coordinate system $\alpha\beta$ is as follows:

$$G_{im}(s) = \frac{U_{F\alpha\beta m}(s)}{E_{F\alpha\beta}(s)} = \underline{K}_i(jm\omega_1)\frac{1}{s - jm\omega_1},\tag{12}$$

in which the complex-value integral gain, shown in the scheme in Figure 3, is given by:

$$\underline{K}_i(jm\omega_1) = \frac{K_p}{T_i\underline{G}_{cp}(jm\omega_1)}.\tag{13}$$

where the controller integration time was assumed equal to $T_i = 0.5/f_1 = 10$ ms.

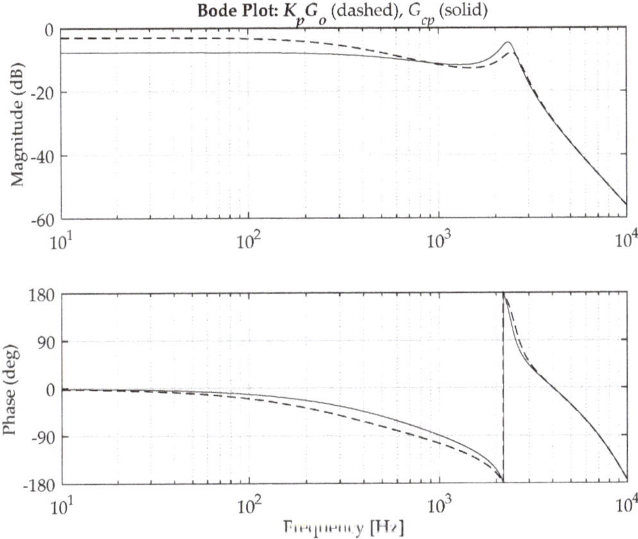

Figure 4. Bode plots of open-loop and closed-loop control systems with proportional controller.

The gains of the integral terms have complex values and ensure proper phase correction of particular output signals $u_{F\alpha\beta m}$, without additional trigonometric function calculations.

5. Laboratory Results

To verify the operation of the S-APF control algorithm, a series of experimental tests were performed. The task of the current controller was to suppress current harmonics of the following orders: -1, ± 3, $\pm(6n \pm 1)$ for $n = 1, 2, \ldots, 6$. Firstly, the steady-state operation of the current control system was tested. The obtained results were compared with the data recorded for two remaining configurations of the converter system. Figure 5 shows sample oscillograms of supply currents and their amplitude spectra recorded at nominal load: (1) for only 18-pulse rectifier (Figure 5a,b), (2) for rectifier with additional series reactor L_R (Figure 5c,d), and (3) for rectifier with series active filter (S-APF) (Figure 5e,f). The waveforms of the phase currents with their amplitude spectra and THD (total harmonic distortion) values were recorded and calculated using the Precision Power Analyzer LMG670 made by Zes Zimmer. The amplitude spectra of the supply currents are given in logarithmic scale.

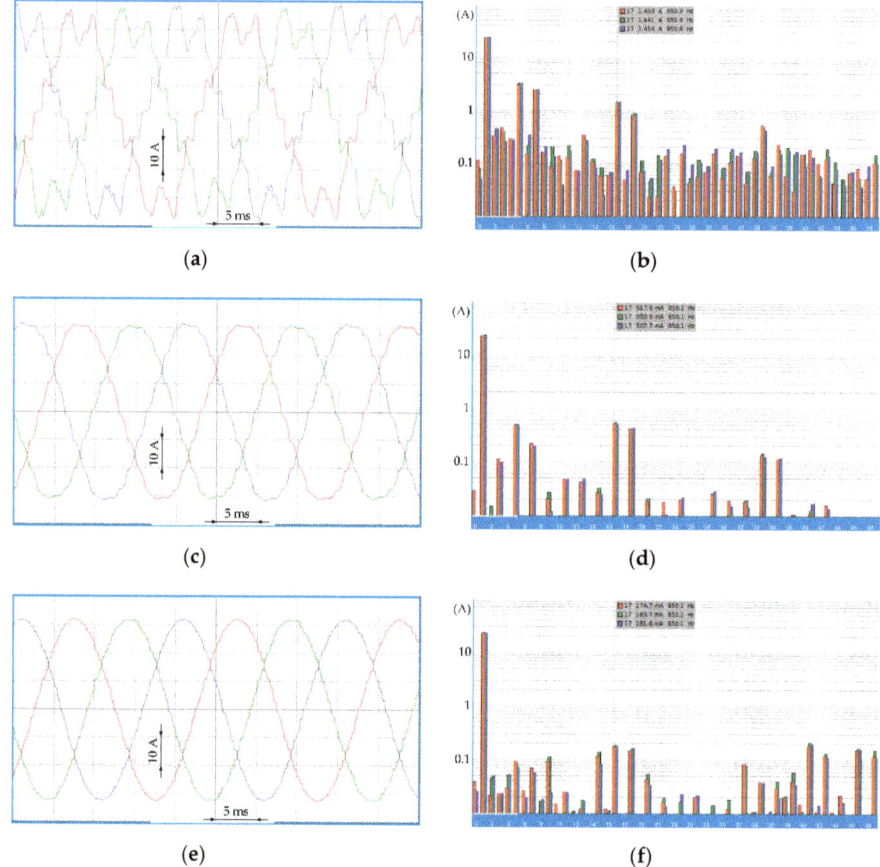

Figure 5. Oscillograms and spectra of converter supply currents at nominal load: (**a,b**) system without reactor L_R and S-APF; (**c,d**) system with reactor L_R and without S-APF; (**e,f**) system without reactor L_R and with S-APF.

The amplitudes of the dominant harmonics are the lowest when using S-APF. In each spectrum shown in Figure 5, the amplitudes of the 17-th harmonic (about 850 Hz) are marked for three phase currents. The application of the series reactor reduced the values of this harmonic from the level of 1.433 A to 0.525 A, while the use of S-APF to the level of 0.180 A. In general, all dominating harmonics considered in the current controller with S-APF were reduced, as compared to the remaining configurations.

Figure 6 shows the results of measurements of supply current THD and output voltage for three system configurations. The introduction of a series reactor alone has already reduced significantly the harmonic distortion of supply currents as compared to the autonomous operation of the 18-pulse rectifier. Replacing the reactor L_S with the S-APF system improves the quality of supply currents within the entire output power range. For the configuration with S-APF operating at nominal load, the THD of supply currents remains at the approximate level of 2%. Significant improvement in quality of supply currents can be observed during converter system operation at low load. Compared to the configuration with additional series reactor, the system with S-APF also ensures slightly higher output voltage.

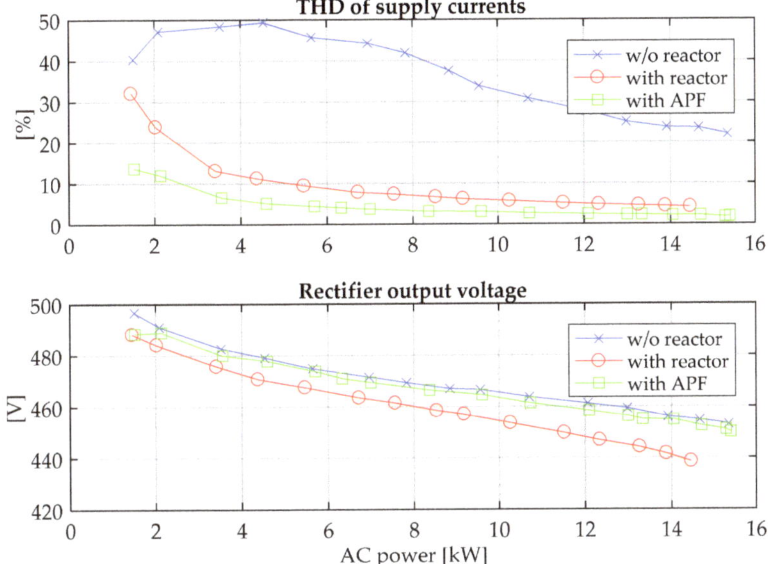

Figure 6. THD of supply currents (**top**) and rectifier output voltage (**bottom**) as functions of input power for three converter system configurations.

Figures 7–9 show sample results of converter system examination in dynamic states. The values of supply network currents were measured and recorded in the converter control system unit. To illustrate the quality of the controller's performance, THD values of the supply current are shown, which were obtained from harmonic values calculated in moving window with fixed width corresponding to frequency 50 Hz:

$$THD[k] = \frac{\sqrt{\sum_{m=-40}^{-1} \left| I_{S\alpha\beta m}[k] \right|^2 + \sum_{m=2}^{40} \left| I_{S\alpha\beta m}[k] \right|^2}}{\left| I_{S\alpha\beta 1}[k] \right|}, \tag{14}$$

where $I_{S\alpha\beta m}[k]$ is the amplitude of m-th harmonic of the supply current converted to the coordinate system $\alpha\beta$, calculated in the moving window, and $I_{S\alpha\beta 1}[k]$ is amplitude of fundamental harmonic of the supply current.

Figure 7 shows the waveforms of supply network currents and THD values after control algorithm activation when the system is loaded with half of the nominal power. During two supply network voltage periods, significant reduction in the level of harmonic distortion is observed, from about 36% to 14.5%, with further slower reduction to the approximate level of 3.7%.

Figures 8 and 9 show sample transient states related with load change at converter system output. Figure 8 presents the step increase of load from 33% to 100%, while Figure 9 shows the reverse situation, i.e., rapid load drop.

In both situations, the converter system operation is stable and leads to the reduction of supply current harmonic distortions after the transient state. Unfortunately, during transients the THD values are not a meaningful indicator and reach disproportionately high values in relation to the level of distortion visible in the current waveforms.

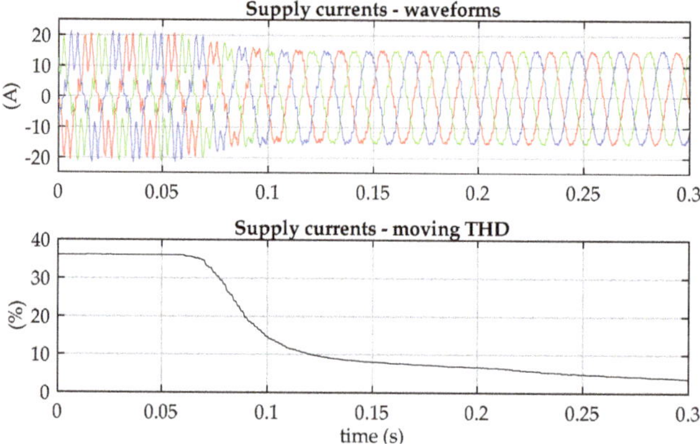

Figure 7. Supply current waveforms and THD values recorded after initiating the current control algorithm.

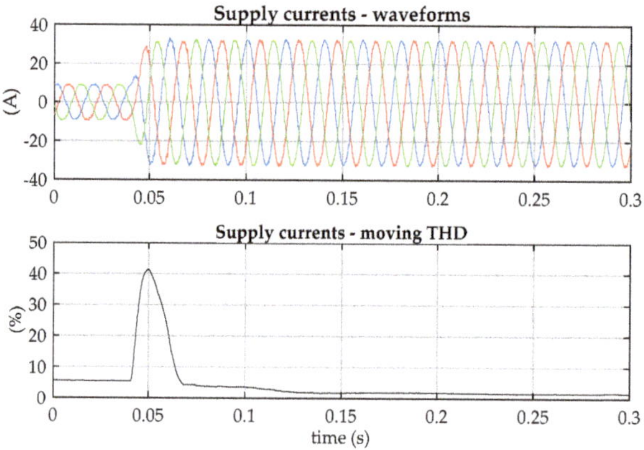

Figure 8. Supply current waveforms and THD values recorded after load increase from 33% to 100%.

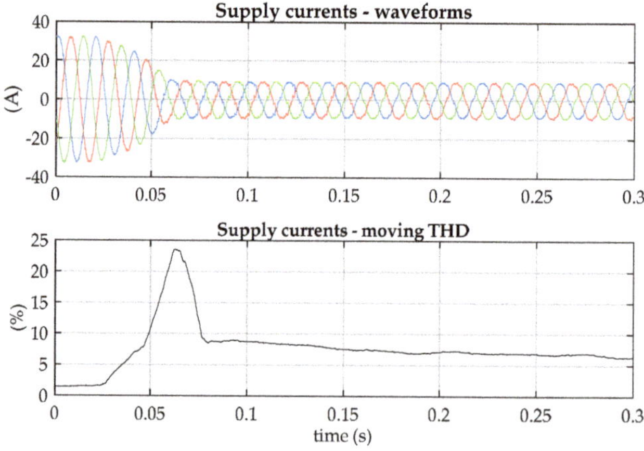

Figure 9. Supply current waveforms and THD values recorded after load decrease from 100% to 33%.

6. Conclusions

The article proposed a current harmonic controller for a series active filter integrated with 18-pulse diode rectifier with coupled reactors. The ac sides of S-APF inverters were coupled with booster transformers via LC filters, which enables significant reduction of booster voltage ripples, but may lead to unstable operation of the converter system.

To suppress undesired harmonics in supply currents, a proportional controller was used with integral terms implemented in multiple coordinate systems, rotating synchronously with angular frequencies of the dominant harmonics. The use of integral gains with complex values, ensures proper phase correction of integral's output signals and stable operation of the converter system.

The proposed current controller suppresses dominating harmonics up to the order of 37. For nominal load, the controller can reduce the THD coefficient from 22% to about 2%.

Author Contributions: Conceptualization, methodology, validation, W.S., A.C. and P.M.; formal analysis, investigation, visualization, W.S. and A.C.; resources, project administration, W.S. and P.M.; software, data curation, writing—original draft preparation, supervision, W.S.; writing—review and editing, A.C. and P.M.; funding acquisition, P.M. All authors have read and agreed to the published version of the manuscript.

Funding: This research was partially funded by the Polish National Centre for Research and Development, grant number PBS1/A4/5/2012.

Conflicts of Interest: The authors declare no conflict of interest.

References

1. Singh, B.; Gairola, S.; Singh, B.N.; Chandra, A.; Al-Haddad, K. Multipulse AC-DC Converters for Improving Power Quality: A Review. *IEEE Trans. Power Electron.* **2008**, *23*, 260–281. [CrossRef]
2. Mysiak, P. *Multi-Pulse Diode Rectifiers with Higher Current Harmonic Blocking Reactors*; Publishing House of Gdynia Maritime University: Gdynia, Poland, 2010.
3. Lian, Y.; Yang, S.; Ben, H.; Yang, W. A 36-Pulse Diode Rectifier with an Unconventional Interphase Reactor. *Energies* **2019**, *12*, 820. [CrossRef]
4. Ryndzionek, R.; Sienkiewicz, Ł. Evolution of the HVDC Link Connecting Offshore Wind Farms to Onshore Power Systems. *Energies* **2020**, *13*, 1914. [CrossRef]
5. Mysiak, P.; Strzelecki, R. A robust 18-pulse diode rectifier with coupled reactors. *Bull. Pol. Acad. Sci. Tech. Sci.* **2012**, *59*, 541–550. [CrossRef]
6. Iwaszkiewicz, J.; Mysiak, P. Supply System for Three-Level Inverters Using Multi-Pulse Rectifiers with Coupled Reactors. *Energies* **2019**, *12*, 3385. [CrossRef]
7. Strzelecki, R.; Mysiak, P. A hybrid, coupled reactors based 18-pulse diode rectifier with active power filter. In Proceedings of the 2014 IEEE International Conference on Intelligent Energy and Power Systems (IEPS), Kiev, Ukraine, 2–6 June 2014; pp. 94–101.
8. Fujita, H.; Akagi, H. An approach to harmonic current-free AC/DC power conversion for large industrial loads: The integration of a series active filter with a double-series diode rectifier. *IEEE Trans. Ind. Appl.* **1997**, *33*, 1233–1240. [CrossRef]
9. Srianthumrong, S.; Fujita, H.; Akagi, H. Stability analysis of a series active filter integrated with a double-series diode rectifier. *IEEE Trans. Power Electron.* **2002**, *17*, 117–124. [CrossRef]
10. le Roux, A.D.; du Mouton, H.T.; Akagi, H. DFT-Based Repetitive Control of a Series Active Filter Integrated With a 12-Pulse Diode Rectifier. *IEEE Trans. Power Electron.* **2009**, *24*, 1515–1521. [CrossRef]
11. Mysiak, P.; Sleszynski, W.; Cichowski, A. Experimental test results of the 150kVA 18-pulse diode rectifier with series active power filter. In Proceedings of the 2016 10th International Conference on Compatibility, Power Electronics and Power Engineering (CPE-POWERENG), Bydgoszcz, Poland, 29 June–1 July 2016; pp. 380–383.
12. Śleszyński, W.; Cichowski, A.; Mysiak, P. Current harmonic controller in multiple reference frames for series active power filter integrated with 18-pulse diode rectifier. *Bull. Pol. Acad. Sci. Tech. Sci.* **2018**, *66*, 699–704.
13. Dixon, J.W.; Venegas, G.; Moran, L.A. A series active power filter based on a sinusoidal current-controlled voltage-source inverter. *IEEE Trans. Ind. Electron.* **1997**, *44*, 612–620. [CrossRef]

Energies **2020**, *13*, 6060

14. Boillat, D.O.; Friedli, T.; Mühlethaler, J.; Kolar, J.W.; Hribernik, W. Analysis of the design space of single-stage and two-stage LC output filters of switched-mode AC power sources. In Proceedings of the 2012 IEEE Power and Energy Conference at Illinois, Champaign, IL, USA, 24–25 February 2012; pp. 1–8.

15. McGrath, B.P.; Holmes, D.G.; Galloway, J.J.H. Power converter line synchronization using a discrete Fourier transform (DFT) based on a variable sample rate. *IEEE Trans. Power Electron.* **2005**, *20*, 877–884. [CrossRef]

16. Lascu, C.; Asiminoaei, L.; Boldea, I.; Blaabjerg, F. High Performance Current Controller for Selective Harmonic Compensation in Active Power Filters. *IEEE Trans. Power Electron.* **2007**, *22*, 1826–1835. [CrossRef]

17. Lascu, C.; Asiminoaei, L.; Boldea, I.; Blaabjerg, F. Frequency Response Analysis of Current Controllers for Selective Harmonic Compensation in Active Power Filters. *IEEE Trans. Ind. Electron.* **2009**, *56*, 337–347. [CrossRef]

18. Buso, S.; Mattavelli, P. Digital Control in Power Electronics, 2nd Edition. *Synth. Lect. Power Electron.* **2015**, *5*, 1–229. [CrossRef]

Article

The Unified Power Quality Conditioner Control Method Based on the Equivalent Conductance Signals of the Compensated Load

Andrzej Szromba

Faculty of Electrical and Computer Engineering, Cracow University of Technology, 31-155 Cracow, Poland; aszromba@pk.edu.pl

Received: 31 October 2020; Accepted: 24 November 2020; Published: 29 November 2020

Abstract: This paper proposes a new control method for a Unified Power Quality Conditioner (UPQC). This method is based on the load equivalent conductance approach, originally proposed by Fryze. It can be useful not only for compensation for nonactive current and for improving voltage quality, but it also allows one to perform some unconventional functions. This control method can be performed by extending the orthodox notion of 'static' load equivalent conductance into a time-variable signal. It may be used to characterize energy changes in the whole UPQC-and-load circuitry. The UPQC can regulate energy flow between all sources and loads being under compensation. They may be located as well for UPQC's AC-side as DC-side. System works properly even if they switch their activity to work either as loads or generators. The UPQC can operate also as a buffer, which can store/share the in-load generated energy amongst other loads, or it can transmit this energy to the source. Therefore, in addition to performing the UPQC's conventional compensation tasks, it can also serve as a local energy distribution center.

Keywords: power quality; power distribution; reverse power flow; compensation for nonactive current; voltage regulation; UPQC

---

1. Introduction

It is well known that distortions of supply voltage and load current cause power quality degradation, diminish the power factor of the supply system and may result in disruption of sensitive loads [1]. Shunt and series active power filters can alleviate these problems. Shunt filters are intended to compensate for load non-active current whereas series ones can improve supply voltage quality. The Unified Power Quality Conditioner (UPQC) can integrate advantages of both shunt and series active filters in order to achieve control over load voltage and source (line) current [2,3].

A wide review on UPQC configurations can be found in [3]. According to this classification the discussed UPQC can be classified as intended for a single-phase supply system that is based on a two-H-bridge converters employing the same DC-link capacitor. Depending on the point of injecting the compensating current with respect to the injecting transformer (Figure 1), the UPQC under study can be implemented as well in the UPQC-R (right-side shunt) as UPQC-L (left-side shunt) configuration. It is also classified as UPQC-P—It compensates source voltage sags using active power of supply sources. UPQCs can obtain reference signals on the base of frequency or time domain detection methods. Some researchers argue that "Harmonic current estimation is the key technology of power electronics systems to generate a harmonics reference current for harmonic control" and propose extensive and flexible solutions in this field [4]. Other researchers develop time-domain techniques [5,6]. Since the considered UPQC calculates references for load voltage and line current directly on the base of time variable quantities it can be classified as operating using time-domain signal analysis. In particular, this method refers to Fryze's concept of the load equivalent conductance [7].

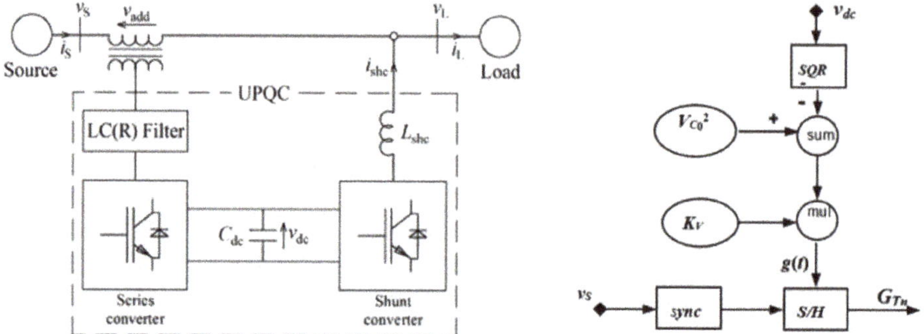

Figure 1. UPQC power circuitry diagram and scheme of obtaining conductance signal on the base of DC-link capacitor voltage v_{dc}, according to Equation (6). The S/H block is an sample-and-hold module that is synchronized with source voltage waveform using *sync* block.

Commonly used control methods of UPQC involves continuous measurement of source voltage and load voltage and current in order to calculate reference signals for UPQC action. Such control methods are known as direct control techniques. However, UPQC can be steered using a somewhat different scenario, which may be classified as the indirect control method [8–12]. A type of the indirect method, which has been dubbed the conductance signal control method, has been successfully implemented to control the shunt active power filter (SAPF) action [9]. However, there is no implementation of this technique for UPQC so far. Applying the conductance signal control method references required to control the UPQC operation are obtained on the base of measuring the voltage across the UPQC's DC-link capacitor. Since for this method the capacitor is employed as "a sensor" of load active power its voltage must not be controlled to be constant. On the contrary, the "freewheeling" capacitor voltage is measured and processed in order to obtain an equivalent (or hypothetical) conductance element that characterizes the consumption of the total active power taken from the source [8]. Since the load active power may vary in time, this conductance element should also be considered as time-dependent one. The ongoing information on conductance of this element may be referred to as the conductance signal. The first part of this paper shows that the conductance signal can be used to control the UPQC action.

As it turns out, if the conductance signal method is used some noteworthy additional functionalities of UPQC can be obtained. In particular, the UPQC can also be used to control the flow of energy between the supply source and the AC or DC passive or active elements of the network being connected with the particular UPQC. It can be said that in addition to perform the UPQC's conventional tasks, it can also serve as a local energy distribution center operating with high power factor. This center may serve as a spot improving grid's energetic efficiency. The second part of this paper describes these extra UPQC's functionalities.

There are many other technical problems related to UPQC extended operation in smart grids. From this perspective problems of UPQC real-time control [13–16] and their optimal sizing and siting are considered very important [13–19].

2. Control of UPQC with the Use of Signal of Load Equivalent Conductance

2.1. Basic Scheme of UPQC

From the point of view of the studied control method the UPQC can be considered as composed of: (1) the shunt converter, which shapes source current i_S to be active and of amplitude required to supply the load with the required active power and maintains the UPQC's DC-link capacitor voltage in the assumed range, and (2) of the series converter, which tracks the supply voltage v_S and—If needed—injects suitable voltage corrections v_{add} (Figure 1). As the result load voltage v_L is shaped to be sinusoidal and of nominal amplitude. In the vast majority of cases the control circuitry of the UPQC's shunt converter steers also the operation of the entire UPQC filter. The same rule is applied in this paper. A comprehensive review on the possible shunt converter control techniques can be found in [20,21].

2.2. Principle of UPQC's Shunt Converter Control

The control unit of the shunt converter processes the voltage signal of the UPQC's DC-link capacitor C_{dc}, Figure 1, in order to obtain the conductance signal. This signal gives crucial information needed to produce the reference for the source current. The conductance signal can be used to obtain the current reference signal as well for the shunt active power filter (SAPF) as for UPQC's shunt converter.

In general, the compensated load may be nonlinear, time variable, passive or active, of single or polyphase structure with or without the neutral conductor, unbalanced, etc. From this perspective applying an universal method for obtaining the signal of equivalent conductance would be beneficial. Such an universal method, which allows this signal calculation as a function of amount of energy stored in the active filter's reactance elements, has been proposed in [8]:

$$g(t) = \frac{(W_{APF0} - w_{APF}(t))(N_{SF} + 1)}{T_{st} V_S^2} \tag{1}$$

where: $g(t)$ is the instantaneous load equivalent conductance signal; W_{APF0} is initial amount of energy, which has been stored in all UPQC's reactance elements during UPQC initialization procedure; $w_{APF}(t)$ is amount of energy stored in these elements at instant t; N_{SF} is the ratio of amount of energy delivered to the load from the supply source with respect to energy that is simultaneously delivered to the load from UPQC's reactance elements—After each instant of change of load active power until the moment of achieving a new stead state by UPQC; T_{st} is a user dependent parameter that may be utilized to define UPQC time response on change of load active power; V_S is source voltage rms.

The Equation (1) can be simplified to the form that only information on a part of total amount of energy stored in the UPQC is taken into account: namely that is stored in its DC-link capacitor:

$$g(t) = \frac{C_{dc}\left(V_{C0}^2 - v_{dc}^2(t)\right)(N_{SF} + 1)}{2T_{st} V_S^2} = K_V\left(V_{C0}^2 - v_{dc}^2(t)\right)(N_{SF} + 1) \tag{2}$$

where C_{dc} is capacity of DC link capacitor, V_{C0} is its initial (i.e., after UPQC initialization procedure, see also W_{APF0} in Equation (1)) voltage and $v_{dc}(t)$ is its voltage at instant t, and where:

$$K_V = \frac{C_{dc}}{2T_{st} V_S^2} \tag{3}$$

It is characteristic for the discussed control technique that the DC-link capacitor voltage is not controlled to be constant. On the contrary, the "freewheeling" capacitor voltage is an input signal for obtaining the conductance signal. The K_V factor gives a proportion between a signal related to the DC-link capacitor voltage and the conductance signal. The K_V factor has a practical meaning: it may be used as the gain coefficient of a simple P-type regulator in the active filter's control unit. No other signal converters of DC-link capacitor voltage are needed to obtain the conductance signal (Equation (2)).

There is a parameter T_{st} in the denominator of Equation (3). By changing this parameter the user can control the UPQC's shunt converter inertial response on any change of load active power. By increasing/decreasing magnitude of this parameter more/less energy of each change of load active power can be buffered by DC-link capacitor. In other words, the T_{st} parameter can be used to regulate the energy flow between the source and the load in order to stabilize (or average) the source active power.

Having the conductance signal the reference for source current can be determined by the relationship:

$$i_S^*(t) = g(t)v_{1S}(t) \tag{4}$$

where $v_{1S}(t)$ is fundamental component signal of source voltage. This component can be obtained in many ways (e.g., using filtration or PLL based techniques).

A variable component may appear in conductance signal (Equation (2)) if the load current contains a non-active component. Since UPQC compensates such component with the use of energy stored in its reactance elements (Equation (1)) this cause an oscillating component in DC-link capacitor voltage (Equation (2)). This component can distort the reference (Equation (4)). In order to eliminate impact of this component on the reference (Equation (4)) the continuous signal (Equation (2)) should be transformed into the stepwise waveform. To do this the signal (Equation (2)) is sampled at the very end of each subsequent period T_m of source voltage cycle. Then each sample is hold for the next period T_{m+1}, [8]. Application of such sample-and-hold procedure causes a "step-by-step" UPQC's shunt converter action in that every change in load active power is practically entirely buffered with energy stored in the DC-link capacitor. For such method of full-buffering of energy flow the N_{SF} parameter, see Equations (1) and (2), should be set to zero. As the result the source-to-load flow of energy is delayed for one period T and the stepwise form of the conductance signal applied for a T_m period is given by:

$$G_{T_m} = \frac{C_{dc}\left(V_{C0}^2 - v_{dc}^2(T_{m-1})\right)}{2T_{st}V_S^2} \tag{5}$$

where: $v_{dc}(T_{m-1})$ is capacitor C_{dc} voltage at the end of $(m\text{-}1)$th period T.

Finally, on the base of Equations (4) and (5) the source current reference signal i_S^* for period T_m is:

$$i_{S,T_m}^*(t) = G_{T_m}v_{1S}(t) \tag{6}$$

It should be emphasized that during compensation the following inequality has to be satisfied:

$$v_{dc}(t) >> v_S(t) \tag{7}$$

If this condition is not satisfied the UPQC dynamics can be insufficient. In an extreme case, when $v_{dc}(t) < v_S(t)$, the UPQC action may become even harmful.

2.3. Principle of UPQC's Series Converter Control

Source voltage waveform may deviate from its fundamental component due to wide range of physical phenomena existing in the grid. They may be considered as voltage harmonics, flicker, swell or sag, or pulse transients. There are specialized devices to overcome power quality problems that are related to voltage disturbances. The dynamic voltage restorer (DVR) seems to be the most economical solution in this field, [22]. However, UPQC's series converter can maintain the load voltage v_L to be close to the fundamental component v_{1S} of the source voltage v_S.

Independently of the reason of voltage distortion its shape bettering can be performed with the use of the same conductance signal-based control method considered. In other words, there is no need to identify the reason or spectrum of the source voltage distortion. In any case it is sufficient to inject the adequate voltage correction v_{add} in series with the source voltage v_S (Figure 1). To produce appropriate voltage correction v_{add} the series converter generates (using energy stored in the DC-link capacitor) the current flow through the converter's side winding of the injecting transformer. The hysteresis controller compares load voltage to its reference, i.e., the source voltage fundamental component, and steer switches action of the series converter in order to keep this voltage near this reference. As the result the required voltage v_{add} appears across the grid side winding of the injecting transformer.

Voltage and current distortion components may be considered as nonactive ones. Therefore, while compensating and being in the steady state, both UPQC converters impact the compensated voltage/current runs using nonactive power only, i.e., without change of mean magnitude of DC-link capacitor energy (if skip energy loss in the UPQC circuitry). In such situation the load conductance signal is still constant and, consequently, source current amplitude is constant as well. This observation is important from the perspective of the considered control method.

However, for the control method considered each change of load active power cause change of the conductance signal. Also energy losses in UPQC circuitry influence the total load-and-UPQC active power, so they impact the conductance signal (Equation (5)). It can be then said that there are no changes of signals (Equations (5) and (6)) when load active power is constant and the UPQC's series converter compensates only for higher harmonics of source voltage. On the contrary, the signals (Equations (5) and (6)) get new magnitude when the series converter counteracts change of source voltage rms, or if there is a change in source voltage harmonic content. As a result the source is loaded higher/lower in order to maintain constant voltage rms across load terminals.

Finally, as an important conclusion it can be said, that all energy relations between the UPQC's series converter and the rest of the system considered can be supervised by the control unit of the UPQC's shunt converter and there is no operational incompatibility between both converters.

3. Studies for UPQC Standard Operation

The considered control method has been extensively verified by means of computer simulation. The IsSpice software (Intusoft, San Pedro, CA, USA) has been used. During some analyses performed the deformation of source voltage and load current often went beyond the voltage and current runs encountered in practice. They caused strong overload of UPQC circuitry. This approach, attractive in simulation studies, allows to assess the usability area of the considered UPQC control method.

In this paper simulation studies are divided into two parts. The first one, Section 3, considers UPQC standard operations, i.e., compensation for nonactive current and improving the voltage waveform on load terminals. The second part, Section 4, describes additional UPQC functionalities that arise if the conductance signal control method is used. In particular, this section considers the possibility of using UPQC as a distribution center for locally generated power.

For all analyses performed the same supply source characteristic and UPQC circuitry were used:

(1) Supply source. Supply voltage waveform, v_S in Figure 1, is composed of fundamental harmonic of rms 230 V/50 Hz and of two higher harmonics: rms 32 V/250 Hz and rms 32 V/350 Hz. Internal resistance and inductance of the supply voltage source is 2 mΩ and 100 µH, respectively.

(2) UPQC's shunt converter. Energy phenomena related to DC-link capacitor are essential for obtaining the UPQC reference signal. The capacity of UPQC's DC-link capacitor C_{dc} is rated with respect to the maximal magnitude P_{Lmax} of the load active power and the C_{dc} capacitor maximal-to-minimal voltage ratio (compare Equation (7)). Thus, on the base of Equation (5) the needed capacity can be estimated as $C_{dc} = 2P_{Lmax}/(V_{C0}^2 - v_{dcT1}^2)$. Finally, the initial voltage V_{C0} of 600 V and the C_{dc} capacity of 8 mF were used. A band-bang regulator operating with ΔI loop equal to 1 A has been used to force the reference current (Equation (6)). The inductance of shunt converter series inductor is set to 5 mH in order to limit the converter's switching frequency to about 40 kHz. Switching elements of the converter are modeled to work similarly to IGBT transistors.

(3) UPQC's series converter. A hysteresis bang-bang regulator operating with ΔV equal to 5 V has been used to shape the load voltage waveforms near its reference signal: The fundamental harmonic component of source voltage. The inductance and capacity of the series converter's LC(R) smoothing filter are designed so that the switching frequency of the serial converter is about 15 kHz.

(4) Switching elements of the shunt converter are modeled in such a way that they work similarly to IGBT transistors.

(5) A high-pass passive LC(R) filter has been added in parallel to the UPQC's shunt converter in order to diminish flow of high-frequency component of the compensating current into the supply source branch.

3.1. Basic Examination of the Control Method. Turning UPQC and Load On and Off

Achieving the steady state after turning the UPQC or load on and off is considered in this Subsection. Analyses of selected signals of the network after switching the load on and off gives information on correctness of source-and-load energy balancing performed by UPQC. This energy balancing is crucial for the studied UPQC control method.

For all analyses carried out in this subsection the load is a thyristor power controller. It is composed of a 10 Ω resistor in series with two thyristors in antiparallel connection. Both thyristors are fired symmetrically with phase angle of $\pi/2$. This load brings in abrupt load power changes and wide harmonic spectrum for load current that can be difficult to compensate by SAPF and UPQC devices.

3.1.1. Turning UPQC's Shunt Converter On

Since the shunt converter controls the source current it is also responsible for the total power delivered to the whole UPQC-and-load network. In particular, it controls also the source current component that is related to energy dissipated in all UPQC's circuitry. For this reason the shunt converter should be turned on no later than the series one. The process of turning the shunt converter on is shown in Figures 2 and 3, and then characterized with the use of basic electrical load and UPQC parameters collected in Tables 1 and 2.

In Figure 2 source voltage $v_S(t)$, source current $i_S(t)$ and the conductance signal $G(T_n)$—According to Equation (5), are shown as waveforms 1, 2 and 3, respectively. Before the time instant $t = 400$ ms the source-and-load circuit acts in the steady state. The UPQC is inactive yet and does not impact the source-load energy transmission. Then the shunt converter is turned on at instant $t = 400$ ms. Until this moment the DC-link capacitor voltage is of its initial magnitude V_{C0}, so the conductance signal (Equation (6)) is null. For that reason for the whole period T starting at this moment the load is powered using energy stored in UPQC reactance elements, practically solely from its DC-link capacitor. Therefore the source current is practically null. At the end of this period T, i.e., at $t = 420$ ms, the load equivalent conductance related to this period can be calculated and then used to produce the reference signal (Equation (6)) for the next period T, i.e., for time period 420 ms–440 ms.

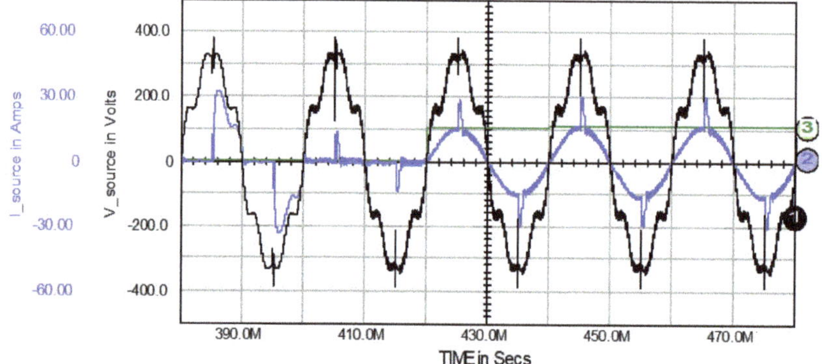

Figure 2. Shunt converter turning on. Source voltage: run 1, source current: 2, conductance signal: 3. Y scale for conductance signal is 45 mS/div.

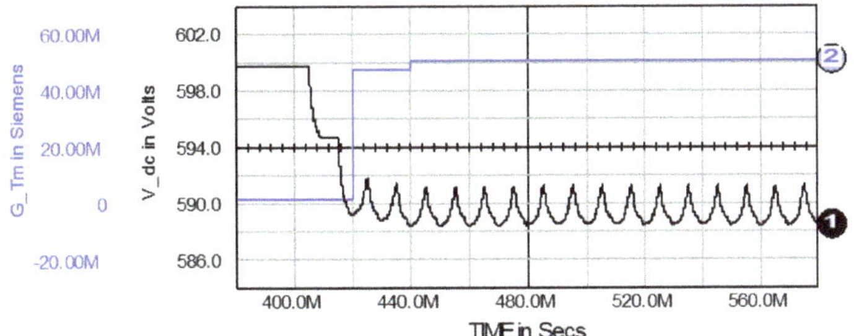

Figure 3. Shunt converter turning on process. DC-link capacitor voltage: waveform 1, and load equivalent conductance signal: waveform 2.

Table 1. Compensated load basic electrical parameters just before (for 380 ms–400 ms) and after (for 400 ms–460 ms) the instant (at 400 ms) of turning the shunt converter on.

Time (ms) Parameter	380–400	400–420	420–440	440–460
V_{Load} (V)	229	234	234	234
I_{Load} (A)	15.6	16.6	17.0	17.0
S_{Load} (VA)	3572	3884	3978	3978
P_{Load} (W)	2439	2743	2885	2879
PF_{Load}	0.68	0.71	0.73	0.73
W_{Load} (J)	48.8	54.9	57.7	57.6
G_{Load} (mS)	46.6	50.1	52.7	52.6

V_{Load} and I_{Load} are voltage and current rms on load terminals, S_{Load} and P_{Load} are load apparent and active powers, PF_{Load} is load power factor, W_{Load} is energy consumed by load, G_{Load} is load equivalent conductance equal to $P_{Load}/(V_{Load})^2$.

Table 2. UPQC-and-load subcircuit basic electrical parameters before (380 ms–400 ms) and after (400 ms–460 ms) the instant (at 400 ms) of turning the shunt converter on.

Time (ms) Parameter	380–400	400–420	420–440	440–460
V_{Source} (V)	232	232	232	232
I_{Source} (A)	15.5	2.8	12.1	12.9
I_{Source} THD (%)	59	146	17	14
$S_{UPQC+Load}$ (VA)	596	650	2807	2993
$P_{UPQC+Load}$ (W)	449	277	2717	2899
$PF_{UPQC+Load}$	0.68	0.42	0.97	0.97
$W_{UPQC+Load}$ (J)	49.0	5.5	54.3	58.0
ΔW_{DCCap} (J)	0.0	−49.5	−4.2	0.0
G_{Signal} (mS)	1.2	1.2	42.3	59.0

V_{Source} and I_{Source} are voltage and current rms on UPQC terminals, I_{Source} THD is source current THD factor, $S_{UPQC+Load}$ and $P_{UPQC+Load}$ are UPQC-and-load apparent and active powers seen on UPQC terminals, $PF_{UPQC+Load}$ is UPQC-and-load subcircuit power factor, $W_{UPQC+Load}$ is energy delivered to UPQC-and-load subcircuit from source, ΔW_{DCCap} is change of energy stored in DC-link capacitor, G_{Signal} is signal of load equivalent conductance calculated accordingly to Equation (5).

The whole "static" change of the capacitor voltage takes place in two steps: the first step from 599.7 V (sample taken at time $t = 400$ ms), down to 589.3 V (sample taken at $t = 420$ ms) and then the second step from 589.3 V for sample taken at $t = 420$ ms down to 588.4 V at $t = 440$ ms. There are two main reasons of this two-step process of energy balancing:

- the first reason is due to intentional increasing the T_{st} parameter, see Equation (5). Setting this parameter to be longer then the period T introduces some inertia to the UPQC response on a change of source voltage or load power, making the UPQC action more stable. Here T_{st} has been fixed as increased by 2.2% with respect to the period $T = 20$ ms.
- the second reason is that energy stored only in the DC-link capacitor has been taking into account when calculating the conductance signal. In such case a comparatively small amount of non-considered energy, which is stored in other UPQC's reactance elements, affects the conductance signal making it a little oscillating. This effect can be neglected because of possible load power changes for time-variable loads as well as possible random variations in source voltage.

Finally, after the two-step updating of the conductance signal magnitude the whole network achieves the steady state.

There are basic electrical parameters characterizing load and UPQC action, related to waveforms shown in Figures 2 and 3, collected in Tables 1 and 2. They characterize the load and source work, respectively.

The load work is buffered through the UPQC. Therefore, variations in load action are "seen modified" from the perspective of the supply source. In particular, there are two major changes, which are seen from this perspective, that occur in whole load-and-UPQC circuitry action at $t = 400$ ms and then 20 ms later. These changes are related to the periodical updating of the conductance signal, see Equation (5) and comments to Figures 2 and 3.

It is noteworthy, that there is no difference in UPQC operation if the order of turning on the load and UPQC is reversed, that is, when UPQC is already active at the moment when the load begins its work. Similarly, at this moment the DC-link capacitor voltage equals its initial magnitude and the going magnitude of the conductance signal is zero. Consequently, the energy flow is buffered by UPQC with all consequences on UPQC action already described above.

3.1.2. Turning UPQC's Series Converter On

The series converter starts its action at $t = 800$ ms (Figure 4), when the network operates in the steady state and the shunt converter is already active. Since for the control method considered no time is needed for analysis of source voltage spectrum its distortion can be compensated immediately. This compensation cause change in harmonic components of DC-link capacitor voltage, waveform 2 in Figure 4. Although compensation for harmonics does not require the use of active power an increase of mean magnitude of the capacitor voltage can be observed. This is due to change of load active power being result of elimination of higher harmonics power from the total active power of the load.

Change in mean value of capacitor voltage causes change in magnitude of conductance signal (Equation (5)). Therefore, appropriate change of source current amplitude can be also observed. Changes of source-UPQC-load system parameters, related to Figures 4 and 5, are collected in Tables 3 and 4.

Just after turning the series converter on, at $t = 800$ ms, there is a drop of load active power resulting from eliminating load voltage distortion. As a consequence a part of source current turns out to be "excessive" with respect to going load and UPQC energy demand. Energy of this excessive current is stored in the DC-link capacitor increasing its voltage/energy. Therefore, on the base of Equation (5) the conductance signal is recalculated to a new magnitude at the beginning of the next T period. From this instant, i.e., for $t = 820$ ms, possibly delayed by the corrections described in the commentary next to Figure 2, the whole network reaches a new steady state.

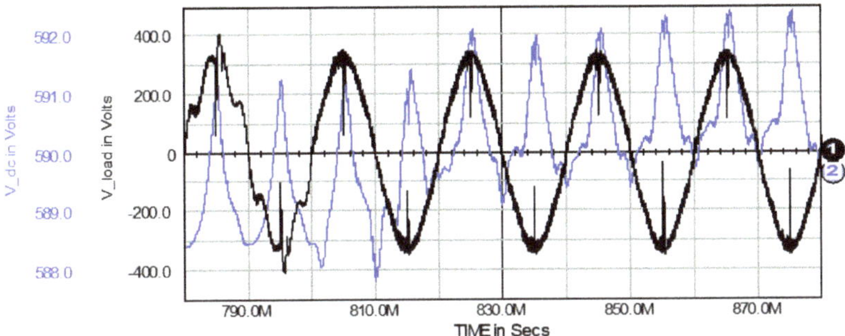

Figure 4. Series converter turning on. Load voltage: waveform 1, and DC-link capacitor voltage: waveform 2.

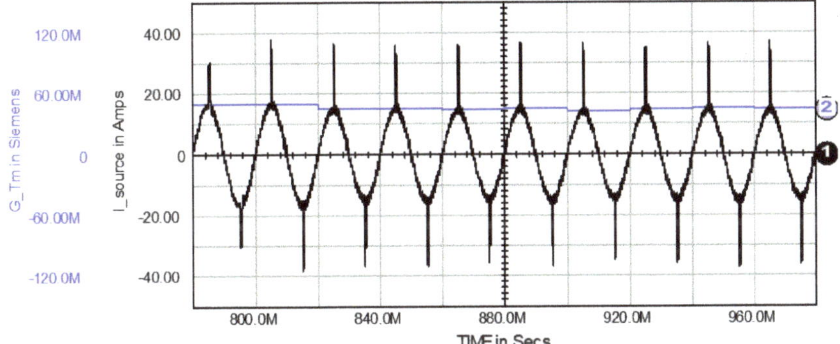

Figure 5. Series converter turning on process. Source current: waveform 1, and conductance signal: waveform 2.

Table 3. Basic parameters describing load action before (780 ms–800 ms) and after (800 ms–860 ms) the instant (800 ms) of turning the series converter on.

Time (ms)				
Parameter	780–800	800–820	820–840	840–860
V_{Load} (V)	234	229	229	229
V_{Load} THD (%)	15	3.3	3.1	3.5
I_{Load} (A)	16.9	16.1	16.1	16.0
S_{Load} (VA)	3955	3687	3686	3664
P_{Load} (W)	2873	2589	2608	2563
PF_{Load}	0.73	0.86	0.71	0.70
W_{Load} (J)	57.5	51.8	52.2	51.3
G_{Load} (mS)	52.4	49.5	49.9	49.0

V_{Load} THD [%] is load voltage THD factor and other parameters are defined as for Table 1.

Table 4. Basic parameters describing load action before, (780 ms–800 ms) and after (800 ms–860 ms) the instant (800 ms) of turning the series converter on. All parameters are defined as for Table 2.

Time (ms) Parameter	780–800	800–820	820–840	840–860
V_{Source} (V)	232	232	232	232
I_{Source} (A)	12.9	13.1	12.0	12.1
I_{Source} THD (%)	14	13	14	14
$S_{UPQC+Load}$ (VA)	2993	3039	2784	2807
$P_{UPQC+Load}$ (W)	2908	2919	2659	2692
$PF_{UPQC+Load}$	0.97	0.96	0.96	0.96
$W_{UPQC+Load}$ (J)	58.2	58.4	53.2	53.8
ΔW_{DCCap} (J)	0.5	5.2	−0.9	1.9
G_{Signal} (mS)	51.0	50.8	45.8	46.3

3.1.3. Turning the Load Off

Before the load is turned off the whole network operates in the steady state. Then, at the moment $t = 1400$ s, the load is turned off (Figure 6).

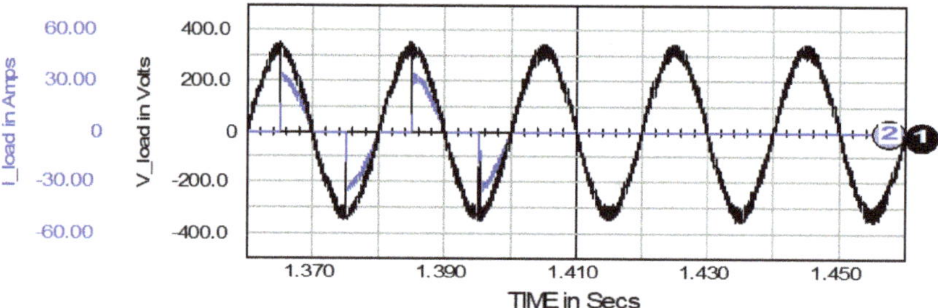

Figure 6. Load turning off process. Load voltage: waveform 1, and load current: waveform 2.

Just after the load turning off moment the load current stops immediately. However, because of the energy flow buffering the flow of source current has been extended for the next period T, i.e., for the time 1400 s–1420 s (Figures 7 and 8). As the result the whole circuit achieves zero steady state, i.e., with no load current/power, and UPQC is ready to the next compensation if load is turned on again.

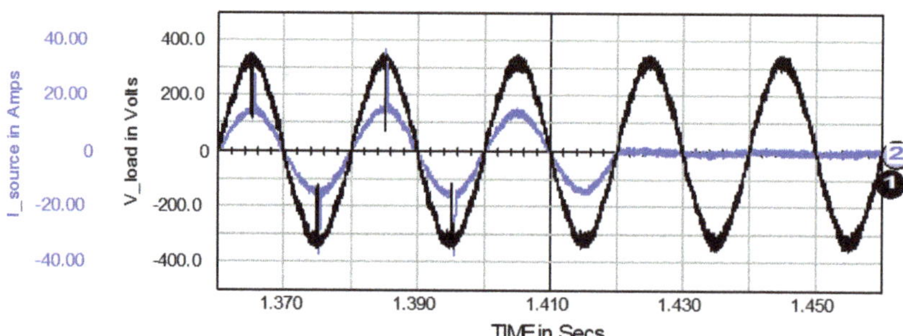

Figure 7. Load turning off process. Load voltage: waveform 1, and source current: waveform 2.

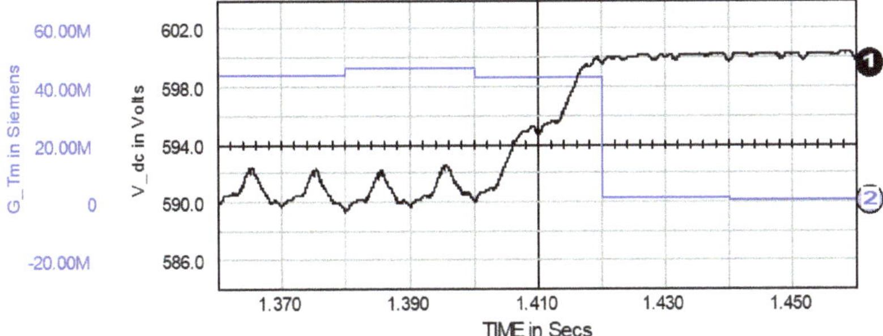

Figure 8. Load turning off process. DC-link capacitor voltage: waveform 1, and load conductance signal: waveform 2.

3.2. Compensation for Source Voltage and Current Distortions

Beside compensation for voltage deformation from higher harmonics the UPQC should maintain sinusoidal load voltage even if source voltage is influenced by irregular or unexpected distortions. Source voltage swells and sags, flicker, and pulse-type distortions can be enumerated in this context.

In Section 3.2. the load consists of two branches in parallel. The first one contains a thyristor power controller, consists of a 15 Ω resistor in series with two thyristors in antiparallel connection operating with phase angle equal to $\pi/4$. The second one consists of 15 Ω resistor in series with 32 mH inductor. This load branch introduces reactive power into the network.

3.2.1. Compensation for Source Voltage Swell

The source voltage is composed initially of fundamental frequency component of rms 230 V and also of two harmonics: 32 V/250 Hz and 32 V/350 Hz. Then, beginning from $t = 120$ ms there is a swell of the source voltage fundamental component by 50%. The most important waveforms and electrical quantities are shown in Figures 9–11, and then in Tables 5 and 6.

The load operates in the steady state when at instant $t = 80$ ms UPQC's shunt and series converters are activated simultaneously. For the first period T of UPQC operation, i.e., for time period 80 ms–100 ms, the load is powered practically solely with the use of energy stored in the DC-link capacitor, i.e., without drawing energy from the supply source. This cause decrease of DC-link capacitor voltage. Then, the first non-zero magnitude of conductance signal (Equation (5)) can be obtained at the very end of this time period, i.e., at instant $t = 100$ ms, waveform 3 in Figure 10.

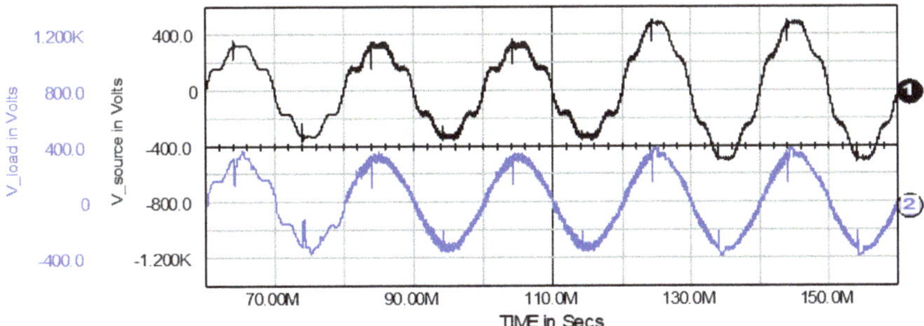

Figure 9. Source voltage swell compensation. Source voltage: waveform 1, and load voltage: waveform 2.

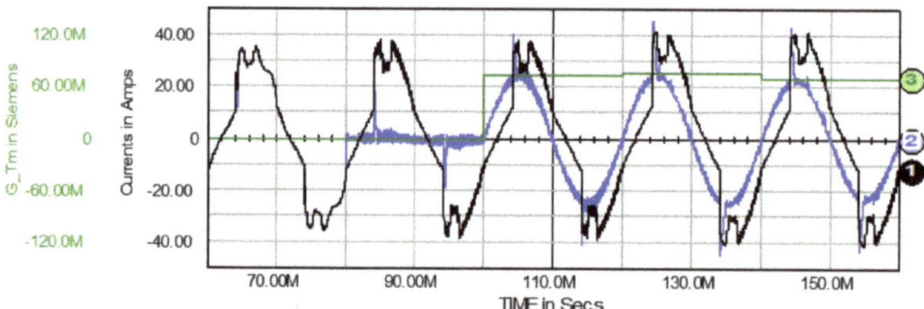

Figure 10. Source voltage swell. Load current, source current and conductance signal: waveforms 1, 2 and 3.

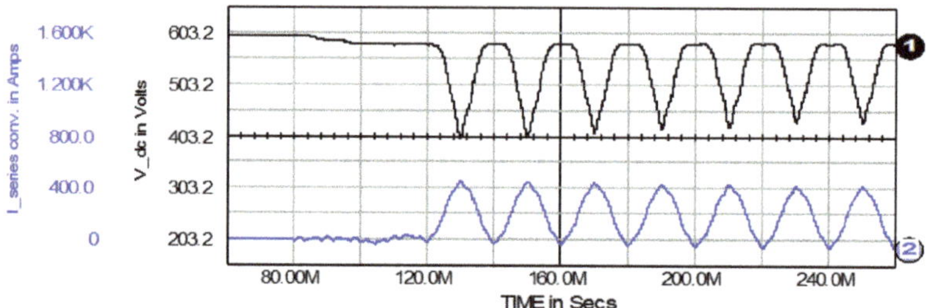

Figure 11. Source voltage swell. DC-link capacitor voltage and series converter current: waveforms 1 and 2.

Table 5. Basic parameters characterizing load work before and during source voltage swell. The parameter definitions are the same as for Table 3.

Time (ms) Parameter	0–80	80–100	100–120	120–140	180–200	240–260
V_{Load} (V)	225	229	228	247	248	247
I_{Load} (A)	22.3	22.4	22.3	24.6	24.6	24.8
S_{Load} (VA)	5018	5130	5084	6076	6101	6126
P_{Load} (W)	4203	4278	4241	5145	5147	5214
PF_{Load}	0.84	0.83	0.83	0.85	0.84	0.85
W_{Load} (J)	84.0	85.6	84.8	102.9	102.9	104.3
G_{Load} (mS)	83.0	81.6	81.6	84.3	83.7	85.5

Table 6. Basic parameters characterizing UPQC-and-load subcircuit operation for the voltage swell. The parameter describing is the same as for Table 2.

Time (ms) Parameter	60–80	80–100	100–120	120–140	180–200	240–260
V_{Source} (V)	231	232	232	346	346	346
I_{Source} (A)	22.2	2.6	18.2	19.0	17.7	17.7
$S_{UPQC+Load}$ (VA)	5128	603	4222	6574	6124	6124
$P_{UPQC+Load}$ (W)	4210	254	4146	6458	6014	6009
$PF_{UPQC+Load}$	0.82	0.42	0.98	0.98	0.98	0.98
$W_{UPQC+Load}$ (J)	84.2	5.1	82.9	129.2	120.3	120.2
ΔW_{DCCap} (J)	−1.0	−81.8	−2.8	7.0	−0.5	0.0
G_{Signal} (mS)	0.2	0.3	74.4	76.8	71.1	71.4

At t = 120 ms the amplitude of source voltage fundamental component rises from 325 V up to 487 V, Figure 9. Since for the control method considered there is no time needed for analysis of source voltage spectrum this source voltage increase can be compensated immediately. The load voltage amplitude rises, but only to 349 V, Figure 9 and Table 5.

In order to compensate for this voltage increase a voltage correction is generated using energy stored in the DC-link capacitor. Depending on the voltage swell magnitude the current of the converter side of the series converter can reach significant values. This can result in significant increase of energy loss in this converter (Figure 11) and compare P_{Load} and $P_{UPQC+Load}$ in Tables 5 and 6.

3.2.2. Compensation for Source Voltage Sag

The load operates in the steady state when the UPQC is turning on at t = 80 ms. At t = 120 ms the amplitude of source voltage fundamental frequency component decreases to 163 V. Due to the UPQC action the load voltage can be maintained on the amplitude about 295 V, Table 7 and Figure 12.

Table 7. Basic parameters characterizing load work before and during source voltage sag. The parameter describing is the same as for Table 3.

Time (ms) Parameter	60–80	80–100	100–120	120–140	200–220	380–400
V_{Load} (V)	225	229	228	214	209	208
I_{Load} (A)	22.3	22.4	22.3	20.9	20.7	20.2
S_{Load} (VA)	5018	5130	5084	4473	4326	4121
P_{Load} (W)	4203	4278	4241	3735	3653	3487
PF_{Load}	0.84	0.83	0.83	0.84	0.84	0.85
W_{Load} (J)	84.1	85.6	84.8	74.7	73.1	69.7
G_{Load} (mS)	83.0	81.6	81.6	81.6	83.6	83.4

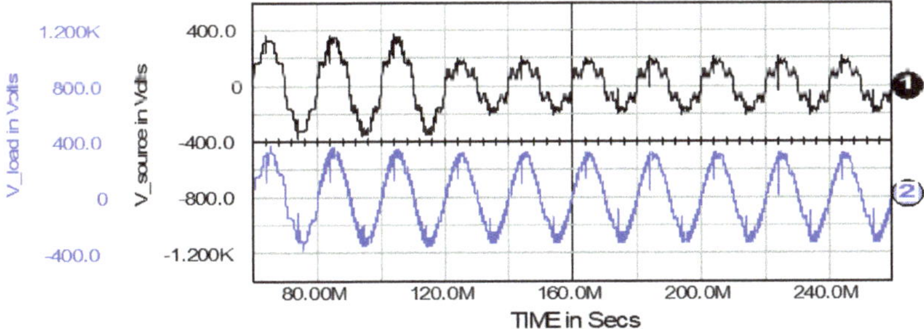

Figure 12. Source voltage sag compensation. Source voltage: waveform 1 and load voltage: waveform 2.

The source voltage drop is compensated with the use of energy drown from the supply source. This is performed as follows. The deficiency of source energy, caused by voltage sag, is balanced using of energy stored in the DC-link capacitor. As the result its voltage decreases. It causes an increase in the conductance signal (Equation (5)) and an increase in amplitude of the source current reference (Equation (6)). Source current rises increasing source active power (Figure 13). This "additional" source power is utilized to increase load voltage to be near its nominal magnitude.

Similarly to the case of voltage swell the compensation for voltage sag may require large current of the series converter. This imply a high magnitude of variable component of the DC-link capacitor voltage. If the instantaneous capacitor voltage falls close to the source instantaneous voltage, then UPQC may lose the possibility of correct operation. Therefore it can be said that in the analyzed case UPQC operates at the limit of its capabilities. This is illustrated in Figure 14, waveform 1.

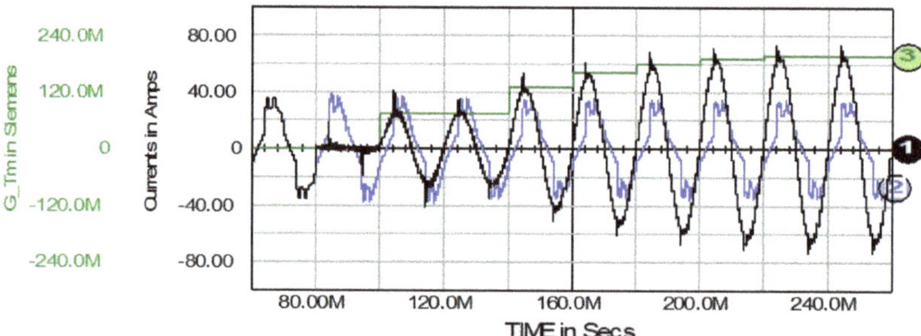

Figure 13. Source voltage sag compensation. Source current: waveform 1, load current: waveform 2, and conductance signal: waveform 3.

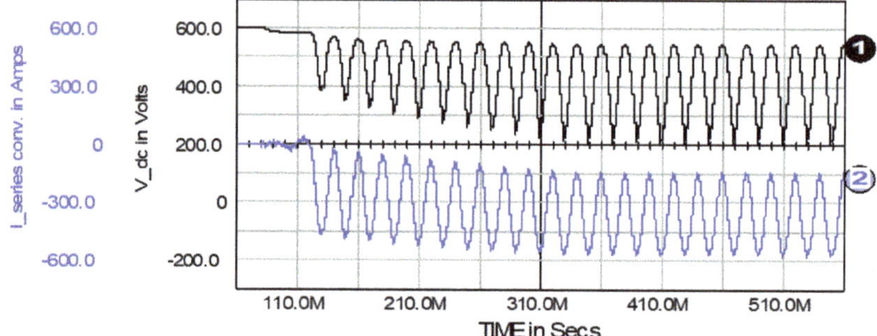

Figure 14. Source voltage sag compensation. DC-link capacitor voltage and load series converter current: waveform 1 and 2, respectively.

It should be also noticed a problem of energetic cost of compensation for the source voltage sag. During compensation high current of the series converter causes a large dissipation of energy in its power circuitry. Such power loss can be estimated on the basis of a comparison of the active power at the load terminals against the active power at the UPQC's input terminals: compare the parameter P_{load} in Table 7 with the parameter $P_{UPQC+Load}$ in Table 8.

Table 8. Basic parameters describing UPQC-and-load subcircuit action before and during the source voltage sag. The parameters shown were defined in Table 2.

Time (ms) Parameter	60–80	80–100	100–120	120–140	140–160	200–220	280–300	380–400
V_{Source} (V)	231	232	232	120	120	120	120	120
I_{Source} (A)	22.2	2.6	18.2	18.1	30.3	44.7	46.5	46.6
$S_{UPQC+Load}$ (VA)	5128	603	4222	2172	3636	5364	5580	5592
$P_{UPQC+Load}$ (W)	4210	254	4146	2086	3492	5123	5364	5369
$PF_{UPQC+Load}$	0.82	0.42	0.98	0.96	0.96	0.96	0.96	0.96
$W_{UPQC+Load}$ (J)	84.2	5.1	82.9	41.7	69.8	102.5	107.3	107.4
ΔW_{DCCap} (J)	−0.48	−82.8	−2.3	−57.6	−33.0	−7.1	−10.5	−0.4
G_{Signal} (mS)	0.2	0.3	74.4	76.8	129.6	192.6	200.0	200.0

3.2.3. Compensation for Source Voltage Fluctuations

Compensation for source voltage fluctuations within the range of magnitude of 0.9 to 1.1 of its nominal value and at frequency of 8 Hz is considered in this section. Such voltage fluctuation may be classified as flicker. Flickers cause a number of adverse effects in electrical circuits operation, for example for electric motors action electronic devices operation or lighting installations.

For the discussed UPQC control method, the way and effect of reducing flicker-type of source voltage fluctuations is identical with the method of compensating for source voltage swells and sags. In addition, because of smaller disturbances in source voltage amplitude they are easier for compensation. However, the flicker-type voltage oscillations can be considered as repetitive run in a narrow range of low frequencies. Therefore, it seems to be convenient to present here a distinctive way of the UPQC action which can be useful for such periodical-like voltage or current disturbances. In particular, the option of choosing a convenient value for the T_{st} parameter, see Equations (1) and (5), is utilized here. This possibility has been used to increase the inertia in UPQC operation against changes in load active power. Voltage fluctuations cause changes in this power. In general, the greater the magnitude of the T_{st} parameter the closer the conductance signal (Equation (5)) run with respect to its multi-period mean. As a result the source current (and load voltage) can be stabilized, so that flicker is easier to be reduced in the whole grid.

It should be noted, that if increase the T_{st} time the DC-link capacitor operates with lowered voltage. For this reason the condition (Equation (7)) may not be met. This can reduce UPQC dynamics or even cause UPQC operation to failure.

The UPQC operation when the time T_{st} is set to be equal to the source voltage period T, and then when it has been increased to $3T$ is analyzed. The conductance signal (Equation (5)) and DC-link capacitor voltage are shown in the same scale in Figure 15, respectively. It can be observed that for the T_{st} parameter increased to $3T$ fluctuation of conductance signal is reduced: Waveform 3 in contrast with waveform 1. Simultaneously the UPQC operates with lowered DC-link capacitor voltage: waveform 4 with respect to waveform 2.

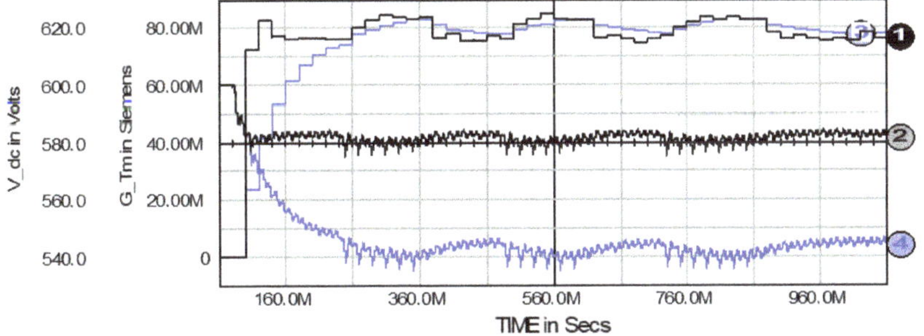

Figure 15. Source voltage flicker compensation. DC-link capacitor voltage for $T_{st} = T$ and for $T_{st} = 3T$: waveforms 2 and 4, respectively, and conductance signal for $T_{st} = T$ and for $T_{st} = 3T$: waveforms 1 and 3, respectively.

The effect of flicker (and still existing harmonics) compensation for $T_{st} = 3T$ is shown in Figure 16. The compensation can be considered sufficient. Taking into account that the load voltage amplitude is maintained to be constant, it can be concluded that the flicker compensation is sufficient.

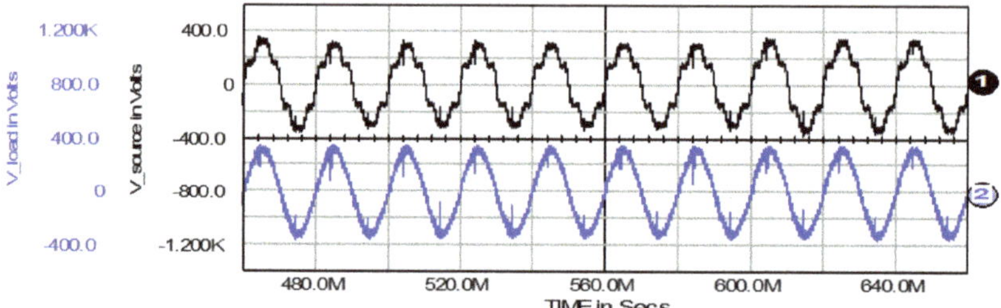

Figure 16. Source voltage flicker compensation. Source voltage: waveform 1, and harmonics-compensated and amplitude-levelled load voltage: waveform 2.

4. Studies for UPQC Extended Operation

The extended functionality of UPQC is understood here as the possibility of using it also to control the energy exchange between all—Being influenced by given UPQC device action—Elements of the network. These elements may be of passive or active type or may be changeable from this point of view. There is no restriction on location in the network of these elements. They may be covered by UPQC extended action as well being located on the AC-side as on the DC-side of UPQC device. Therefore, beside compensating for undesirable components of grid voltage and current runs the UPQC can also serve as a local energy distribution center that can operate with high power factor. There is no change in UPQC circuitry parameters and in source voltage waveform components with respect to those introduced in Section 3. However, in order to highlight UPQC's extended capabilities the load is composed to be nonlinear, time variable and of changeable passive or active kind.

4.1. UPQC Operation with Switched Passive/Active Work of AC-Side Load Elements

In general, there are two main possibilities of energy flow management when some nominally passive elements of load become generators and the amount of energy being generated in the load exceeds energy being consumed there:

(a)   the "excessive" amount of energy is transmitted up stream to the supply source or
(b)   the "excessive" amount of energy is stored in the DC-link capacitor.

The case (a) may be realized in a full form, when all amount of the "excessive" energy is transmitted up stream to the source, or in a partial form, when some portion of the "excessive" energy is accumulated in the DC-link capacitor. Waveforms related to the (a) strategy are shown in a general outlook in Figures 17–19, and then, in more precise look and with detailed comments, are presented in Figures 20–22. Then, waveforms related to the (b) strategy are shown in general outlook in Figures 23 and 24, and then are detailed in Figures 25–29.

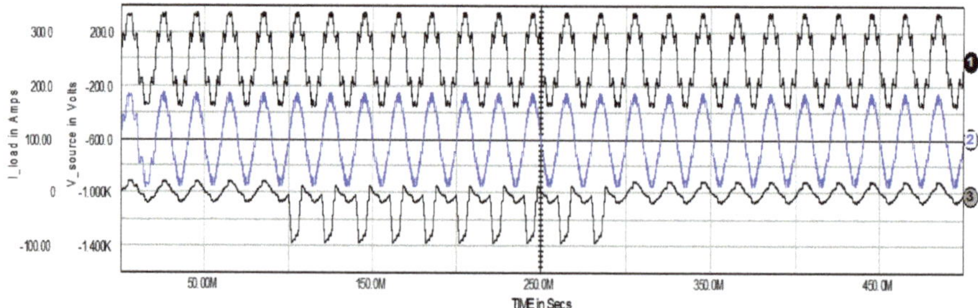

Figure 17. Source voltage: waveform 1, load voltage: waveform 2 and load current: waveform 3. Whole network action, general view.

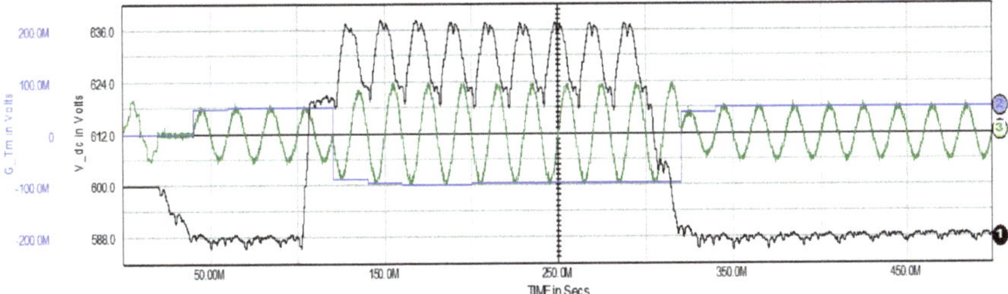

Figure 18. DC-link capacitor voltage: waveform 1, conductance signal: waveform 2 and source current: waveform 3 with Y scale of 54 mS/div. Whole network action, case 4.1 (a).

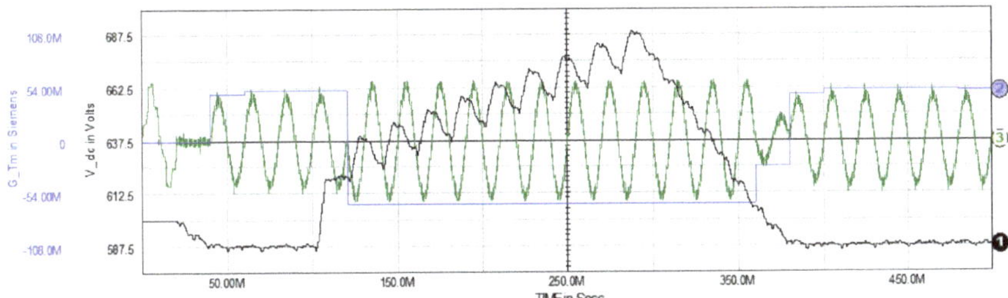

Figure 19. DC-link capacitor voltage: waveform 1, conductance signal: waveform 2 and source current: waveform 3 with Y scale of 10 A/div. Whole network action, case 4.1 (b).

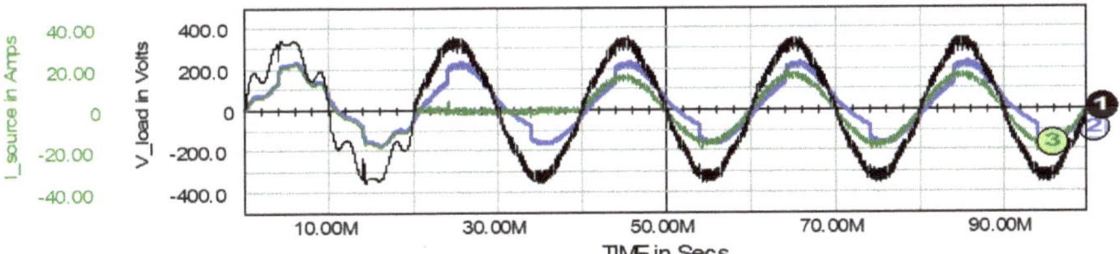

Figure 20. Load voltage: waveform 1, load current: waveform 2 and source current: waveform 3.

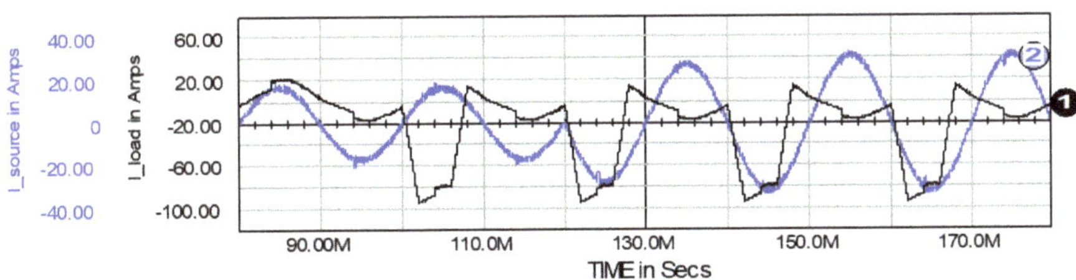

Figure 21. Load current: waveform 1 and source current: waveform 2.

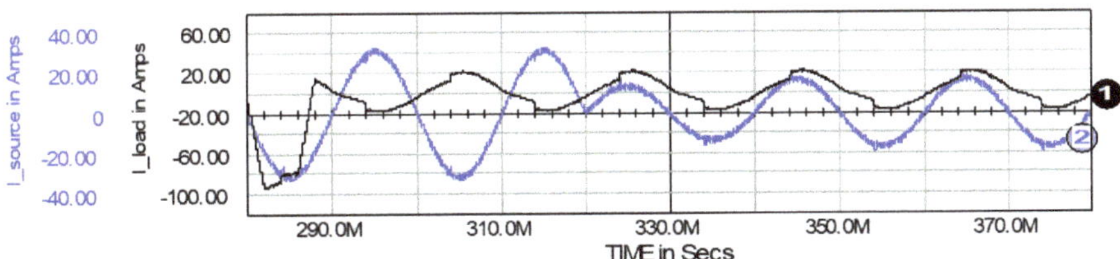

Figure 22. Load current: waveform 1 and source current: waveform 2.

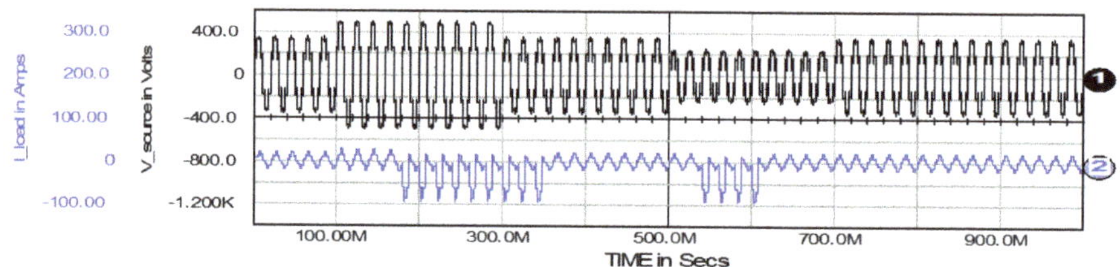

Figure 23. Source voltage: waveform 1 and load current: waveform 2.

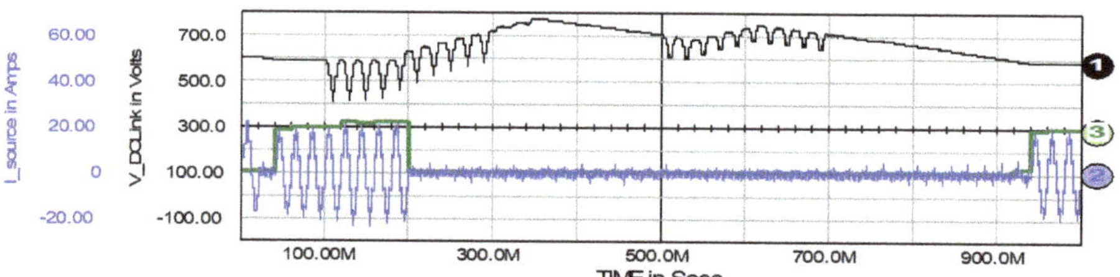

Figure 24. DC-link capacitor voltage: waveform 1, source current: 2 and conductance signal: 3 with Y scale of 27 mS/div.

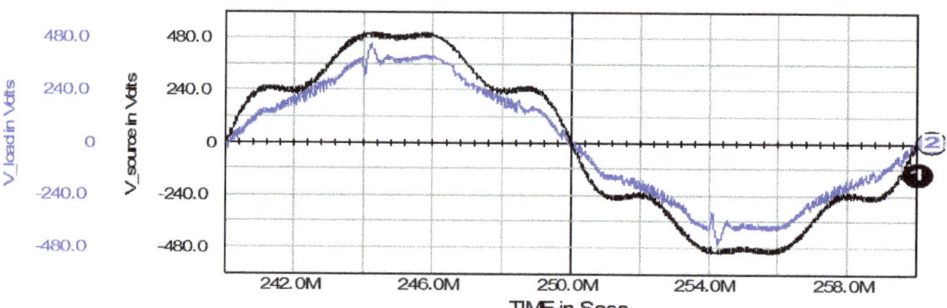

Figure 25. Source and load voltages: waveform 1 and 2. Theirs RMS and THD parameters are 348 V and 14.7%, and then 268 V and 7.7%, respectively.

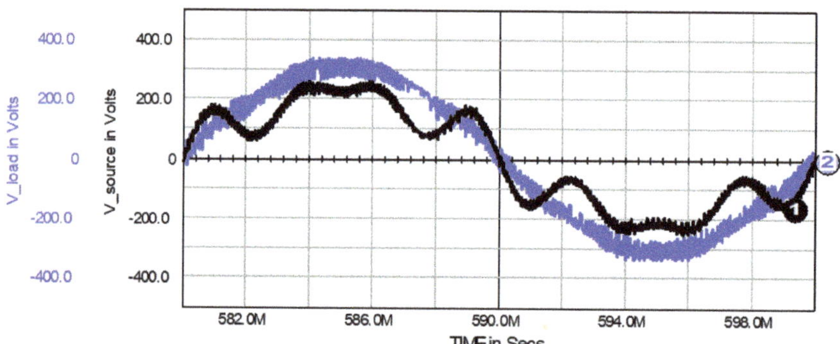

Figure 26. Source and load voltages: waveform 1 and 2. Theirs RMS and THD parameters are 162 V and 32.7%, and then 219 V and 4.1%, respectively.

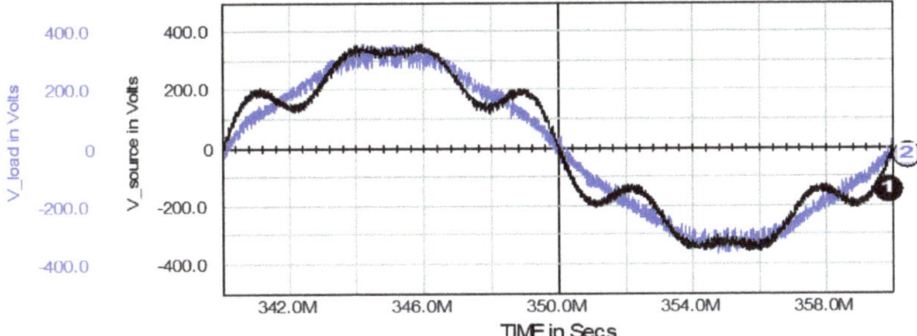

Figure 27. Source and load voltages: waveform 1 and 2. Theirs RMS and THD parameters are 235 V and 21.4%, and then 231 V and 4.3%, respectively.

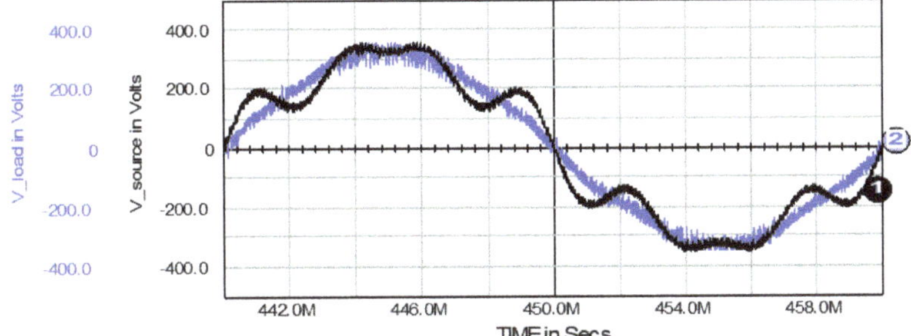

Figure 28. Source and load voltages: waveform 1 and 2. Theirs RMS and THD parameters are 235 V and 21.7%, and then 231 V and 4.7%, respectively.

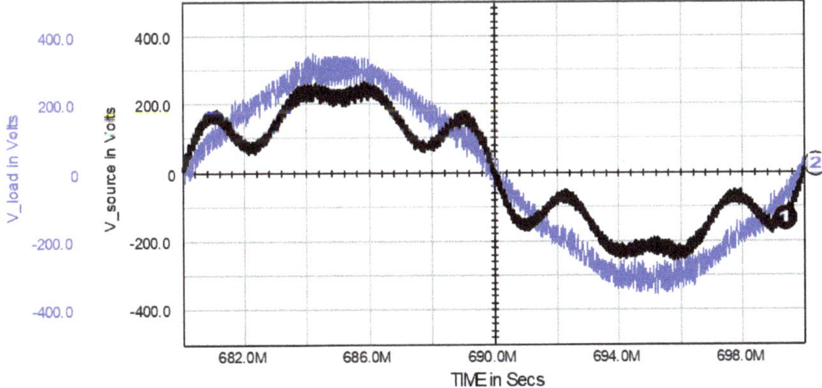

Figure 29. Source and load voltages: waveform 1 and 2. Theirs RMS and THD parameters are 162 V and 32.6%, and then 219 V and 4.2%, respectively.

Figure 20 shows load voltage and current, and then source current during the first 100 ms of whole network action. This time period corresponds with the same time interval in Figures 17–19, and then in Figures 23 and 24.

The waveform 2 in Figure 20 is formed to represent total current of several loads, where some of them are nonlinear and time variable. The current is highly distorted, having also an inductive and DC components. Initially, in the time period T of 0 ms–20 ms the load active and apparent powers are

2.8 kW and 3.0 kVA, respectively, and 2.7 kW and 2.9 kVA during time period *T* of 20 ms–40 ms when UPQC starts its action.

Then, during the time interval 100 ms–300 ms, a new load-sided element was activated and therefore the load current run changes, see waveform 1 in Figure 21 (and waveform 3 in Figure 17). The load current is now composed to be almost unrealistically strongly distorted in order to show high and extended performance of UPQC. In particular a relatively large negative constant component of 17–28 A appears in the load current spectrum, so the fundamental frequency component is inverted relative to the fundamental component of the source voltage. Thus, the load, taken as a whole, works now as a source of energy. Because the in-load generated power exceeds both the power consumed in the load and dissipated in the UPQC the instantaneous DC-link capacitor voltage rises above its initial magnitude. As a result, the conductance signal changes its sign to negative, see waveform 3 in Figure 18. Consequently, the source current, still being purely active, begins to be controlled by the UPQC's shunt converter in order to carry some amount of the in-load generated power to the source, waveform 2 in Figure 21.

Estimation of energy balance in the network when it is practically in the steady state (here in the time period 200 ms–220 ms, when there is no change in the load power and in the static DC-link capacitor voltage, see Figures 17 and 18) is as follows: in-load generated power: 7.85 kW, in-load consumed power: 2.72 kV, power transmitted from the load to the source: 5.07 kW, power dissipated in UPQC's shunt converter: 43 W and power dissipated in UPQC's series converter: 226 W. From this energy balance results that the in-load generated power feeds passive elements of the compensated load and covers UPQC's energy loss. The remaining "excessive" amount of in-load generated power is transmitted to the source.

After switching off the in-load generating element the DC-link capacitor voltage diminishes below its initial magnitude. For this reason the conductance signal becomes positive, consequently source current polarity has been inverted and load draws energy from the source again. This is shown in Figures 18 and 22.

During the time period about 100 ms–350 ms the UPQC operates as a compensator as well as an energy distributor. Note, during this time load voltage and source current were maintained to be sinusoidal, irrespective of energy flow direction (Figures 17 and 18).

4.1.1. Split Storing/Distributing of the In-Load Generated Energy

As it was already stated the energy generated in the active part of load may be decomposed into the portion consumed immediately and into the "excessive" portion. In turn this "excessive" portion may be split into a piece to be transmitted immediately to the grid and a piece to be stored in the DC-link capacitor. In other words, a power limit may be imposed on energy transmission to the grid (or, if needed, a voltage limit may be imposed on maximal DC-link capacitor voltage).

Such possibility of limited back-transmission is illustrated in Figure 19. In this example the maximal back-transmission power is bounded to 2.7 kW. Because the in-load generated power is greater than the allowed power limit of the back-transmission the DC-link capacitor voltage (energy) rises. It lasts till the moment of switching-off the generating element that is located in the load. Then the "excessive" energy portion, which were stored in the capacitor, is discharging partially to the source and partially to UPQC—Being dissipated in its power circuitry. After achieving the balance between load and source active powers, what can be seen in Figure 19 about *t* = 380 ms, the source takes over powering the load.

4.1.2. Full Storing of the In-Load Generated Energy

Characteristic waveforms of currents and voltages for the case of the full energy-storing mode (without upstream energy transmission) are shown in general outlook in Figures 23 and 24, and then some critical time areas are zoomed in Figures 25–29.

The waveform 1 in Figure 24 demonstrates the process of accumulating/discharging the in-load generated "excessive" energy in the DC-link capacitor. The energy accumulation process takes place in time periods 180 ms–340 ms and then 540 ms–620 ms, whereas the energy discharging can be seen during time periods 340 ms–520 ms and 620 ms–940 ms. These time intervals fall within the wider 180 ms–940 ms range in which UPQC's shunt converter blocks any current flow to-or-from the supply source. In other words, during the time 180 ms–940 ms the load works in an energetically autonomous way, that means without any energy drawing from or giving to the supply source. Power fluctuations of all elements of the network are buffered by UPQC. Simultaneously all UPQC's conventional compensation tasks are fully fulfilled.

The most critical areas occurs about 200 ms–300 ms and then about 550 ms–600 ms. There is a cumulation of strong source voltage harmonics distortions with voltage swell and later with voltage sag. There are also large load current deformations, including high energy negative pulses that generates "asymmetrical" active power, occurring only during positive half-waves of source voltage.

The 240 ms–260 ms T period was chosen to show the effect of UPQC action (Figure 25).

For this period the RMS and THD parameters of source voltage run have been reduced at load terminals from the "swelled" magnitude of 348 V down to 268 V and from 14.7% to 7.7%, respectively. Unfortunately, it can be said that despite a significant improvement of load voltage parameters, the voltage quality requirements have not been met.

Within the second critical area the 580 ms–600 ms T period was chosen to show the effect of UPQC action (Figure 26). For this period the RMS and THD parameters of source voltage waveform have been improved at load terminals from the "sagged" magnitude of 162 V up to 219 V and from 32.7% down to 4.1%, respectively. It can be said that these parameters may be considered satisfactory.

The source and load voltages for non-critical areas are presented in next three figures. The 340 ms–360 ms, 440 ms–460 ms and 680 ms–700 ms T periods were chosen to show these waveforms in Figures 27–29, respectively. The load voltage RMS and THD parameters can be accepted as sufficient from the voltage quality point of view. In particular, there is no significant difference in the content of harmonics when working with or without compensation for source voltage sag.

4.1.3. A Side Effect of Conductance Signal Control Method: Alleviating and Catching Energy of Source Voltage Spike Distortion

From time to time an impulsive voltage transient (voltage spike) can appear in the grid, e.g., as a result of lighting stroke. Lighting arresters can be used to stop the transient. Fortunately, it follows from the principle of the considered control method that UPQC can alleviate, or even store and then utilize some amount of energy of such voltage distortion.

The network operates in the steady state when a voltage spike appears at $t = 105$ ms, Figure 30. Parameters characterizing this spike are 3.3 kV in magnitude, 1.2 µs rise time, 10 µs peak voltage duration and 200 µs fall time.

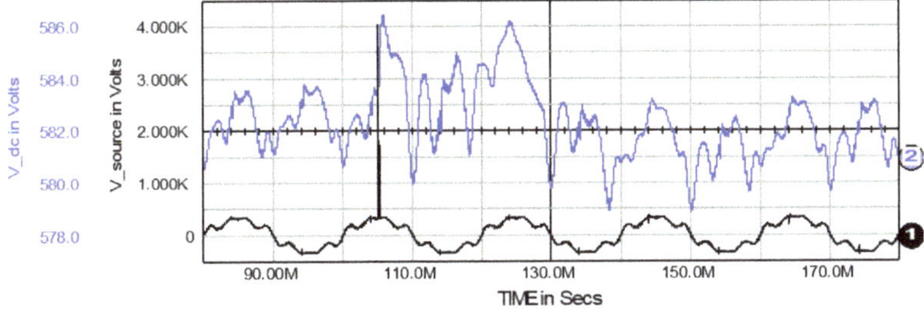

Figure 30. Source voltage with spike distortion: waveform 1 and DC-link capacitor voltage: waveform 2.

Energy of this source spike increases amount of energy, which flows through UPQC input terminals, from 89.9 J during the stead state T period 80 ms–100 ms, up to 141.6 J for the next (i.e., hit by the spike) 100 ms–120 ms T period. Some amount of energy of the spike impacts the load immediately, what can be seen as load voltage distortion shown in waveform 2 in Figure 31. As the result, the energy consumed by the load rises from 88.2 J for the 80 ms–100 ms time period up to 125.4 J for the next one. However, some portion of energy of the spike has been caught by UPQC. This energy portion increases energy stored in the DC-link capacitor: note the difference in capacitor voltage between instants t = 120 ms and t = 100 ms in waveform 2 in Figure 30.

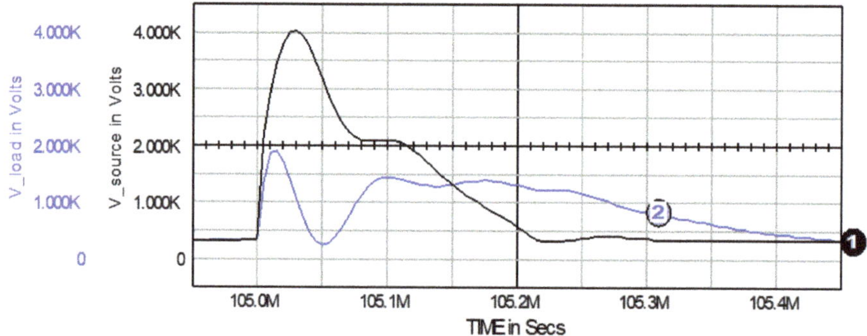

Figure 31. Source voltage: waveform 1 and load voltage: waveform 2.

Because UPQC buffers energy variations between the supply source and the load the energy of the pulse distortion, which reaches load terminals, is lower than the distortion of energy that appears on UPQC input terminals. This is 125.4 J with respect to 141.6 J, respectively. This energy difference increases electric charge stored in the DC-link capacitor and, at the same time, increases its static voltage from 581 V at t = 100 ms up to 584 V at t = 120 ms, waveform 2 in Figure 30. This "additional" voltage (or energy) decreases the conductance signal (Equation (1)) from 82.3 mS for T period 100 ms–120 ms down to 65.7 mS for the next T period 120 ms–140 ms, waveform 1 in Figure 32. This fall of conductance signal (Equation (1)) causes decreasing in source current amplitude from 26.6 A down to 21.3 A, waveform 2 in Figure 32 during the T period 120 ms–140 ms. In other words, during this T period UPQC uses the from-the-distortion energy to power both itself and the load.

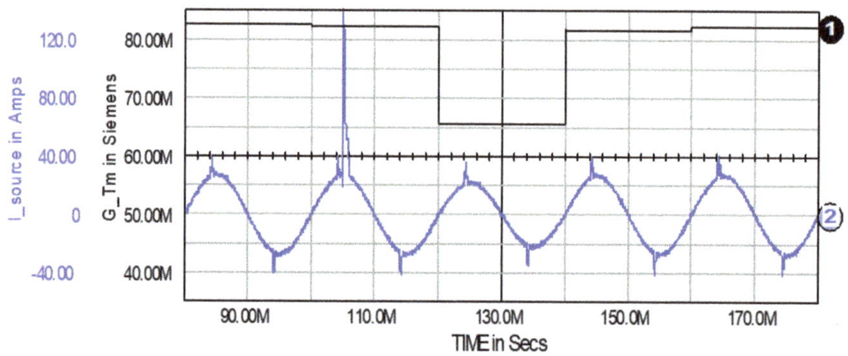

Figure 32. Conductance signal: waveform 1 and source current: waveform 2.

Just after the moment at which energy of the pulse distortion stored in the DC-link capacitor is discharged, the network returns to the steady state, waveform 2 in Figure 30 and both waveforms in Figure 32.

4.1.4. UPQC Operation During Switched Passive/Active Work of DC-Side Load

In order to extend the UPQC usefulness a load as well as a source of energy may be connected to the UPQC's DC-link capacitor. In such a case UPQC can perform some extra tasks. In particular, depending on direction of energy flow, UPQC can act as a high power factor rectifier, which can power DC-side loads with the use of AC-side generated energy, or as an inverter, which can supply AC-side loads with the use of energy generated by DC-side sources. Therefore, UPQC may also serve as an energy bridge and buffer that can control energy flow between all UPQC's AC-side and DC-side loads and sources. Figures 33 and 34 illustrate such extended mode of UPQC operation.

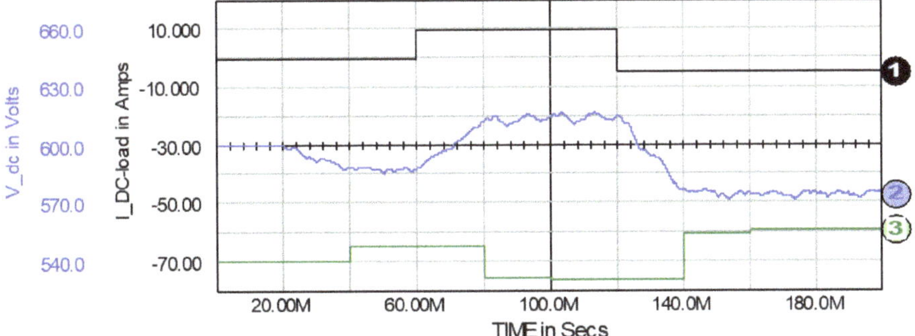

Figure 33. DC-side load current: waveform 1, DC-link capacitor voltage: waveform 2 and conductance signal: waveform 3 (Y scale for this signal is 25 mS/div).

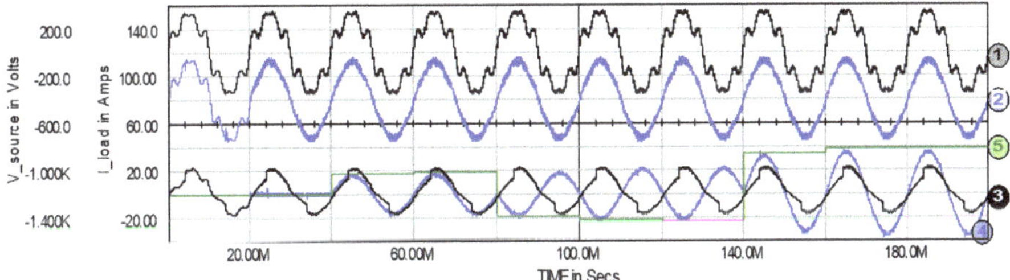

Figure 34. AC-side source voltage: waveform 1, AC-side load voltage: waveform 2, AC-side load current: waveform 3, AC-side source current: waveform 4 and conductance signal: waveform 5, where Y scale for the conductance signal is 57 mS/div.

In Figure 33 UPQC operates in the steady state when a DC-side current of 10 A magnitude begins to charge the DC-link capacitor at the moment $t = 60$ ms, waveform 1 in Figure 33. Due to the increase in DC-link capacitor voltage, the conductance signal changes both its magnitude and sign to be negative, waveform 3 in Figure 33 and waveform 5 in Figure 34.

As a result energy of this charging DC-side current is transferring to the AC-side network. This energy can power AC-side loads and, concurrently, its surplus may be transmitted upstream to the grid. This to-the-grid energy transferring process starts at $t = 80$ ms, waveform 4 in Figure 34.

Then, at time instant $t = 120$ ms, the DC-side current begins to discharge the DC-link capacitor. The conductance signal is recalculated into a new magnitude that depends on sum of AC-side and DC-side active powers. As a result the UPQC, still compensating for voltage and current disturbances, operates concurrently as a high power factor rectifier, i.e., a rectifier with sinusoidal input current, waveform 4 in Figure 34.

Figures 35–39 characterize the UPFC work for when voltage/current runs to be compensated are highly and variously distorted. An additional element, which can operate either as a load or a source of energy and which power can vary in time both in magnitude and in sign, is connected on the DC-link capacitor. Such a network can be considered as generating a kind of worst case of voltage and current runs to be improved. Parameters of these voltage/current runs are specified in the comment that follows Figure 36.

The distorted AC-side source voltage and then the compensated AC-side load voltage are shown in Figure 35. The AC source voltage is distorted by the same higher harmonics and swell/sag disturbances compared to that shown in Figure 23, but this time the whole system is also influenced by energy consumption/generation by the DC-side circuitry. The current run of the DC-side load/source element is shown in Figure 36. Positive polarity of this current means that this element generates energy and vice versa.

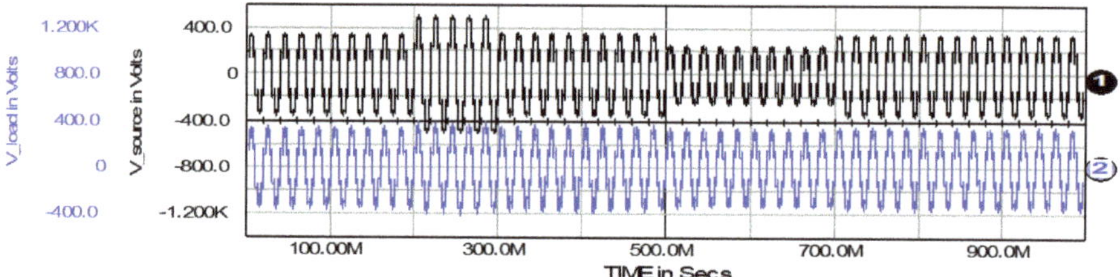

Figure 35. Global view on AC-side source voltage: waveform 1 and on AC-side load voltage: waveform 2.

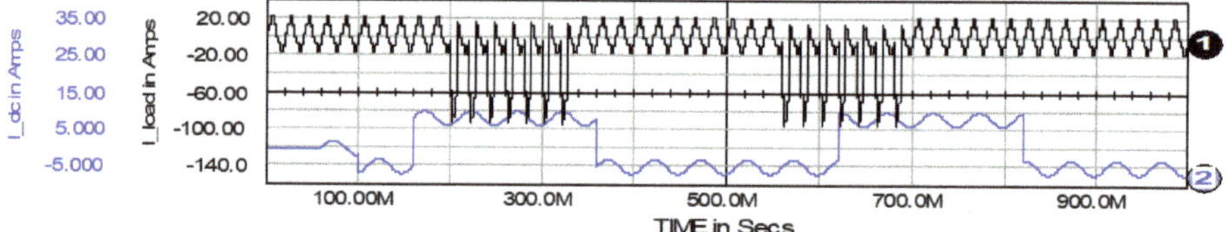

Figure 36. Global view on AC-side load current: waveform 1 and on DC-side load current: waveform 2.

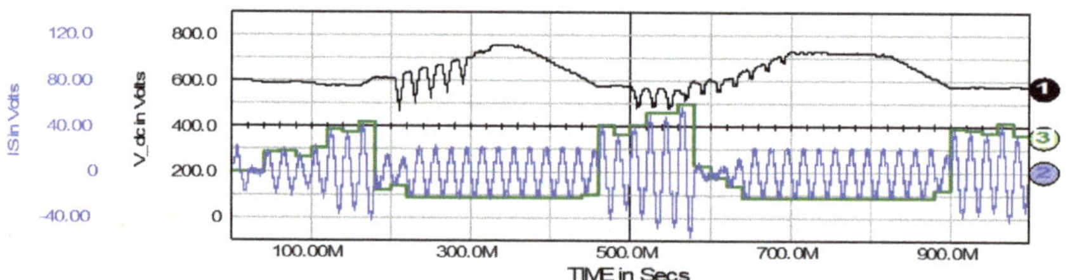

Figure 37. Global view on DC-link capacitor voltage: waveform 1, AC-side source current: waveform 2 and conductance signal: waveform 3 with Y scale of 56 mS/div.

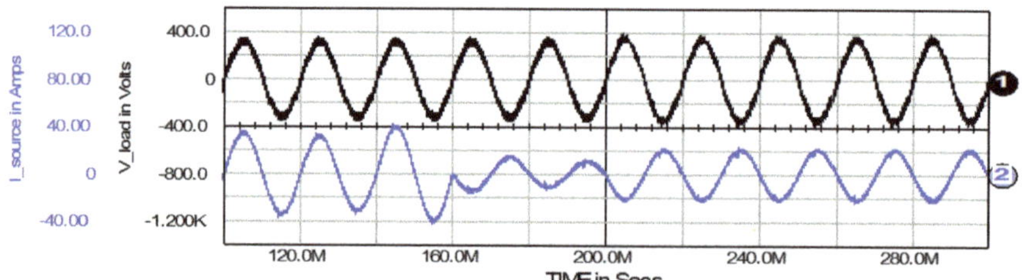

Figure 38. Critical time period 100 ms–300 ms. AC-load voltage: waveform 1 and AC-source current: waveform 2.

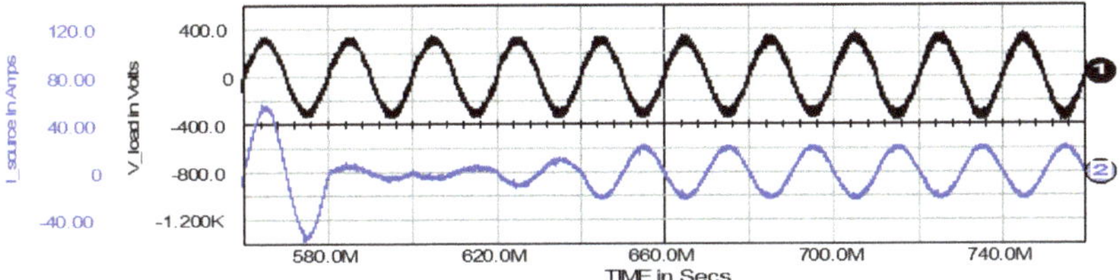

Figure 39. Critical time period 560 ms–760 ms. AC-load voltage: waveform 1 and AC-source current: waveform 2.

DC-link capacitor voltage, AC-side source current and conductance signal (shown as the envelope of source current) are presented in Figure 37. These signals are essential for discussing the UPQC extended operation. In particular, energy balancing between all sources and loads of the network as well as UPQC's "internal energy effort" in order to compensate for nonactive voltage and current components can be identified when analyzing the waveform of DC-link capacitor voltage. This voltage run contains the in-T-period oscillations that relate to compensation for nonactive components of AC-side current and voltage. In the same time the DC-link capacitor voltage increases or decreases statically, i.e., from T_n period to T_{n+1} one, when UPQC balances active powers of all loads and energy sources of the network.

The most critical conditions for UPQC action appear around 200 ms–300 ms and 500 ms–700 ms. They may be treated as a base for a general assessment of UPQC performance.

(a) The first critical time period 200 ms–300 ms: There exist a 50% increase in the amplitude of fundamental component of AC source voltage, a large harmonic deformation with heavy DC-component in AC-side load current and successive energy generation/consumption with variable power on UPQC's DC-side circuitry. The related UPQC's output runs, i.e., AC-side load voltage and AC-side source current, are shown in Figure 38. For the steady state of the grid waveforms, T period 220 ms–240 ms, parameters of UPQC's input and output runs are:

- source voltage RMS/THD of 348 V/14.5% have been transformed into load voltage 247 V/3.8%, respectively;
- load current THD of 82.3% and its DC component of magnitude −28.2 A have been transformed into source current THD and DC component of 1.9% and 0.0 A, respectively.

(b) The second critical time period 500 ms–700 ms. There is a 33% decrease in the amplitude source voltage fundamental component, other parameters are the same as in (a). The UPQC output signals are shown in Figure 39. For the steady state of grid runs within this region, T period 660 ms–680 ms, parameters of UPQC's input and output signals are as follows:

- source voltage RMS of 162 V and THD of 33.2% have been transformed into load voltage RMS and THD of 218 V and 4.6%, respectively;
- load current THD and DC component of 79.5% and −28.5 A has been transformed into source current THD and DC component of 2.6% and 0.1 A, respectively.

It can be stated that in both critical areas of the UPQC operation the disturbed input voltage and current runs have been compensated satisfactory.

5. Conclusions

The paper presents the possibility of compensation for nonactive voltage/current components with the use of compensators that are controlled using a conductance signal. In general, the conductance signal results from the active power of the compensated load. This signal can be calculated based on two variables to be measured:

(a)   signal of load power—But obtained indirectly, by measuring energy stored in the reactive elements of the compensator;

(b)   signal of supply voltage.

By using the conductance signal as the reference for the compensator action it is possible to omit the technically complicated methods of analysis of voltage/current waveforms by their decomposition into plurality of components. The same idea of avoiding the harmonic analysis is visible here if one compares the considered method to the p-q instantaneous power theory and its use to compensators control. Depending on planned purpose of compensation, such a solution may well be an advantage or a disadvantage.

The conductance signal is obtained based on observation of energy balance in the circuit consisting of the source, the UPQC compensator and the load. Any energy imbalance between the power required by the load and supplied by the source is buffered by the UPQC. In other words the network aims the steady state under control of the UPQC.

Using the conductance signal control method extends the functionality of UPQC. There is the possibility of controlling the flow of energy between all active and passive components of the network. This can help to increase the efficiency of the network.

The use of conductance signal in order to control the UPQC action enables bi-directional energy transmission with unity power factor both from the source to the load and in the opposite direction when the load can also generate energy. Both UPQC's AC- and DC-side generated or consumed energy may be handled and exchanged in such bi-directional way. This opens the possibility of using UPQC also as a local energy buffering-and-distribution center that can be useful for smart microgrids, increasing their energy efficiency.

Funding: This research received no external funding.

Conflicts of Interest: The author declares no conflict of interest.

References

1.   Singh, B.; Chandra, A.; Al-Haddad, K. *Power Quality: Problems and Mitigation Techniques*; John Wiley & Sons: Hoboken, NJ, USA, 2014.

2.   Akagi, H.; Watanabe, E.; Aredes, M. *Instantaneous Power Theory and Applications to Power Conditioning*; IEEE Press: Piscataway, NJ, USA, 2017.

3.   Khadkikar, V. Enhancing electric power quality using UPQC: A comprehensive overview. *IEEE Trans. Power Electron.* **2012**, *27*, 2284–2297. [CrossRef]

4.   Chen, D.; Xiao, L. Novel Fundamental and Harmonics Detection Method Based on State-Space Model for Power Electronics System. *IEEE Access* **2020**, *8*, 170002–170012. [CrossRef]

5.   Harirchi, F.; Simoes, M.G. Enhanced instantaneous power theory decomposition for power quality smart converter applications. *IEEE Trans. Power Electron.* **2018**, *33*, 9344–9359. [CrossRef]

6.   Pinto, J.O.; Macedo, R.; Monteiro, V.; Barros, L.; Sousa, T.; Alfonso, J. Single-phase shunt active power filter based on a 5-Level converter topology. *Energies* **2018**, *11*, 1019. [CrossRef]

7.   Fryze, S. Wirk-, Blind-, und Scheinleistung in Elektrischen Stromkreisen mit nichtsinusformigem Verlauf von Strom und Spannung. *Elektrotechnische Z.* **1932**, *25*, 33.

8.   Szromba, A. Conductance-controlled global-compensation-type shunt active power filter. *Arch. Electr. Eng.* **2015**, *64*, 259–275. [CrossRef]

9.   Szromba, A.; Mysiński, W. Voltage-source-based conductance-signal controlled shunt active power filter. In Proceedings of the 2017 European Conference on Power Electronics an Applications, Warsaw, Poland, 11–14 September 2017.

10.   Orts-Grau, S.; Gimeno-Sales, F.J.; Segui-Chilet, S.; Abellen-Garcia, A.; Alcaniz-Fillol, M.; Masot-Periz, R. Selective compensation in four-wire electric systems based on a new equivalent conductance approach. *IEEE Trans. Ind. Electron.* **2009**, *56*, 2862–2874. [CrossRef]

11.   Montero, M.I.; Cadaval, E.R.; Gonzalez, F.B. Comparison of control strategies for shunt active power filters in three-phase four-wire systems. *IEEE Trans. Power Electron.* **2007**, *22*, 229–236. [CrossRef]

12. Moreno, V.; Pigazo, A. Modified FBD method in active power filters to minimize the line current harmonics. *IEEE Trans. Power Deliv.* **2006**, *22*, 735–736. [CrossRef]
13. Hafezi, H.; D'Antona, G.; Dede, A.; Giustina, D.D.; Faranda, R.; Massa, G. Power quality conditioning in LV distribution networks: Results by field demonstration. *IEEE Trans. Smart Grid* **2017**, *8*, 418–427. [CrossRef]
14. Panda, A.K.; Patnaik, N. Management of reactive power sharing and power quality improvement with SRF-PAC based UPQC under unbalanced source voltage condition. *Int. J. Electr. Power Energy Syst.* **2017**, *84*, 182–194. [CrossRef]
15. Ambati, B.B.; Khadkikar, V. Optimal sizing of UPQC considering VA loading and maximum utilization of power-electronics converters. *IEEE Trans. Power Deliv.* **2014**, *29*, 1490–1498. [CrossRef]
16. Ye, J.; Gooi, H.B.; Wu, F. Optimization of the size of UPQC system based on data-driven control design. *IEEE Trans. Smart Grid* **2018**, *9*, 2999–3008. [CrossRef]
17. Moghbel, M.; Masoum, M.A.; Fereidouni, A.; Deilami, S. Optimal sizing, siting and operation of custom power devices with STATCOM and APLC functions for real-time reactive power and network voltage quality control of smart grid. *IEEE Trans. Smart Grid* **2018**, *9*, 5564–5575. [CrossRef]
18. Campanhol, L.B.; da Silva, S.A.; de Oliveira, A.A.; Bacon, V.D. Power flow and stability analyses of a multifunctional distributed generation system integrating a photovoltaic system with Unified Power Quality Conditioner. *IEEE Trans. Power Electron.* **2019**, *34*, 6241–6256. [CrossRef]
19. Kaur, A.; Kaushal, J.; Basak, P. A review on microgrid central controller. *Renew. Sustain. Energy Rev.* **2016**, *55*, 338–345. [CrossRef]
20. Asiminoaei, L.; Blaabjerg, F.; Hansen, S. Detection is key. Harmonic detection methods for active power filter applications. *IEEE Ind. Appl. Mag.* **2007**, *13*, 22–33. [CrossRef]
21. Tareen, W.; Mekhilef, S.; Seyedmahmoudian, M.; Horan, B. Active power filter (APF) for mitigation of power quality issues in grid integration of wind and photovoltaic energy conversion system. *Renew. Sustain. Energy Rev.* **2017**, *70*, 635–655. [CrossRef]
22. Farhadi-Kangarlu, M.; Babaei, E.; Blaabjerg, F. A comprehensive review of dynamic voltage restorers. *Electr. Power Energy Syst.* **2017**, *92*, 136–155. [CrossRef]

Publisher's Note: MDPI stays neutral with regard to jurisdictional claims in published maps and institutional affiliations.

Article

Effective Permeability of Multi Air Gap Ferrite Core 3-Phase Medium Frequency Transformer in Isolated DC-DC Converters

Piotr Dworakowski [1,*], Andrzej Wilk [2], Michal Michna [2], Bruno Lefebvre [1], Fabien Sixdenier [3] and Michel Mermet-Guyennet [1]

[1] Power Electronics & Converters, SuperGrid Institute, 69100 Villeurbanne, France; bruno.lefebvre@supergrid-institute.com (B.L.); michel.mermet-guyennet@alstomgroup.com (M.M.-G.)
[2] Faculty of Electrical and Control Engineering, Gdańsk University of Technology, 80-233 Gdansk, Poland; andrzej.wilk@pg.edu.pl (A.W.); michal.michna@pg.edu.pl (M.M.)
[3] Univ Lyon, Université Claude Bernard Lyon 1, INSA Lyon, ECLyon, CNRS, Ampère, 69100 Villeurbanne, France; Fabien.sixdenier@univ-lyon1.fr
* Correspondence: piotr.dworakowski@supergrid-institute.com

Received: 4 February 2020; Accepted: 11 March 2020; Published: 14 March 2020

Abstract: The magnetizing inductance of the medium frequency transformer (MFT) impacts the performance of the isolated dc-dc power converters. The ferrite material is considered for high power transformers but it requires an assembly of type "I" cores resulting in a multi air gap structure of the magnetic core. The authors claim that the multiple air gaps are randomly distributed and that the average air gap length is unpredictable at the industrial design stage. As a consequence, the required effective magnetic permeability and the magnetizing inductance are difficult to achieve within reasonable error margins. This article presents the measurements of the equivalent $B(H)$ and the equivalent magnetic permeability of two three-phase MFT prototypes. The measured equivalent $B(H)$ is used in an FEM simulation and compared against a no load test of a 100 kW isolated dc-dc converter showing a good fit within a 10% error. Further analysis leads to the demonstration that the equivalent magnetic permeability and the average air gap length are nonlinear functions of the number of air gaps. The proposed exponential scaling function enables rapid estimation of the magnetizing inductance based on the ferrite material datasheet only.

Keywords: average air gap length; dc-dc power converters; gapped magnetic core; magnetic permeability; magnetizing inductance; medium frequency transformer

---

1. Introduction

The medium frequency transformer (MFT) is one of the key components in the isolated dc-dc converters [1–4] related to: smart grids [5], photovoltaic power plants [6], wind power plants [7], and electric vehicle charging [8,9]. The three-phase topology is considered for high power applications where the high power density and high efficiency are required. In [10,11], an analytical approach was proposed to compare multi-phase dc-dc topologies. In [12], the single-phase and three-phase topologies were compared. A 10 kVA 1 kHz three-phase MFT prototype was reported in [13], and a 2 kVA 100 kHz three-phase MFT was reported in [14]. A 5 MW three-phase converter was presented in [15] but using three single-phase MFTs. The general circuit diagram of the three-phase isolated dc-dc converter is composed of two voltage source converters (VSC), and an MFT is presented in Figure 1.

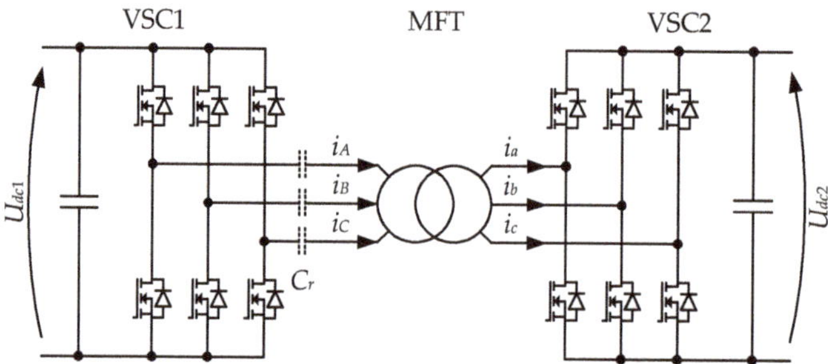

Figure 1. Three-phase isolated dc-dc converter circuit diagram; C_r is the optional resonant capacitor.

The performance of the converter highly depends on the MFT and its equivalent circuit parameters. The leakage inductance has a significant influence on the operation of the converter and the specified value is usually well achieved in the MFT development process. In the LLC resonant dc-dc (LLC) converter [1,16,17], the magnetizing inductance has a significant effect on the zero voltage switching (ZVS) [18–20], but it may be difficult to achieve within reasonable error margins [21]. The maximum value of the magnetizing inductance should take into account the drain-source capacitance C_{ds} of the MOSFET (or other power semiconductor switch). It should ensure the magnetizing current sufficient to charge and discharge the C_{ds} during the dead time of a VSC leg. In the dual active bridge (DAB) converter [2,22], the magnetizing inductance should not increase the VSC current and it should be considered at low operating power.

The operating frequency of the 100 kilowatt class isolated dc-dc converters is considered in the range from few kilohertz to tens of kilohertz [23–25]. The voltage and current fundamentals and harmonics influence the design of the MFT magnetic core and windings. The choice of MFT magnetic core material should be done according to the material properties and cost. The performance factor, which is defined as a product of the frequency and flux density at a specified core loss density, is used to compare different types of core materials [26,27]. The amorphous and especially nanocrystalline materials are preferred in the low and medium frequencies due to the high flux density [28,29]. On the other hand, the main advantage of ferrite cores is their low power loss, which makes them an attractive material for the construction of medium and high frequency transformers [30,31]. The ferrite also offers low cost in terms of material and transformer assembly. In [32], the ferrite core MFT was considered for an optimized dc-dc converter operating at a few kHz. Finally, the ferrite seems as a good candidate for the short term industrialization of the high power three-phase MFT. However, the construction of a ferrite magnetic core for high power MFT requires an assembly of type "I" cores since the C-cores or E-cores do not exist for large transformers. This results in a multi air gap structure of the magnetic core.

The influence of the air gap on the transformer magnetic properties in LLC converters was analysed in [33,34]. It was assumed that the air gap length was known and controlled in the MFT design process. The considered air gaps had relatively large size in order to reduce the slope of the $B(H)$ curve and to minimize the influence of magnetic saturation on the magnetizing inductance value. The influence of the air gap length on the equivalent magnetic permeability, magnetic reluctance and magnetizing inductance in ferrite core transformers was analysed in [35–38]. The influence on the core and winding power loss was studied in [39–41]. All the analysed cases considered a single and uniform air gap of a known length. The analysis of a single but non-uniform air gap in toroidal cores was presented in [42,43]. The influence of the number of uniform air gaps with a controlled size on the magnetic properties and transients in a current transducer was considered in [44].

In the transformer core structure characterized by a construction periodicity (ferromagnetic material—air gap, ferromagnetic material—diamagnetic material, etc.), it is possible to utilize the homogenization technique or multiscale methods in the description of magnetic properties (reluctance of homogenized core, equivalent magnetic permeability, equivalent $B(H)$, etc.). The use of the homogenization technique in the finite element method (FEM) analysis of step-lap joints in steel sheet transformers was proposed in [45]. The homogenization technique was further developed in 2D FEM of steel sheet cores [46–48] and amorphous cores [49]. The multiscale methods were proposed in the analysis of the magnetic properties of transformer cores in [50]. In order to increase the accuracy of magnetic computations, a higher order FEM [51] and a step-wise method were proposed [52].

In all the presented references, it was assumed that the air gap length or the diamagnetic material dimensions were known. However, during the core assembly, the core experiences different mechanical constraints, which are required to ensure its integrity. This impacts the magnetic properties [53] and changes the core structure near the air gaps. In many cases, these changes are difficult to determine, especially once the core is assembled.

According to the authors' knowledge, a study of multiple air gaps in the ferrite core transformers, enabling an efficient MFT design for the isolated dc-dc converters, has not been reported. In this article, it is proposed the analysis of the number of air gaps on the equivalent $B(H)$ and the equivalent magnetic permeability. It is considered that different MFTs have a similar probability distribution of the average air gap length. The authors propose an experimental approach to the determination of the equivalent $B(H)$, implying that the physical phenomena as: nonlinearity, fringing effect, structure dissymmetry, technological aspects, etc. are taken into account.

The novel aspects of this work includes:

- Determination of the equivalent $B(H)$ and the equivalent magnetic permeability in a three-phase multi air gap ferrite core MFT.
- Demonstration that the equivalent magnetic permeability and the average air gap length of the multi air gap ferrite core MFT are nonlinear functions of the number of air gaps.
- Proposal of an exponential scaling function, enabling a rapid estimation of the magnetizing inductance based on the ferrite material datasheet only.

The multi air gap medium frequency transformer prototype is presented in Section 2. The measurement of the magnetic flux in the function of the magnetizing current and the calculation of the equivalent $B(H)$ and the equivalent magnetic permeability are presented in Section 3. The finite element simulation of the MFT no load test, using the measured equivalent $B(H)$, is presented in Section 4. The FEM simulation result is compared with an experimental measurement on a 100 kW 1.2 kV 20 kHz dc-dc converter in Section 5. The results are analysed and discussed in Section 6, where the influence of the number of air gaps on the equivalent permeability and the average air gap length are presented. A scaling function enabling a rapid estimation of the magnetizing inductance is proposed.

2. High Power Medium Frequency Transformer

2.1. MFT Prototypes

The authors have developed two three-phase MFT prototypes for a 100 kW 1.2 kV 20 kHz dc-dc converter. The dc-dc converter is presented in details in [54]. The three-phase structure is still novel in the MFT applications with very little demonstrators. The specifications of two MFT prototypes T1 and T2 are presented in Table 1. The MFT T1 can operate in delta and star vector groups whereas the T2 in star only. The winding of both transformers is made of the same litz wire composed of 3870 strands of 0.1 mm diameter.

Table 1. Specification of the medium frequency transformer prototypes for the nominal operating conditions.

Parameter	T1 Dd	T1 Yy	T2 Yy
Phase voltage (V)	980	566	566
Phase current (A)	36	65	65
Core flux density (T)	0.22	0.15	0.27
Winding current density (A/mm^2)	1.2	2.1	2.1
Dimensions of active parts (cm)	$67 \times 20 \times 35$		$45 \times 20 \times 30$
Total weight (kg)	57		36

The MFT T2 is presented in Figure 2 and its design is detailed in [55]. In particular, a significant difference between the calculated and measured magnetizing inductance is highlighted. This shows the important influence of the parasitic air gaps on the magnetizing inductance.

Figure 2. Medium frequency transformer prototype T2 showing primary winding terminals: 1*-1, 2*-2, 3*-3, secondary winding terminals: 4*-4, 5*-5, 6*-6, three columns A, B, C, and additional auxiliary coils AUX1 and AUX2 for flux measurement (blue wire around the yoke).

2.2. Magnetic Core

The magnetic core of the MFT prototypes is made of MnZn ferrite 3C90 from Ferroxcube. The core is assembled with I-cores measuring 25 mm $\times$ 25 mm $\times$ 100 mm each. The core assembly is presented in Figure 3. In this core design, the I-cores are not interleaved. It can be seen that the core involves multiple parasitic air gaps. Moreover, due to manufacturing tolerances, the I-core is not an ideal rectangular cuboid and its dimensions vary from one sample to another. This causes the non-uniform parasitic air gaps in the core. There are at least two types of parasitic air gaps: perpendicular and longitudinal to the axis of the magnetic flux path. The authors claim that the parasitic air gap size is unpredictable at the industrial design stage and that it cannot be modelled precisely. In Appendix A, some example views of the ferrite core assembly are presented. It can be seen that the air gap length varies from almost zero to about 0.5 mm. Consequently, the use of material datasheet in the calculation of effective magnetizing inductance leads to significant errors. However, the magnetizing inductance or the equivalent $B(H)$ can be measured on the transformer prototype. Such a measurement can be helpful in a new transformer design with a similar core assembly.

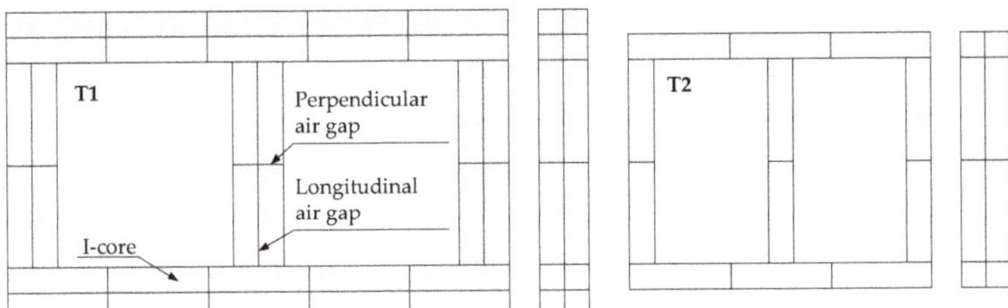

Figure 3. Medium frequency transformer core assembly composed of elementary I-cores: T1 (**left**) and T2 (**right**).

3. Equivalent B(H) Measurement

3.1. Measurement Setup

The nonlinear magnetic properties of core material are represented by the magnetic permeability, which relates the magnetic flux density B with the magnetic field strength H. The nonlinear magnetic properties of a transformer core can be described by the current-dependent flux linkage characteristics $\Psi(i)$ using the experimental approach. From the flux linkage characteristics, the $B(H)$ curve can be determined under certain simplifying assumptions. The measurement of $\Psi(i)$ hysteretic characteristics for inherently asymmetric three-phase transformer with three columns was proposed in [56]. In this approach to determine $\Psi(i)$ characteristics for each winding, only two phases are excited in a special manner.

A dedicated static $B(H)$ measurement setup was developed as presented in Figure 4. It is composed of a high current AC power supply, oscilloscope and probes. The primary and secondary windings of each phase were connected in series in order to achieve a high magnetomotive force (MMF). The windings of two columns were connected in anti-parallel so that their MMFs add together. Two additional auxiliary coils (AUX1 and AUX2) were placed on the yoke allowing the measurement of the magnetic flux in the core (see the blue wire in Figure 2) and minimizing the magnetic coupling in the air. The voltage of the remaining winding (so-called zero-coil) is measured in order to verify that the magnetic flux coupled with this winding is close to zero.

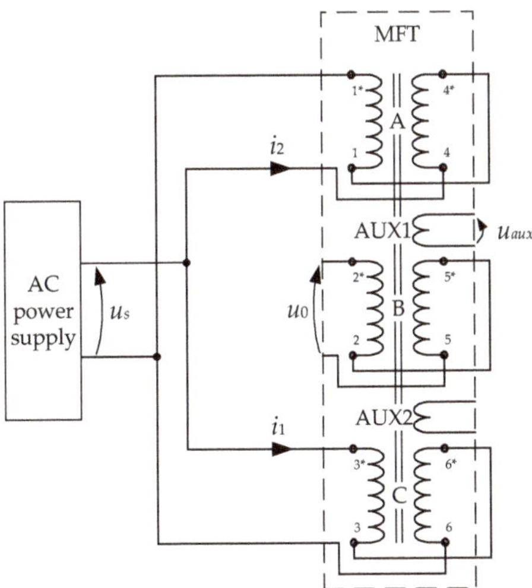

Figure 4. Circuit diagram of the equivalent $B(H)$ measurement setup where the windings C and A are supplied.

For each MFT prototype, three measurements were performed according to the winding configurations presented in Table 2. The frequency of the power supply in the static $B(H)$ measurement setup was set to 100 Hz. This value was considered in order to minimize the effect of eddy currents (considering a high frequency material as ferrite) and to achieve good performance of the available power supply.

Table 2. Winding configurations of the equivalent $B(H)$ mearement circuits.

u_s	u_{aux}	u_0	Magnetic Flux Path
A + B	AUX1	C	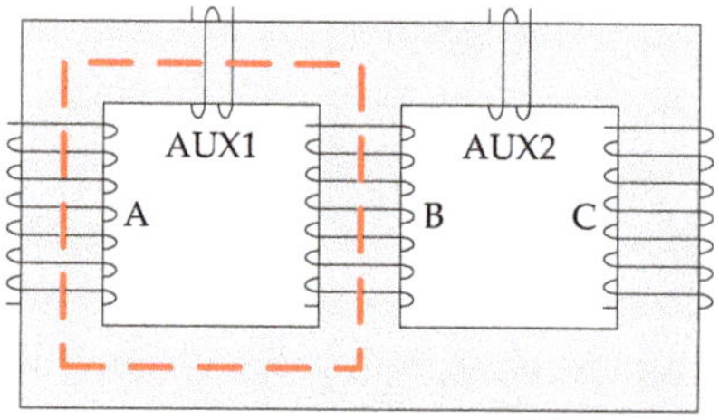
B + C	AUX2	A	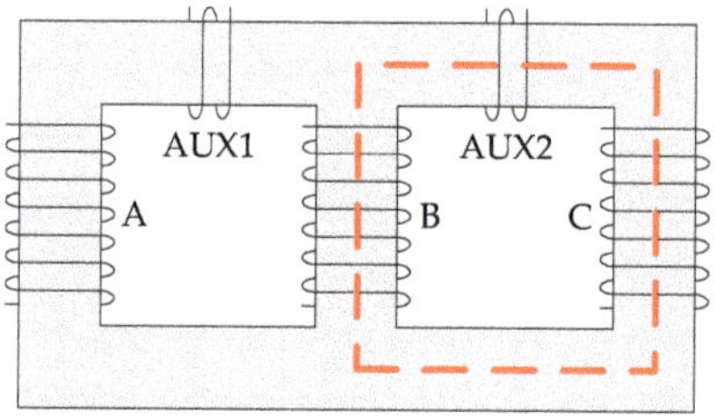
C + A	AUX1 or AUX2	B	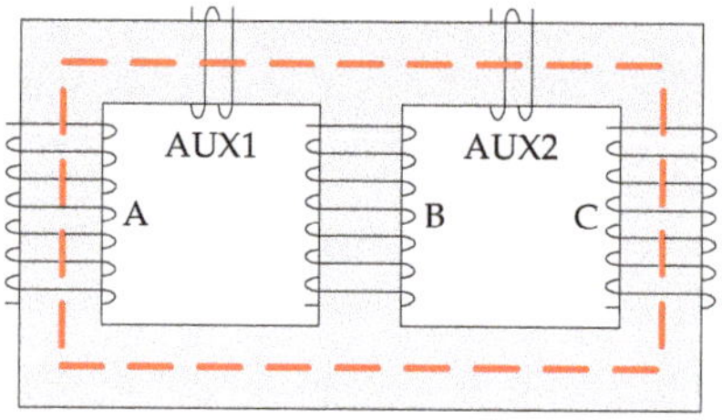

The waveforms of the magnetic flux density $B(t)$ and the magnetic field strength $H(t)$ were calculated with:

$$H(t) = \frac{N_{exc}[i_1(t) + i_2(t)]}{l_m} \tag{1}$$

$$\Phi(t) = \int_0^T u_{aux}(t)dt \tag{2}$$

$$B(t) = \frac{\Phi(t)}{N_{aux}A_c}. \tag{3}$$

where i_1 and i_2 are the current of the first and second excitation winding respectively, N_{exc} is the number of turns of each excitation winding, l_m is the average magnetic circuit length (visualized in Table 2), u_{aux} is the voltage of the auxiliary coil placed on the yoke, T is the period of the excitation voltage, Φ is the core magnetic flux, N_{aux} is the number of turns of the auxiliary coil, and A_c is the average cross-section of the core.

3.2. Measurement Results

The measured waveforms for the example case where the C and A windings of T2 are supplied are presented in Figure 5a. The measurement was performed with the transformer temperature equal to ambient at 25 °C. It can be observed that the supply voltage is close to sinusoidal. The currents in two excitation windings show the core saturation. Their amplitudes are slightly different due to a difference in winding impedance. The amplitude of the zero-coil voltage is relatively low.

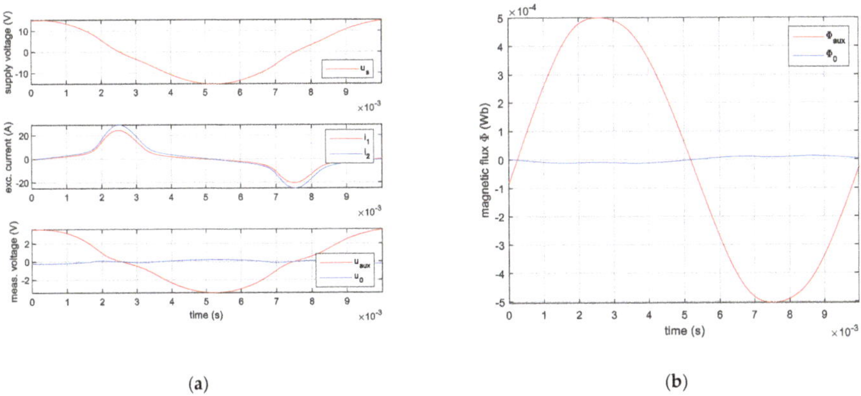

(a)   (b)

Figure 5. Waveforms of the T2 supplied with C and A windings: (**a**) measured supply voltage u_s, excitation currents i_1 (C) and i_2 (A), auxiliary coil voltage u_{aux} (AUX1) and zero coil voltage u_0 (B); (**b**) magnetic flux of the auxiliary coil Φ_{aux} (AUX1) and magnetic flux of the zero coil Φ_0 (B).

Figure 5b presents the waveforms of the magnetic flux calculated according to (2). The Φ_{aux} correspond to the main magnetic flux in two side columns and two yokes. The Φ_0 corresponds to the magnetic flux in the central column. It is observed that the magnetic flux in the central column is below 5% of the main flux so it seems fair to neglect it.

Thanks to (1) and (3), the magnetic field strength H and the magnetic flux density B are calculated. In Figure 6, the resulting $B(H)$ is plotted for the positive values of H. The $B(H)$ is separated into the upward and downward curves, which are then interpolated with piecewise linear functions in order to facilitate the data analysis. The anhysteretic $B(H)$ curve is calculated as the average of the interpolated upward and downward curves and further filtered to achieve a smooth curve adequate for further processing. Moreover, the coercive field H_c and remanent flux density B_r can be captured.

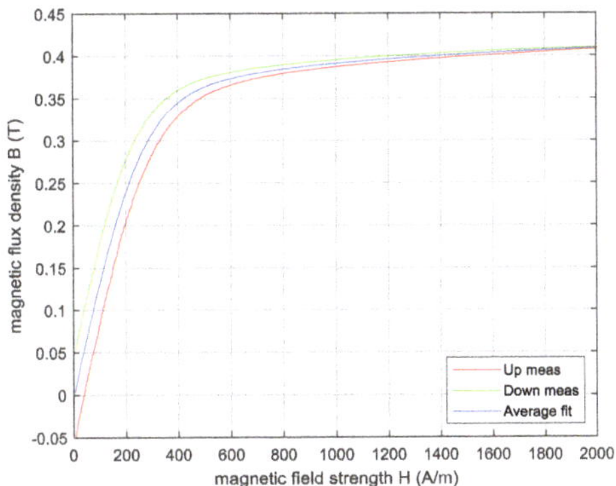

Figure 6. Measured equivalent $B(H)$ of the T2 supplied with C and A windings: upward curve (red), downward curve (green) and interpolated anhysteretic curve (blue).

3.3. Synthesis of Equivalent B(H) Measurement

The measurement process presented in the previous section was repeated for the MFT T1 and T2 for the cases with the supply of windings: A and B, B and C, and C and A, according to Table 2. The measured equivalent anhysteretic $B(H)$ and relative permeability $\mu_r(H)$ are presented in Figure 7. The 3C90 datasheet curves [57] are plotted for comparison. As expected, a significant difference between the datasheet and the measurement is observed. There is a difference between T1 and T2 since they have a different core assembly, T1 having more parasitic air gaps than T2 (see Figure 3). For each MFT, the equivalent $B(H)$ differs slightly for different measurement circuits. This proves that the parasitic air gaps are randomly distributed in the core assembly. For each transformer, the authors arbitrarily select the solid line curve (CA) as the reference $B(H)$ for the whole core.

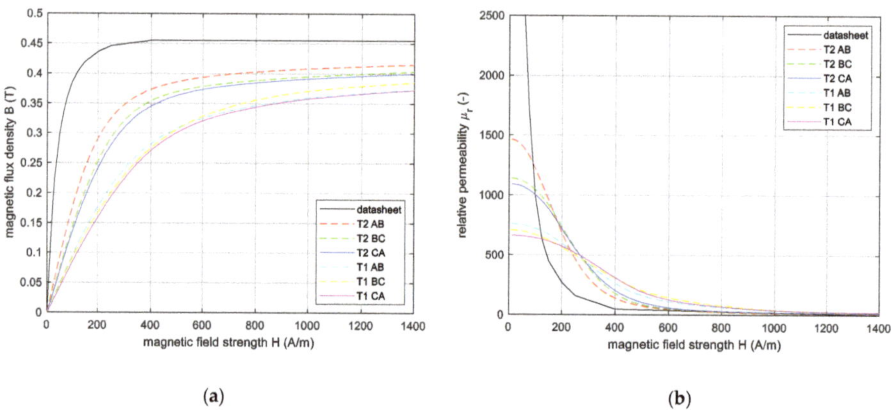

(a)   (b)

Figure 7. Synthesis of equivalent $B(H)$ measurement: (**a**) equivalent anhysteretic $B(H)$; (**b**) equivalent relative permeability μ_r; curves based on 3C90 datasheet (black) and measurement: T2 supply of A and B windings (red), T2 supply of B and C windings (green), T2 supply of C and A windings (blue)—the same as in Figure 6, T1 supply of A and B windings (cyan), T1 supply of B and C windings (yellow), T1 supply of C and A windings (magenta).

4. Finite Element Simulation

4.1. Finite Element Model

A 3D MFT T2 model was developed in Ansys Maxwell. A simplified transformer geometry was considered. The model was divided into three computational domains as shown in Figure 8. The Ω_1 domain is the volume of the windings, the Ω_2 domain is the volume of the core, and the Ω_3 domain consists of the air surrounding the MFT. In this model, it is assumed that the magnetic core is homogenized. It means that the core components: ferrite, air gaps and also glue, impregnation resin, etc. form a homogenous material. In a similar manner, the winding is also homogenized.

The Maxwell's equations for the defined domains have the form:

$$\nabla \times \mathbf{H} = \begin{cases} \mathbf{j} \text{ in } \Omega_1 \\ \overleftrightarrow{\sigma}\mathbf{E} \text{ in } \Omega_2 \\ 0 \text{ in } \Omega_3 \end{cases} \tag{4}$$

$$\nabla \times \mathbf{E} = -\frac{\partial \mathbf{B}}{\partial t}; \nabla \bullet \mathbf{B} = 0; \mathbf{B} = \nabla \times \mathbf{A} \tag{5}$$

where $\overset{\leftrightarrow}{\sigma}$ is the electrical conductivity tensor:

$$\overset{\leftrightarrow}{\sigma} = \begin{bmatrix} \sigma_{xx}(x,y,z) & 0 & 0 \\ 0 & \sigma_{yy}(x,y,z) & 0 \\ 0 & 0 & \sigma_{zz}(x,y,z) \end{bmatrix} \tag{6}$$

The permeability tensor, which for nonlinear properties describes the relation between dB and dH in the constitutive equation, can be expressed as:

$$\overset{\leftrightarrow}{\mu} = \begin{cases} \mu_0 \text{ in } \Omega_1 \\ \overset{\leftrightarrow}{\mu}_{core} \text{ in } \Omega_2 \\ \mu_0 \text{ in } \Omega_3 \end{cases} \tag{7}$$

where $\overset{\leftrightarrow}{\mu}_{core}$ is the magnetic permeability tensor:

$$\overset{\leftrightarrow}{\mu}_{core} = \begin{bmatrix} \mu_{xx}(x,y,z) & 0 & 0 \\ 0 & \mu_{yy}(x,y,z) & 0 \\ 0 & 0 & \mu_{zz}(x,y,z) \end{bmatrix} \tag{8}$$

It was assumed that the ferrite core has isotropic electrical and magnetic properties. Hence, the electrical conductivity and magnetic permeability tensors have the form:

$$\overset{\leftrightarrow}{\sigma} = \begin{bmatrix} \sigma_c & 0 & 0 \\ 0 & \sigma_c & 0 \\ 0 & 0 & \sigma_c \end{bmatrix} ; \overset{\leftrightarrow}{\mu}_{core} = \begin{bmatrix} \mu_c & 0 & 0 \\ 0 & \mu_c & 0 \\ 0 & 0 & \mu_c \end{bmatrix} \tag{9}$$

where $\sigma_c = 0.25$ S/m (at 25 °C) and $\mu_c = dB/dH$ are defined in the previous section (Figure 7b, curve T2 CA). In Ansys Maxwell, the material conductivity enables the calculation of eddy current effects. However, it can be noticed that the ferrite conductivity is low so the eddy current effects do not have a significant impact on the magnetic field and core power loss.

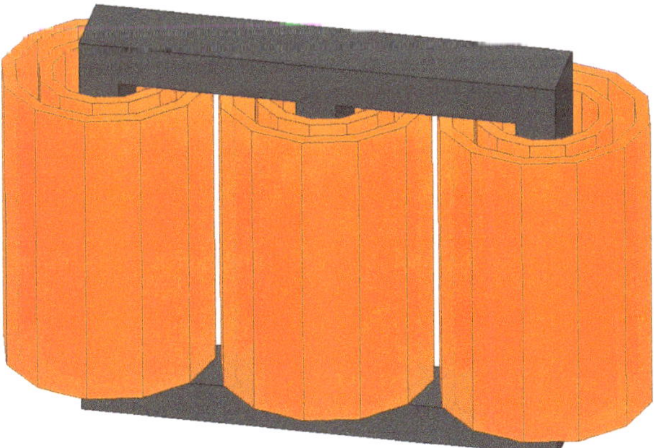

Figure 8. 3D MFT model divided into three computational domains: Ω_1 volume of the windings (orange), Ω_2 volume of the homogenized core (grey) and Ω_3 air surrounding the MFT (white).

4.2. Magnetic Simulations

In order to perform a magnetic transient simulation, the finite element model was coupled with an equivalent circuit model. A no load test was considered, as presented in Figure 9. The coupling between the finite element model and the equivalent circuit model is done through the nonlinear inductances L_1, L_2 and L_3, which correspond to the primary winding. The voltage sources model the VSC square output voltage, and R_p is the primary winding resistance.

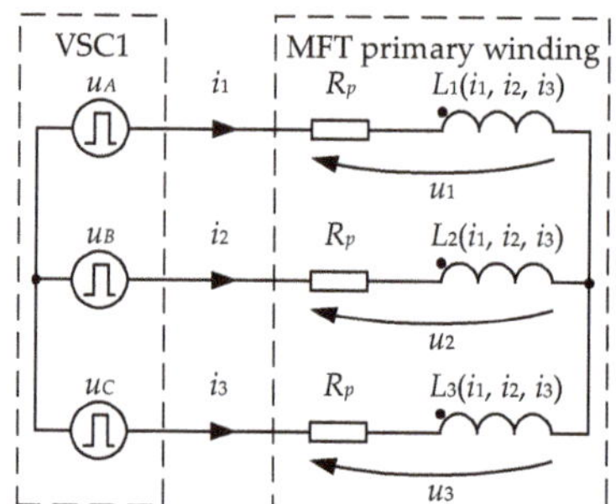

Figure 9. MFT no load test equivalent circuit model coupled with the finite element model through the nonlinear inductances L_1, L_2 and L_3.

The magnetic transient simulation result is presented in Figure 10. The MFT phase voltage is presented, being a typical VSC output voltage waveform. The MFT primary current is presented in steady-state. This result will be further used to validate the measured equivalent $B(H)$.

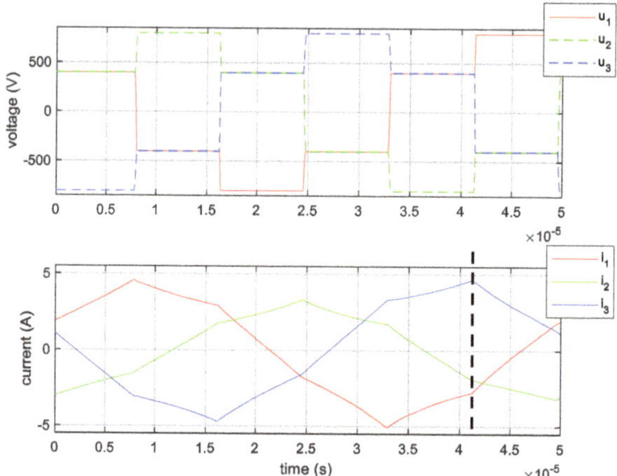

Figure 10. MFT no load test magnetic transient simulation result: primary phase voltage (**top**) and primary current (**bottom**); the dashed vertical line indicates the time instant for the magnetostatic simulation.

In Figure 11, the magnetostatic simulation result corresponding to the time instant defined by the dashed line in Figure 10 is presented. The magnitude of the flux density is plotted on the core surface and the maximum value of 0.27 T is observed, as expected. In Figure 12, the magnetic field strength and the magnetic flux density are plotted along the path defined by the dashed line in Figure 11. The different values of quotient $B/(\mu_0 H)$ in the central and the right column can be observed due to the nonlinearity of the $B(H)$ curve.

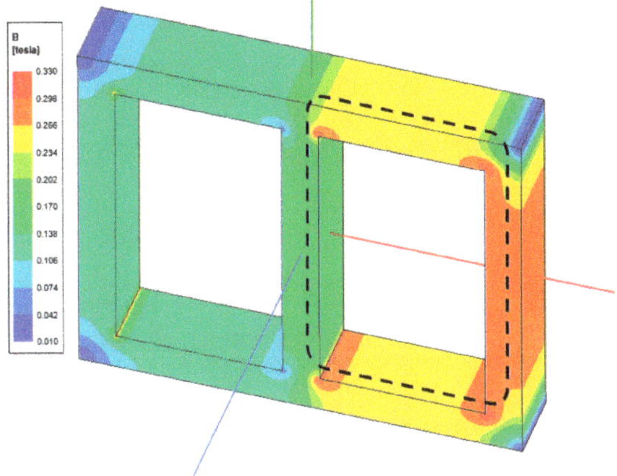

Figure 11. Magnetic flux density B magnitude on the core surface with the current excitation $i_1 = -2.76$ A, $i_2 = -1.93$ A, $i_3 = 4.69$ A; the dashed line indicates the magnetic flux path in the centre of the core.

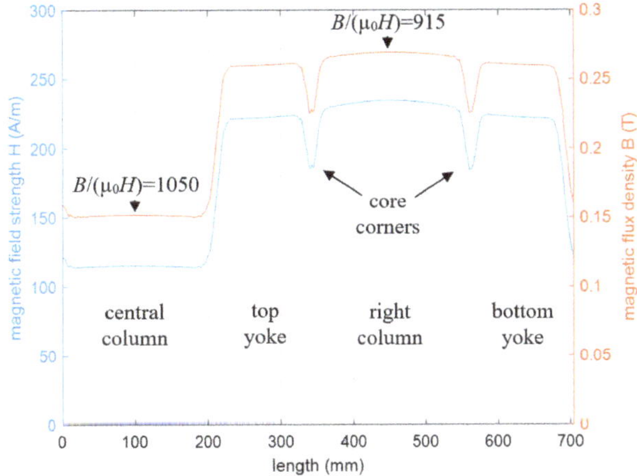

Figure 12. Magnetic field strength H and magnetic flux density B along the path in the centre of the core passing through the central column, top yoke, right column, and bottom yoke; the values of static permeability $B/(\mu_0 H)$ are presented.

5. Experimental Verifications

5.1. Converter Test Bench

The power converter test bench was developed for the 100 kW dc-dc converter, as presented in Figure 13. A MFT no load test was considered in order to evaluate the magnetizing inductance. In the no load test, the VSC1 operates normally with 1200 Vdc input voltage and the AC terminals of the VSC2 are disconnected. The circuit diagram of the experimental setup is equivalent to the one used in the simulation that is presented in Figure 9. The test was performed at an ambient temperature of 25 °C. The MFT temperature was measured nearly equal to the ambient as the test lasts for a few minutes only.

Figure 13. 100 kW three-phase isolated dc-dc converter test bench implementation.

5.2. No Load Test Experimental Results

The measured waveform of the MFT T2 no load current is presented in Figure 14. The simulated no load current from the previous section is plotted for the comparison. Generally, quite a good fit between the simulation and the measurement is observed. Some minor differences are discussed hereafter.

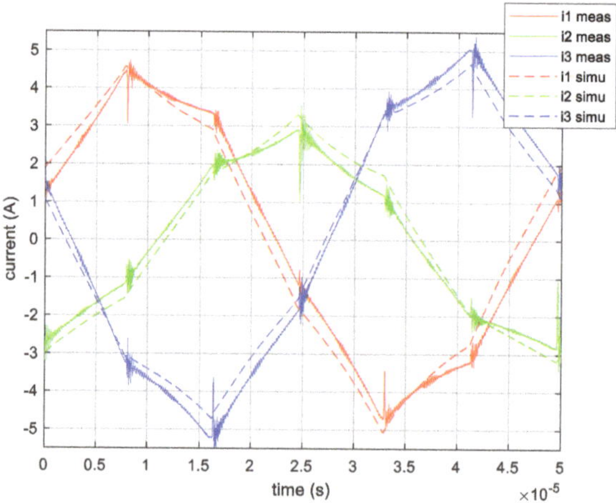

Figure 14. MFT T2 no load test primary current: experimental result (solid line), magnetic transient simulation result (dashed line).

There are some high frequency oscillations present in the measurement. They are due to the parasitic capacitance of the windings that have not been modelled. This could be improved by adding the winding self and mutual capacitances into the model. However, the simulation time would increase significantly.

There are some differences in the current amplitude of different phases. As it has been presented in Figure 7, the $B(H)$ is not strictly the same for the whole core. Since in the simulation, the authors have assumed a single equivalent $B(H)$, then it seems normal to observe some differences in the measured currents.

Moreover, there might be some differences due to the fact that the simulation model assumes the anhysteretic $B(H)$. In Figure 6, one can see that the measured equivalent $B(H)$ is hysteretic, thus it may influence the shape of the current waveform, in particular, the corresponding ascending and descending slopes of the current.

Finally, the RMS current error is within 10% and the authors consider this acceptable. If the datasheet $B(H)$ was used (Figure 7), then the RMS current error would reach approximately 500%. This experimental result proves the validity of the measured equivalent $B(H)$.

6. Scaling of Relative Permeability

The approach presented in the previous paragraphs has limited usage in the MFT design process since it is based on the measurement on a physical device. This limits the practical usage to post-manufacturing analysis or to a new design of a similar transformer. In this section, an approach based on a simple count of perpendicular parasitic air gaps is proposed.

In the MFT design process from scratch, when evaluating the performance of isolated dc-dc converters, one is usually interested in the magnetizing inductance at the nominal $B(H)$ operating point. This is usually below the $B(H)$ saturation, so the anhysteretic curve from Figure 7 can be linearized as presented in Figure 15.

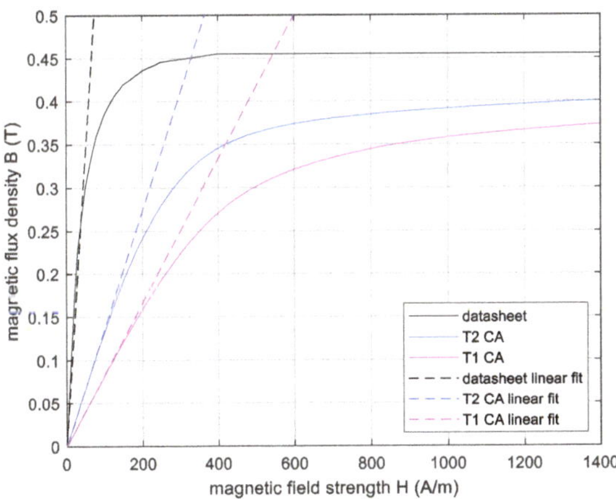

Figure 15. Equivalent anhysteretic $B(H)$: datasheet and measurement (solid line), linear interpolation (dashed line).

From Figure 3, we can count the number of perpendicular parasitic air gaps along the magnetic path. This equals to 10 and 14 for T2 and T1 respectively. The core used for the datasheet measurement had zero air gaps. The value of datasheet linearized relative permeability, which equals $\mu_{r0} = 5300$, is read from Figure 15. Thus, the equivalent relative permeability ratio K_μ of the multi air gap core can be calculated with:

$$K_\mu = \frac{\mu_r}{\mu_{r0}} \tag{10}$$

where μ_r is the equivalent relative permeability defined in Figure 15 for T1 or T2. The equivalent relative permeability ratio is plotted in Figure 16 as a function of a number of parasitic air gaps.

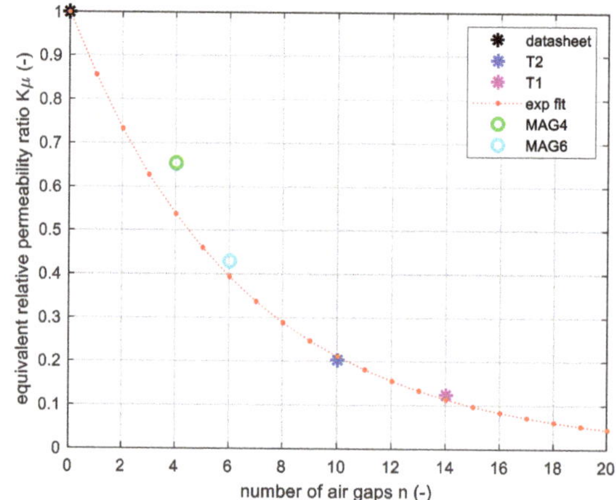

Figure 16. Equivalent relative permeability ratio K_μ in the function of a number of parasitic air gaps n: datasheet, T2 and T1 measurement (stars), exponential interpolation (red dashed line), and single-phase multi air gap transformer MAG4 and MAG6 measurement (circles).

In addition, an exponential interpolation is proposed allowing to estimate the equivalent relative permeability for any high power ferrite core MFT with a similar core assembly. The exponential interpolation function is defined as:

$$K_\mu(n) = e^{-0.155n} \tag{11}$$

where n is the number of perpendicular parasitic air gaps along the magnetic flux path.

This function was validated with the experimental $B(H)$ measurement on two single-phase multi air gap (MAG) transformers presented in Appendix B. The MAG4 transformer has four air gaps and MAG6 has six air gaps. Both use the same I-cores as T1 and T2. The resulting ratios are displayed in Figure 16 and it can be seen that for MAG4 the ratio is slightly higher than the exponential interpolation. This is normal because for this transformer the I-cores were carefully selected to minimize the parasitic air gaps and the core assembly is simpler compared to the three-phase MFT. However, a general trend of the equivalent relative permeability ratio is clearly observed even if the four MFT prototypes involve different technologies and different manufacturers.

Furthermore, a simple reluctance model of the magnetic core neglecting the fringing effect is considered according to [35]. The total magnetic circuit reluctance can be related to the sum of the I-core and air gap reluctances as:

$$\frac{l_m}{\mu_0 \mu_r A_c} = n\frac{l_I}{\mu_0 \mu_{r0} A_c} + \frac{l_a}{\mu_0 A_c} \tag{12}$$

where l_m is the average magnetic circuit length, l_I is the length of the I-core, l_a is the average air gap length, and A_c is the average cross-section of the core. Assuming that the average magnetic circuit length l_m is equal to $n \cdot l_I$, then it can be found the relative average air gap length l_a/l_m defined as:

$$\frac{l_a}{l_m} = \frac{1}{\mu_r} - \frac{1}{\mu_{r0}} \tag{13}$$

Considering an ideal core assembly, where the average air gap length l_a equals n times the known individual air gap length l_g, the relative average air gap length l_a/l_m is a linear function of n:

$$\frac{l_a}{l_m} = \frac{l_g}{l_m}n \tag{14}$$

In Figure 17, these linear functions are presented for four transformers T2, T1, MAG4, and MAG6. It was verified that the individual air gap length l_g changes between prototypes. However, considering the proposed exponential interpolation (11), the effective relative average air gap length is a nonlinear function of n as presented in Figure 17. This is due to the fact that the I-core is not an ideal rectangular cuboid and its dimensions vary from one sample to another. As a consequence, the mechanical assembly of the core gets more difficult when a large number of I-cores is assembled.

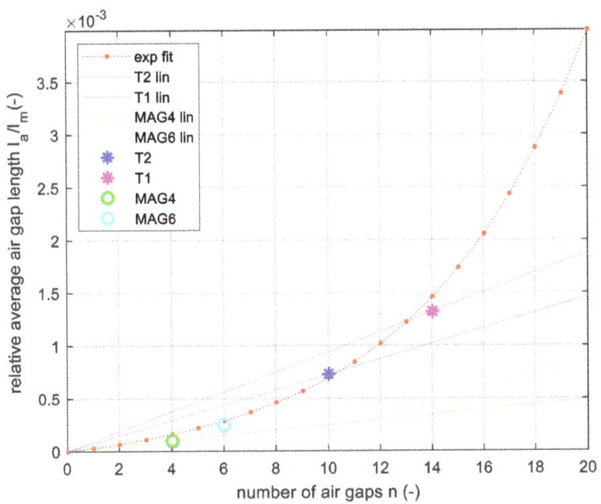

Figure 17. Relative average air gap length l_a/l_m in the function of a number of parasitic air gaps n: T2, T1, MAG4 and MAG6 measurement (stars/circle), the corresponding idealized reluctance model (solid lines), and the relative average air gap length calculated based on the proposed exponential interpolation (red dashed line).

The proposed approach can be used in scaling the datasheet $B(H)$ for a finite element simulation, in the rapid estimation of transformer magnetizing inductance or in evaluating the size of the average air gap length. The magnetizing inductance can be estimated based on the magnetic reluctance model according to:

$$L_m = K_\mu(n) \frac{\mu_0 \mu_{r0} N^2 A_c}{l_m} \tag{15}$$

where N is the primary/secondary number of turns. It shall be mentioned that the proposed estimation is meant to provide an order of magnitude of the magnetizing inductance. This shall be sufficient when evaluating the performance of isolated dc-dc converters. However, the proposed scaling function could be further validated with a large number of MFT prototypes with different types of I-cores and a different number of parasitic air gaps.

7. Conclusions

The analysis of the effective permeability and average air gap length in multi air gap ferrite core three-phase medium frequency transformer was presented. The calculation of the magnetizing inductance in multi air gap ferrite core MFT based on core material datasheet leads to significant errors. This may impact the design of isolated dc-dc converters as the magnetizing inductance influences their performance.

The measurement of the equivalent $B(H)$ and the equivalent permeability for two three-phase MFT prototypes was presented. The measured equivalent $B(H)$ was used in a finite element simulation, giving good results when compared to a 100 kW dc-dc converter no load operation. The use of the anhysteretic $B(H)$ gives satisfactory results within 10% error compared to the experiment.

This article demonstrates that the equivalent magnetic permeability and the average air gap length of the multi air gap ferrite core MFT are nonlinear functions of the number of air gaps. An empirical scaling function is proposed for the rapid estimation of the magnetizing inductance in the multi air gap MFT. In fact, the relative average air gap length increases with the number of parasitic air gaps due to the increasing difficulty in mechanical assembly of the core. The proposed scaling function can be used in the design of isolated dc-dc converters using 25 mm × 25 mm × 100 mm I-cores or similar, based on the core material datasheet and a number of parasitic air gaps.

The measured or scaled equivalent $B(H)$ can also be used in the equivalent circuit simulation instead of finite element simulation. This would allow more convenient simulations as well including the winding capacitance. The measured $B(H)$ can be further utilized in the simulations taking into account the hysteresis. This work shall be further extended taking into account the influence of the temperature since the ferrite relative permeability depends on the temperature. The proposed scaling function could be further validated with a large number of MFT prototypes with different types of I-cores and a different number of parasitic air gaps in order to determine the uncertainty range. The experimental validation of the influence of the magnetizing inductance on the performance of three-phase isolated dc-dc converters is recommended.

Author Contributions: Conceptualization, P.D. and A.W.; methodology, P.D., A.W. and M.M.; software, P.D.; validation, P.D.; formal analysis, P.D., A.W. and M.M.; investigation, P.D., A.W. and M.M.; resources, P.D., A.W. and M.M.; data curation, P.D.; writing—original draft preparation, P.D., A.W. and M.M; writing—review and editing, P.D., A.W., M.M. and F.S.; visualization, P.D., A.W. and M.M; supervision, A.W., M.M., B.L., M.M.-G.; project administration, B.L. and M.M.-G.; funding acquisition, M.M.-G. All authors have read and agreed to the published version of the manuscript.

Funding: This work was supported by a grant overseen by the French National Research Agency (ANR) as part of the "Investissements d'Avenir" Program (ANE-ITE-002-01) and the LINTE^2 Laboratory at the Gdansk University of Technology.

Acknowledgments: The authors would like to thank Caroline Stackler, Martin Guillet and Alexis Fouineau for their valuable support in writing this article.

Conflicts of Interest: The authors declare no conflict of interest.

Appendix A

In Figure A1, two example views of the ferrite core assembly are presented where the perpendicular and longitudinal parasitic air gaps can be observed up to about 0.5 mm. In each assembly, 4 randomly selected I-cores are aligned along a calliper on a flat surface. The I-cores are assembled tight together so that even if there is an air gap on the visible surface then there is somewhere a direct contact between the neighbour I-cores. In the 3-phase MFT core assembly, composed of tens of I-cores, the air gaps are even larger due to the cumulating I-core misalignments.

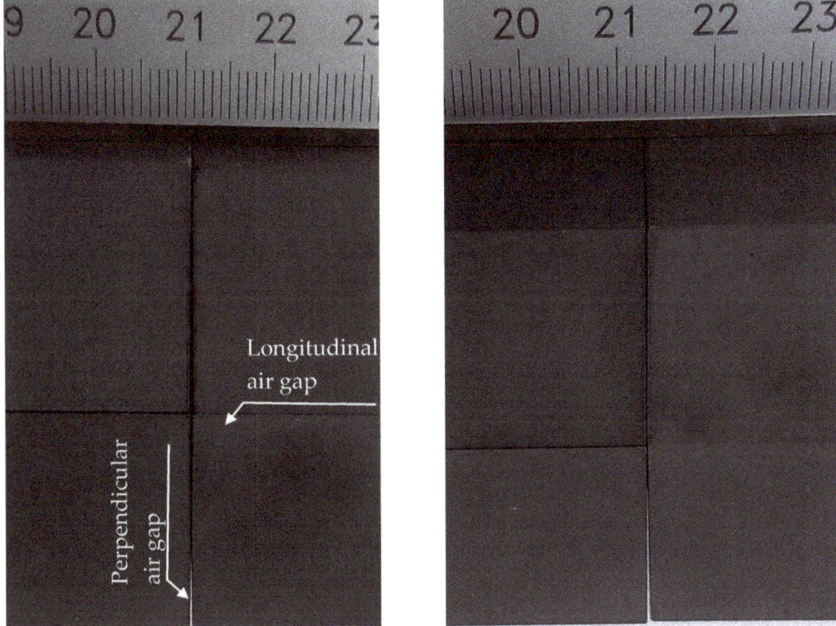

Figure A1. Two assemblies of 4 randomly selected 3C90 ferrite I-cores showing the perpendicular parasitic air gap and the longitudinal parasitic air gap measuring up to about 0.5 mm.

Appendix B

The core assemblies of the single-phase multi air gap MFTs are presented in Figure A2. The MAG4 has 4 and the MAG6 has 6 perpendicular parasitic air gaps along the magnetic flux path. The MAG4 is detailed in [58].

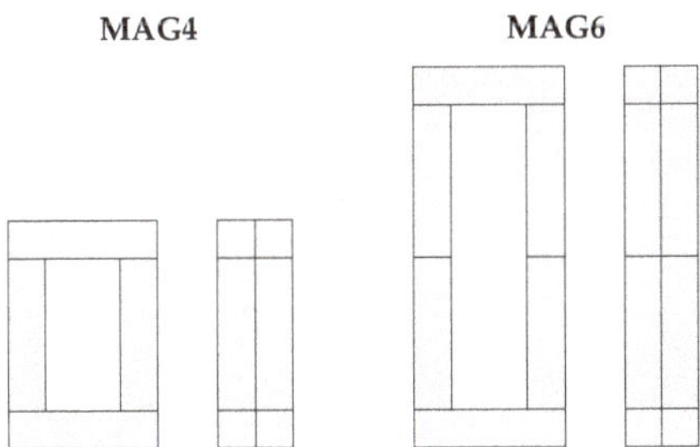

Figure A2. Single-phase multi air gap MFT core assembly: MFT4 with 4 air gaps (**left**) and MFT6 with 6 air gaps (**right**).

References

1. Yang, B.; Lee, F.C.; Zhang, A.J. Guisong Huang LLC resonant converter for front end DC/DC conversion. In Proceedings of the APEC. Seventeenth Annual IEEE Applied Power Electronics Conference and Exposition (Cat. No.02CH37335), Dallas, TX, USA, 10–14 March 2002; Volume 2, pp. 1108–1112.
2. De Doncker, R.W.A.A.; Divan, D.M.; Kheraluwala, M.H. A three-phase soft-switched high-power-density DC/DC converter for high-power applications. *IEEE Trans. Ind. Appl.* **1991**, *27*, 63–73. [CrossRef]

3.   Schwarz, F.C.; Klaassens, J.B. A Controllable 45-kW Current Source for DC Machines. *IEEE Trans. Ind. Appl.* **1979**, *IA-15*, 437–444. [CrossRef]

4.   Mweene, L.H.; Wright, C.A.; Schlecht, M.F. A 1 kW 500 kHz front-end converter for a distributed power supply system. *IEEE Trans. Power Electron.* **1991**, *6*, 398–407. [CrossRef]

5.   Adamowicz, M. Power Electronics Building Blocks for implementing Smart MV/LV Distribution Transformers for Smart Grid. *Acta Energetica* **2014**, *4*, 6–13. [CrossRef]

6.   Walker, G.R.; Sernia, P.C. Cascaded DC-DC converter connection of photovoltaic modules. *IEEE Trans. Power Electron.* **2004**, *19*, 1130–1139. [CrossRef]

7.   Dincan, C.G.; Kjaer, P.; Chen, Y.-H.; Sarrá-Macia, E.; Munk-Nielsen, S.; Bak, C.L.; Vaisambhayana, S. Design of a High-Power Resonant Converter for DC Wind Turbines. *IEEE Trans. Power Electron.* **2019**, *34*, 6136–6154. [CrossRef]

8.   Du, Y.; Lukic, S.; Jacobson, B.; Huang, A. Review of high power isolated bi-directional DC-DC converters for PHEV/EV DC charging infrastructure. In Proceedings of the 2011 IEEE Energy Conversion Congress and Exposition, Phoenix, AZ, USA, 17–22 September 2011; pp. 553–560.

9.   Ruffo, R.; Khalilian, M.; Cirimele, V.; Guglielmi, P.; Cesano, M. Theoretical and experimental comparison of two interoperable dynamic wireless power transfer systems for electric vehicles. In Proceedings of the 2017 IEEE Southern Power Electronics Conference (SPEC), Puerto Varas, Chile, 4–7 December 2017; pp. 1–6.

10.   Garcia-Bediaga, A.; Villar, I.; Rujas, A.; Etxeberria-Otadui, I.; Rufer, A. Analytical Models of Multiphase Isolated Medium-Frequency DC–DC Converters. *IEEE Trans. Power Electron.* **2017**, *32*, 2508–2520. [CrossRef]

11.   Lagier, T. Convertisseurs Continu-Continu Pour Les Réseaux d'électricité à Courant Continu. Doctoral Thesis, INPT, Toulouse, France, 2016.

12.   Xue, J.; Wang, F.; Boroyevich, D.; Shen, Z. Single-phase vs. three-phase high density power transformers. In Proceedings of the 2010 IEEE Energy Conversion Congress and Exposition, Atlanta, GA, USA, 12–16 September 2010; pp. 4368–4375.

13.   Lee, Y.; Vakil, G.; Watson, A.J.; Wheeler, P.W. Geometry optimization and characterization of three-phase medium frequency transformer for 10 kVA Isolated DC-DC converter. In Proceedings of the 2017 IEEE Energy Conversion Congress and Exposition (ECCE), Cincinnati, OH, USA, 1–5 October 2017; pp. 511–518.

14.   Noah, M.; Kimura, S.; Endo, S.; Yamamoto, M.; Imaoka, J.; Umetani, K.; Martinez, W. A novel three-phase LLC resonant converter with integrated magnetics for lower turn-off losses and higher power density. In Proceedings of the 2017 IEEE Applied Power Electronics Conference and Exposition (APEC), Tampa, FL, USA, 26–30 March 2017; pp. 322–329.

15.   Soltau, N.; Stagge, H.; De Doncker, R.W.; Apeldoorn, O. Development and demonstration of a medium-voltage high-power DC-DC converter for DC distribution systems. In Proceedings of the 2014 IEEE 5th International Symposium on Power Electronics for Distributed Generation Systems (PEDG), Galway, Ireland, 24–27 June 2014; pp. 1–8.

16.   Kim, E.-S.; Oh, J.-S. High-Efficiency Bidirectional LLC Resonant Converter with Primary Auxiliary Windings. *Energies* **2019**, *12*, 4692. [CrossRef]

17.   Bouvier, Y.E.; Serrano, D.; Borović, U.; Moreno, G.; Vasić, M.; Oliver, J.A.; Alou, P.; Cobos, J.A.; Carmena, J. ZVS Auxiliary Circuit for a 10 kW Unregulated LLC Full-Bridge Operating at Resonant Frequency for Aircraft Application. *Energies* **2019**, *12*, 1850. [CrossRef]

18.   Morel, F.; Stackler, C.; Ladoux, P.; Fouineau, A.; Wallart, F.; Evans, N.; Dworakowski, P. Power electronic traction transformers in 25 kV/50 Hz systems: Optimisation of DC/DC Isolated Converters with 3.3 kV SiC MOSFETs. In Proceedings of the PCIM Europe 2019; International Exhibition and Conference for Power Electronics, Intelligent Motion, Renewable Energy and Energy Management, Nuremberg, Germany, 7–9 May 2019; pp. 1–8.

19.   Dujic, D.; Steinke, G.K.; Bellini, M.; Rahimo, M.; Storasta, L.; Steinke, J.K. Characterization of 6.5 kV IGBTs for High-Power Medium-Frequency Soft-Switched Applications. *IEEE Trans. Power Electron.* **2014**, *29*, 906–919. [CrossRef]

20.   Huber, J.E.; Rothmund, D.; Wang, L.; Kolar, J.W. Full-ZVS modulation for all-SiC ISOP-type isolated front end (IFE) solid-state transformer. In Proceedings of the 2016 IEEE Energy Conversion Congress and Exposition (ECCE), Milwaukee, WI, USA, 18–22 September 2016; pp. 1–8.

21. Mogorovic, M.; Dujic, D. Medium Frequency Transformer Design and Optimization. In Proceedings of the PCIM Europe 2017; International Exhibition and Conference for Power Electronics, Intelligent Motion, Renewable Energy and Energy Management, Nuremberg, Germany, 16–18 May 2017; pp. 1–8.

22. Everts, J. Design and Optimization of an Efficient (96.1%) and Compact (2 kW/dm3) Bidirectional Isolated Single-Phase Dual Active Bridge AC-DC Converter. *Energies* **2016**, *9*, 799. [CrossRef]

23. Mogorovic, M.; Dujic, D. 100 kW, 10 kHz Medium-Frequency Transformer Design Optimization and Experimental Verification. *IEEE Trans. Power Electron.* **2019**, *34*, 1696–1708. [CrossRef]

24. Villar, I.; Mir, L.; Etxeberria-Otadui, I.; Colmenero, J.; Agirre, X.; Nieva, T. Optimal design and experimental validation of a Medium-Frequency 400 kVA power transformer for railway traction applications. In Proceedings of the 2012 IEEE Energy Conversion Congress and Exposition (ECCE), Raleigh, NC, USA, 15–20 September 2012; pp. 684–690.

25. Ortiz, G.; Biela, J.; Bortis, D.; Kolar, J.W. 1 Megawatt, 20 kHz, isolated, bidirectional 12 kV to 1.2 kV DC-DC converter for renewable energy applications. In Proceedings of the 2010 International Power Electronics Conference—ECCE ASIA, Sapporo, Japan, 21–24 June 2010; pp. 3212–3219.

26. Villar, I.; Garcia-Bediaga, A.; Viscarret, U.; Etxeberria-Otadui, I.; Rufer, A. Proposal and validation of medium-frequency power transformer design methodology. In Proceedings of the 2011 IEEE Energy Conversion Congress and Exposition, Phoenix, AZ, USA, 17–22 September 2011; pp. 3792–3799.

27. Hurley, W.G.; Merkin, T.; Duffy, M. The Performance Factor for Magnetic Materials Revisited: The Effect of Core Losses on the Selection of Core Size in Transformers. *IEEE Power Electron. Mag.* **2018**, *5*, 26–34. [CrossRef]

28. Ruiz-Robles, D.; Ortíz-Marín, J.; Venegas-Rebollar, V.; Moreno-Goytia, E.; Granados-Lieberman, D.; Rodríguez-Rodriguez, J. Nanocrystalline and Silicon Steel Medium-Frequency Transformers Applied to DC-DC Converters: Analysis and Experimental Comparison. *Energies* **2019**, *12*, 2062. [CrossRef]

29. Ruiz-Robles, D.; Venegas-Rebollar, V.; Anaya-Ruiz, A.; Moreno-Goytia, E.L.; Rodríguez-Rodríguez, J.R. Design and Prototyping Medium-Frequency Transformers Featuring a Nanocrystalline Core for DC–DC Converters. *Energies* **2018**, *11*, 2081. [CrossRef]

30. Bahmani, M.A. Design considerations of medium-frequency power transformers in HVDC applications. In Proceedings of the 2017 Twelfth International Conference on Ecological Vehicles and Renewable Energies (EVER), Monte Carlo, Monaco, 11–13 April 2017; pp. 1–6.

31. Stojadinović, M.; Biela, J. Modelling and Design of a Medium Frequency Transformer for High Power DC-DC Converters. In Proceedings of the 2018 International Power Electronics Conference (IPEC-Niigata 2018 -ECCE Asia), Niigata, Japan, 20–24 May 2018; pp. 1103–1110.

32. Stackler, C.; Morel, F.; Ladoux, P.; Fouineau, A.; Wallart, F.; Evans, N. Optimal sizing of a power electronic traction transformer for railway applications. In Proceedings of the IECON 2018—44th Annual Conference of the IEEE Industrial Electronics Society, Washington, DC, USA, 21–23 October 2018; pp. 1380–1387.

33. Keuck, L.; Schafmeister, F.; Boecker, J.; Jungwirth, H.; Schmidhuber, M. Computer-Aided Design and Optimization of an Integrated Transformer with Distributed Air Gap and Leakage Path for LLC Resonant Converter. In Proceedings of the PCIM Europe 2019; International Exhibition and Conference for Power Electronics, Intelligent Motion, Renewable Energy and Energy Management, Nuremberg, Germany, 7–9 May 2019; pp. 1–8.

34. Noah, M.; Endo, S.; Kimura, S.; Yamamoto, M.; Imaoka, J.; Umetani, K.; Hiraki, E. An investigation into a slight-variation of the transformer effective permeability in LLC resonant converter. In Proceedings of the 2017 19th European Conference on Power Electronics and Applications (EPE'17 ECCE Europe), Warsaw, Poland, 11–14 September 2017.

35. Ayachit, A.; Kazimierczuk, M.K. Sensitivity of effective relative permeability for gapped magnetic cores with fringing effect. *IET Circuits Devices Syst.* **2017**, *11*, 209–215. [CrossRef]

36. Salas, R.A.; Pleite, J. Simulation of the Saturation and Air-Gap Effects in a POT Ferrite Core With a 2-D Finite Element Model. *IEEE Trans. Magn.* **2011**, *47*, 4135–4138. [CrossRef]

37. Balakrishnan, A.; Joines, W.T.; Wilson, T.G. Air-gap reluctance and inductance calculations for magnetic circuits using a Schwarz-Christoffel transformation. *IEEE Trans. Power Electron.* **1997**, *12*, 654–663. [CrossRef]

38. Stenglein, E.; Albach, M. A Novel Approach to Calculate the Reluctance of Air-Gaps in Ferrite Cores. In Proceedings of the PCIM Europe 2017; International Exhibition and Conference for Power Electronics, Intelligent Motion, Renewable Energy and Energy Management, Nuremberg, Germany, 16–18 May 2017; pp. 1–8.

39. Ayachit, A.; Kazimierczuk, M.K. Steinmetz Equation for Gapped Magnetic Cores. *IEEE Magn. Lett.* **2016**, *7*, 1–4. [CrossRef]

40. Komma, T.; Gueldner, H. The effect of different air-gap positions on the winding losses of modern planar ferrite cores in switch mode power supplies. In Proceedings of the Automation and Motion 2008 International Symposium on Power Electronics, Electrical Drives, Ischia, Italy, 11–13 June 2008; pp. 632–637.

41. Albach, M.; Rossmanith, H. The influence of air gap size and winding position on the proximity losses in high frequency transformers. In Proceedings of the 2001 IEEE 32nd Annual Power Electronics Specialists Conference (IEEE Cat. No.01CH37230), Vancouver, BC, Canada, 17–21 June 2001; Volume 3, pp. 1485–1490.

42. Zurek, S. FEM Simulation of Effect of Non-Uniform Air Gap on Apparent Permeability of Cut Cores. *IEEE Trans. Magn.* **2012**, *48*, 1520–1523. [CrossRef]

43. Stenglein, E.; Kuebrich, D.; Albach, M. Analytical calculation of the current depending inductance of a stepped air gap inductor. In Proceedings of the 2016 18th European Conference on Power Electronics and Applications (EPE'16 ECCE Europe), Karlsruhe, Germany, 5–9 September 2016; pp. 1–10.

44. Lesniewska, E.; Jalmuzny, W. Influence of the number of core air gaps on transient state parameters of TPZ class protective current transformers. *IET Sci. Meas. Technol.* **2009**, *3*, 105–112. [CrossRef]

45. Nakata, T.; Takahashi, N.; Kawase, Y. Magnetic performance of step-lap joints in distribution transformer cores. *IEEE Trans. Magn.* **1982**, *18*, 1055–1057. [CrossRef]

46. Pietruszka, M.; Napieralska-Juszczak, E. Lamination of T-joints in the transformer core. *IEEE Trans. Magn.* **1996**, *32*, 1180–1183. [CrossRef]

47. Hihat, N.; Napieralska-Juszczak, E.; Lecointe, J.-P.; Sykulski, J.K.; Komeza, K. Equivalent Permeability of Step-Lap Joints of Transformer Cores: Computational and Experimental Considerations. *IEEE Trans. Magn.* **2011**, *47*, 244–251. [CrossRef]

48. Gyselinck, J.; Melkebeek, J. Two-dimensional finite element modelling of overlap joints in transformer cores. *COMPEL Int. J. Comput. Math. Electr. Electron. Eng.* **2001**, *20*, 253–268. [CrossRef]

49. Shin, P.S.; Lee, J. Magnetic field analysis of amorphous core transformer using homogenization technique. *IEEE Trans. Magn.* **1997**, *33*, 1808–1811. [CrossRef]

50. Hollaus, K.; Schöbinger, M. Multiscale finite element method for perturbation of laminated structures. In Proceedings of the 2017 International Conference on Electromagnetics in Advanced Applications (ICEAA), Verona, Italy, 11–15 September 2017; pp. 1262–1263.

51. Hauck, A.; Ertl, M.; Schöberl, J.; Kaltenbacher, M. Accurate magnetostatic simulation of step-lap joints in transformer cores using anisotropic higher order FEM. *COMPEL Int. J. Comput. Math. Electr. Electron. Eng.* **2013**, *32*, 1581–1595. [CrossRef]

52. Da Luz, M.V.F.; Dular, P.; Leite, J.V.; Kuo-Peng, P. Modeling of Transformer Core Joints via a Subproblem FEM and a Homogenization Technique. *IEEE Trans. Magn.* **2014**, *50*, 1009–1012. [CrossRef]

53. Penin, R.; Parent, G.; Lecointe, J.-P.; Brudny, J.-F.; Belgrand, T. Impact of Mechanical Deformations of Transformer Corners on Core Losses. *IEEE Trans. Magn.* **2015**, *51*, 1–5. [CrossRef]

54. Lagier, T.; Chédot, L.; Ghossein, L.; Wallart, F.; Lefebvre, B.; Dworakowski, P.; Mermet-Guyennet, M.; Buttay, C. A 100 kW 1.2 kV 20 kHz DC-DC converter prototype based on the Dual Active Bridge topology. In Proceedings of the 2018 IEEE International Conference on Industrial Technology (ICIT), Lyon, France, 20–22 February 2018; pp. 559–564.

55. Dworakowski, P.; Wilk, A.; Michna, M.; Lefebvre, B.; Lagier, T. 3-phase medium frequency transformer for a 100 kW 1.2 kV 20 kHz Dual Active Bridge converter. In Proceedings of the IECON 2019—45th Annual Conference of the IEEE Industrial Electronics Society, Lisbon, Portugal, 14–17 October 2019; Volume 1, pp. 4071–4076.

56. Fuchs, E.F.; You, Y. Measurement of λ-1 characteristics of asymmetric three-phase transformers and their applications. *IEEE Power Eng. Rev.* **2002**, *22*, 69–70. [CrossRef]

57. Ferroxcube 3C90 Material Specification. Available online: https://www.ferroxcube.com/upload/media/product/file/MDS/3c90.pdf (accessed on 19 January 2020).

58. Wilk, A.; Michna, M.; Dworakowski, P.; Lefebvre, B. Influence of air gap size on magnetizing current and power losses in ferrite core transformers—Experimental investigations. In Proceedings of the EPNC 2018 Twenty-Fifth Symposium on Electromagnetic Phenomena in Nonlinear Circuits, Arras, France, 26–29 June 2018.

Article

SiC-Based Power Electronic Traction Transformer (PETT) for 3 kV DC Rail Traction

Marek Adamowicz [1],* and Janusz Szewczyk [2]

[1]   Faculty of Electrical and Control Engineering, Gdańsk University of Technology, Narutowicza 11/12, 80-233 Gdańsk, Poland
[2]   MMB Drives Ltd., Maszynowa 26, 80-298 Gdańsk, Poland; j.szewczyk@mmb-drives.com.pl
*   Correspondence: marek.adamowicz@pg.edu.pl; Tel.: +48-58-348-6076

Received: 17 September 2020; Accepted: 22 October 2020; Published: 24 October 2020

Abstract: The design of rolling stock plays a key role in the attractiveness of the rail transport. Train design must strictly meet the requirements of rail operators to ensure high quality and cost-effective services. Semiconductor power devices made from silicon carbide (SiC) have reached a level of technology enabling their widespread use in traction power converters. SiC transistors offering energy savings, quieter operation, improved reliability and reduced maintenance costs have become the choice for the next-generation railway power converters and are quickly replacing the IGBT technology which has been used for decades. The paper describes the design and development of a novel SiC-based DC power electronic traction transformer (PETT) intended for electric multiple units (EMUs) operated in 3 kV DC rail traction. The details related to the 0.5 MVA peak power medium voltage prototype, including the electrical design of the main building blocks are presented in the first part of the paper. The second part deals with the implementation of the developed SiC-based DC PETT into a regional train operating on a 3 kV DC traction system. The experimental results obtained during the testing are presented to demonstrate the performance of the developed 3 kV DC PETT prototype.

Keywords: silicon carbide; dual active bridge dc-dc converter; power electronic traction transformer; 3 kV DC railway traction; electric multiple unit

---

1. Introduction

Rail is among the most efficient and lowest emission modes of transport. It is under constant pressure to increase the accessibility of connections and quality of passenger services to enhance their competitiveness with other transport modes. For rail to compete more effectively with other modes of transport and attract more passengers, it needs the next generation of passenger trains that will be lighter, more energy-efficient and cost-effective. Traditional passenger trains consisting of a locomotive and several carriages penalize performance in terms of acceleration since they have a limited number of available drive wheelsets. Modern electric multiple units (EMUs) do not require a locomotive, as hey consist of several self-propelled units in a fixed assembly. Traction is distributed along the length of the train and the motors are housed by bogies of different carriages [1]. The demand for EMUs used for local, regional and intercity transport, grows globally every year. Thanks to the use of common bogies for the central wagons, EMUs are shorter and lighter than trains pulled by locomotives. This translates into lower movement resistance, better acceleration and lower energy consumption [2]. Moreover, EMUs better implement the specificity of passenger traffic and station service as they have a greater number of doors and provide better exchange of travelers.

Quantitatively, the most important part of energy drawn from the overhead traction line by an EMU is consumed by the traction propulsion systems. Traction drive is crucial for the EMU's efficiency, reliability and availability. The on-board propulsion converters, which for years have been traditionally installed in separate technical compartments, are currently mounted on the roof [3] or under the floor [4,5] of the EMU train. This corresponds to the growing requirements for the number of passenger seats due to the limited length of the passenger train.

The most common R&D works carried out by the manufacturers of traction drives are focused on improving efficiency and reducing the size of the on-board propulsion and power systems [3–6]. Another important topic is the reduction of noise emission [3]. Noise from mechanical and electrical components is troublesome to the human ear and significantly reduces travel comfort. A noticeable source is the working traction motor, whose noise depends on the frequency of switching transistors and the control method used in the drive inverter [7,8].

In the case of a drive inverter for rolling stock for 3 kV DC traction, it would be difficult to find any significant changes in the construction and control technology in the last decades [2,9,10]. The 3 kV dc-line voltage is a challenging task for power electronics. Due to a significant voltage drop on the overhead line, the traction power supply works at a higher voltage, usually 3.6 kV or even higher. This results in a requirement where the operation of a vehicle is often expected from 2.4 kV up to more than 4 kV, with the nominal power available from 2.8 kV or 3 kV DC [9]. A power conversion chain operating under direct current catenary is reduced to a bulky input filter and a voltage source inverter (VSI). Nowadays, EMUs powered by 3 kV DC voltage are equipped with two-level VSIs constructed of six 6.5 kV insulated gate bipolar transistors (IGBTs) generating pulse width modulated (PWM) voltage for supplying medium voltage asynchronous traction motors.

A small number of EMUs are equipped with three-level neutral point clamped (NPC) VSIs using 3.3 kV IGBTs [11]. A typical traction drive with a two-level VSI is shown in Figure 1a. In such a system, the LC filter capacitor also takes on the role of energy storage and is cyclically charged and discharged with impulse currents. The line filter choke, which is the most massive element of the inverter and weighs several hundred kgs, serves to limit high frequency, high di/dt pulse currents flowing between the catenary and the traction vehicle when the capacitor is recharged.

The power supply of the inverter has a large impact on both noise and the electromagnetic and thermal properties of traction motors. Table 1 shows the switching frequencies of HV IGBTs used in propulsion inverters for 3kV DC traction.

The harmonic voltages caused by PWM inverters operated with frequency ranging from individual hundreds of Hz to 2 kHz are the cause of significant current heat losses in the traction motor winding resistances [7,8]. The additional losses due to inverter supply can amount in the range of a few percent of the rated motor power and also occur in the laminated core of the traction motor. The winding and core losses most notably entail thermal problems, increasing the traction motor temperature by an average 30–50 K compared with a sinusoidal supply [8], but also lead to reductions in efficiency. Moreover, the negative effects of the low frequency PWM operation of the inverter feeding the traction motor are increased mechanical loads and increased noise emission. Typical voltage and current transients during the switching of the 6.5 kV IGBT are shown in Figure 1b. The main reason for low IGBT switching frequency are the relatively long switching times of the order of several microseconds and a characteristic current tail during turning off. The increase in the switching frequency of IGBTs would be at the expense of an unacceptable increase in the switching losses. Due to the above-mentioned reasons, current research and development conducted by manufacturers of the traction drives for rolling stock are focused on the utilization of new generation power transistors made of silicon carbide (SiC). Thanks to the exceptional properties, such as: significant achievable energy savings, quieter operation, improved reliability, and reduction in maintenance costs, SiC MOSFETs are ideal for traction power converters new designs instead IGBTs—which were in use in the rolling stock industry for decades [3–6,10]. The most obvious benefit of SiC metal-oxide semiconductor field-effect transistors (MOSFETs) compared to IGBTs is a significant reduction in switching losses, up to 55%, and total power losses, up to 80% [6].

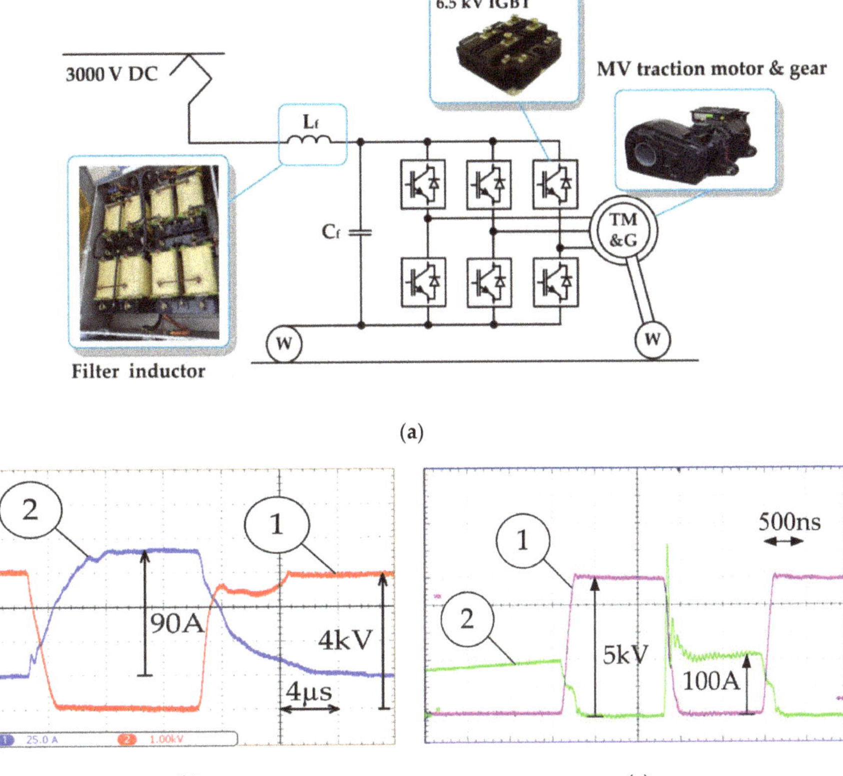

Figure 1. Conventional electric drive of 3 kV DC traction rolling stock with asynchronous traction motor and 6.5 kV IGBT-based two-level inverter: L_f, C_f—inductance and capacitance of the traction line filter, TM&G—traction motor and gear, w—wheels (**a**); Typical collector-emitter voltage (1) and collector current (2) transients during switching of the 6.5 kV IGBT, 4µs/div (**b**); Typical drain-source voltage (1) and drain current (2) transients during switching of the 10 kV SiC MOSFET, 500 ns/div (**c**) (comparison in the text).

Table 1. Switching frequencies of HV IGBTs used in rail inverters for 3kV DC traction.

Inverter Type	Manufacturer	Topology	IGBT Ratings	Switching Frequency	Brochure Publication Year
BORDLINE® CC750	ABB	three-level	3.3 kV/1200 A	2 kHz	2018
FT-800-3000-LQC	MEDCOM	two-level	6.5 kV/600 A	1 kHz	2018
TF09 Traction converter	INGETEAM	two-level	6.5 kV/600A	500 Hz	2012

Typical voltage and current transients during switching of the 10 kV SiC MOSFET are shown in Figure 1b. As it can be seen from Figure 1, in the case of an SiC-MOSFET there is in principle no tail current, and so without this contribution, clearly the switching loss is extremely smaller. Table 2 shows the overall losses distribution between the switching and conductive losses for HV SiC MOSFET and HV IGBT technology [12].

Table 2. Loss comparison between IGBT and SiC MOSFET: 5 kV, 33A/cm^2, 50% duty [12].

Device	Breakdown Voltage	$P_{switching}$ 500 Hz	$P_{switching}$ 5 kHz	$P_{switching}$ 20 kHz	$P_{conduction}$ 100 °C
Si IGBT 5SMX 12M6500	2 × 6.5 kV *	72.5 W/cm^2	725 W/cm^2	2900 W/cm^2	182 W/cm^2
SiC n-IGBT CREE	12 kV	6.5 W/cm^2	65 W/cm^2	260 W/cm^2	100 W/cm^2
SiC MOSFET CREE/Powerex	10 kV	4 W/cm^2	40 W/cm^2	160 W/cm^2	100 W/cm^2

* two 6.5 kV IGBTs in series have been selected as device closest to HV SiC IGBT and SiC MOSFET.

The IGBT switching frequency is limited by the power dissipation, which increases the IGBT junction temperature. To preserve the reliability of operation, the total power dissipation density is limited at the value of 200 W/cm^2 which maintains the chip temperature of 125 °C. The packaging technology further reduces the switching losses of the IGBT to 100 W/cm^2, which is the reason for limiting the operating frequency of the 6.5 kV IGBT-based two-level inverters to below 1000 Hz.

Lower switching losses of HV SiC MOSFETs give the possibility of increasing the PWM frequency of the propulsion converters towards tens of kilohertz. This, in turn, makes it possible to reduce the size of passive components of the traction inverter-primarily the bulky traction line filter. The weight of the filter inductor for the conventional traction inverter shown in Figure 1a is about 500 kg. The unique properties of SiC transistors allow to reduce the weight of the input filter inductor proportionally to the increase in the switching frequency. The issue supporting the reduction of dimensions of the input filter is the use of an additional high-efficiency active-front end (AFE) stage at the input of the traction inverter, which provides regulation of the current drawn from the grid. Due to the lower losses of SiC devices, the need for cooling is also substantially reduced [5]. Reducing the size of the cooling system leads to a reduction in the volume, weight and costs of the entire traction drive [6,8–10].

The higher resolution of the PWM generation in SiC-based traction inverters has a positive effect on reducing harmonic losses of the traction motors-making the whole traction systems more efficient. As stated in [4], the use of SiC power modules combined with expanding the control region outputting regeneration torque have made it possible to reduce the energy consumption rate of the railbound vehicle operated on suburban line by more than 37% compared to conventional systems. By increasing the PWM frequency above 20 kHz the acoustic noise level on the platform and in the cabin may be pushed beyond the audible range. In the case of railway inverters, the switching frequency of high voltage (3.3 kV and 10 kV) MOSFET SiC transistors is several kHz [4–6]. In order to obtain noiseless operation of MV drives, multi-level topologies are used with the use of low-voltage (1.2 kV or 1.7 kV) SiC transistors. Among others, power electronic traction transformers (PETTs), built with power electronic building blocks (PEBB) with built-in input/output isolation, realized by small size and high power density medium frequency transformers (MFT) are of particular interest [13–19].

So far, the primary purpose of SiC-based PETTs were primarily traction drives powered from the AC lines: 15 kV/16.7 Hz [13,14,18,19] and 25 kV/50 Hz [15–17]. AC powered locomotives and EMUs have a complex and voluminous conversion chain including a step-down transformer, rectifier, low-frequency filter and traction inverter. Conventional line frequency transformers (LFTs) used in AC EMUs occupy up to 12% of the weight of rail vehicle and a corresponding part of the train space [2]. Lightweight PETTs using MFTs, if they replaced bulky LFTs, could be easily installed on the roof, resulting in more space for passengers inside the train. AC PETTs offer not only a way to reduce the weight of the on-board electric equipment, but also to add additional functionalities and improve energy efficiency [19].

Strictly speaking, the AC PETT replaces the system of an LFT and an active rectifier and the AC PETT topology does not include a propulsion inverter, which stays the same as in a classical propulsion system and is connected to DC output terminals of the PETT [3,13–19].

As it has been highlighted in [13], the presently most promising medium frequency MF topologies for traction feature on their input-side a large number of series connected four-quadrant converters, with an overall switching frequency in the range of 5–8 kHz, modulated with Phase-Shifted PWM.

The results presented in [13] have shown that the use of a passively-damped LCL filter for the input-side results in a significantly lighter solution, compared to a single-pole type (single inductor) reported in most cases. The work [14] shows the benefits of using the originally 15 kV/16.7 Hz-type PETT in a traction drive supplied from a 3kV DC network. The main innovation aspect of the PETT proposed in [14] is the possibility to be operated with a HV-AC electric system, as well as after reconfiguration on a MV-DC catenary. A three-stage technology have been studied. Four-quadrant converters (4QC) connected in cascade and forming the PETT input stage provide, via insulated dual active bridge (DAB) DC-DC converters, voltage stabilization at the output DC terminals of the PETT, while limiting the ripple of current drawn from the DC traction network. Two solutions have been presented: the first solution using hard switching techniques, using a three-times silicon-conversion, and a second solution based on ZVT/ZCS switching of silicon semiconductor devices. The DC PETT described in [14] provides a higher quality of drive operation during DC voltage disturbances which normally occur in 3 kV DC traction, than conventional propulsion systems with two-level and three-level inverters. Prototype modules have been realized with association of diodes and IGBTs in order to provide reverse-blocking devices. However, the concept was verified only on a small scale test rig. The works [15] and [16] present the method of optimal sizing of 25 kV/50 Hz-type PETTs using 3.3 kV SiC MOSFETs that leads to the best efficiency at rated power in a given volume. The authors emphasize that the efficiency cannot be the only parameter to make a clear choice of an isolated DC-DC converter included in a PETT. An important parameter should be also the acoustic noise, even if it is difficult to take it into account in preliminary studies. The results obtained in [16] for 3.3 kV SiC MOSFET-based resonant DC-DC converters show that designing converters with a switching frequency in the range of 20 kHz would not lead to a dramatic reduction of the PETT efficiency. In [19] the design and development of the 1.2 MVA medium voltage PETT prototype for 15 kV/16.7 Hz traction applications have been presented. The use of MFTs developed in [19] allows for weight reduction and power density improvement (0.5–0.75 kVA/kg) compared to conventional traction chains (0.2–0.35 kVA/kg). The authors highlight the new possibilities in terms of management of the railway networks due improved power quality and grid compliance due to the multilevel input voltage waveform and high apparent switching frequency seen from the grid side. The application of PETT to 3 kV DC rail traction means that the entire modular converter operates at much lower maximum voltage than in the case of the AC PETT, which encourages the integration of the propulsion inverter within the structure of the DC PETT with the use of unified low-voltage cells. The use of low-voltage (1.2 kV or 1.7 kV) SiC devices for this purpose seems to be an excellent solution.

The presence of built-in galvanic isolation in low voltage cells, realized utilizing MFTs, allows for series connection of the cells on the catenary side to obtain DC rail traction voltage and, independently, the configuration of the output terminals of the cells as a three-phase medium voltage multi-level inverter on the traction motor side. However, the use of low voltage power devices and electronics in combination with high insulation requirements, imposed by the application to 3 kV DC rail traction, is a challenging research topic regarding the insulation strength of the cells, in particular SiC power devices and MFTs.

The world's first and only traction unit equipped with the roof-mounted SiC-based 3 kV DC PETT is shown in Figure 2. It is the PESA 308 EN81 series electric passenger railcar (construction number 308B-007) with an originally rated power of 560 kW that operates in Polish regional passenger rail transport since 2007. The work presented in this paper covers some of the recent research and development efforts to develop, design, test, commission, and install of the 335 kVA (500 kVA peak) SiC-based 3kV DC PETT on the PESA 308 test traction unit for a field trial.

Figure 2. The EN81 series 3 kV DC electric passenger railcar used for the field trial with the roof-mounted SiC-based DC power electronic traction transformer (PETT) prototype.

2. Configuration of the 3 kV DC PETT Topology

The general description of the 3kV DC traction propulsion system with the SiC-based DC PETT is shown in Figure 3.

The key parts of the system include: nine power electronic cells (1) each consisting of eight commercially available 1.2 kV SiC MOSFET power modules (2) and medium frequency transformers (3) placed in separate air-cooled chambers; a compact input traction LCL filter of a small size has an area of a small chamber (4); a medium voltage asynchronous traction motor integrated with the gear unit (7). The DC PETT has, essentially, a three-stage construction: (1) DC input conversion stage; (2) AC output conversion stage and (3) AC-link, which is an especially designed medium frequency power transformer. The DC input conversion stage is used to adapt constant frequency DC traction voltage to a medium frequency required for conversion, while the AC output stage adapts medium frequency of the AC link to the output AC voltage useful for controlling the traction motor. The input terminals of the nine power electronic cells are connected in series which means that the DC PETT input stage is configured as a 19-level cascaded H-bridge four-quadrant converter (4QC) supporting the 3 kV DC railway traction network. The output terminals are connected in series, three pairs of terminals per phase, creating a three-phase output stage, which means that the traction motor is supplied by a three-phase seven-level cascaded H-bridge (CHB) traction inverter. Table 3 lists the overall system requirements and specifications.

The proposed SiC-based DC PETT takes full advantage of the SiC technology while meeting the requirements of train manufacturers. During DC PETT operation, the energy drawn from the 3kV overhead contact line is processed in discrete portions, with full DC voltage divided into 19 levels, which helps meet stringent electromagnetic compatibility (EMC) standards, including EN 50121-2, EN 50121-3-1, EN 50121-3-2 and EN 50238. The use of a cascade system of series-connected 1.2kV SiC MOSFETs to switch the full voltage of the railway traction enables a much higher operating frequency of the power converter than using HV counterparts. At the same time, the selection of low-voltage transistors with high switching frequency can provide noiseless operation with comparable or even smaller total switching losses of the power converter [19]. The internal terminals of the power electronic cells are configured as nine DAB DC-DC converters connected in series and forming the insulation stage of the DC PETT. Each of the nine 4QC-DAB-DC/AC power electronic cells assembled in this configuration is rated for 38 kVA (56 kVA peak). Each cell, in addition to the galvanic isolation realized with the MFT, contains two energy storage elements, which is a characteristic feature of PET devices and enables the independent implementation of various control tasks at the input and output of the device.

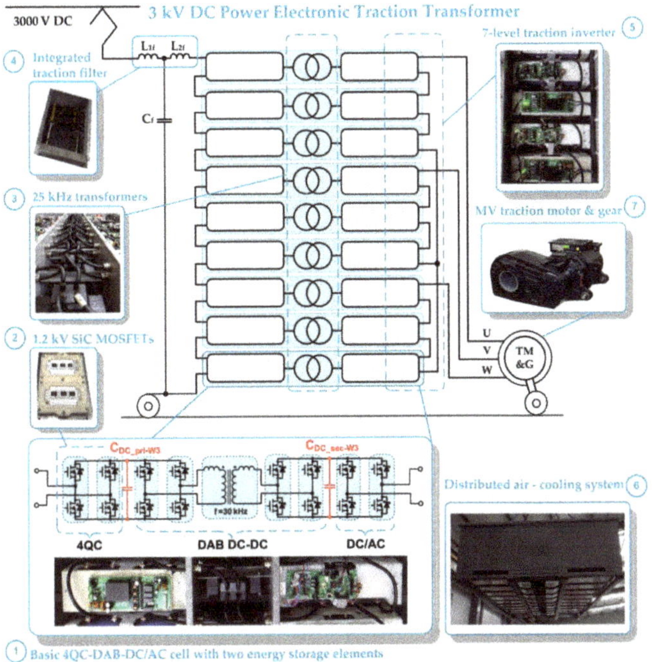

Figure 3. General scheme of the SiC-based 3 kV DC PETT: (1) Basic 4QC-DAB-DC/AC power electronic cell with two energy storage elements; (2) 1.2 kV SiC MOSFET power modules assembled in a 9 kV insulation frame; (3) medium frequency transformers; (4) compact input LCL traction filter; (5) output stage configured in the form of seven-level CHB inverter; (6) medium voltage traction motor integrated with the gear; (7) distributed air cooling system; U, V, W—traction motor phases.

Table 3. 3kV DC PETT system specification.

Parameter	Description
Input stage circuit structure	19-level 4QC
Output stage circuit structure	3 phase seven-level CHB inverter
Rated input voltage	3 kV V_{DC}/2.2 kV V_{AC}
Operating DC traction voltage	2 kV DC … 3.9 kV DC
Operating power	335 kVA
Rated output current	88 A (RMS)
Rated output frequency	60 Hz
LCL traction filter parameters	L_{1f} = 2 mH, L_{2f} = 1 mH, C_f = 10 µF
Filter inductors total weight	60 kg
MFT and DAB cells switching frequency	30 kHz
4QC cells switching frequency	20 kHz
CHB inverter cells switching frequency	20 kHz

As the heat generation is concentrated around the MFTs and SiC power modules, a distributed air cooling system (6), when uses PWM controlled fans and eighteen cooling ducts is provided in order to deal with the heat. The use of low voltage SiC MOSFETs makes the construction of the DC PETT more competitive in price than two-level traction inverters using HV SiC MOSFETS due to very high costs of high voltage semiconductor power devices, high voltage capacitors and other high voltage electronic components. The price of CAS120M12BM2-type 1.2 kV/193 A SiC MOSFET power module starts at about $300, the price of CAB450M12XM3-type 1.2 kV/450 A SiC MOSFET starts at about $700, while for the 3.3 kV nHPD2-type SiC MOSFET modules prices start at about $9500. The 10 kV and 6.5 kV SiC MOSFET power modules are not available on the market yet.

The modular construction of the DC PETT ensures even weight distribution on the roof of the EMU train. The center of gravity of the device is distributed symmetrically on both sides of the longitudinal axis of the vehicle. The distribution of the basic cells of the PETT on the EMU roof surface is shown in Figure 4. The operation of current-controlled nine-level four-quadrant converter connected to the 3 kV DC traction enables the minimization of the integrated input LCL traction filter and the mitigation of the resonances coming from the railway grid at a level that cannot be obtained in a conventional two-level inverter. As it can be seen from Table 3, for the investigated EN81 railcar, the weight of the traction filter chokes was reduced by nearly 10 times.

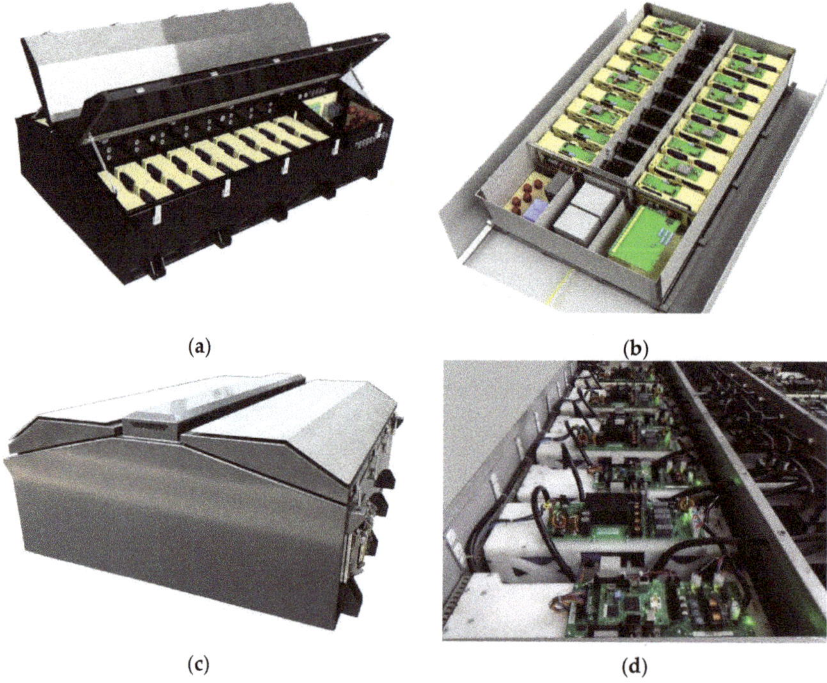

(a)   (b)

(c)   (d)

Figure 4. Assembly of power electronic cells of the DC PETT on the roof surface: (**a**) overall view of the whole device from the CAD program; detailed view of the 4QC-DAB-DC/AC power electronic cells and the LCL filter in the middle at the bottom (**b**); photos of the prototype (**c**) and (**d**).

Moreover, modular construction enables quick replacement of individual cells in case of failure. Therefore, the minimization of the MTTR (Mean Time to Repair), which is one of the basic indicators of reliability required in the railway industry [20,21] can be provided.

3. Low Voltage SiC Power Modules

According to numerous recent publications, e.g., [20,21], cascaded cells converter systems, such as the CHB topology, are a very attractive solution to interface power electronic systems to MVA-scale medium voltage applications. For the systems connected to medium voltage grids the choice of fast switching and low voltage drop 1.2 kV or 1.7 kV devices (giving up to 15 cells per phase stack) can be the most suitable trade-off between efficiency and power density, with efficiencies above 99% at a power density of about 5 kW/dm^3 [22].

Higher switching frequency obtainable at lower blocking voltages allows to reduce loss and volume contributions of the grid filter inductances and the cooling system.

With the wide availability of low voltage SiC MOSFET power modules on the market their performance in the range of maximum working currents offered is constantly increasing along with the progressive heat dissipation capacity. Figure 5 shows a comparison of the dimensions of the 1.2 kV SiC MOSFET power module, part no. CAS120M12BM2, used in this project, with a rated drain current of 138 A (T_C = 90 °C), $R_{DS(on)}$ = 23 mΩ, junction to case thermal resistance R_{thJC} = 0.125 °C/W, maximum dissipated power P_D = 450 W (T_C = 90 °C) characterizing commutation circuit series inductance of 15nH and a base surface of 65.4 cm^2, and new generation 1.2 kV SiC MOSFET power module, part no. CAB450M12XM3, available on the market from mid-2019, with 3 times higher rated drain current I_D = 409 A (T_C = 90 °C), $R_{DS(on)}$ = 4.6 mΩ, R_{thJC} = 0.11 °C/W, P_D = 750 W (T_C = 90 °C), and characterizing commutation circuit series inductance more than twice as low, L_{stray} = 6.7 nH and 36% less footprint.

Figure 5. The 1.2 kV SiC MOSFET power module, part no. CAS120M12BM2, used in the project, with a rated current of 138 A (left) and a 36% smaller 1.2 kV SiC MOSFET power module, part no. CAB450M12XM3, with a rated current of 409A, available on the market from mid-2019.

The use of SiC MOSFET 1.2 kV power modules in the design of the proposed 3 kV DC PETT requires higher insulation strength than the original housing offers to withstand the full operating voltage of the converter. To increase the voltage strength, an additional frame made of insulating material has been developed and mounted between each transistor module and the heat sink, which is shown in Figure 6. A flexible, thermally conductive silicone film with thermal conductivity of 5 W/mK and a breakdown voltage of 9 kV AC was used between the base of the power modules and the heat sink.

There are several design challenges associated with switching performance of the SiC power modules which need to be carefully addressed including the minimization of ringing and overshoots caused by the parasitic loop inductance. These loop inductances together with SiC MOSFET output capacitance create a resonant tank, which is the source of unwanted EMI emissions [23,24]. To overcome the ringing, parasitic parameters converter should be minimized by a careful converter design.

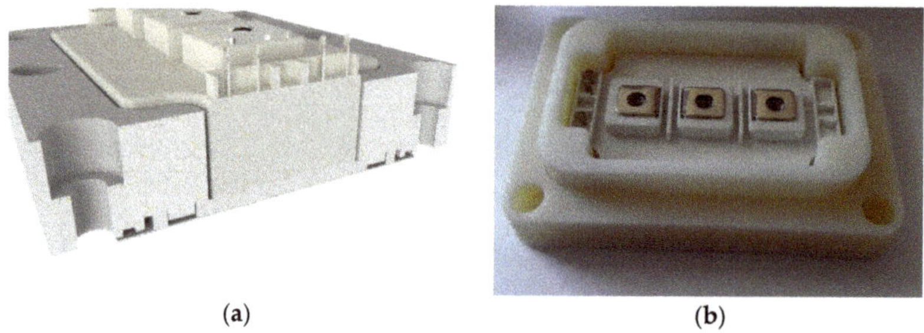

(a)   (b)

Figure 6. General view and cross-section of the frame of insulating material, ensuring increased voltage strength of the SiC MOSFET transistor module housing and heat sink insulation: (**a**) overall view from the CAD program; (**b**) photo of the prototype.

Figure 7 shows the drain-source voltage (U_{DS}) of the SiC MOSFET and the primary side transformer current of the developed DAB DC-DC converter measured in the developed SiC-based 3 kV DC PETT during commutation of 700 V DC voltage within 80 ns. A single standard PPA2124150-type, 1.5 uF snubber capacitor was attached to the DC bus of each SiC MOSFET power module. As can be seen in Figure 7, thanks to the obtained minimization of the switching loop inductance the measured magnitude of the voltage ringing is about 40 V (5.7%), which is an acceptable level.

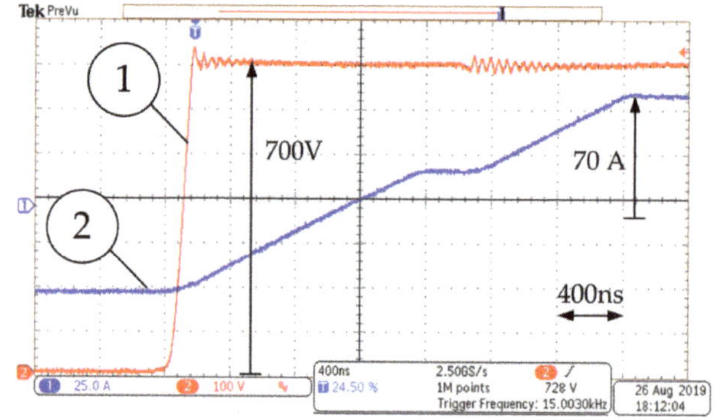

Figure 7. SiC MOSFET drain-source voltage (1), 100 V/div and the primary MFT current (2), 25 A/div of the developed DAB DC-DC converter during commutation of 700V DC voltage, 400 ns/div.

4. Design of MFT for 3kV DC PETT

The proposed 3kV DC PETT topology requires individual isolated DC-DC converters for the feeding of each H-bridge of the three-phase seven-level traction converter. The primary function is to provide galvanic insulation of each supply voltage [17]. The isolation stage is ensured by nine DAB DC-DC converters [25–32], which, in the same time, are core elements of the entire system (Figure 3). The main idea of the DC-DC DAB converter design is to combine the advantages of hard-switching pulse-width modulation (PWM) topologies characterizing a large dynamic range and no (reactive) circulating power operation and resonant topologies, which can operate in a soft-switching manner with a reduced electromagnetic interference (EMI) and effective utilization of transformer parasitics but, with strongly increased circulating power and a limited dynamic range.

The performance of DAB DC-DC converters and the entire DC PETT system is strongly affected by the design of the MFTs [26]. The key element enabling each DAB DC-DC converter to transfer energy between input DC stage and output AC stage is the series inductance L_s, which acts as a decoupling element between the square-wave voltages and influences the conducted currents and switched currents of all the semiconductors of the active bridges of the DAB DC-DC converter. Series inductance can either be implemented as a separate component, using its own magnetic core, or can be built into the MFT. Using the MFT leakage inductance as the series inductance, $L_s = L_\sigma$, simplifies the mechanical design, eliminates the losses and volume resulting from the interconnection of the external inductor, and enables the achievement of higher power densities. However, in some justified cases, the use of additional auxiliary inductance may be unavoidable [33].

The insulation between the primary and secondary side of the MFT must withstand the rated voltage of the DC railway overhead line. Achieving the desired MFT design that will maximize power density and efficiency while maintaining space and weight restrictions requires a complicated optimization procedure [25]. Therefore the following considerations have been taken into account at the design stage:

- Interleaving of windings was not considered in order to avoid voltage isolation problems,
- The windings have been designed to provide very low coupling capacity between primary and secondary side to avoid capacitive coupling currents,
- The coil-formers have been designed in order to implement the required leakage inductance L_σ, at the same time to comply with minimum clearance and creepage distances to the core and to maximize the cooling surface of the windings.

The working principle of the DAB DC-DC converter lies in the phase-shift δ, introduced between the rectangular AC voltages generated by the two active bridges [26,28]. The AC current flowing through the MFT is introduced by the phase shift of the active bridges AC voltages and depends on the difference of the primary and secondary DC voltages V_{dc1} and V_{dc2} and the value of the series inductance L_s. For rectangular AC voltages characterizing the same duty cycle $D = 0.5$, the transferred power P_{DAB} is adjusted by controlling the phase angle δ, according to the following formula:

$$P_{DAB} = \frac{V_{dc1}V_{dc2}|\delta|(\pi - |\delta|)}{2\pi^2 f_s L_s} \tag{1}$$

where L_s is the primary-referred leakage inductance [26]. In principle, the designed transformer had to meet the thermal requirements and insulation distances required to achieve the desired leakage inductance. To design and manufacture of two prototype transformers, two types of magnetic material, i.e., amorphous material characterizing high saturation level and N87 ferrite core have been used. Both materials, amorphous and ferrite, are suitable for high power and high-frequency applications and it can be of interest to investigate their performance on two transformer prototypes.

Figure 8 shows the dimensions of the windings and view of the first developed MFT prototype consisting of two high performance iron-based amorphous alloy (Fe-Cu-Nb-Si-B) wound cores with a rectangular shape (core length 222 mm, core width 118 mm, core height 30 mm, core build 35 mm). The classical shell-type structure with two uniform, concentric windings has been selected for the design. The leakage inductance has been set by arranging the position of the primary and secondary windings.

Uncut cores have been used, which can help reduce noise emissions from the transformer [27] and are valuable for rail vehicles. Moreover, the absence of an interlayer impregnation, which can be found in cut cores, eliminates additional mechanical stress to the lamination. Both, the primary and secondary windings consist of 1400 × 0.2 mm Litz-wire with 14 turns. As can be seen from Figure 8b, to facilitate the construction of the winding, plastic formers have been used. Since the magnetic core is uncut, an insulating tape was used to assemble the fragments of the former—so that they can be mounted around the central limb of the core.

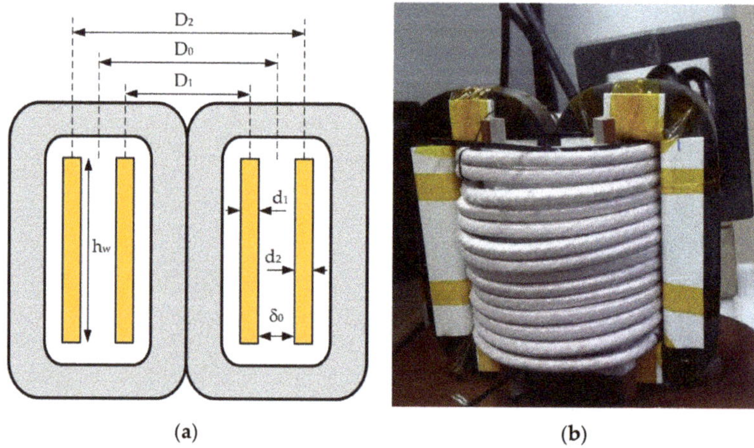

Figure 8. Main dimensions (**a**); and the overall view (**b**) of the first developed MFT with concentric windings of cylindrical shape (auxiliary inductor visible in the right top corner).

The leakage inductance of two uniform, concentric windings, of equal height, for which the leakage field inside the windings can be assumed to be axial, can be determined from Rogowski approximation method [34] using the mean length per turn for the whole arrangement of coils l_m:

$$L_\sigma = \mu_0 N_P^2 \frac{l_m\left(\frac{d_1+d_2}{3} + \delta_0\right)}{h_w} k_\sigma \tag{2}$$

where $\mu_0 = 4\pi \times 10^{-9}$ H/cm is the vacuum magnetic permeability, N_P is the number of turns in one winding, d_1 and d_2 are the radial sizes of internal and external windings, δ_0 is the width of the channel between the windings, h_w is the windings height and k_σ is Rogovskii's coefficient:

$$k_\sigma = 1 - \frac{d_1 + d_2 + \delta_0}{\pi h_w} \tag{3}$$

or, alternatively, using the area of reduced leakage channel S_L:

$$L_\sigma = \mu_0 N_P^2 \frac{S_L}{h_w} k\sigma \tag{4}$$

The area of the reduced leakage channel for concentric windings from Figure 8 can be calculated from [34]:

$$S_L = \frac{\pi}{6}\left(D_2^2 - D_1^2 + 2\delta_0 D_0\right), \tag{5}$$

where D_1 and D_2 are mean diameters of the primary and secondary winding and D_0 is mean diameter of the clearance ring between windings. Using k_σ, real concentric windings with height h_w are replaced by conditional windings of height h_w/k_σ, which reach the yokes. This permits one to replace a real leakage field that is not convenient for calculations with an ideal one in which all field lines are parallel to the winding axis [34]. For the developed prototype $N_P = 14$; $h_w = 14.6$ cm; $D_1 = 9$ cm; $D_2 = 13.7$ cm; $D_0 = 11.4$ cm; $d_1 = d_2 = 0.9$ cm; $\delta_0 = 1.5$ cm; $k_\sigma = 0.92$. Specifications of the first MFT prototype are listed in Table 4.

The first MFT prototype from Figure 8 has been operated with a DAB DC-DC converter. The dual phase-shift (DPS) control which uses the phase-shift between output voltages of the bridges along with the pulse width variation of both bridges output voltages has been applied to the DAB DC-DC converter. Although the applied DPS modulation helps to minimize the reactive power and thus maximize the active power transmitted by the DAB DC-DC converter, the desired output active power of 40 kW could not be achieved due to too low obtained leakage inductance of the prototype transformer of about 5.5 µH per side. Hence, two auxiliary series inductors of 5 µH have been added to increase the resultant series inductance value and keep the transformer running at rated power.

Table 4. First MFT prototype specification.

Parameter	Description
Feeding voltage	$V_{DC1n} = V_{DC2n} = 700$ VDC
Output current	$I_{DC2} > 70$ A
Rated output power	40 kW
Operating frequency	20 kHz
Magnetic material	Amorphous alloy (Fe-Cu-Nb-Si-B)
Maximum induction	$B_{sat} = 1.56$ T
Specific Losses @ 0.1 T, 100 kHz	0.2 kW/kg
Diameter of a strand of a Litz wire	0.2 mm
Number of strands in a Litz wire	1400
Effective surface of a Litz wire	44 mm^2
Number of windings	$N_1 = N_2 = 14$
Leakage inductance (sum of primary and secondary)	$L_\sigma = 11.5$ µH
Auxiliary series inductors	5 µH (x2)

Thermal management to dissipate power losses is key to achieving high power density strongly required in the roof-mounted power electronic converters. During laboratory tests, particular attention was paid to the mechanism of formation of local temperature hot spots. In order to carry out transformer thermal measurements a thermal camera has been used to show the temperature distribution in the MFT. Figure 9a shows experimental waveforms of the developed first prototype transformer operating at half rated power: primary current and the collector-emitter voltage of the SiC MOSFET of the DAB DC-DC converter operating with 20 kHz switching frequency and the DPS modulation while Figure 9b describes the thermal characterization of the transformer operating without forced cooling at rated power of 40 kW. Excessive heating of the core was measured, which can be seen in Figure 9b. Local temperature increase far above 100 °C was observed during the tests, even at partial load, which was associated with: local heat-up of the core and the proximity effect losses in the transformer windings. The used laminated amorphous cores are wound from a strip of material of several tenths of micrometric thickness. Local heating of the amorphous alloy core was observed in the strip-end area, which was attributed to eddy current losses due to normal flux components in the zone of the amorphous strip-end. On the other hand, the reason for the observed proximity effect in windings was the time-varying flux density field in a conductor caused by a current flowing in another conductor nearby. Non-uniform current density of a conductor section, caused by the proximity effect, leads to higher effective resistance which in turn increases winding losses and the total MFT losses and is directly responsible for the hot-spot temperature gradients. The hottest observed places of the first prototype transformer occurred around the middle limbs of the transformer core and in the center of the primary winding, inside of the leakage layer. The maximal temperature of windings exceeded 109 °C while the maximum measured temperature of the core reached 200 °C in the strip-end area. Additionally, a disadvantageous fact was that the presence of two additional auxiliary inductors has significantly increased the volume of the magnetic circuit of the DAB DC-DC converter. For the above reasons, an improved version of the MFT was produced with a split planar litz-wire windings placed coaxially one above the another in the disc arrangement. Split windings construction enables obtaining a higher leakage flux density and correspondingly higher leakage inductance compared to the cylindrical transformer with the same number of turns [35]. In the disc type transformer the winding height h_w is measured in the direction of the core window width, while windings width d_1 and d_2 are in the direction of the core window height [26], which is shown in Figure 10. Therefore, the trapezoidal field distribution occurs in a direction perpendicular to the direction that occurred in the first prototype with the concentric windings. In the second prototype the N87 ferrite material with four times lower saturation level was used, which required increasing the core cross-section A_c of the second prototype, according to the relationship below:

$$A_c = \frac{V_{rms1}}{k_f k_c N_1 B_m f_s}$$ (6)

where V_{rms1} is the RMS value of the primary voltage, k_c is the filling factor of the core, N_1 is the number of primary turns, and f_s is the fundamental frequency. The coefficient k_f in Equation (6) depends on the duty cycle D of the phase shift modulation of the DC-DC DAB converter:

$$k_f = \frac{2\sqrt{2D}}{D}$$ (7)

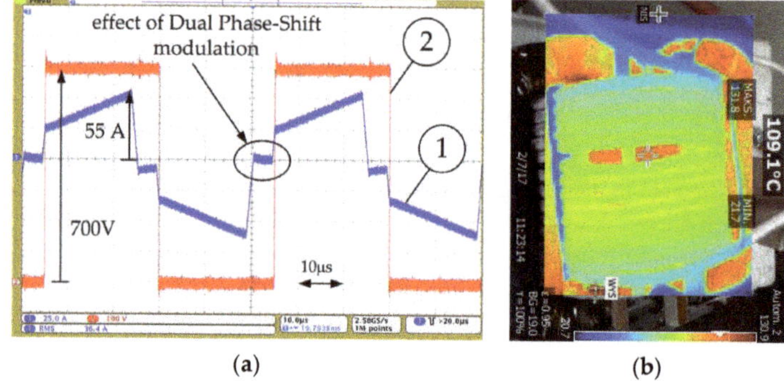

(a)   (b)

Figure 9. Primary MFT current i_{TR} (25 A/div) and drain-source voltage v_{DS} (100 V/div) of the SiC MOSFET of the DAB DC-DC converter operating with 20 kHz switching frequency and the DPS modulation (20 µs/div) (**a**); temperature measurement of the amorphous alloy core and observed heating of individual coils as a result of the proximity effect at rated power (**b**).

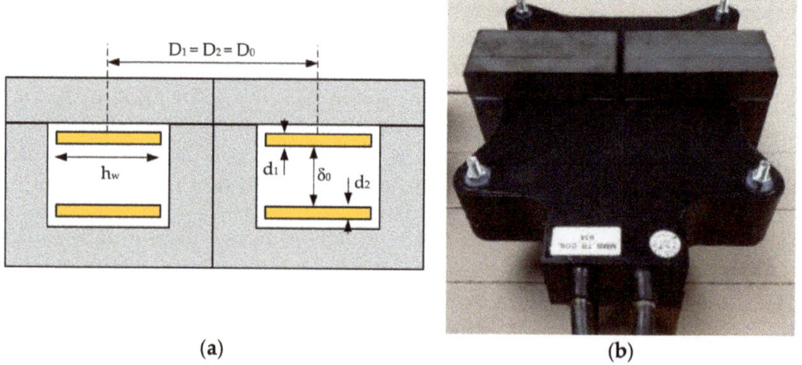

(a)   (b)

Figure 10. Main dimensions (**a**); and the overall view (**b**) of the second developed MFT with a split planar litz-wire windings placed coaxially one above the another in the disc arrangement.

To minimize the desired A_c of the second MFT prototype, the switching frequency of the DC-DC DAB converter has been increased from 20 kHz to 30 kHz. Six pairs of U126/91/20 and I126/20 cores, i.e., three core stacks and litz wires comprised of 700 strands with a thickness of 0.2 mm have been used in the second MFT prototype. The effective surface of the used litz wires is 22 mm^2. The transformers were placed in separate chambers inside the DC PETT housing and are cooled by air blown by fans. The used forced cooling enables the maximum permissible current density of 4 A/mm^2 [26] and the windings of the second prototype can carry a maximum current of 88 A. The used planar winding technology enables to adjust the leakage inductance of the transformer very precisely and reproducibly by using insulation distances between the primary and secondary windings.

The application of standard U-type and I-type ferrite core profiles for the core construction enables the use of prefabricated windings and supporting trays made of insulation material. To achieve isolation level of 9 kV, which is two times of the maximum DC-rail traction voltage level, casting the windings by the epoxy resin has been applied. The primary and secondary coils of the second prototype have the same height h_w, measured in the direction of the core window width. Hence it fulfils the precondition for the application of the method of Rogowski (Equation (2)) for prediction of leakage inductance [36]. Specifications of the second MFT prototype are listed in Table 5. Figure 10 shows the dimensions of the windings and view of the first developed MFT prototype.

Table 5. Second MFT prototype specification.

Parameter	Description
Feeding voltage	$V_{DC1n} = V_{DC2n} = 700$ VDC
Output current	$I_{DC2} > 70$ A
Rated output power	38 kW
Operating frequency	30 kHz
Magnetic material	N87 ferrite
Maximum induction	$B_{sat} = 0.4$ T
Specific Losses @ 0.1 T, 100 kHz	0.009 kW/kg
Diameter of a strand of a Litz wire	0.2 mm
Number of strands in a Litz wire	700
Efective surface of a Litz wire	22 mm^2
Number of windings	$N_1 = N_2 = 8$
Leakage inductance (sum of primary and secondary)	$L_\sigma = 24.5$ µH
Isolation voltage	$U_{Ni} = 9$ kV *

* The isolation voltage is defined as two times of the maximum DC-rail traction voltage level.

The windings were cast with a resin of good thermal conductivity and high mechanical strength, thanks to which the transformer can be placed in the so-called dirty area of the DC PETT housing. The final volume of the whole transformer is:

$$V_t = 307 \ mm \times 270 \ mm \times 131 \ mm = 10.85 \ dm^3$$

With a power density of 3.5 kW/dm^3 ($\approx$5 kW/dm^3 peak). In order to carry out transformer thermal measurements a thermal camera has been used to determine the core and winding temperature distribution. The transformer has been running for 2 h at rated load. Figure 11 describes the thermal characterization of the transformer operating at a power of 45 kW without forced cooling.

In of the second MFT prototype, with the split windings, the dimension of both the primary and the secondary windings are equal, which provides presenting equal dc resistances, while in the first prototype with the concentric windings the length difference between interior and exterior windings was considerable. The realized 30 kHz MFT prototype has been successfully tested at various operating conditions in a full power rated DAB DC-DC converter, which will be presented in detail in Section 5.3.

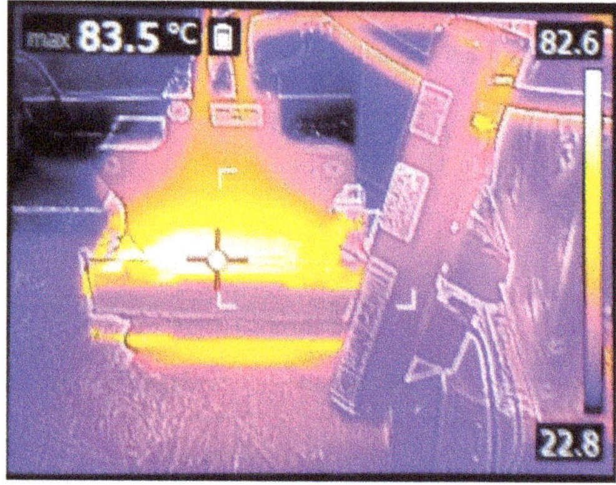

Figure 11. Temperature measurement of the second developed MFT with split windings at power of 45 kW without forced cooling.

5. Control Strategy and Controller Hardware

5.1. Main Control Tasks

As mentioned in Section 2, DC PETT carries out independent control tasks at its output and input, which consist in precisely generating the PWM voltage supplying the traction motor and maintaining full control over the current drawn from the railway overhead contact line. The PWM voltage which feeds the asynchronous motor must be generated in accordance with current tasks of the drive system: control of the electromagnetic torque, which determines the driving dynamics, and control of the magnetic excitation of the motor, which determines the energy consumption.

5.2. Control of the 19-Level H-Bridge 4QC

The presence of galvanic isolation in the DC intermediate circuit and two energy storage elements in each of the power electronic cells, enables independent control of the SiC MOSFET H-bridges on the primary (railway traction) side and secondary (traction motor) side. As mentioned, the nine SiC MOSFET H-bridges of the DC PETT input stage are connected in series and configured as 19-level H-bridge 4QC ensuring a low ripple amplitude of the current drawn from the 3 kV DC overhead contact line. Each H-bridge of the input stage has a capacitor at the output, which plays a role on the DC voltage source: v_{DC_pri-U1}, v_{DC_pri-U2}, $\cdots$, v_{DC_pri-V1}, $\cdots$, v_{DC_pri-W3} for nine individual DAB DC-DC converters. The proposed control scheme of the DC PETT input stage is illustrated in Figure 12.

The 19-level H-bridge 4QC controller has two closed control loops: the voltage controller to deal with the primary DC voltage v_{DC_pri} of the nine SiC MOSFET H-bridges and the current controller for the purpose to control the input current i_{L2f}. The set value for the voltage controller is the desired DC voltage value at the individual capacitors of the H-bridges. The voltage regulator amplifies and integrates the deviation between the voltage set point on a single capacitor and the average value of the voltages measured across the capacitors of all H-bridges. The voltage regulator output signal is the set point value to the current regulator. The output of the current regulator corrected by the actual value of the traction voltage $v_{DC\_traction}$, measured on the traction current collector, constitutes the set signal to the PWM modulator. The control method described in [36] was used to control individual SiC MOSFET H-bridges of the 19-level 4QC from Figure 12. The essence of the operation of the entire 19-level H-bridge 4QC is that the energy from individual capacitors C_{DC_pri-U1}, C_{DC_pri-U2}, $\cdots$, C_{DC_pri-V1}, $\cdots$, C_{DC_pri-W3} is transferred through isolated DAB DC-DC converters to the three-phase seven-level CHB traction inverter supplying the traction motor. The energy flow from primary DC capacitor of the input H-bridge through the DAB DC-DC converter to the output H-bridge of the

U, V or W phase of the CHB traction inverter reduces the voltage on this capacitor. During each control program execution sequence, the individual H-bridges of the input stage, whose intermediate DC circuit voltage reaches the lowest values, are activated, causing the primary DC capacitors to charge and the voltage on these capacitors to rise. The sequence of switching individual cells on and off depends on the current DC voltage levels of individual capacitors $C_{DC\_pri\text{-}U1}$, $C_{DC\_pri\text{-}U2}$, $\cdots$, $C_{DC\_pri\text{-}V1}$, $\cdots$, $C_{DC\_pri\text{-}W3}$. On the other hand, in the case of energy recuperation, the bridges with the highest voltage value on the DC capacitors are connected to the overhead contact line.

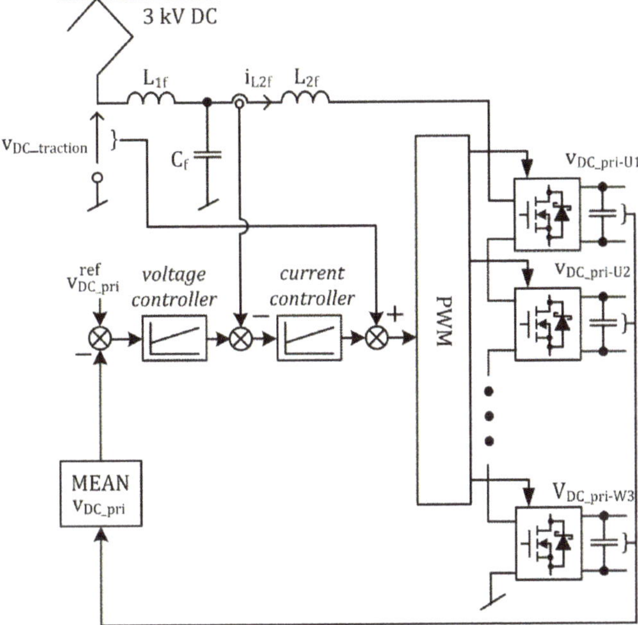

Figure 12. The overall control scheme for the DC PETT input stage.

All the electronic circuits of the 4QC-DAB-DC/AC power electronic cells are designed with the floating ground. The ground planes of the electronic circuits of each 4QC H-bridge shown in Figure 12 is connected to the middle points of the DC-links adjacent H-bridges through the resistive dividers. These resistive dividers acts simultaneously as the bleeder resistors for the DC-link capacitors of the power electronic cells. Thanks to this, the full voltage appearing on the electronic circuits will never exceed several hundred volts and the use of optically isolated op-amps with maximum working insulation voltage of 1 kV is sufficient. All electronic circuits are powered from the train's on-board 24 V DC auxiliary power converter via isolated DC-DC power supplies. These isolated DC-DC power supplies must be able to withstand the full voltage of the 3 kV DC overhead traction line. The measurement of the DC traction voltage $v_{DC\_traction}$ required in the control system from Figure 12 is performed using voltage divider placed on a separate insulated electronic board. Measured signal is transmitted to the MASTER controller via optical fiber. The voltage measurement electronic board is supplied by the on-board 24 V DC auxiliary power converter via an isolated DC-DC power supply. By using high-voltage insulation of the cable connecting the L_{1f} and L_{2f} input filter chokes, the i_{L2f} input current measurement is performed using a conventional current transducer. The measuring signal from the current transducer is transmitted directly to the MASTER controller interface board.

5.3. Control of DAB DC-DC Converters

As can be seen from Figure 3 in Section 2, nine DAB DC-DC converters are used to transfer the electrical energy between the primary DC links (C_{DC_pri-U1}, C_{DC_pri-U2}, $\cdots$, C_{DC_pri-V1}, $\cdots$, C_{DC_pri-W3}) of the 19-level 4QC and the secondary DC links (C_{DC_sec-U1}, C_{DC_sec-U2}, $\cdots$, C_{DC_sec-V1}, $\cdots$, C_{DC_sec-W3}) of the three-phase seven-level CHB traction inverter. Individual DAB DC-DC converters are controlled independently and no information exchange about the control process with the 19-level 4QC, seven-level CHB traction inverter and other DAB DC-DC converters is needed. The single DAB DC-DC converter is shown in Figure 13. The control system of the DAB DC-DC converter is designed to obtain the same voltages on the primary DC-link capacitor and secondary DC-link capacitor $v_{DC\_sec} = v_{DC\_pri}$. As a result, each DAB DC-DC converter equalizes the individual DC-link voltages of the 19-level 4QC and the seven-level CHB traction inverter [36].

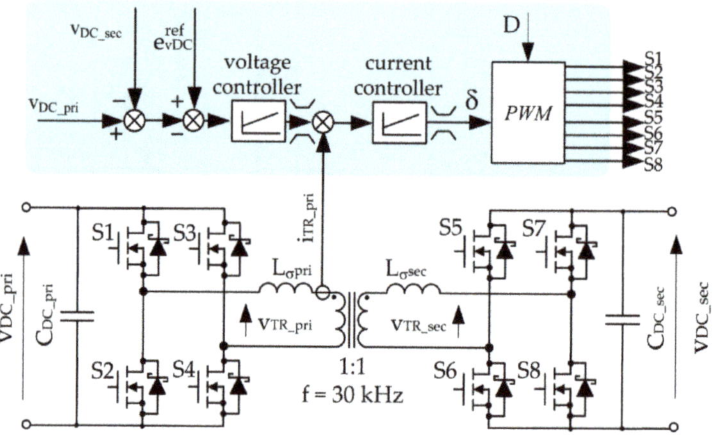

Figure 13. The single DAB DC-DC converter and its control system.

The DC voltages: v_{DC_pri} on the primary side and v_{DC_sec} on the secondary side are converted into high frequency rectangular pulses v_{TR_pri} and v_{TR_sec}, with constant ($D = const$) or modulated ($D = var$) pulse width. The transferred power depends on the mutual phase shift ratio δ between primary and secondary voltages v_{TR_pri} and v_{TR_sec}. For simple phase-shift control transferred power P_{DAB} is defined by (1). Tests in the DAB DC-DC converter with the second MFT prototype were done at 640/640 V and powers of 10 kW, 38 kW and 45 kW for switching frequency of 30 kHz and a dead time 500 ns. Figure 14a shows the characteristic waveforms of the developed DAB DC-DC converter operating at rated power of 38 kW: the primary transformer current i_{TR_pri} (25 A/div), the primary transformer voltage v_{TR_pri} (500 V/div) and the secondary transformer voltage v_{TR_sec} (500 V/div). Figure 14b shows the impact of the dead time on the time duration of the voltage pulses of the primary and secondary transformer voltages v_{TR_pri} and v_{TR_sec}. Although both voltages are controlled with the same constant value of the duty cycle $D = 0.96$, the voltage pulses of the secondary voltage v_{TR_sec} are longer than voltage pulses of v_{TR_pri}. It can be seen from Figure 14, that in the time period when $v_{TR\_pri} = v_{TR\_sec} = 0$ there is no resultant voltage forcing the dynamics of the current and the dynamics of transformer current changes decreases for a fraction of a microsecond. This phenomenon was also recorded when the tested DAB DC-DC converter was operated with a reduced power of 10 kW (Figure 15a). After applying the correction of the duty cycle and taking into account the dead time effect, the above did not occur any more—which can be seen in Figure 15b describing the primary and secondary transformer voltages and current of the DAB DC-DC converter operating with a power of 45 kW.

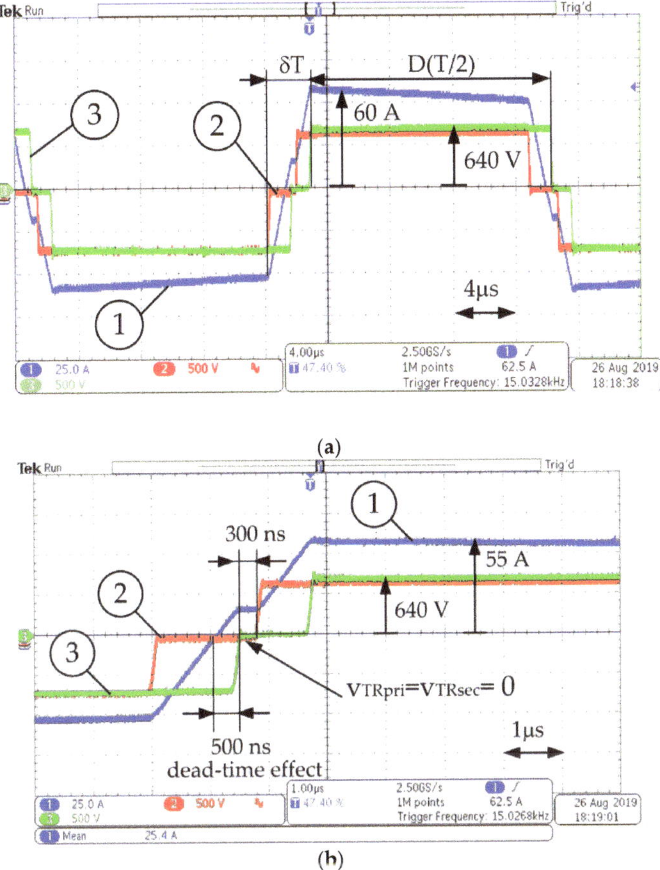

Figure 14. Measured waveforms of the DAB DC-DC converter with the second MFT prototype: primary transformer current i_{TR_pri} (1), 25 A/div; primary voltage v_{TR_pri} (2), 500 V/div; secondary voltage v_{TR_sec} (3), 500 V/div at the rated load. Time scale 4 μs/div. (**a**) and 1 μs/div. (**b**).

Figure 15a shows the characteristic waveforms of the developed DAB DC-DC converter operating at partial power of 10 kW: the primary transformer current i_{TR_pri} (25 A/div), the primary transformer voltage v_{TR_pri} (1 kV/div) and the secondary transformer voltage v_{TR_sec} (1 kV/div). Figure 15b shows the primary transformer current i_{TR_pri} (50 A/div), the primary transformer voltage v_{TR_pri} (1 kV/div) and the secondary transformer voltage v_{TR_sec} (1 kV/div) for the DAB DC-DC converter with the second MFT prototype overloaded with a power of 45 kW. The efficiency of the DAB DC-DC converter has been calculated using the voltages and currents measurements and the math functions of the digital oscilloscope Tektronix DPO4104. The measured resistances of the primary and secondary windings of the MFT was 11 mΩ and 12 mΩ, respectively. The experimental efficiency results are presented in Figure 16. The developed DAB DC-DC converter characterizes peak efficiency above 98 % and has efficiency around 97.5 % in a wide range of the output power.

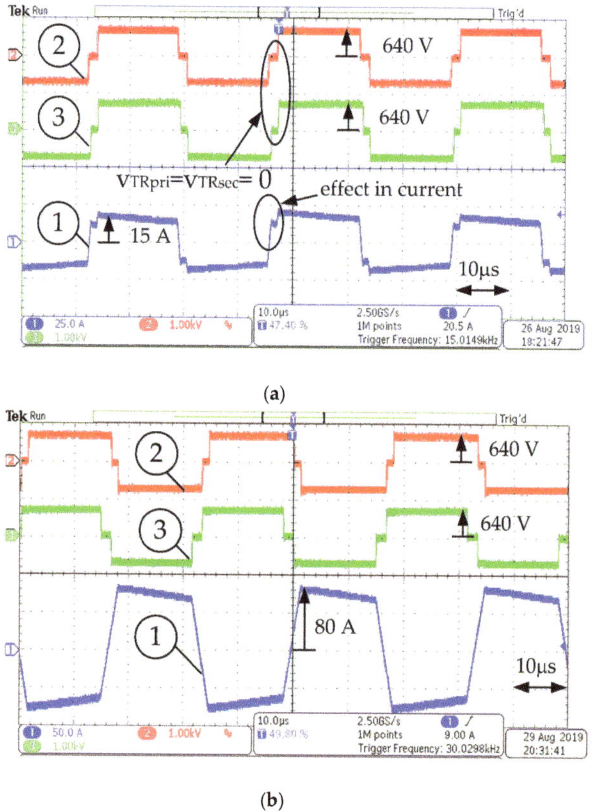

(a)

(b)

Figure 15. Measured waveforms of the DAB DC-DC converter with the second MFT prototype: primary transformer current i_{TR_pri} (1), 25 A/div; primary voltage v_{TR_pri} (2), 500 V/div; secondary voltage v_{TR_sec} (3), 500 V/div at partial power of 10 kW (**a**) and the corresponding waveforms at power of 45 kW (**b**).

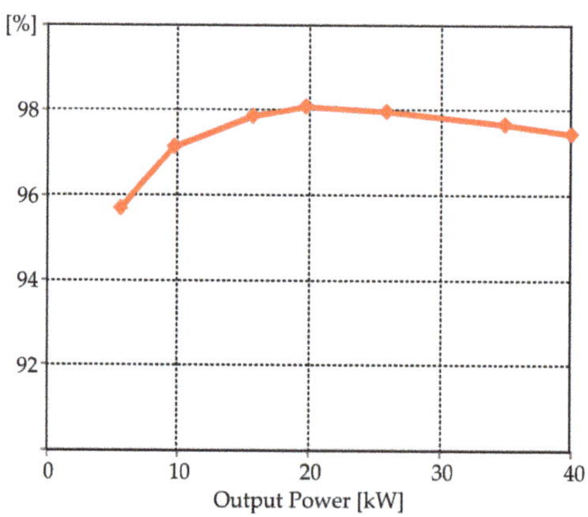

Figure 16. Measured efficiency versus output power of the DAB DC-DC converter with the second MFT prototype.

5.4. Control of Seven-Level CHB Traction Inverter

As it can be deduced from Figure 3, the seven-level CHB traction inverter is composed of three H-bridges connected in series in any of the phases. The traction inverter is controlled using space vector PWM (SVPWM) by successively activating one H-bridge per phase until the reference voltage vector is reached [36]. The CHB topology enables the operation with advantageously high modulation indexes of individual H-bridges. The DC links of the individual H-bridges are coupled with nine DAB DC-DC converters. If the obtained output voltage is different from the reference motor voltage, the next H-bridge is activated in each phase. At each switching sequence only one H-bridge per phase provides a modulated output voltage, while the others are negatively/positively connected or bypassed [37]. Since their transistors do not switch, they do not generate commutation loses. For the above reason, the control system can consider the topology of the seven-level CHB converter as a set of 3 three-level CHB converters connected in series, which simplifies the control strategy. Each of them is then composed using three H-bridges (one H-bridge in each phase of the inverter) and can be controlled using simple SVPWM patterns [37].

5.5. Control of the Traction Motor

The precision of PWM voltage generation resulting from the seven-level topology of the three-phase traction inverter and the adopted transistor switching frequency of 20 kHz, allows the use of advanced traction motor control algorithms, not previously used in rolling stock. The multiscalar model based control [38,39] has been used to control the torque and excitation of the traction motor. According to the multiscalar model concept, the motor torque is defined as the state variable instead of the current vector component in the q axis that occurs in conventional Field Oriented Control (FOC). The complete multiscalar model (or *natural variables* [40]) of the induction motor is received after the nonlinear transformation of the stator current and rotor flux vector components occurring in the classic vector model. The multiscalar variables of the induction motor model are selected as follows:

$$x_{11} = \omega_r, \tag{8}$$

$$x_{12} = \psi_{r\alpha} i_{s\beta} - \psi_{r\beta} i_{s\alpha}, \tag{9}$$

$$x_{21} = \psi_{r\alpha}^2 + \psi_{r\beta}^2, \tag{10}$$

$$x_{22} = \psi_{r\alpha} i_{s\alpha} + \psi_{r\beta} i_{s\beta} \tag{11}$$

where $i_{s\alpha}$, $i_{s\beta}$, $\psi_{r\alpha}$, $\psi_{r\beta}$ are the stator current and rotor flux vector components, x_{11} denotes traction motor rotor speed, x_{12} is proportional to electromagnetic torque, x_{21} is square of the magnitude of rotor flux vector and represents the excitation of the traction motor, and x_{22} is a multiscalar variable with no direct physical interpretation and proportional to the reactive power consumption. Figure 17 shows the basic structure of the multiscalar model based control system for the traction motor. The rotor fluxes $\psi_{r\alpha}$, $\psi_{r\beta}$, rotor speed (8) and remaining multiscalar variables (9)–(11) are estimated in a speed observer. The variables estimated in the speed observer denoted by Λ are used in the control system. Nonlinear feedback is applied to the system of first-order multiscalar model equations obtained from nonlinear transformation of the multiscalar model [38,39].

The approach of using the multiscalar variables (8)–(11) instead of d–q components of the stator current and rotor flux vectors, advantageously eliminates the need for continuous synchronization of the rotating reference frame with the rotating rotor flux vector, which is absolutely required in the FOC method. The use of the linearizing feedback allows to obtain a linear relationship between the outputs and inputs of the multiscalar model and enables decoupled control of the mechanical subsystem of the induction motor, related to the dynamics of the shaft rotational speed x_{11}, and the electromagnetic subsystem, related to the dynamics of the square of the rotor flux vector module x_{21}. Hence, the x_{21} reference value for the control system from Figure 17 can be modulated according to the arbitrary chosen efficiency optimizing formulation, which ensures the improved efficiency of the traction drive [40]. A detailed analysis of the multiscalar control of the traction motor is beyond the scope of this article and will be discussed in more detail in the forthcoming papers.

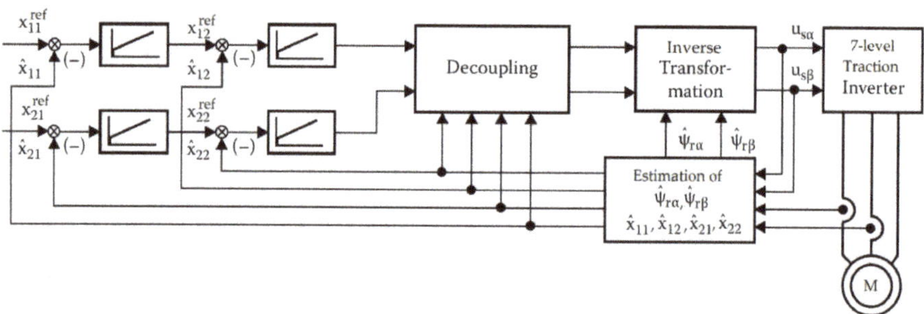

Figure 17. The basic structure of the multiscalar model based control system for the traction motor.

5.6. Controller Hardware

For implementing the proposed control strategy a DSP-based MASTER controller board and ten ARM processor-based SLAVE controller boards have been designed as shown in Figure 18 with the features of advanced functionalities and fast execution time. As shown in the Figure 18a, the MASTER controller board consists of a digital signal processor (ADSP-21363) control card, including a FPGA (FPGA-CYCLONE II EP2C8F256) and additional ARM Cortex-M4 32b processor (STM32F407IGT6) control card. The ADSP-21363 floating-point signal processor (3 Mb SRAM, 333 MHz, 2GFLOPS) implements the traction motor torque and excitation control, and the traction line current and voltage control, while the STM32F407IGT6 processor (MCU + FPU, 210DMIPS, 1MB Flash/192 + 4KB RAM, USB OTG HS/FS, Ethernet) realizes human-machine interfacing (HMI) and the communication with other STM32F407IGT6 ARM Cortex-M4 32b processors of nine power electronic cells.

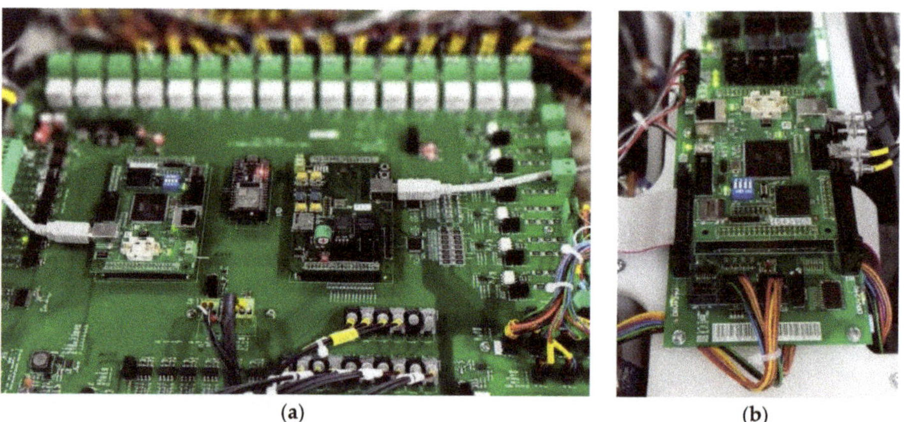

(a)　　　　　(b)

Figure 18. MASTER controller board consisting of main ADSP-21363 based control card and the auxiliary STM32F407IGT6 Arm Cortex based control card (**a**); one of nine STM32F407IGT6 Arm Cortex based SLAVE controller boards controlling nine 4QC-DAB-DC/AC power electronic cells (**b**).

The DSP-based MASTER controller board receives the measurements from 4QC-DAB-DC/AC power electronic cells and implements the traction motor control and traction line voltage and current control algorithms. The FPGA receives the input DC voltage and three-phase output voltage commands from the DSP and implements the PWM algorithm and outputs nine command signals to ARM-based SLAVE controller boards, shown in Figure 18b, via optical fibers. The individual ARM-based SLAVE controller boards receive a command from the MASTER controller board and implement PWM times and outputs gate signals to the transistors. Ten SLAVE controller boards output gate signals to 72 SiC MOSFET dual power modules. To ensure complete isolation between the input stage and the output stage of the DC PETT, one SLAVE controller controls the SiC MOSFET H-bridges of the input sides of the two 4QC-DAB-DC/AC power electronic cells, while the other SLAVE controller controls the transistor bridges of the output sides of these two 4QC-DAB-DC/AC power electronic cells. Both SLAVE controllers communicate with each other using fiber optics.

The MASTER controller board and all SLAVE controller boards contain signal conditioning circuits designed to receive individual voltage and current sensors signals and send them to the DSP or ARM processors respectively. The isolated communication interfaces are realized by a serial Controller Area Network (CAN) interface port.

The overall MASTER control commands of the DSP can be received in two modes: from the driver's console in the train driver's cab (train running mode) and from the operator's PC through the DSP CAN port (service mode). The execution period of one cycle of the control scheme in the main DSP is 150 µs, while the FPGA on the MASTER control card works with a three times shorter execution period of the PWM algorithm. Individual FPGAs on SLAVE controller boards operate with the execution period equal to 33.33 µs. The accuracy of the PWM time counting on the SLAVE controller board results from the used 150 MHz clock.

The control relay outputs on the main-board shown in Figure 18a are used to control the train's individual switching devices, such as the circuit breaker circuit or the contactor circuits in the on-board high-voltage switchgear.

6. On-Track Testing

The developed DC PETT prototype for the railway applications has been assembled as shown in Figure 4 in Section 2. Due to the direct availability of the 3 kV DC railway network on the railway siding, a number of experiments were conducted with the developed DC PETT prototype mounted on the roof of the EN81 series electric passenger railcar shown in Figure 2 in the introductory section. During the tests, only one drive set of the traction motor powered by the developed 3kV DC PETT was running, which made it possible to obtain measurement results specifically for one complete drive system. The second traction motor of the EN81 series electric passenger railcar was not running during the tests. Due to the limitation of the track length, the maximum achievable speed during the experiments was, however, limited and did not reach the operating speed of 120 km/h given by the manufacturer of the rail vehicle. The route of the test runs with a length of 600 m on the Bydgoszcz-Towarowa railway siding used during on-track testing is shown in Figure 19.

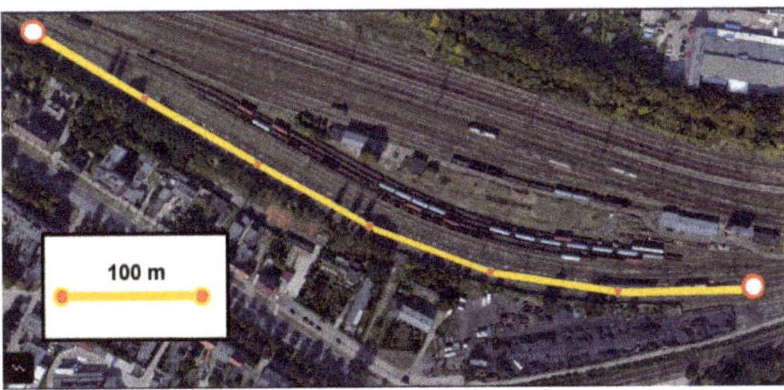

Figure 19. The route of test runs with a length of 600 m on the Bydgoszcz-Towarowa railway siding.

Specifications of the DKLBZ 0910-04 type traction motor of the EN81 series electric passenger railcar are listed in Table 6.

Table 6. DKLBZ 0910-04 type traction motor specification.

Parameter	Description
Rated stator voltage	$U_{sN} = 2200$ V
Rated stator current	$I_{sN} = 88$ A
Rated power	$P_N = 300$ kW
Rated torque	$M_N = 1506$ Nm
Power factor	$\cos\varphi_N = 0.89$
Efficiency	$\eta = 94.3\%$
Max. voltage during breaking	$U_{sMAX} = 2300$ V
Max. power during breaking	$P_{break} = 320$ kW
Max. torque	$M_{MAX} = 2160$ Nm

The results of the track test run on 600 m railway siding are shown in Figure 20. The recorded waveforms were saved on a PC connected to the DSP-based MASTER controller card via USB.

Figure 20 illustrates recorded waveforms during accelerating of the train with the DC PETT based asynchronous traction drive to the speed of 45 km/h and immediate braking. Before the connection of the DC PETT to the overhead traction line voltage there is no energy stored in the DC PETT and all DC links are empty. Similar to AC PETT reported in [19], initial charging is performed from the DC side utilizing the startup resistor that is bypassed later on. Moreover, the developed DC PETT has a second mode of pre-charging the intermediary circuits from the on-board battery bank using a set of DC-DC converters. The latter mode enables the DC PETT to be connected to the overhead traction line without any inrush current.

Referring to Figure 20, after the train pantograph is on, the uncontrolled voltage on each of the primary DC links equals the overhead line voltage divided by nine. Then, after starting DC PETT, the primary DC-link voltage is controlled to 520 V and the secondary DC-link voltage is controlled to be equal primary DC-link voltage. During DC PETT operation the input current drawn from the overhead traction line is controlled according to the control scheme shown in Figure 14. In all experiments, the power recovered during braking was transferred back to the traction network. As can be seen from Figure 20, the delivery of braking power of the order of 200 kW to the traction network did not significantly change the voltage of the traction network.

Figure 21 shows the characteristic traction motor stator current and the PWM output voltage waveforms obtained during three different operation modes of the DC PETT. Figure 21a shows the stator current waveform during start-up and the acceleration of the traction motor (50 A/div; 1 s/div). Thanks to the use of high-performance torque control, mentioned in Section 5, the stator current magnitude is limited at the desired value and its amplitude does not exceed 100 A at all times.

As can be seen, there are no undesirable oscillations in the current waveform. Figure 21b shows the stator current waveform during the final braking phase of the train from 15 km/h until the train stops (50 A/div; 1 s/div). Figure 21c shows the stator current and the PWM voltage at the DC PETT output in the steady-state operation of the traction motor (50 A/div; 1 kV/div; 10 ms/div). As it is shown in Figure 21c, the applied three-phase, seven-level CHB topology provides an almost sinusoidal PWM voltage, which has not been demonstrated in any other 3kV DC rail traction inverter so far. By using CHB technology, the instantaneous maximum value of the voltage switched by the SiC MOSFET transistors does not exceed a few hundred volts. At high switching speeds of SiC MOSFET transistors, this will significantly facilitate compliance with the stringent requirements of railway electromagnetic compatibility (EMC) standards, which will be the focus of the planned continuation of the work carried out by the authors.

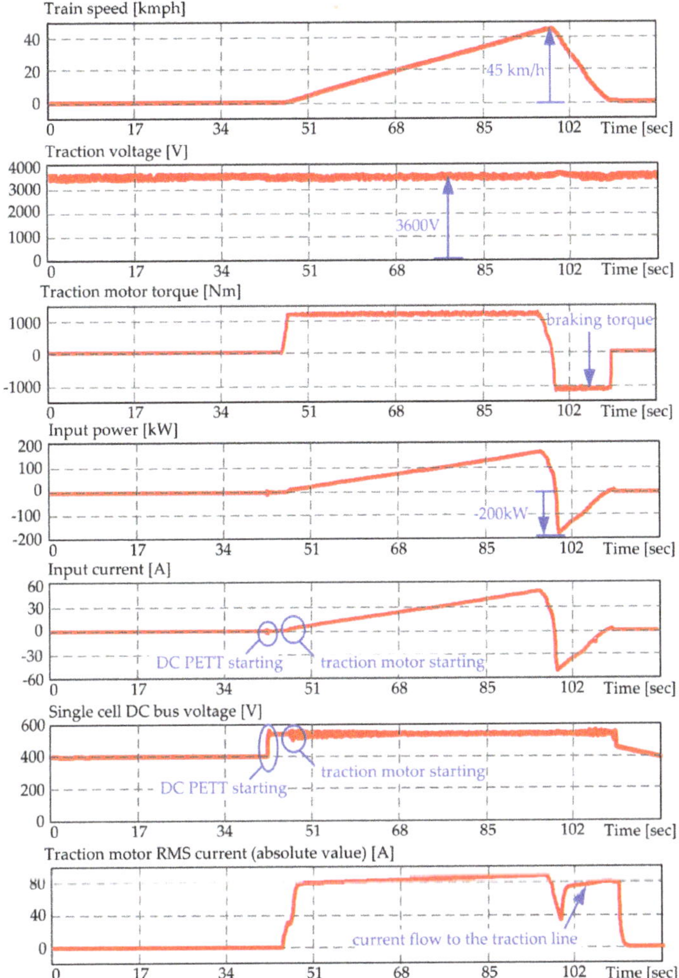

Figure 20. Track test results: Recorded waveforms during start-up of the train with the DC PETT based asynchronous traction drive to a speed of 45 km/h and immediate braking. Top to bottom: (1) train speed; (2) traction motor torque; (3) DC PETT input power; (4) DC PETT input current; (5) traction DC voltage; (6) primary DC bus voltage of a single 4QC-DAB-DC/AC power electronic cell.

The efficiency of the SiC-based 3kV DC PETT prototype has been measured. As in [19], the power consumption of the auxiliary converters supporting the cooling system and the control system has not been included in the efficiency calculations. The plots of efficiencies versus output power are shown in Figure 22.

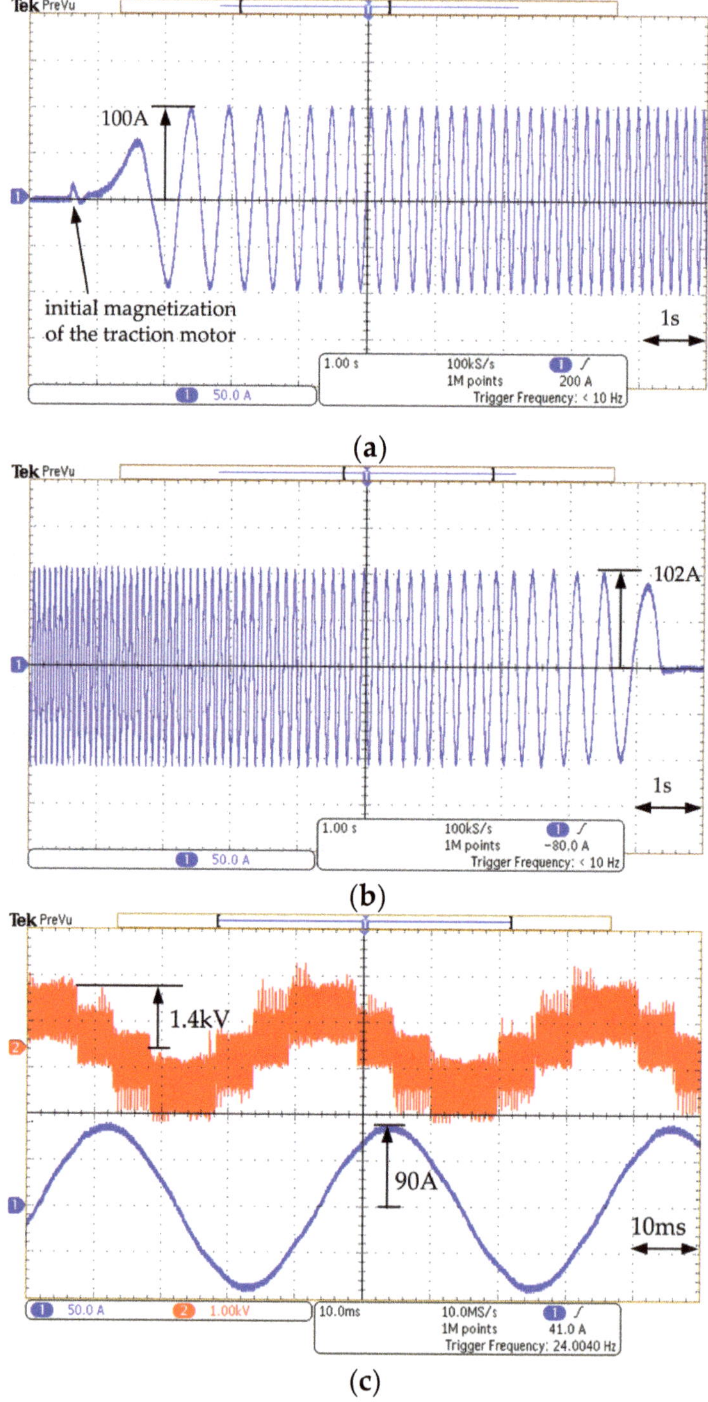

Figure 21. Traction motor stator current and output PWM voltage waveforms obtained during three operation modes of the DC PETT: start-up and acceleration mode (**a**); braking (**b**); phase voltage and phase current of the traction motor steady-state operation (**c**).

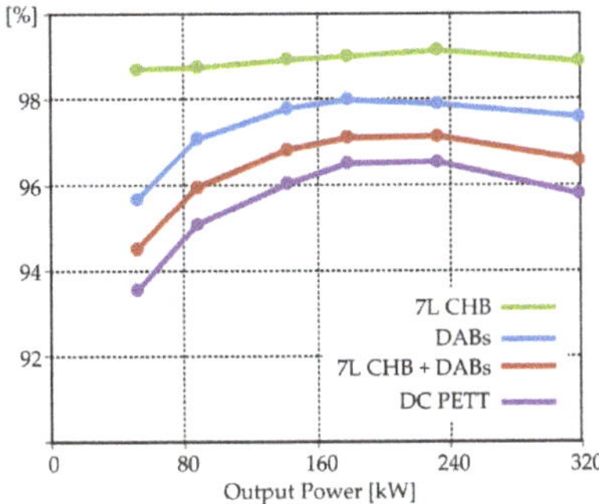

Figure 22. Measured efficiency versus output power of the three power stages and the entire 3kV DC PETT prototype.

As can be seen from Figure 22, the three-phase seven-level CHB traction inverter (7L CHB), considered separately, characterizes peak efficiency around 99%, which is comparable to the corresponding SiC based MV multilevel CHB inverters constructed from low-voltage SiC MOSFET transistors, presented in the literature [22]. The efficiency of the 4QC input stage is slightly higher than that of the seven-level CHB traction inverter, because when cascaded H-bridges operate with a constant voltage of 3kV DC overhead traction line, actually only two SiC MOSFETs work in each H-bridge. However, thanks to the use of the full H-bridges in the DC PETT input stage instead of half bridges, it is possible that the developed SiC-based DC PETT can also work with an AC input voltage, as is the case with multi-system locomotives. However, the operation of the proposed SiC-based DC PETT in an AC voltage system is beyond the scope of this paper and will be the subject of future publications. As can be seen in Figure 22, the isolation stage ensured by nine DAB DC-DC converters has the greatest impact on the efficiency of the entire SiC-based DC PETT. The SiC-based DC PETT prototype has an efficiency of around 96% in a wide range of output power and the peak efficiency around 96.5%.

7. Conclusions

The design and development of the SiC-based DC PETT intended for EMUs operated in 3 kV DC rail traction have been presented in this paper. The developed DC PETT has been implemented into the PESA 308 EN81 series electric passenger railcar that operates in Polish regional passenger rail transport. The conducted experimental tests during train runs on the trial confirm the full functionality of the developed device.

As with the MV PETT for AC traction [19], the proposed DC PETT offers a number of advantages that make it very attractive for rolling stock operating in 3 kV DC traction. First of all, in the era of widespread striving to design highly efficient and ultra-quiet drive converters from SiC semiconductor devices, the proposed solution has a number of advantages if one compares it to high-voltage SiC traction inverters with a classic design being currently at the stage of analyzes and preliminary tests. Conventional two-level voltage source traction inverters of the working voltage in the catenary 3 kV DC would contain SiC MOSFET transistors with a voltage blocking of 6 kV, and the conventional three-level inverter voltage would contain SiC MOSFET transistors with a voltage blocking of 3 kV. It is already known that obtaining high voltage switching frequencies SiC MOSFET above 5 kHz is energy inefficient and the management of electromagnetic disturbances at such high switched energies is quite a challenge. The component modules of the proposed DC PETT, in the form of nine 4QC-DAB-DC/AC power electronic cells, are made with the use of low-voltage SiC technology (1200 V). The applied high switching frequency: 30 kHz to 1.2 kV SiC MOSFETs used in DAB DC-DC converters and 20 kHz to 1.2 kV SiC MOSFETs used in SiC MOSFET H-bridges of the input and output stage, do not cause as significant energy losses as it would be in the case of high voltage (>3 kV) SiC MOSFET technology. The applied high switching frequency allows for favorable elimination of noise from the converter operation. Moreover, the use of multi-level topology made it possible to follow the command voltage from the control system with very high precision and, therefore, enables the application of the high precision control of the traction motor. Moreover, compared to classical topologies, the applied active input stage with the regulator of the current drawn from the overhead line enables the minimization of the input LC filter and, thus, minimization of the total volume and cost, which the authors intend to make the subject of detailed analysis in future publications.

The MFT design path, discussed in detail in the article, shows that the important factors influencing the power density of the developed transformers are the provision of appropriate insulation gaps to ensure galvanic isolation at the level of 9 kV DC and the provision of structural gaps between the windings to obtain the desired transformer leakage inductance. In the case of the traction drive investigated in the paper, with a relatively small power of 325 kVA (500 kVA in peak), the power density of the designed 38 kW MFT was 3.5 kW/dm^3 ($\approx$5 kW/dm^3 peak). This allows the authors to reasonably hope that for a higher power MFT, the power density obtained will also be higher. At the present stage, it is difficult to compare the power density of the developed 3kV DC PETT prototype with a built-in lightweight LCL input traction filter with 3 kV DC roof-mounted traction inverters available on the market because the solutions known to the authors have a heavy external traction filter mounted in a separate container, which is not taken into account by manufacturers to estimating the power density of the traction converter. The developed 3kV SiC-based DC PETT prototype, thanks to the built-in 4QC power input stage, is immediately ready for cooperation with the AC traction network in a multi-system EMUs. The proposed modular DC PETT structure, composed of the same repeatable power electronic cells, could ensure lower maintenance costs, short inspection and repair times for potential faults, and thus high availability - required in the rolling stock.

Author Contributions: Conceptualization, M.A.; software, J.S.; validation, M.A., J.S.; investigation, M.A., J.S.; writing—original draft preparation, M.A.; writing—review and editing, M.A., J.S.; visualization, M.A.; All authors carried out the theoretical analysis and contributed to writing the paper. All authors have read and agreed to the published version of the manuscript.

Funding: The authors gratefully acknowledge financial support from the European Union under the Polish Operational Programme Smart Growth, The National Centre for Research and Development, Grant no. POIR.01.02.00-00-0193/16-00. The APC was funded by Gdańsk University of Technology.

Acknowledgments: The authors acknowledge the commitment and resources provided by MMB Drives Ltd., H.Cegielski-Energocentrum Ltd. and PESA Bydgoszcz S.A. during the research. The authors would like to thank Jędrzej Pietryka, Sebastian Giziewski and Patryk Kruk for their valuable support in writing this article.

Conflicts of Interest: The authors declare no conflict of interest.

References

1. Brenna, M.; Foiadelli, F.; Zaninelli, D. *Electrical Railway Transportation Systems*; IEEE Press Series on Power Engineering; John Wiley & Sons, Inc.: Hoboken, NJ, USA, 2018.
2. Ronanki, D.; Singh, S.A.; Williamson, S.S. Comprehensive topological overview of rolling stock architectures and recent trends in electric railway traction systems. *IEEE Trans. Transp. Electrif.* **2017**, *3*, 724–738. [CrossRef]
3. Falco, M.; Desportes, G.; Piton, M. Innovative Railway Traction System. In Proceedings of the 7th Transport Research Arena (TRA 2018), Vienna, Austria, 16–19 April 2018; pp. 1–10.
4. Ishikawa, K.; Terasawa, K.; Sakai, T.; Sugimoto, S.; Nishino, T. Development of rolling stock inverters using SiC. *Hitachi Rev.* **2017**, *2*, 155–160.
5. Rujas, A.; Lopez, V.M.; Villar, I.; Nieva, T.; Larzbal, I. SiC-hybrid based railway inverter for metro application with 3.3kV low inductance power modules. In Proceedings of the IEEE Energy Conversion Congress and Exposition (ECCE 2019), Baltimore, MD, USA, 29 September–3 October 2019; pp. 1992–1997.
6. Lindahl, M.; Velander, E.; Johansson, M.H.; Blomberg, A.; Nee, H.P. Silicon Carbide MOSFET Traction Inverter Operated in the Stockholm Metro System Demonstrating Customer Values. In Proceedings of the IEEE Vehicle Power and Propulsion Conference (VPPC 2018), Chicago, IL, USA, 27–30 August 2018; pp. 1–6.

7. Schmidt, E.; Müllner, F.; Neudorfer, H. Modelling and Precalculation of Additional Losses of Inverter Fed Asynchronous Induction Machines for Traction Applications. In Proceedings of the International Aegean Conference on Electrical Machines and Power Electronics and Electromotion, Istanbul, Turkey, 8–10 September 2011; pp. 1–6. [CrossRef]

8. Dangl, F.; Neudorfer, H. Impact of Additional Losses due to Inverter Supply on the Thermal Behaviour of Induction Machines for Traction Drives. *Elektrotechnik Inf.* **2016**, *1*, 27–35.

9. Bakran, M.M.; Eckel, H.G. Power Electronics Technologies for Locomotives. In Proceedings of the IEEE Power Conversion Conference (PCC 2007), Nagoya, Japan, 2–5 April 2007; pp. 1362–1368. [CrossRef]

10. Soltau, N.; Wiesner, E.; Hatori, K.; Uemura, H. 3.3 kV Full SiC MOSFETs—Towards High-Performance Traction Inverters. *Bodos Power Syst.* **2018**, *1*, 22–24.

11. Steczek, M.; Szeląg, A.; Chatterjee, D. Analysis of disturbing effect of 3 kV DC supplied traction vehicles equipped with two-level and three-level VSI on railway signalling track circuits. *Bull. Pol. Acad. Sci.* **2017**, *65*, 663–674. [CrossRef]

12. Das, M.K.; Capell, C.; Grider, D.E.; Leslie, S.; Ostop, J.; Raju, R.; Schutten, M.; Nasadoski, J.; Hefner, A. 10 kV, 120 A SiC half H-bridge power MOSFET modules suitable for high frequency, medium voltage applications. In Proceedings of the IEEE Energy Conversion Congress and Exposition (ECCE 2011), Phoenix, Arizona, 17–22 September 2011; pp. 2689–2692.

13. Aceiton, R.; Weber, J.; Bernet, S. Input filter for a power electronics transformer in a railway traction application. *IEEE Trans. Ind. Electron.* **2018**, *65*, 9449–9458. [CrossRef]

14. Rufer, A.; Schibli, N.; Chabert, C.; Zimmermann, C. Configurable front-end converters for multicurrent locomotives operated on 16 2/3 Hz AC and 3 kV DC systems. *IEEE Trans. Power Electron.* **2003**, *18*, 1186–1193. [CrossRef]

15. Stackler, C.; Morel, F.; Ladoux, P.; Fouineau, A.; Wallart, F.; Evans, N. Optimal sizing of a power electronic traction transformer for railway applications. In Proceedings of the 44th Annual Conference of the IEEE Industrial Electronics Society (IECON 2018), Washington, DC, USA, 20–23 October 2018; pp. 1380–1387. [CrossRef]

16. Morel, F.; Stackler, C.; Ladoux, P.; Fouineau, A.; Wallart, F.; Evans, N.; Dworakowski, P. Power electronic traction transformers in 25 kV/50 Hz systems: Optimisation of DC/DC Isolated Converters with 3.3 kV SiC MOSFETs. In Proceedings of the International Exhibition and Conference for Power Electronics, Intelligent Motion, Renewable Energy and Energy Management (PCIM Europe 2019), Nurnberg, Germany, 7–9 May 2019; pp. 1–9.

17. Tremong, J.-F.; Vulturescu, B.; Chamaret, A.-P. Evaluation of the energy saving potential of a Power Electronic Transformer for rolling stock under 25kV, 50Hz. In Proceedings of the 7th Transport Research Arena (TRA 2018), Vienna, Austria, 16–19 April 2018; pp. 1–9.

18. Farnesi, S.; Marchesoni, M.; Passalacqua, M.; Vaccaro, L. Solid-State Transformers in Locomotives Fed through AC Lines: A Review and Future Developments. *Energies* **2019**, *12*, 4711. [CrossRef]

19. Zhao, C.; Dujic, D.; Mester, A.; Steinke, J.K.; Weiss, M.; Lewdeni-Schmid, S.; Chaudhuri, T.; Stefanutti, P. Power electronic traction transformer—Medium voltage prototype. *IEEE Trans. Ind. Electron.* **2014**, *61*, 3257–3268. [CrossRef]

20. Huber, J.E.; Kolar, J.W. Optimum number of cascaded cells for high-power medium-voltage AC–DC converters. *IEEE J. Emerg. Sel. Top. Power Electron.* **2017**, *5*, 213–232. [CrossRef]

21. Magro, M.C.; Savio, S. Reliability and availability performances of a universal and flexible power management system. In Proceedings of the IEEE International Symposium on Industrial Electronics (ISIE 2010), Bari, Italy, 4–7 July 2010; pp. 2461–2468. [CrossRef]

22. Pan, J.; Ke, Z.; Al Sabbagh, M.; Li, H.; Potty, K.; Perdikakis, W.; Na, R.; Zhang, J.; Wang, J.; Xu, L. 7-kV, 1-MVA SiC-Based Modular Multilevel Converter Prototype for Medium-voltage Electric Machine Drives. *IEEE Trans. Power Electron.* **2020**, *35*, 10137–10149. [CrossRef]

23. Engelmann, G.; Sewergin, A.; Neubert, M.; De Doncker, R.W. Design Challenges of SiC Devices for Low- and Medium-Voltage DC-DC Converters. In Proceedings of the IEEE International Power Electronics Conference (IPEC-Niigata 2018 -ECCE Asia), Niigata, Japan, 20–24 May 2018; pp. 3979–3984. [CrossRef]

24. Adamowicz, M.; Giziewski, S.; Pietryka, J.; Krzeminski, Z. Performance comparison of SiC Schottky diodes and silicon ultra fast recovery diodes. In Proceedings of the 7th IEEE International Conference-Workshop Compatibility and Power Electronics (CPE 2011), Tallin, Estonia, 1–3 June 2011; pp. 144–149.

25. Leibl, M.; Ortiz, G.; Kolar, J.W. Design and experimental analysis of a medium-frequency transformer for solid-state transformer applications. *IEEE J. Emerg. Sel. Top. Power Electron.* **2017**, *5*, 110–123. [CrossRef]
26. Villar, I. Multiphysical Characterization of Medium-Frequency Power Electronic Transformers. Ph.D. Thesis, École Polytechnique Fédérale de Lausanne, Lausanne, Switzerland, 2010.
27. Shua, P.; Biela, J. Influence of material properties and geometric shape of magnetic cores on acoustic noise emission of medium-frequency transformers. *IEEE Trans. Power Electron.* **2017**, *32*, 7916–7931. [CrossRef]
28. Jain, A.K.; Ayyanar, R. PWM control of dual active bridge: Comprehensive analysis and experimental verification. *IEEE Trans. Power Electron.* **2011**, *26*, 1215–1227. [CrossRef]
29. Villar, I.; Mir, L.; Etxeberria-Otadui, I.; Colmenero, J.; Agirre, X.; Nieva, T. Optimal design and experimental validation of a medium-frequency 400 kVA power transformer for railway traction applications. In Proceedings of the IEEE Energy Conversion Congress and Exposition (ECCE), Raleigh, NC, USA, 15–20 September 2012; pp. 684–690. [CrossRef]
30. Everts, J. Modeling and Optimization of Bidirectional Dual Active Bridge ac-dc Converter Topologies. Ph.D. Thesis, Arenberg Doctoral School Faculty of Engineering Science, KU Leuven, Leuven, Belgium, 2014.
31. Adamowicz, M. Power Electronics Building Blocks for implementing SmartMV/LV Distribution Transformers for Smart Grid. *Acta Energetica* **2014**, *4*, 6–13. [CrossRef]
32. Saeed, M.; Rogina, M.R.; Rodriguez, A.; Arias, M.; Briz, F. SiC-Based High Effciency High Isolation Dual Active Bridge Converter for a Power Electronic Transformer. *Energies* **2020**, *13*, 1198. [CrossRef]
33. Haneda, R.; Akagi, H. Design and Performance of the 850-V 100-kW 16-kHz Bidirectional Isolated DC–DC Converter Using SiC-MOSFET/SBD H-Bridge Modules. *IEEE Trans. Power Electron.* **2020**, *35*, 10013–10025. [CrossRef]
34. Kantor, V.V. Methods of Calculating Leakage Inductance of Transformer Windings. *Russ. Electr. Eng.* **2009**, *80*, 224–228. [CrossRef]
35. Prieto, R.; Cobos, J.A.; Garcia, O.; Alou, P.; Uceda, J. Taking into account all the parasitic effects in the design of magnetic components. In Proceedings of the IEEE Thirteenth Annual Applied Power Electronics Conference and Exposition (APEC'98), Anaheim, CA, USA, 15–19 February 1998; pp. 400–406.
36. Doebbelin, R.; Teichert, C.; Benecke, M.; Lindemann, A. Computerized calculation of leakage inductance values of transformers. *Piers Online* **2009**, *5*, 721–726.
37. Lewicki, A.; Morawic, M. Structure and the space vector modulation for a medium-voltage power-electronic-transformer based on two seven-level cascade H-bridge inverters. *IET Electr. Power Appl.* **2019**, *13*, 1514–1523. [CrossRef]
38. Krzemiński, Z. Nonlinear control of induction motor. *IFAC* **1987**, *20*, 357–362. [CrossRef]
39. Wilamowski, B.M.; Irwin, J.D. (Eds.) *The Industrial Electronics Handbook. Power Electronics and Motor Drives*, 2nd ed.; CRC Press, Taylor & Francis Group: Boca Raton, FL, USA, 2011.
40. Dong, G.; Ojo, O. Efficiency optimizing control of induction motor using natural variables. *IEEE Trans. Ind. Electron.* **2006**, *53*, 1791–1798. [CrossRef]

Publisher's Note: MDPI stays neutral with regard to jurisdictional claims in published maps and institutional affiliations.

Article

Feasibility Study GaN Transistors Application in the Novel Split-Coils Inductive Power Transfer System with T-Type Inverter

Viktor Shevchenko [1,*], Bohdan Pakhaliuk [1,2], Oleksandr Husev [1,3,*], Oleksandr Veligorskyi [1], Deniss Stepins [4] and Ryszard Strzelecki [2,*]

[1] Chernihiv Power Electronics Laboratory, BRAS Department, Educational-Scientific Institute of Electronic and Information Technologies, Chernihiv National University of Technology, 14039 Chernihiv, Ukraine; bohdan.pakhaliuk@gmail.com (B.P.); alexveligorsky@gmail.com (O.V.)

[2] Faculty of Electrical and Control Engineering, Gdansk University of Technology, 80-233 Gdansk, Poland

[3] Power Electronics Research Group, Tallinn University of Technology, 12616 Tallinn, Estonia

[4] Institute of Industrial Electronics and Electrical Engineering, Riga Technical University, 12/K1 Azenes Street, LV-1658 Riga, Latvia; stepin2@inbox.lv

* Correspondence: shevaip1990@gmail.com (V.S.); oleksandr.husev@gmail.com (O.H.); profesor1958@gmail.com (R.S.); Tel.: +380-93-957-4211 (V.S.); +380-96-985-5012 (O.H.)

Received: 17 July 2020; Accepted: 28 August 2020; Published: 1 September 2020

Abstract: A promising solution for inductive power transfer and wireless charging is presented on the basis of a single-phase three-level T-type Neutral Point Clamped GaN-based inverter with two coupled transmitting coils. The article focuses on the feasibility study of GaN transistor application in the wireless power transfer system based on the T-type inverter on the primary side. An analysis of power losses in the main components of the system is performed: semiconductors and magnetic elements. System modeling was performed using Power Electronics Simulation Software (PSIM). It is shown that the main losses of the system are static losses in the filter inductor and rectifier diodes on the secondary side, while GaN transistors can be successfully used for the wireless power transfer system. The main features of the Printed Circuit Board (PCB) design of GaN transistors are considered in advance.

Keywords: wireless power transfer; inductive power transmission; multilevel converter; AC-DC power converters; T-type inverter; GaN-transistors; electromagnetic coupling

---

1. Introduction

Interest in inductive wireless power transmission is constantly growing due to the increasing interests of both low-power wireless chargers for mobile and wireless charging stations of medium and high power for electric bikes and electric vehicles. Such chargers transfer the electric energy wirelessly from primary to secondary inductor by means of inductive coupling [1]. Inductive wireless power transfer systems consist of a transmitting part (contains an inverter, compensation circuit and primary inductor) and a receiving part (receiving inductor, compensation circuit, rectifier) [1]. The researchers have already analyzed the main possible topologies of compensation schemes, their advantages and disadvantages, and described the general recommendations for their implementation. It is well known that Wireless Power Transfer (WPT) systems have some limitations, such as short transmission distance(centimeters or dozens of centimeters at acceptable levels of transmission efficiency) [2,3], sensitivity to the exact positioning of the receiving coil relative to the transmission coil [2,4], size and cost of the system.

Among existing limitations, the issue of the size and cost of the WPT system is one of the most important. Researchers are still looking for the optimal system configurations and topologies of power converters that would best meet the above requirements.

Different types of switches are utilized in the power electronics converters [5–8]. The conventional Insulated Gate Bipolar Transistors (IGBTs) are gradually going out of use in industrial circuits of WPT systems due to their low switching capability [9]. The reverse blocking voltage capability of the conventional IGBT is very low; there are relatively large power losses [10]. It is well known that the use of wideband gap semiconductors (such as GaN-transistors) instead of classical Si power switches can significantly reduce the power losses that lead to the increasing of the system efficiency or significantly increase switching frequency reducing size of passive elements [11–13]. It is advisable to use GaN transistors for T-type topologies [5–8,14]. The GaN features fast switching, low parasitic charges, reverse conductivity with zero recovery charge and low driving power losses and dynamic losses; compared to Si-IGBTs and SiC-MOSFETs [5,6,15–18], higher efficiency, low parasitic output capacitance [16–18] can be achieved. The advantage of GaN over Si is mostly visible at higher frequencies in dynamic losses [15,19]. However, the conduction losses are comparable with the SIC semiconductors [18,19].

The main goal of the article is to study the feasibility of GaN transistor application in the proposed non-traditional (non-classical) WPT system. This will be based on the loss analysis of the main components of the circuit.

The paper, consisting of seven sections, proposes a new solution of the wireless power transfer system based on two parallel single-phase T-type GaN-based invertors (dual T-type inverter) with two transmitting coils on one ferrite core (coupled transmitting inductances). The case study system description and advantages of such solution are represented in Section 2 of the paper. According to previous research [20], more than 70% of the losses in WPT systems for various cases of power, loads and working frequencies depend on semiconductors and inductors [20,21]. Therefore, the contribution of such parameters was taken into account in calculations in this paper. Confirmation of the advantages of the proposed solution, made mainly by power losses analysis, described in Sections 3–5. Sections 3 and 4 proposes the losses models of the GaN transistors and coil inductors, respectively. Simulation and experimental verification of the proposed solution is described in Section 5, with conclusions and list of patents devoted to the proposed WPT system on Sections 6 and 7, respectively.

2. Case Study System Description

Figure 1 depicts the proposed circuit of a multi-level converter for WPT. The primary converter consists of a full-bridge three-level T-type inverter connected to bidirectional auxiliary semiconductor switches. The GaN-based T-type inverter is first proposed for use in a WPT system together with two coupled inductors. It provides a number of advantages over existing analogues.

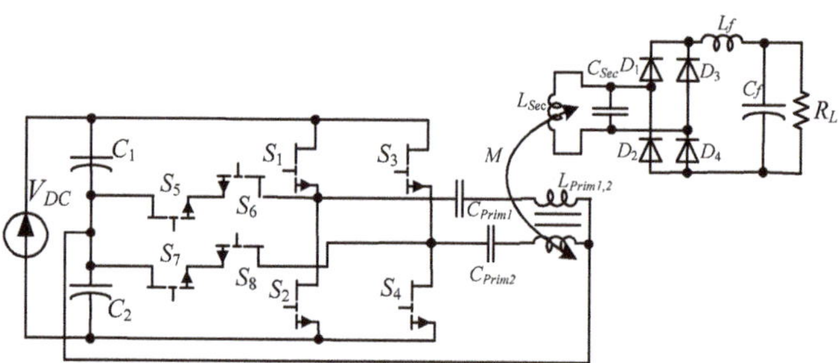

Figure 1. Proposed Inductive Power Transfer IPT converter.

The DC source is applied to the T-type inverter. The energy is transmitted to the secondary side through the primary coils with an air gap (Figure 1). The output current is rectified by the passive full-bridge rectifier, filtered and supplied to the load.

In the scheme, V_{dc} is the source of the input dc voltage; C_1, C_2—input capacitors; S_1–S_4—switches of the single-phase full-bridge inverter; S_5, S_6 and S_7, S_8—auxiliary bidirectional switches; $L_{prim1,2}$—coupled primaries inductances; C_{prim1}, C_{prim2}—primaries compensating capacitances; L_{sec}—secondary inductance; C_{sec}—secondary compensating capacitance; D_1–D_4—bridge rectifier; C_f—filter capacitor; L_f—filter inductance; R_L—output resistive load.

First of all, the multi-level inverters have a number of advantages over conventional H-bridge inverters, including better Electromagnetic Compatibility EMC and higher efficiency [14,22–24], which are extremely important for wireless power transmission systems [24]. However, analysis of the existed publications shows that using of multi-level inverters for WPT systems are just at the beginning stage [25].

Each multi-level circuit has its advantages and disadvantages, but among these types, T-type has some advantages over other types: smaller size, simpler operation principles, lower THD, lower conduction losses and smaller number of semiconductors [10,14,23]. The most important advantage of T-type solution in the WPT application is that only half of dc-link voltage applied to the primary side coil which in turn reduces the primary inductance and size of the coil [14]. The equation (1) for calculating the primary inductance L_{prim} at SP compensation analytically confirms this fact [20]:

$$L_{Prim} \approx L_{Sec}\left(\frac{8}{\pi^2 k_{nom}}\frac{V_{in}}{V_{out}}\right)^2,\tag{1}$$

where L_{sec}-secondary side self-inductance, k_{nom}—nominal coupling, V_{in} and V_{out}—input and output voltages, respectively.

Finally, the splitting of the transmitting coils will reduce the conduction and overall diameter of the primary inductance. The application of two coupled transmission coils of inductance on a single ferrite core reduces the total dimensions of the magnetic components on the primary side (and the losses in copper and ferrite). Multiple magnetic resonant coils lead to higher transmission efficiency and longer transmission distances [26]. In addition, the coupling coefficient between them is considered constant, which simplifies certain calculations of the system. Furthermore, this solution reduces the current through each coil. This leads to a lower overall resistance of the transmission coils, which, in turn, increases the Q factor and the energy transfer efficiency.

It is expected that wireless power transmission systems based on of multi-coil circuits with GaN-based T-type inverters in the transmission part is a promising solution, joining the advantages of the parts which already existed. The proposed solution does not have any heatsinks, has reduced the size of primary coils and can be considered for industrial application. Such a system can be used directly in power supply systems for transmission on different power levels.

Compensating capacitors are required to compensate the leakage inductance on the primary and secondary sides in the WPT systems. In this case, the possible distance between the coils increases [2]. Systems with Series-Parallel (SP) compensation work efficiently with a wide range of loads in addition to the advantages for middle-power and low-power applications and allow reducing dimensions of the receiver coil [20]. Series-Series (SS) compensation does not depend on the change of magnetic coupling and load. These compensation topologies are most widely used for wireless charging. This solution is very well investigated; its benefits and drawbacks are well known [2,27–29].

3. Losses Models of the GaN Transistors

The main sequence of the calculations is presented in this section and Section 5. Initially, the control signals and the shape of voltage and current signals in the inverter were derived from simulation and are used for calculations of power losses in semiconductors.

3.1. T-type Inverter Operation Mode

Two auxiliary switches S_5 and S_6 (Figure 1) are turned on in the T-type topology (Figure 2). Switching states and the voltage are shown for the 3-level T-type Neutral Point Clamped NPC inverter in Table 1 for Figure 1 (for one leg).

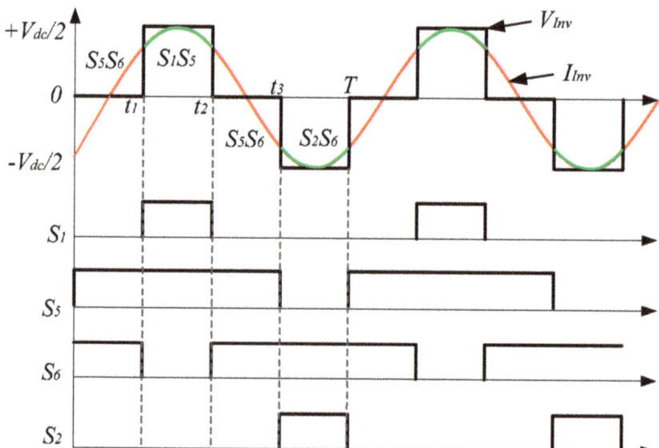

Figure 2. Control signal sequences and shapes of the voltage and the current of the IPT converter. Static losses are calculated, knowing the current through each transistor at a certain interval.

Table 1. Switching States and Voltage for the 3-Level T-Type NPC Inverter [23].

Level	S_1	S_5	S_6	S_2	Voltage
Positive (+)	1	1	0	0	$+V_{dc}/2$
Neutral (N)	0	1	1	0	0
Negative (−)	0	0	1	1	$-V_{dc}/2$

Figure 2 shows the control signal sequence of the transistor (shown for one phase), the voltage (V_{Inv}) and the inverter output current (I_{Inv}).

3.2. Losses Model of the GaN Transistor under Compensation Condition

It is well known that the conduction losses are a significant component to estimate the total losses in the transistors [19]. In addition, the dynamic losses in GaN transistors should be taken into account. The total transistor power losses are determined by the sum of the static and dynamic losses [17].

$$P_{Total} = P_{Cond} + P_{Dyn},\tag{2}$$

where P_{Cond}—static losses (conduction losses), P_{Dyn}—dynamic losses.

The equation that determines the conduction losses (when the transistor is fully on) is as follows:

$$P_{Cond} = I_{Drms}{}^2 + R_{DS(on)},\tag{3}$$

where I_{Drms}—rms current value through drain, $R_{DS(on)}$—on-resistance of a transistor's drain-source.

The current I_{Drms} is determined at each interval (Figure 2) for the positive half wave as follows:

$$I^2_{S1S5} = \frac{1}{T}\int_{t1}^{t2} (i_m \sin(\omega t))^2 dt,\tag{4}$$

where i_m—amplitude value of drain current, ω—angular frequency, T—period. For the negative half wave:

$$I^2_{S2S6} = \frac{1}{T} \int_{t3}^{T} (i_m \sin(\omega t))^2 dt. \tag{5}$$

Two sections are defined for the zero state (Figure 2):

$$I^2_{S5S6} = \frac{1}{T} \int_{t0}^{t1} (i_m \sin(\omega t))^2 dt + \frac{1}{T} \int_{t2}^{t3} (i_m \sin(\omega t))^2 dt. \tag{6}$$

Currents are added from all sections and multiplied by the resistance of the transistor over a period of time due to (3) to calculate the conduction losses of the transistor.

Total dynamic power losses [17]:

$$P_{Dyn} = P_{SW(on)} + P_{SW(off)} + P_G + P_{rcl} + P_{oss} + P_{RR}, \tag{7}$$

where $P_{SW(on)}$ and $P_{SW(off)}$—switching losses, P_G—gate charge losses, P_{rcl}—power loss due to the reverse conduction voltage through the body diode; P_{RR}—reverse recovery loss; P_{OSS}—power loss due to the output capacitance.

Gate charge power losses of a transistor:

$$P_G = Q_G + V_{dr} f_{sw}, \tag{8}$$

where Q_G—total gate charge of a transistor, V_{dr}—driving voltage, f_{sw}—switching frequency.

The equations show, that charge losses are increasing at high switching frequency.

Reverse recovery loss is caused by the charge stored in the junction of the internal body diode of a transistor in the T-type inverter [19]:

$$P_{RR} = Q_{RR} \frac{V_{In}}{2} f_{sw}, \tag{9}$$

where Q_{RR} is the reverse recovery charge, V_{in}—input voltage of the inverter. In the transistor datasheet, GaN transistors do not contain the internal body diode, so, reverse recovery loss is not present in these devices.

Power losses due to the output capacitance [17] are the following:

$$P_{OSS} = f_{sw} \int_{0}^{V_{in}} (V_{DS} C_{OSS}(V_{DS})) dv_{DS}, \tag{10}$$

where C_{OSS}—output capacitance of the transistor (determined by the dependencies in the datasheet). These losses are independent of power and are insignificant at the increase of power, but the contribution of this type of losses is significant at low power and high switching frequency.

The power losses due to the reverse conduction voltage through the body diode (or dead-time losses):

$$P_{rcl} = V_{REV} I_D t_{dead} f_{sw}, \tag{11}$$

where t_{dead}—length of the dead-time (reverse diode conduction time), V_{REV}—reverse voltage drop in a GaN transistor.

The exact equations from [17,19], take into account several values (values of switching time, level of corresponding voltages, etc.) from the simulation data or the experimental data. It is not possible to accurately determine the switching time of the transistors from the model in PSIM and at high frequency. The approximate switch-on time is 4.9 ns, the switch-off time is 3.4 ns according to the datasheet. At the same time, the duration of the on- and off-transistors of each of the shoulders in the model is much larger.

Therefore, a simplified equation is used to estimate the magnitude of switching losses quantitatively [18]:

Turn-on switching losses of a transistor:

$$P_{SW(on)} = \int_0^{t\_rise} (V_{DS}I_D)dt, \tag{12}$$

Turn-off switching losses of a transistor:

$$P_{SW(off)} = \int_0^{t\_fall} (V_{DS}I_D)dt, \tag{13}$$

where t_{rise} and t_{fall}—transistor's time for turn-on and turn-off (values from datasheet), V_{ds}—drain-source voltage, I_D—drain current.

According to [19], at frequencies below 100 kHz, switching losses of transistors are very low. Switching losses are increasing at frequencies up to 500 kHz but they have no significant effect on the total loss estimation [19]. Thus, most transistor losses are conduction losses and power losses due to the output capacitance.

3.3. Losses Model of Rectifier Diode Losses

The losses in rectifying diodes are also significant, especially at high switching frequency and high current through diodes. The parameters of high-speed Schottky diodes were used for modeling and in the experimental verification.

It is known that the static losses in a diode are determined by the current flow through the diode multiplied by the voltage drop across the diode. Two diodes simultaneously conduct current in the case of a diode bridge:

$$P_{Cond.D} = 2I_{forv}V_{drop}, \tag{14}$$

where I_{forv} is the forward current value through the diode, V_{drop}—voltage drop per diode.

Dynamic losses in the diode are switching losses. All diodes in the diode bridge are involved in the process of rectification. If a Q_{RR} value (this is the reverse recovery charge) is given in the datasheet, then the equation is as follows:

$$P_{SW\_D} = 4Q_{rr}V_{rev}f_{sw}. \tag{15}$$

Reverse recovery charge value is defined as a product of $C_{junction}$ (junction capacitance) by V_{rev} (the reverse voltage on the diode). The result in (16) is obtained by substituting this product into Equation (15):

$$P_{SW\_D} = 4V_{rev}^2 C_{junction}f_{sw}. \tag{16}$$

The total losses in all semiconductors in this scheme are higher than the losses in the transmitting and receiving coils.

4. Design and Losses Models of Coils Inductors

Coupled primary coils were calculated with a nominal value of 90 μH each and a receiving coil with a nominal value of 24 μH (Table 2).

Table 2. Parameter Table for Experiments.

Symbol	Description	Value
V_{in}	Input voltage	300 V
S_1-S_8	GaN transistors	GS66508T
F_{sw}	Switching frequency	150; 200 kHz
C_{sn}	Snubber capacity	100; 470 pF
D_1-D_4	Rectifying diodes	RB228NS100TL
$L_{prim1,2}$	Primary inductances	90 µH
L_{sec}	Secondary inductance	24 µH
k	Coupling coefficient	0.7; 0.9

4.1. Design of the Transmitter and Receiver Coils

The simulation was performed in almost the same order as described in [4]. The model of coils was carried out in ANSYS Electrimagnetic Suite and designed using the Finite Elements Modeling (FEM) method. The transmission coils are on the top and the receiving coil is on the bottom (Figure 3).

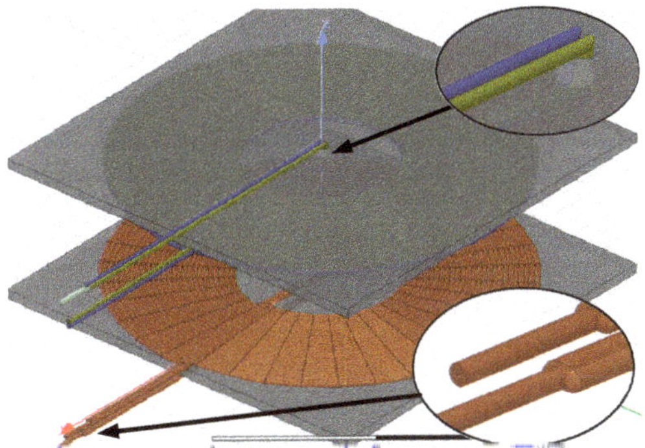

Figure 3. Designed model of coils for WPT.

The primary coil consists of two coils connected in parallel at one lead. One end of the coil pins is drawn through a hole in the ferrite. They do not interfere with the coils as close as possible to each other and do not distort the magnetic flux with this solution. Both primary coils have equal turns each. They are arranged in two layers, one above the other. This solution reduces losses in copper and ferrite and also the total dimensions of the magnetic components on the primary side. Multiple magnetic resonant coils lead to higher transmission efficiencies and longer transmission distances.

The secondary coil consists of double turns (shown in the figure with an enlarged fragment) in one layer. Coil winding with double turns reduces the coil's own resistance at the same value of the inductance itself and increases the quality factor. It increases also the maximum current that the secondary coil misses.

4.2. Losses Model of the Coils Inductors under Compensation Condition

Certain simplifications are allowed in the calculations of losses in the inductors. It is quite a complex mathematical problem to determine the core losses and eddy current losses, especially including the skin and proximity effects of inductors [30]. It is not always possible to achieve acceptable accuracy of calculations even when those losses are determined. The challenges are caused by the complex physical nature of these phenomena.

The core is mainly intended for shielding the magnetic induction flux in the WPT system [4]. Losses in the core are determined by the modified bulky Steinmetz equation at non-ideal sinusoidal voltage [20,30]. However, it is difficult to determine the ferrite coefficients for the equation since an experimental procedure is needed for a specific material under right conditions with high quality equipment. These factors are given rarely by the manufacturer. Therefore, determination of the value of core losses by other methods for this material is not accurate and has no scientific validity.

Most of the available FE tools do not support Litz wire modeling. In addition, magnetic field H differs from turn to turn at determining the proximity effect [4]. Hence, H must be evaluated in the center of each turn individually to calculate the proximity loss for each turn [20]. Conduction losses are usually added to this value at determining a skin effect [20,30]. The Litz wire reduces the skin effect and proximity effect significantly. Furthermore, the proximity effect is minimal between the transmitting and the receiving coils due to the large air gap.

The value of DC conduction losses is sufficient to understand the effect of the geometrical parameters of the coils on the losses value in transistors and coils.

Conduction (ohmic) losses in the primary coils:

$$P_{Lprim12} = I_{Lprim\_rms}^{2}(2R_{Lprim}), \tag{17}$$

where I_{Lprim_rms}—current through primary coils; R_{Lprim}—resistance of one primary coil. Both paired transmitting coils are the same and have the same resistance in this case.

Conduction losses in the secondary coil are as follows:

$$P_{L\,sec} = I_{L\,sec\_rms}^{2} R_{L\,sec}, \tag{18}$$

where I_{Lsec_rms}—current through the secondary coil.

Power losses in the primary coils mostly depend on the inverter current, together with the own resistance of the primary coil.

5. Results of Experiments and Simulation

An experimental model was made to check the feasibility of using GaN transistors in the described scheme (Figure 4). Transistors GS66508Twere used with maximum drain current of 30 A, maximum drain-to-source voltage 650 V, drain-to-source on resistance at 25 °C equal to 50 mΩ. The transistor has zero reverse recovery loss, fast fall and rise times, low inductance and low thermal resistance in a small package. The GS66508T is a top-side cooled transistor that offers very low junction-to-case thermal resistance. Transistors are located at the bottom of the Printed Circuit Board (PCB) without additional radiators. These features combine to provide very high efficiency of power switching.

The PCB consists of four copper layers, divided into signal and power parts. In the signal part of the board top and bottom layers are devoted for signal traces and internal layers for power supply voltage and ground polygons. In contrast, all layers, external as well as internal, are used for power traces. Some special techniques were used on the PCB aiming increase the efficiency and decrease power losses. First of all, high-current power traces were repeated on all four layers and stitched with via matrix to reduce parasitic resistance of such traces. GaN transistors, as it was mentioned above, were placed on the bottom side of the board in accordance to the producer's recommendations [31]. Taking into account that GaN transistors can operate on high frequencies, EMC considerations was implemented on the board. Image of the bottom side of the board with power GaN transistors and other components of one half of T-type circuit is shown on Figure 5. It should be noted that component designators on Figure 5 corresponds to designators on Figure 1. Highlighted components representing current flow in the circuit for switching states marked as *"Positive (+)"* in the Table 1. As it can be seen, such placement of the components on PCB provides as low as possible square of current loops.

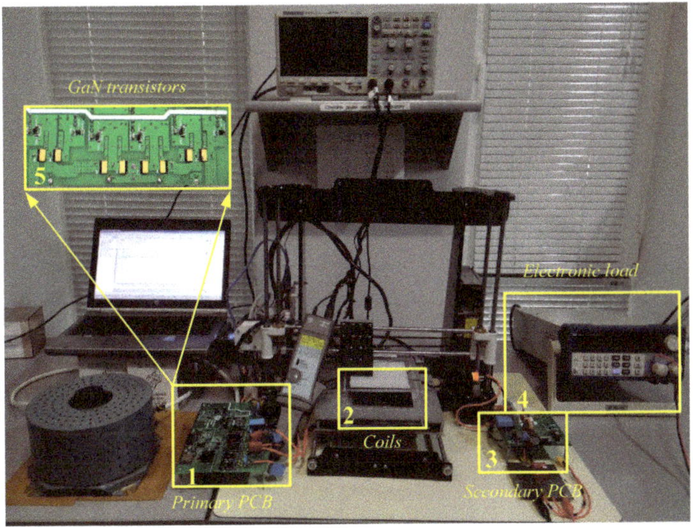

Figure 4. Experimental laboratory GaN-based WPT system: (**1**)—primary side Printed Circuit Board (PCB); (**2**)—transmitting and receiving coils; (**3**)—secondary side PCB; (**4**)—electronic dc load; (**5**)—GaN transistors located on the bottom of the primary PCB.

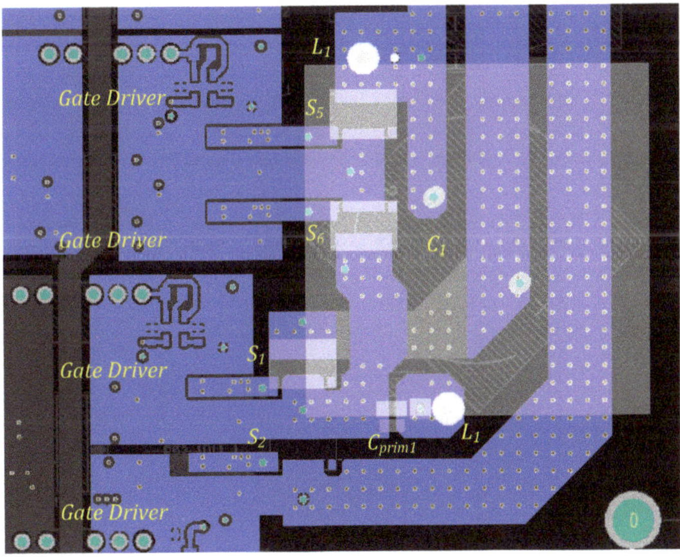

Figure 5. Bottom side of the power part of primary PCB.

The similar square of current loops on PCB are also provided for other switching states, presented in the Table 1. Therefore, such small squares of the current loops in all possible operation modes of the inverter provide low electromagnetic interferences of the power converter, improving EMC of WPT.

The experiments were performed at a distance between the coils of 1 cm (coupling coefficient = 0.9) and at a distance of 2 cm ($k = 0.7$)—Table 2. The dependences of the output parameters on the operating frequency (150 and 200 kHz) were investigated at different load resistances for each of these distances.

A wirewound resistor was connected for power distribution since the available electronic load has a maximum power of 300 W.

Combinations of the two frequencies described above and the two coupling coefficients were investigated. Figure 6, for example, shows the cases at a coupling coefficient of 0.7.

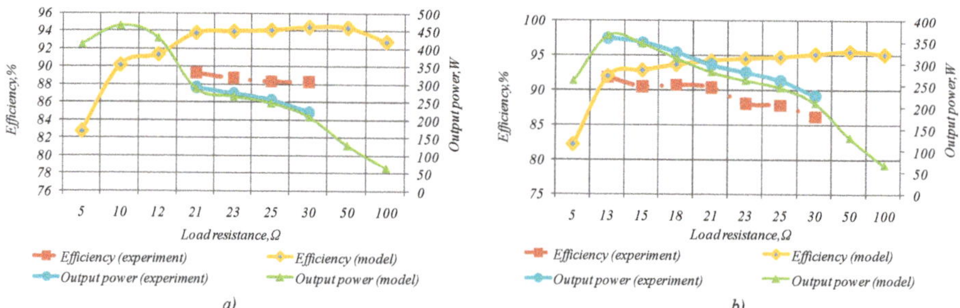

Figure 6. Experimental and simulation dependencies at the load resistance variable at $k = 0.7$: (a) efficiency and output power on the load resistance for $f = 150$ kHz; (b) efficiency and output power on the load resistance for $f = 200$ kHz.

Figure 6 shows that, in general, the efficiency in the model was slightly higher than in the experiments. However, the experimental efficiency also reached more than 90%. The unevenness of the experimental graphs is explained by the fact that the voltage in the grid can vary constantly during the experiment, while in the model, the desired value is specified.

The investigated maximum transmitted power was 360 W during the experiment under the operating frequency of 200 kHz and load resistance of 15 Ω (Figure 6b). The measured temperatures are shown in Figure 7. The temperature on the transistors surface and coils has almost not changed (40 °C and 42 °C, respectively) during a long-term operation of the circuit at this power. This indicates to the power reserve in these elements (they can withstand more power). The temperature increased to 89 °C on the rectifier diodes. Obviously, the losses in the diodes are the largest of the whole scheme, which confirms the power losses calculations. The heating temperature can be reduced by using a larger radiator or forced air cooling. As a conclusion, the diode losses are the main limitation factor of the further switching frequency increasing.

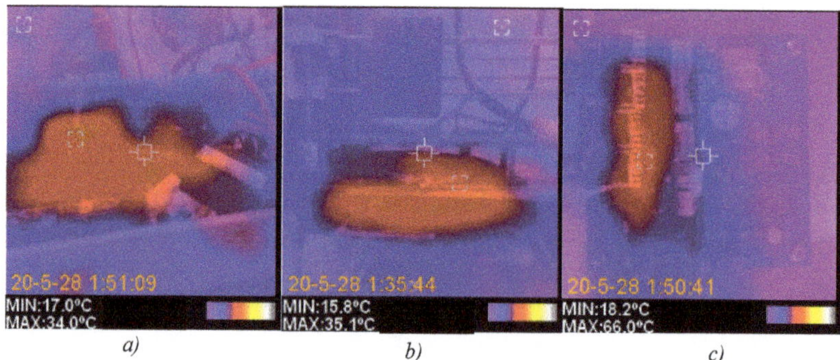

Figure 7. Pictures from the thermal camera at $R_L = 15$ Ω, $k = 0.7$, $f = 200$ kHz, $V_{in} = 300$ V: (a) transistors temperature; (b) coils temperature; (c) temperature of rectifying diodes.

Large coupling coefficients has been studied to obtain more power (Figure 8). Higher power with a larger duty cycle is of course due to the longer time when the transistors are open and conducting.

The duration of the zero voltage level is minimal under such conditions (Figure 9). The change in the duty cycle has little effect on the efficiency. The dead-time of the transistors was selected taking into account the maximum efficiency of energy transfer. Dead-time in the transistors was set less than 5% of the period.

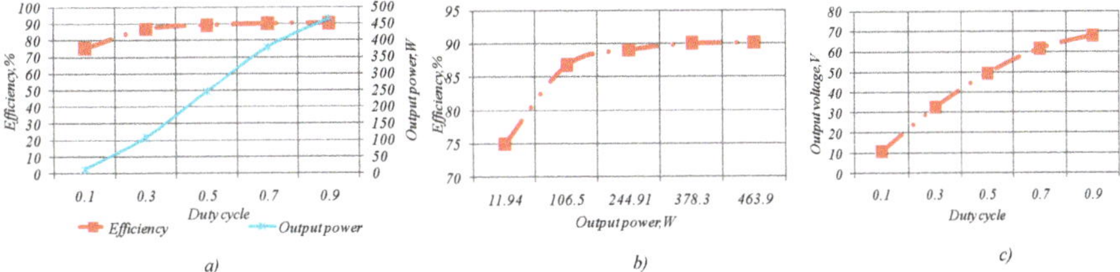

Figure 8. Experimental dependencies at $R_L = 10\ \Omega$, $k = 0.9$, $f = 150$ kHz, $V_{in} = 300$ V: (**a**) efficiency and output power on the duty cycle; (**b**) efficiency on the output power with the duty cycle change; (**c**) output voltage on the duty cycle.

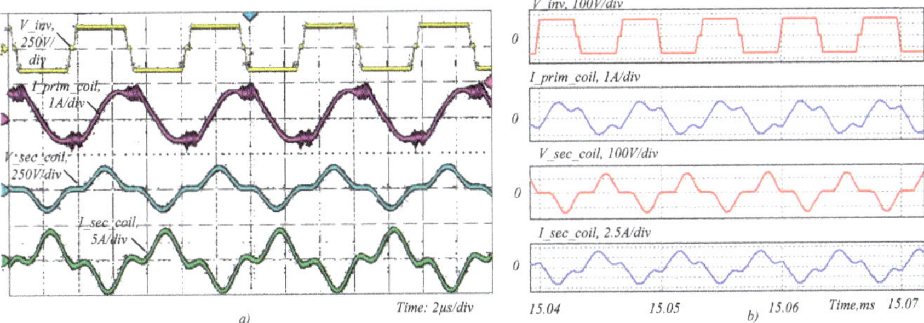

Figure 9. The waveforms on the primary and secondary sides at $R_L = 30\ \Omega$, $k = 0.9$, $f = 150$ kHz, $V_{in} = 300$ V: (**a**) experiment; (**b**) simulation.

The shape of the signals in the experiment is shown in Figure 9 for one of the used cases. In particular, the voltage at the output of the inverter (V_inv), the current through the primary coil (I_prim_coil), the voltage (V_sec_coil) and current on the secondary coil (I_sec_coil) were recorded.

The shape and the magnitude of the currents and the voltages are very close. The resonance condition is fulfilled.

The influence of the snubber capacity on the power and energy transfer efficiency was also investigated experimentally (Figure 10).

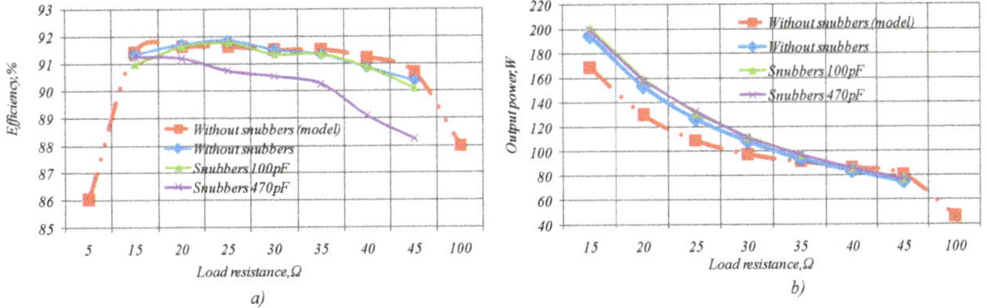

Figure 10. Experimental and simulation dependencies at the load resistance variable at $k = 0.9$, $f = 150$ kHz, $V_{in} = 300$ V: (**a**) efficiency on the load resistance; (**b**) output power on the load resistance.

The snubber capacitance was designed to reduce voltage peaks when the transistor is turned on and off, especially at voltage levels close to critical. Snubber capacitors of different capacities were alternately installed between the drain-source leads of the transistors. The addition of a small snubber capacity in the bridges of the T-type scheme has almost no effect on the energy efficiency (Figure 10a). The effect was more noticeable with an increasing capacitor value and an increasing operating frequency of the transistors.

The distribution of losses is shown at changing the load resistance for the case $k = 0.7, f = 150$ kHz, $V_{in} = 300$ V (Figure 11). The circuit model made in PSIM simulation is an exact copy of the experimental circuit. The data were taken from the model to calculate the losses in the circuit elements.

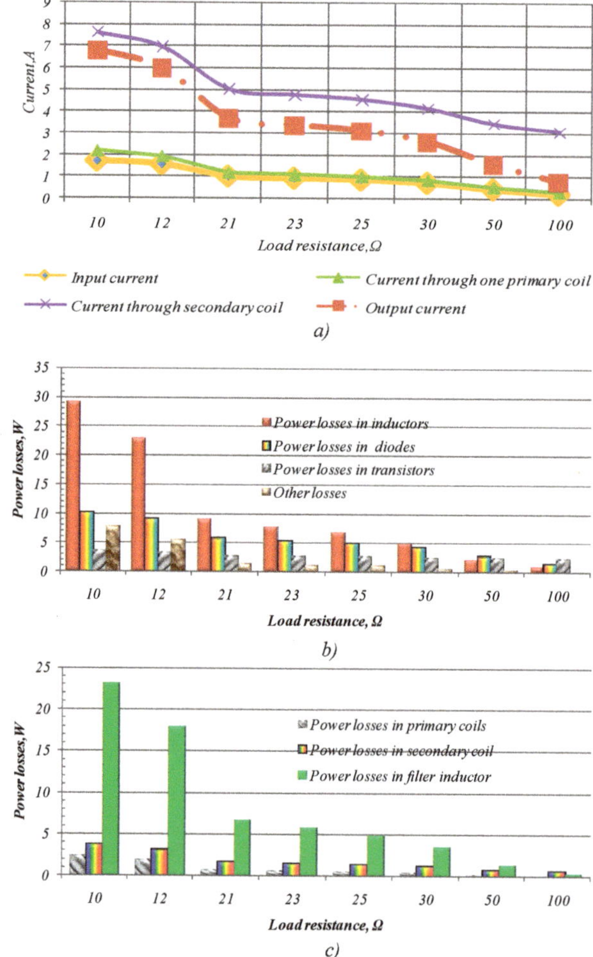

Figure 11. Calculation of power losses at $k = 0.7, f = 150$ kHz, $V_{in} = 300$ V and change of load resistance: (**a**) dependencies of currents through elements on the load resistance; (**b**) distribution of losses by major groups; (**c**) power losses in inductors.

The losses in the inductors and diodes are decreasing significantly with an increasing resistance (and hence with a decreasing current: see Figure 11a. Evidence for this is shown in the charts (Figure 11b,c). Other losses are also decreasing with increasing load resistance for the same reasons.

The total power losses in the transistor remain at approximately the same level in this case. It means that GaN transistors do not reach the maximum current, which they can pass through themselves.

The losses distribution in the semiconductor and magnetic components of the circuit for the same case $k = 0.7, f = 150$ kHz, $V_{in} = 300$ V at a load resistance of 10 Ω and 50 Ω are compared in more detail in Figure 12. The static losses in the inductors and diodes are significant (predominant) (Figure 12a). The reason is that with less resistance there is more input and output current. The total calculated losses are 84% relative to other losses. This ratio is already 95% in Figure 12 that confirms the statement described in [32]. It can be explained due to a significant decrease in the current value through all elements of the circuit, especially the output current.

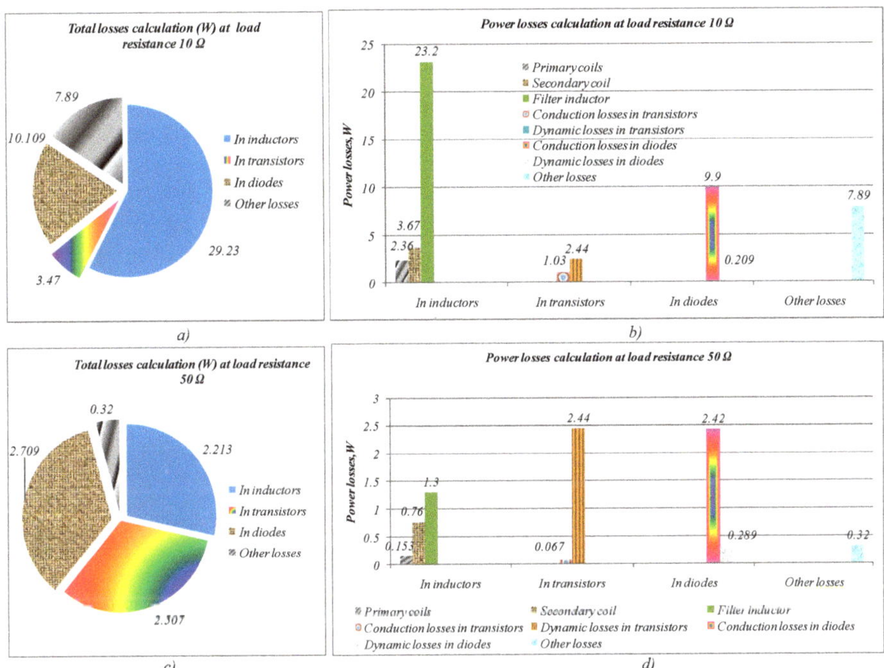

Figure 12. Power losses calculation at $k = 0.7, f = 150$ kHz, $V_{in} = 300$ V: (**a**) distribution of losses by major groups for the load resistance 10 Ω; (**b**)detailed distribution of losses for the load resistance 10 Ω; (**c**) distribution of losses by major groups for the load resistance 50 Ω; (**d**)detailed distribution of losses for the load resistance 50 Ω.

The current in the secondary coil is higher due to the lower resistance in the secondary coil and the value of the self-inductance (turns ratio). This causes greater losses compared to the primary coils (Figure 12b,d).

The effect of dynamic losses in the transistor is less noticeable with an increasing transmission power. It means that the contribution of the transistor dynamic losses to the total power losses will be smaller (because they are almost constant due to the large value of P_{oss}).

The dynamic losses of the diode are not significant (Figure 12b,d). They are dependent on the change of output voltage and the operating frequency according to the expression (15) and do not depend on the current.

In summary, in this topology and these transistors, it is possible to obtain an efficiency of up to 95% and a power greater than that shown in this article. This requires very accurate instruments for measuring data and more careful selection of rectifier diodes.

6. Conclusions

The paper analyzes the feasibility of utilizing of GaN transistor in the three-level T-type NPC inverter with two coupled transmitting coils for WPT. The promising T-type topology of the inverter was selected because of the advantages described in the paper over the classical solutions.

To study the system operation, a detailed analysis of the calculation of static and dynamic losses in the transistors and rectifier diodes and static losses in the inductors was conducted. A series of experiments followed, focused on changing different input and output parameters and calculated power losses in the magnetic components and semiconductors. It was established that the greatest losses have concentrated in the magnetic components and rectifying diodes. These losses are mainly of conduction nature, caused by the significant current through these elements. At the same time, the total losses in the transistors are the smallest compared to all other losses on the circuit elements, which shows that this transistor type is absolutely justified for wireless power transfer in this non-classical circuit. However, these transistors are more critical to overvoltage (surges) on the drain-source than other types of transistors. Due to this fact, the particular attention should be paid to PCB design.

The overall energy transfer efficiency was 90% at the maximum experimentally investigated power of 360 W, which is at the level of industrial samples. The system showed excellent stability at different load resistances and changes in different parameters. At the same time, due to the T-type application, along with GaN transistors, this solution has reduced primary coil and does not have heatsink. It has great potential and can operate at higher power with greater efficiency of wireless power transfer.

7. Patents

Ukrainian patent No. 142050 "Wireless power transfer system based on three-level T-type inverter and two coupled transmission coils".

Author Contributions: Idea, supervising: O.H.; theoretical review and writing: D.S.; analytical calculations, simulations and final writing of paper: V.S.; PCB design: O.V.; software and the experiments: B.P.; final editing, revising: R.S. All authors have read and agreed to the published version of the manuscript.

Funding: This research and paper was funded by Ukrainian Ministry of Education and Science (Grants No. 0117U007260 "A highly effective system for low-voltage charging of low-voltage accumulators" and No. 0118U003865 "High-efficient wireless power transfer systems based on new semiconductor converters topologies") and Ukrainian-Latvian project (Grant No. 0119U102105 "Novel power electronics facilities for wireless power transfer"). Also it was supported by the Estonian Research Council grant PRG675.

Acknowledgments: The authors would like to acknowledge the financial support of the Ukrainian Ministry of Education and Science (Grants No. 0117U007260 and No. 0118U003865) and Ukrainian-Latvian project (Grant No. 0119U102105).

Conflicts of Interest: The authors declare no conflict of interest.

References

1. Patil, D.; McDonough, M.K.; Miller, J.M.; Fahimi, B.; Balsara, P.T. Wireless power transfer for vehicular applications: Overview and challenges. *IEEE Trans. Transp. Electr.* **2018**, *4*, 3–37. [CrossRef]
2. Shevchenko, V.; Husev, O.; Strzelecki, R.; Pakhaliuk, B.; Poliakov, N.; Strzelecka, N. Compensation topologies in IPT systems: Standards, requirements, classification, analysis, comparison and application. *IEEE Access* **2019**, *7*, 120559–120580. [CrossRef]
3. Zahid, Z.U.; Zheng, C.; Chen, R.; Faraci, W.E.; Lai, J.-S.J.; Senesky, M.; Anderson, D. Design and control of a single-stage large air-gapped transformer isolated battery charger for wide-range output voltage for EV applications. In *2013 IEEE Energy Conversion Congress and Exposition*; IEEE: Piscataway, NJ, USA, 2013; pp. 5481–5487.
4. Shevchenko, V.; Husev, O.; Pakhaliuk, B.; Karlov, O.; Kondratenko, I. Coil design for wireless power transfer with series-parallel compensation. In Proceedings of the IEEE 2nd Ukraine Conference on Electrical and Computer Engineering (UKRCON), Lviv, Ukraine, 2–6 July 2019; pp. 401–407.

5. Palmer, P.; Zhang, X.; Shelton, E.; Zhang, T.; Zhang, J. An experimental comparison of GaN, SiC and Si switching power devices. In Proceedings of the IECON 2017-43rd Annual Conference of the IEEE Industrial Electronics Society, Beijing, China, 9 October–1 November 2017; pp. 780–785. [CrossRef]

6. Kumari, K.; Mapa, S.; Maheshwari, R. Loss analysis of NPC and T-type three-level converter for Si, SiC, and GaN based devices. In Proceedings of the 2020 IEEE 9th Power India International Conference (PIICON), Sonepat, India, 28 February–1 March 2020; pp. 1–6.

7. Kuring, C.; Lenth, J.; Boecker, J.; Kahl, T.; Dieckerhoff, S. Application of GaN-GITs in a single-phase T-type inverter. In Proceedings of the PCIM Europe 2018 International Exhibition and Conference for Power Electronics, Intelligent Motion, Renewable Energy and Energy Management, Nuremberg, Germany, 5–7 June 2018; pp. 1–8.

8. Kurumatani, H.; Katsura, S. GaN-HEMT-based three level T-type NPC inverter using reverse-conducting mode in rectifying. In Proceedings of the IEEE 26th International Symposium on Industrial Electronics (ISIE), Edinburgh, UK, 18–21 June 2017; pp. 1941–1946.

9. Anthon, A.; Zhang, Z.; Andersen, M.A.E.; Holmes, D.G.; McGrath, B.; Teixeira, C.A. The benefits of SiC mosfets in a *t*-type inverter for grid-tie applications. *IEEE Trans. Power Electron.* **2017**, *32*, 2808–2821. [CrossRef]

10. Avci, E.; Uçar, M. Analysis and design of grid-connected 3-phase 3-level AT-NPC inverter for low-voltage applications. *Turkish J. Electr. Eng. Comput. Sci.* **2017**, *25*, 2464–2478. [CrossRef]

11. Ueno, H.; Kinoshita, Y.; Yamada, Y.; Suzuki, A.; Ichiryu, T.; Nomura, M.; Fujiwara, H.; Ishira, H.; Hatsuda, T. A 3-phase T-type 3-level inverter using GaN bidirectional switch with very low on-state resistance. In Proceedings of the PCIM Europe 2019, International Exhibition and Conference for Power Electronics, Intelligent Motion, Renewable Energy and Energy Management, Nuremberg, Germany, 7–9 May 2019; pp. 1–4.

12. Pulakhandam, H.; Bhattacharya, S.; Byrd, T. Hybrid operation of a GaN-based three-level T-type inverter for pulse load applications. In Proceedings of the IEEE 7th Workshop on Wide Bandgap Power Devices and Applications (WiPDA), Raleigh, NC, USA, 29–31 October 2019; IEEE: Raleigh, NC, USA; pp. 378–383.

13. Ma, C.-T.; Gu, Z.-H. Review of GaN HEMT Applications in Power Converters over 500 W. *Electronics* **2019**, *8*, 1401. [CrossRef]

14. Zhang, Z.; Wu, X.; Zhang, J. A single phase T-type inverter operating in boundary conduction mode. In Proceedings of the 2016 IEEE Energy Conversion Congress and Exposition (ECCE), Milwaukee, WI, USA, 18–22 September 2016; IEEE: Milwaukee, WI, USA; pp. 1–6.

15. Gurpinar, E.; Castellazzi, A. Single-phase T-type inverter performance benchmark using Si IGBTs, SiC MOSFETs, and GaN HEMTs. *IEEE Trans. Power Electr.* **2016**, *31*, 7148–7160. [CrossRef]

16. Chub, A.; Rabkowski, J.; Blinov, A.; Vinnikov, D. Study on power losses of the full soft-switching current-fed DC/DC converter with Si and GaN devices. In Proceedings of the IECON 2015—41st Annual Conference of the IEEE Industrial Electronics Society, Yokohama, Japan, 9–12 November 2015; pp. 13–18.

17. Lidow, A.; Lidow, A.; Strydom, J.; de Rooij, M.; Reusch, D. *GaN Transistors for Efficient Power Conversion*, 2nd ed.; Wiley: Hoboken, NJ, USA, 2014; p. 250.

18. Wang, B.; Dong, S.; Jiang, S.; He, C.; Hu, J.; Ye, H.; Ding, X.A. Comparative study on the switching performance of GaN and Si power devices for bipolar complementary modulated converter legs. *Energies* **2019**, *12*, 1146. [CrossRef]

19. Aksamit, W.; Rzeszutko, J. Application of GaN transistors to increase efficiency of switched-mode power supplies. *Zeszyty Naukowe Wydziału Elektrotechniki I Automatyki Politechniki Gdańskiej* **2016**, *49*, 11–16.

20. Bosshard, R.; Kolar, J.W.; Mühlethaler, J.; Stevanović, I.; Wunsch, B.; Canales, F. Modeling and η-α-pareto optimization of inductive power transfer coils for electric vehicles. *IEEE J. Emerg. Sel. Top. Power Electron.* **2015**, *3*, 50–64. [CrossRef]

21. Song, K.; Li, Z.; Jiang, J.; Zhu, C. Constant current/voltage charging operation for series–series and series–parallel compensated wireless power transfer systems employing primary-side controller. *IEEE Trans. Power Electr.* **2018**, *33*, 8065–8080. [CrossRef]

22. Kouro, S.; Malinowski, M.; Gopakumar, K.; Pou, J.; Franquelo, L.G.; Wu, B.; Rodriguez, J.; Pérez, M.A.; Leon, J.I. Recent advances and industrial applications of multilevel converters. *IEEE Trans. Ind. Electr.* **2010**, *57*, 2553–2580. [CrossRef]

23. Salem, A.; Abido, M.A. T-type multilevel converter topologies: A comprehensive review. *Arab. J. Sci. Eng.* **2019**, *44*, 1713–1735. [CrossRef]
24. Schweizer, M.; Kolar, J.W. Design and implementation of a highly efficient three-level T-type converter for low-voltage applications. *IEEE Trans. Power Electr.* **2013**, *28*, 899–907. [CrossRef]
25. Liu, J.; Xu, W.; Chan, K.W.; Liu, M.; Zhang, X.; Chan, N.H.L. A three-phase single-stage AC–DC wireless-power-transfer converter with power factor correction and bus voltage control. *IEEE J. Emerg. Sel. Top. Power Electron.* **2020**, *8*, 1782–1800. [CrossRef]
26. Liang, P.; Wu, Q.; Brüns, H.; Schuster, C. Efficient modeling of multi-coil wireless power transfer systems using combination of full-wave simulation and equivalent circuit modeling. In Proceedings of the IEEE International Symposium on Electromagnetic Compatibility and 2018 IEEE Asia-Pacific Symposium on Electromagnetic Compatibility (EMC/APEMC), Singapore, 14–17 May 2018; pp. 466–471.
27. Ni, B.; Chung, C.Y.; Chan, H.L. Design and comparison of parallel and series resonant topology in wireless power transfer. In Proceedings of the IEEE 8th Conference on Industrial Electronics and Applications (ICIEA), Melbourne, Australia, 19–21 June 2013; pp. 1832–1837.
28. Zhang, W.; Mi, C.C. Compensation topologies of high-power wireless power transfer systems. *IEEE Trans. Veh. Technol.* **2016**, *65*, 4768–4778. [CrossRef]
29. Zhang, W.; Wong, S.-C.; Tse, C.K.; Chen, Q. Analysis and comparison of secondary series- and parallel compensated inductive power transfer systems operating for optimal efficiency and load-independent voltage-transfer ratio. *IEEE Trans. Power Electron.* **2014**, *29*, 2979–2990. [CrossRef]
30. Muhlethaler, J. Modeling and Multi-Objective Optimization of Inductive Power Components. Ph.D. Thesis, Swiss Federal Institute Technology in Zurich, ETHZ, Zurich, Switzerland, 2012.
31. GN009 Application Note. PCB Layout Considerations with GaN E-HEMTs, GaN Systems. 2019. Available online: https://gansystems.com/wp-content/uploads/2019/01/GN009-PCB-Layout-Considerations-with-GaN-E-HEMTs_20190118.pdf (accessed on 29 May 2020).
32. Rodriguez, J.; Lai, J.-S.; Peng, F.Z. Multilevel inverters: A survey of topologies, controls, and applications. *IEEE Trans. Ind. Electr.* **2002**, *49*, 724–738. [CrossRef]

Article

Aggregated Conducted Electromagnetic Interference Generated by DC/DC Converters with Deterministic and Random Modulation

Hermes Loschi [1,*], Robert Smolenski [1], Piotr Lezynski [1], Douglas Nascimento [1] and Galina Demidova [2]

[1] Institute of Automatic Control, Electronics and Electrical Engineering, the University of Zielona Gora, 65-417 Zielona Góra, Poland; r.smolenski@iee.uz.zgora.pl (R.S.); p.lezynski@iee.uz.zgora.pl (P.L.); eng.douglas.a@ieee.org (D.N.)

[2] Faculty of Control Systems and Robotics, ITMO University, 197101 Saint Petersburg, Russia; galina.demidova@ieee.org

* Correspondence: eng.hermes.loschi@ieee.org

Received: 9 June 2020; Accepted: 13 July 2020; Published: 21 July 2020

Abstract: The assessment of electromagnetic compatibility (EMC) is important for both technical and legal reasons. This manuscript addresses specific issues that should be taken into account for proper EMC assessment of energy systems that use power electronic interfaces. The standardized EMC measuring techniques have been used in a laboratory setup consisting in two identical DC/DC converters with deterministic and random modulations. Measuring difficulties caused by the low frequency envelopes, resulting from frequency beating accompanying aggregation of harmonic components of similar frequencies, were indicated as a phenomenon that might lead to significant problems during the EMC assessment using currently binding standards. The experimental results describing deterministic and random modulated converters might be useful for practitioners implementing power interfaces in microgrids and power systems as well as for researchers involved in EMC assurance of power systems consisting in multiple power electronic interfaces.

Keywords: conducted electromagnetic interference; electromagnetic compatibility; aggregated electromagnetic interference; power electronic interfaces; frequency beat

1. Introduction

Electromagnetic compatibility (EMC) assessment is demanded for technical as well as legal reasons. EMC evaluation is usually based on the use of the dedicated standards, which determine the permissible limit values for electromagnetic interference (EMI), measurement methods, test equipment and provide classification of products according to their characteristics and electromagnetic environment where they are intended to be used [1]. The shape of conducted EMI depends on the source of the interference as well as complex phenomena accompanying the flow of interference in circuits, including parasitic couplings. In the subject matter literature, some papers emphasize the necessity for assurance of reliable operation of complex energy systems and the need for EMC assurance [2–8]. Furthermore, some papers highlight how approaches concerning deterministic modulation (DetM) and random modulation (RanM), based on the parameters' control of fundamental switching frequency (f_{sw}) and duty cycle (d), may contribute to achieving EMC requirements [9–20]. Indeed, the RanM has been widely used since the 1980s [21]. From the practical viewpoint, beyond the reduction in the maximum level of voltage or current harmonics, the choice for RanM has been considered in order to provide, for instance, reduced of burdensome acoustic noise related to switching frequency [10]. However, some manuscripts have shown that the aggregation of interference in the case of deterministic modulation might be accompanied by low frequency envelopes. This phenomenon may lead to misinterpretations during the EMC assessment [22–25].

According to requirements of the EMC Directive [26] "where apparatus is capable of taking different configurations, the electromagnetic compatibility assessment should confirm whether the apparatus meets the essential requirements in the configurations foreseeable by the manufacturer as representative of normal use in the intended applications". Moreover, the EMC Directive defines responsibility of standard organizations in this context: "The European standardisation organisations should take due account of that objective (including the cumulative effects of the relevant types of electromagnetic phenomena) when developing harmonised standards". Taking into account a global approach to standardization, the issues presented in this paper, concerning aggregation of the conducted electromagnetic interference introduced by power electronic converters with deterministic [27] and random modulation, might constitute a contribution to the elaboration of relevant standards as well as practical information for engineers dealing with assurance of EMC in systems consisting power electronic converters.

As mentioned above, random modulation might contribute to a reduction of maximum levels of EMI spectrum due to more even dispersion of interference over frequency range in comparison with deterministic modulation. Figure 1 shows the EMI measurement of one buck converter topology, with the f_{sw} = 60 kHz, d = 0.5 and with both switch control strategies, DetM and RanM. The EMI measurement presented by Figure 1 was carried out based on the FPGA-based system proposed in [20].

The detailed standard requirements concerning conducted EMI can be found in CISPR 16. Standardized conducted EMI measurements consider the frequency range from 9 kHz to 30 MHz, where the Intermediate Frequency Band Width (IFBW) equal to 200 Hz is applied for the range from 9 kHz to 150 kHz (CISPR A) and IFBW = 9 kHz is applied for the range from 150 kHz to 30 MHz (CISPR B). Since the core concept of the DetM is to provide a f_{sw} constant under the time. The power spectral density is concentrated for frequencies equal to the harmonics of the switching frequency. On the other hand, RanM provides the spreading of interference over frequency range, thus the reduction of maximum observed values is obtained.

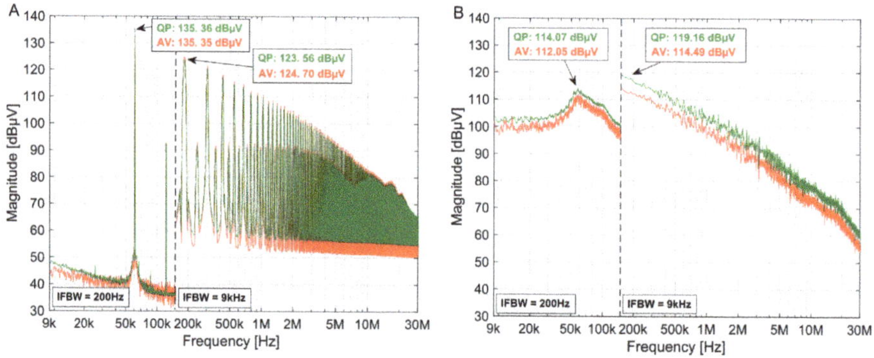

Figure 1. Electromagnetic interference (EMI) measurement of DC/DC converter with the fsw = 60 kHz and d = 0.5: (**A**) for deterministic modulation (DetM) and (**B**) for random modulation (RanM). Results obtained through the FPGA-based system proposed in [20].

The novelty of this paper lies in the presentation of the comparative analysis concerning aggregated interference generated by converters with DetM and RanM. This approach allows us to comprehend the behavior of low-frequency envelopes phenomena beyond the traditional knowledge related with DetM and RanM, i.e., the absence of f_{sw} variation means high disturbance values for the f_{sw} and its harmonics. On the other hand, through the introduction of f_{sw} variation means reduced of disturbances levels. The analyses presented in this paper consider simulations and experimental results based on a standardized testing setup.

2. Simulation Results of Aggregated EMI Generated by DC/DC Converters with Deterministic and Random Modulation

The simulations of DC/DC buck converters with deterministic and random modulation have been run on MatLab software. The function spectrogram was used, and it returns the Short-Time Fourier Transform (STFT) of the aggregated signal with a Hamming window.

Figure 2 shows the results of the simulation in the form of 3D spectrograms. Simulations have been performed for the $f_{sw} = 80$ kHz and $d = 0.5$. The spectrogram (A) shows results for one interference signal generated by a single converter with DetM, while spectrogram (B) shows the aggregated interference introduced by two converters operating in parallel. Since the superimpositions of the switching frequency harmonics can be related to the summation of sinusoidal signals of similar frequency. This process of aggregating sinusoidal components with similar frequencies causes modulation of their amplitudes with low frequency envelopes. This phenomenon is well-known in acoustics as frequency beat. The theory of frequency beats [24] highlights that the sum of the harmonic vibrations with the frequencies f_1 and f_2 of amplitudes equal to 1 can be expressed by:

$$S_2\left(t;\{f_1,f_2\}\right) \;=\; \sin\left(2\pi f_1\,t\right) + \sin\left(2\pi f_2\,t\right) \;=\; 2\cos\left(2\pi\,\frac{f_1-f_2}{2}\,t\right)\sin\left(2\pi\,\frac{f_1+f_2}{2}\,t\right). \quad (1)$$

The frequency beat effect appears when $|f_1 - f_2| \ll f_1 + f_2$. In such conditions, the absolute value

$$\mathrm{Env}_2\left(t;\{f_1,f_2\}\right) = \left|2\cos\left(2\pi\,\frac{f_1-f_2}{2}\,t\right)\right| \quad (2)$$

is the envelope of the aggregated signal. It is also possible to observe that the period of the envelope does not depend on the frequencies of the components, but on the difference between the frequencies of the aggregated signals [24].

The appearance of low frequency envelopes in the case of aggregated interference might cause significant measuring problems.

Additionally, the comparison between spectrograms (A) and (B) in Figure 2 reveals that the maximum observed amplitude is lower in the case of the aggregated interference. However, it should be noted that the power spectral density in a sufficiently wide frequency range and measuring time is higher in the case of aggregated interferences.

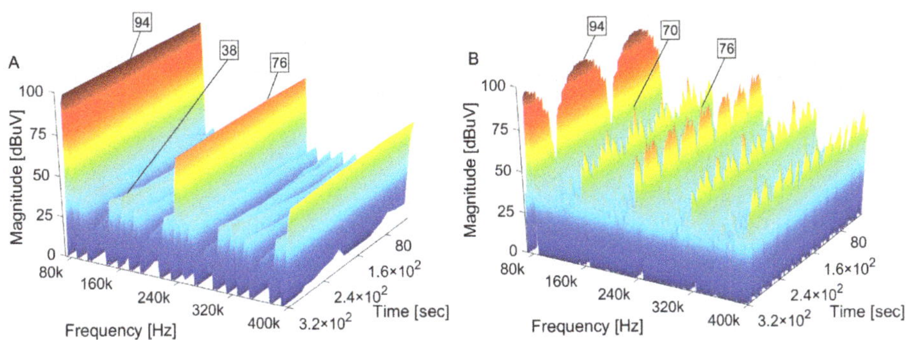

Figure 2. Simulation 3D spectrograms of interference caused by one DC/DC converter with DetM (**A**), and two DC/DC converters with DetM (**B**).

Figure 3 shows the spectrograms corresponding to those presented in Figure 2 with the same parameters, but for random modulation. In both cases of Figure 3, item (A) and (B), the interference power has been spread over the frequency range, and is more even in comparison with DetM, Figure 2. As a result of a more even distribution of interference power, the maximum measured levels have been significantly decreased.

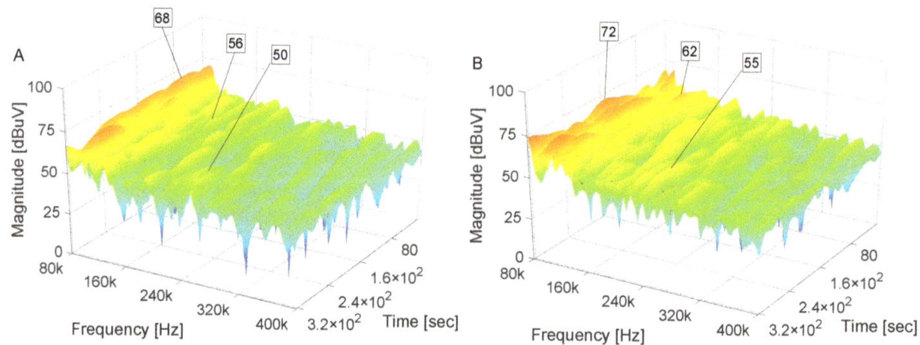

Figure 3. Simulation 3D spectrograms of interference caused by one DC/DC converter with RanM (**A**), and two DC/DC converters with RanM (**B**).

3. Measurements of Aggregated EMI Generated by DC/DC Converters with Deterministic and Random Modulation

In order to confirm the results of the simulation, standardized EMI measurements have been obtained from a laboratory setup fully compliant with EN 55011 based on a voltage probe. Two DC/DC buck-converters constitute the Equipment Under Test (EUT). Both converters are based on a C2-class high speed insulated-gate bipolar transistor (IGBT). The hardware interface for signal and ground are made by the R-Series Multifunction RIO (FPGA PXI-7854R), with VIRTEX-5 LX110. The control signal output (RanM and DetM) is provided at the hardware level by the shielded connector block NI SCB-68A. Figure 4 illustrates the scheme for the measuring testbed.

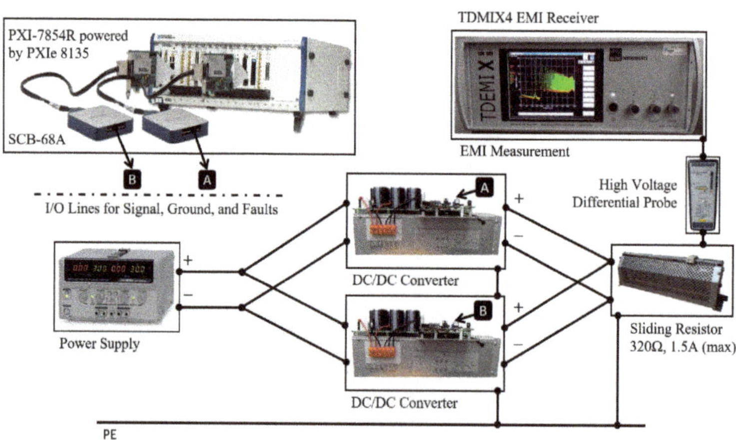

Figure 4. Schematic diagram of measuring testbed.

The schematic diagram presented in Figure 4 shows that both buck-converters are powered by the same regulated laboratory power supply by means of the cables of equal length. In addition, two FPGA control boards were used, and both were powered by controller PXIe 8135 to avoid additional couplings through the power source. A 1.5 A Leybold sliding resistor 320 Ω, was connected as the load on the output of buck-converters (24 V) connected in parallel. In addition, equal length of cables has been applied. The most important parameters of the buck-converter topology have been summarized in Table 1.

Table 1. The main parameters of buck-converter topology.

Component/Function	Specification
Transistors type	IXGH40N60C2D1
I_C(max)	40 A
t_{on}	40 ns
t_{off}	180 ns
Transistor Gate Drivers	HCPL-316J
Converter Power	1800 W (max)
DC capacitors	1500 µF
Max DC voltage	450 V
Load	sliding resistor 320 Ω (max), 1.5 A (max)

The output voltage was measured by a differential voltage probe SI-9010A from Sapphire Instruments (with a 40 dB attenuation level). In both cases, for all presented experimental results, the f_{sw} = 80 kHz and d = 0.5 remained unchanged. Figures 5 and 6 show the measurements obtained using a digital receiver type TDMIX6 EMI, which provides a 3D spectrogram for both Quasi Peak (QP) and Average (AV) detector and CISPR 16-1-1 compliant measurements. In order to increase readability of the figures, measurements have been taken up to 6th harmonic with IFBW = 200 Hz. The experimental results presented in Figure 5 have confirmed the presence of the frequency beat phenomenon observed in simulations. In a case of two converters low frequency envelopes resulting from frequency beat are superimposed on the interference harmonics.

The use of random modulation to disperse interference over the frequency range prevents the frequency beating phenomenon, which appears during aggregation of sinusoidal components of similar frequencies. Thus, in the case of RanM presented in Figure 6 the low frequency envelopes do not appear for aggregated interference introduced by two DC/DC converters connected in parallel Figure 6B.

Generally, the shapes of experimental results, presented in the form of 3D spectrograms, based on data from a laboratory setup, fit well with corresponding 3D spectrograms obtained by simulations. Both simulation and experimental results confirm the theoretical assumptions concerning aggregation of interference for deterministic and random modulation. The obtained results encouraged us to perform multiple measurements according to standard requirements.

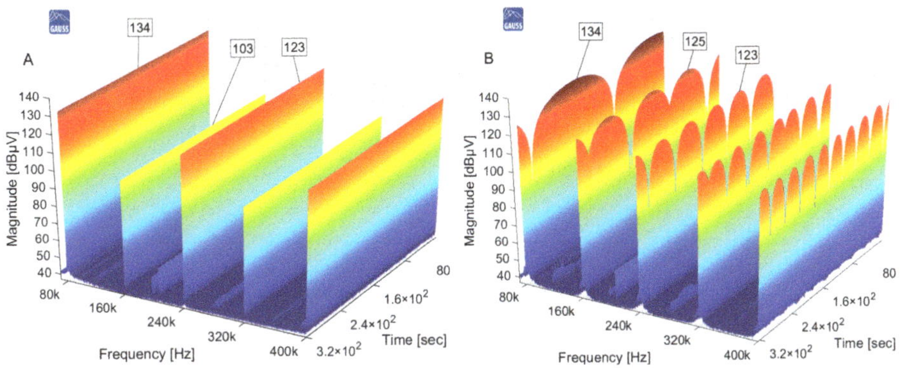

Figure 5. Experimental 3D spectrograms of interference measured using AV detector, caused by one DC/DC converter with DetM (**A**), and two DC/DC converters with DetM (**B**).

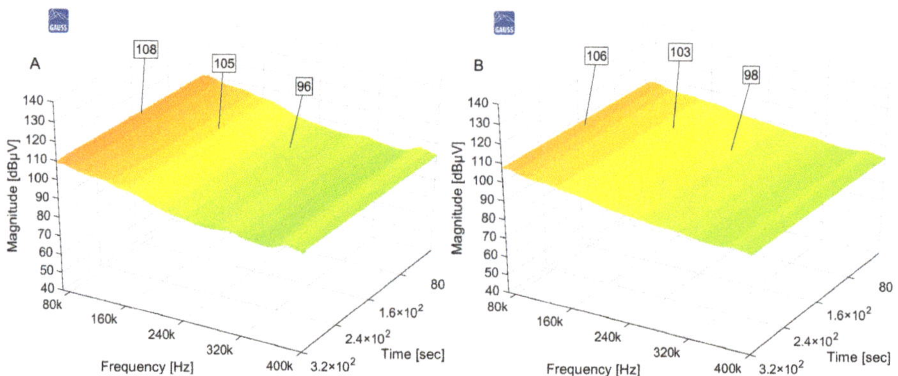

Figure 6. Experimental 3D spectrograms of interference measured using AV detector, caused by one DC/DC converter with RanM (**A**), and two DC/DC converters with RanM (**B**).

4. Statistical Analyses of Aggregated EMI Generated by Converters with Deterministic and Random Modulation Measured According to Standards

In order to present measurement problems connected with the frequency beat phenomenon multiple measurements of the frequency linked with the highest emission were taken. The results of the measurements have been presented in the form of box-and-whisker plots, supplemented with individual values of measured EMI depicted as points. According to standard requirements [28], one final measurement taken during a measuring period equal to 1 s can be compared with the limit line for a presumption of conformity based on harmonized standards. The standards require measurements using QP as well as AV detector. Since the results obtained for both detectors did not differ significantly the presented analyses were based on AV detector measurements only. For each investigated case, 1000 final measurements during 1 s were taken [29].

Figure 7 shows distributions of the results obtained for single DC/DC converters with DetM (**A**) and RanM (**B**). The dispersion of the 1000 results in the case of DetM (**A**) is lower than 0.1 dB. The randomization of the switching frequency caused an increase of the dispersion up to 2 dB. Such distributions of the results confirm that a case of EMI generated by a single DC/DC converter is sufficient for EMI evaluation.

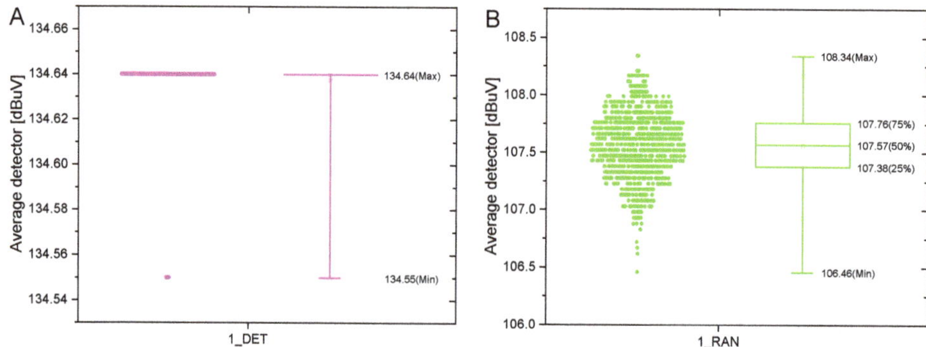

Figure 7. Box-and-whisker plots of 1000 average detector 1 s measurements for one DC/DC converter with: (**A**) deterministic modulation and (**B**) random modulation

The 2 dB dispersion remained unchanged in the case of aggregated interference introduced by two DC/DC converters with random modulation, Figure 8B. However, low frequency envelopes, linked with the frequency beat phenomenon and accompanying aggregation of EMI introduced by converters with deterministic modulation, caused a significant increase in the range of measured levels. The observed differences reached 25 dB (18 times), Figure 8A.

The observations based on the Figures 7 and 8 are confirmed by statistical parameters determined for empirical distributions presented in the figures. The values of variance and standard deviation of measurements in an arrangement consisting of two DC/DC converters are much greater than in other investigated cases (Table 2). The variance and standard deviation calculation, from the EMI measurement viewpoint, represent the dispersion of the measurements of the AV detector, indicating "how far" in general its values are from the expected value. In fact, such dispersion of the results makes evaluation of aggregated EMI, based on one final measurement, in arrangement consisting converters with deterministic modulation unreliable.

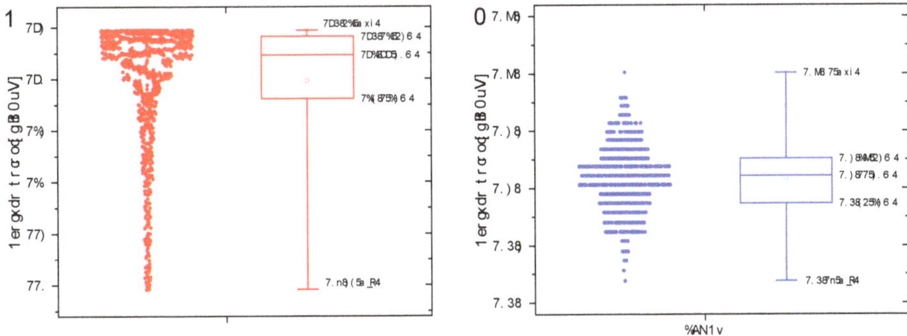

Figure 8. Box-and-whisker plots of 1000 average detector 1 s measurements for two DC/DC converters with: (**A**) deterministic modulation and (**B**) random modulation

Table 2. Statistical parameters of empirical distributions of 1000 final measurements using AV detector.

	Mean	Standard Deviation	Variance	Median	Minimum	Maximum
1_DET	134.64	0.01	0.0002	134.64	134.55	134.64
1_RAN	107.56	0.29	0.08	107.57	106.46	108.34
2_DET	129.82	6.02	36.25	132.33	109.58	134.72
2_RAN	105.09	0.28	0.08	105.11	104.19	106.01

5. Conclusions

In the paper both simulation and experimental results concerning aggregated conducted electromagnetic interference generated by DC/DC converters with deterministic and random modulation have been presented. In the case of deterministic modulation the obtained results have shown that the amplitudes of aggregated interference are modulated with low frequency envelopes caused by the frequency beat phenomenon accompanying summation of sinusoidal components of close frequencies.

The investigation presented in this paper, despite consisting of two identical DC/DC converters, is corroborated by conducted electromagnetic interference in multiconverter systems, as recently investigated by [24], through a real 1 MW photovoltaic power plant. Furthermore, the statistical analyses of large series of final measurement data has confirmed assumptions that low-frequency envelopes might make the standardized EMI tests unreliable.

The research presented has revealed that in the case of random modulation a blurring of instantaneous values of switching frequency contributes to the decreasing of maximum EMI values as well as to the prevention of the frequency beat phenomenon.

Author Contributions: Conceptualization, H.L., R.S., P.L., D.N. and G.D.; methodology, H.L., P.L. and R.S.; validation, formal analysis, visualization H.L. and G.D.; software, investigation, H.L., D.N.; writing—original draft preparation, H.L., R.S., P.L. and G.D.; writing—review and editing, H.L., P.L. and R.S.; supervision, project administration, funding acquisition, R.S. and G.D. All authors have read and agreed to the published version of the manuscript.

Funding: This paper is part of a project that has received funding from the European Union's Horizon 2020 research and innovation program under the Marie Skłodowska-Curie grants agreement No 812391—SCENT, 812753—ETOPIA and in part by the Government of Russian Federation under Grant 08-08.

Conflicts of Interest: The authors declare no conflict of interest.

Abbreviations

The following abbreviations are used in this manuscript:

AV	Average
CISPR	International Special Committee on Radio Interference
DetM	Deterministic Modulation
EM	ElectroMagnetic
EMC	ElectroMagnetic Compatibility
EMI	ElectroMagnetic Interference
EUT	Equipment Under Test
FPGA	Filed-Programmable Gate Array
IEC	International Electrotechnical Commission
IFBW	Intermediate Frequency Band Width
PDF	Probability Density Function
QP	Quasi Peak
RanM	Random Modulation
STFT	Short-Time Fourier Transform

References

1. Braun, S.; Donauer, T.; Russer, P. A real-time time-domain EMI measurement system for full-compliance measurements according to CISPR 16-1-1. *IEEE Trans. Electromagn. Compat.* **2008**, *50*, 259–267. [CrossRef]
2. Loschi, H.; Ferreira, L.; Nascimento, D.; Cardoso, P.; Carvalho, S.; Conte, F. EMC Evaluation of Off-Grid and Grid-Tied Photovoltaic Systems for the Brazilian Scenario. *J. Clean Energy Technol.* **2018**, *6*, 125–133. [CrossRef]
3. Capovilla, C.E.; Casella, I.R.S.; Sguarezi Filho, A.J.; dos Santos Barros, T.A.; Ruppert Filho, E. Performance of a direct power control system using coded wireless OFDM power reference transmissions for switched reluctance aerogenerators in a smart grid scenario. *IEEE Trans. Ind. Electron.* **2014**, *62*, 52–61. [CrossRef]
4. Ma, X.; Zhang, Q.; Luo, Z.; Lin, X.; Wu, G. A novel structure of Ferro-Aluminum based sandwich composite for magnetic and electromagnetic interference shielding. *Mater. Des.* **2016**, *89*, 71–77. [CrossRef]
5. Alam, A.; Mukul, M.K.; Thakura, P. Wavelet transform-based EMI noise mitigation in power converter topologies. *IEEE Trans. Electromagn. Compat.* **2016**, *58*, 1662–1673. [CrossRef]
6. Pyda, D.; Habrych, M.; Rutecki, K.; Miedzinski, B. Analysis of narrow band PLC technology performance in low-voltage network. *Elektron. Elektrotechnika* **2014**, *20*, 61–64. [CrossRef]
7. Shadare, A.E.; Sadiku, M.N.; Musa, S.M. Electromagnetic compatibility issues in critical smart grid infrastructure. *IEEE Electromagn. Compat. Mag.* **2017**, *6*, 63–70. [CrossRef]
8. Smolenski, R.; Bojarski, J.; Kempski, A.; Lezynski, P. Time-domain-based assessment of data transmission error probability in smart grids with electromagnetic interference. *IEEE Trans. Ind. Electron.* **2013**, *61*, 1882–1890. [CrossRef]
9. Lezynski, P.; Smolenski, R.; Loschi, H.; Thomas, D.; Moonen, N. A novel method for EMI evaluation in random modulated power electronic converters. *Measurement* **2020**, *151*, 107098. [CrossRef]
10. Lezynski, P. Random modulation in inverters with respect to electromagnetic compatibility and power quality. *IEEE J. Emerg. Sel. Top. Power Electron.* **2017**, *6*, 782–790. [CrossRef]
11. Lai, Y.S.; Chen, B.Y. New random PWM technique for a full-bridge DC/DC converter with harmonics intensity reduction and considering efficiency. *IEEE Trans. Power Electron.* **2013**, *28*, 5013–5023. [CrossRef]

12. Chatterjee, S.; Garg, A.K.; Chatterjee, K.; Kumar, H. Chaotic PWM spread spectrum scheme for conducted noise mitigation in DC-DC converters. In *2015 International Conference on Energy, Power and Environment: Towards Sustainable Growth (ICEPE)*; IEEE: Piscataway, NJ, USA, 2015; pp. 1–6.

13. Marsala, G.; Ragusa, A. Spread spectrum in random PWM DC-DC converters by PSO&GA optimized randomness levels. In Proceedings of the 2017 IEEE 5th International Symposium on Electromagnetic Compatibility (EMC-Beijing), Beijing, China, 28–31 October 2017; IEEE: Piscataway, NJ, USA, 2017; pp. 1–6.

14. Gamoudi, R.; Chariag, D.E.; Sbita, L. A Review of Spread-Spectrum-Based PWM Techniques—A Novel Fast Digital Implementation. *IEEE Trans. Power Electron.* **2018**, *33*, 10292–10307. [CrossRef]

15. Liou, W.R.; Villaruza, H.M.; Yeh, M.L.; Roblin, P. A digitally controlled low-EMI SPWM generation method for inverter applications. *IEEE Trans. Ind. Informatics* **2013**, *10*, 73–83. [CrossRef]

16. Jing, H.; Weiying, Z.; Jinhong, L. Study on improving EMC of APFC converter with chaotic spread-spectrum technique. In Proceedings of the 2017 IEEE 2nd Advanced Information Technology, Electronic and Automation Control Conference (IAEAC), Chongqing, China, 25–26 March 2017; IEEE: Piscataway, NJ, USA, 2017; pp. 433–437.

17. Arunkumari, T.; Indragandhi, V. An overview of high voltage conversion ratio DC-DC converter configurations used in DC micro-grid architectures. *Renew. Sustain. Energy Rev.* **2017**, *77*, 670–687. [CrossRef]

18. Chen, Y.; Ma, D.B. EMI-Regulated GaN-Based Switching Power Converter With Markov Continuous Random Spread-Spectrum Modulation and One-Cycle on-Time Rebalancing. *IEEE J. Solid-State Circuits* **2019**, *54*, 3306–3315. [CrossRef]

19. Ho, E.N.Y.; Mok, P.K.T. Design of PWM Ramp Signal in Voltage-Mode CCM Random Switching Frequency Buck Converter for Conductive EMI Reduction. *IEEE Trans. Circuits Syst. I: Regul. Pap.* **2013**, *60*, 505–515. [CrossRef]

20. Loschi, H.; Lezynski, P.; Smolenski, R.; Nascimento, D.; Sleszynski, W. FPGA-Based System for Electromagnetic Interference Evaluation in Random Modulated DC/DC Converters. *Energies* **2020**, *13*, 2389. [CrossRef]

21. Legowski, S.; Trzynadlowski, A.M. Hypersonic MOSFET-based power inverter with random pulse width modulation. In Proceedings of the Conference Record of the IEEE Industry Applications Society Annual Meeting, San Diego, CA, USA, 1–5 October 1989; Volume 1, pp. 901–903.

22. Bendicks, A.; Dörlemann, T.; Frei, S.; Hees, N.; Wiegand, M. Active EMI reduction of stationary clocked systems by adapted harmonics cancellation. *IEEE Trans. Electromagn. Compat.* **2018**, *61*, 998–1006. [CrossRef]

23. Smolenski, R.; Bojarski, J.; Kempski, A.; Luszcz, J. Aggregated conducted interferences generated by group of asynchronous drives with deterministic and random modulation. In Proceedings of the 2012 Asia-Pacific Symposium on Electromagnetic Compatibility, Singapore, Singapore, 21–24 May 2012; pp. 293–296.

24. Smolenski, R.; Lezynski, P.; Bojarski, J.; Drozdz, W.; Long, L.C. Electromagnetic Compatibility Assessment in Multiconverter Power Systems-Conducted Interference Issues. *Measurement* **2020**. [CrossRef]

25. Schmidt, S.; Richter, M.; Oberrath, J.; Mercorelli, P. Control oriented modeling of DCDC converters. *IFAC-PapersOnLine* **2018**, *51*, 331–336. [CrossRef]

26. The European Parliament and the Council. Directive 2014/30/EU on the harmonisation of the laws of the Member States relating to electromagnetic compatibility. *Off. J. Eur. Union* **2014**, *L96*, 79–106.

27. Bojarski, J.; Smolenski, R.; Kempski, A.; Lezynski, P. Pearson's random walk approach to evaluating interference generated by a group of converters. *Appl. Math. Comput.* **2013**, *219*, 6437–6444. [CrossRef]

28. International Electrotechnical Commission. Specification for radio disturbance and immunity measuring apparatus and methods—Part 1-1: Radio disturbance and immunity measuring apparatus—Measuring apparatus. In *CISPR 16-1-1:2019*; IEC: Geneva, Switzerland, 2019.

29. Ferrero, A.; Petri, D.; Carbone, P.; Catelani, M. EMC Measurements. In *Modern Measurements: Fundamentals and Applications*; Wiley: Hoboken, NJ, USA, 2015; pp. 317–351.

Article

FPGA-Based System for Electromagnetic Interference Evaluation in Random Modulated DC/DC Converters

Hermes Loschi [1,*], **Piotr Lezynski** [1], **Robert Smolenski** [1], **Douglas Nascimento** [1] **and Wojciech Sleszynski** [2]

[1]   Institute of Automatics, Electronics and Electrical Engineering, the University of Zielona Gora, 65-417 Zielona Gora, Poland; p.lezynski@iee.uz.zgora.pl (P.L.); r.smolenski@iee.uz.zgora.pl (R.S.); eng.douglas.a@ieee.org (D.N.)

[2]   Faculty of Electrical and Control Engineering, Gdansk University of Technology, 80-233 Gdansk, Poland; wojciech.sleszynski@pg.edu.pl

*   Correspondence: eng.hermes.loschi@ieee.org

Received: 20 April 2020; Accepted: 7 May 2020; Published: 11 May 2020

Abstract: Field-Programmable Gate Array (FPGA) provides the possibility to design new "electromagnetic compatibility (EMC) friendly" control techniques for power electronic converters. Such control techniques use pseudo-random modulators (RanM) to control the converter switches. However, some issues connected with the FPGA-based design of RanM, such as matching the range of fixed-point numbers, might be challenging. The modern programming tools, such as LabVIEW, may facilitate the design process, but there are still fixed-point operations and limitations in arithmetic operations. This paper presents the design insights on the FPGA-based EMC friendly control system for DC/DC converter. Probability density functions (PDF) are used to analyse and improve pseudo-random algorithms. The theoretical algorithms, hardware details and experimental results are presented and discussed in terms of conducted electromagnetic interference emission.

Keywords: systems control; electromagnetic compatibility; conducted interference; DC-DC power converters; FPGA; random modulation

---

1. Introduction

Nowadays, with the advent of smart energy environments, the demand for cyber-physical systems [1] and the growth in switch-mode converter applications [2–6], electromagnetic compatibility (EMC) issues are becoming more and more significant [7]. Among many methods of improving EMC of converters, we may use improved control techniques. In addition to the primary function of controlling energy conversion, such a control technique can reduce the level of conducted electromagnetic interference (EMI) by spreading the harmonics on a broader frequency range [6,8]. A Field-Programmable Gate Array (FPGA) may be used for the cyber-physical implementation of such controls. It should be added that FPGA may be much more flexible and may cover more applications than other commonly used control devices such as microcontrollers (μC) or digital signal processors (DSP). In particular, FPGAs allow for the building of hardware circuits of the modified Pulse Width Modulators (PWM), unlike μC or DSP's, where the built-in PWM circuit cannot be changed.

In a typical PWM circuit, the user may set the fundamental switching frequency (fsw) and the duty cycle (D). The parameter D controls the output voltage of the converters and thus affects the energy conversion process. The f_{sw} of the PWM signal is practically irrelevant to the voltage transfer function of the converter, and of the primary energy conversion. However, it affects the losses in the switch-mode converter and the parameters of the reactance elements. Conducted EMI generated by the converter are grouped around the f_{sw} and its harmonics [8,9]. Any modification of f_{sw} leads to changes in EMI emission.

Traditionally, pseudo-random modulators (RanM) can be used as a switch control strategy, which can reduce the level of EMI [8,10]. In such type of modulation, the frequency of the PWM signal is randomly changed in the selected range. As a consequence, disturbance energy is distributed more evenly across a wider spectrum. The development of a random modulation requires a combination of changes in the frequency fsw (or the period) of the PWM signal with a pseudo-random number generator. Therefore, the range of pseudo-random numbers must be adapted to the specific hardware platform. In addition, the probability density function (PDF) of such a pseudo-random stream should be analysed in terms of the emission of conducted disturbances.

The design of RanM in FPGA may be challenging. The modern graphical programming tools such as LabVIEW may facilitate the design process. Nevertheless, in LabVIEW, there is still some inconvenience associated with fixed-point operations, and with the lack of some arithmetic operations, e.g., divide. Therefore, this manuscript demonstrates how to provide an FPGA-based control system for a DC/DC converter that limits the level of conducted EMI. The presented algorithms, based on LabVIEW engineering software, do not change the essential functions or parameters of the converter. During the design of algorithms, we take into account the PDF of frequency changes, and we propose a method of shaping the PDF in FPGA without arithmetic division and using only fixed-point operations. For the presented algorithms, we perform the EMI evaluation in an experimental system. The CISPR-A frequency band was considered as the primary frequency range for tests.

2. FPGA-Based Systems Design

2.1. FPGA Hardware

In the manuscript, we consider National Instruments PXIe-8135 controller as the primary development environment. The PXIe-8135 is an Intel Core i7 embedded controller with the design tool—LabVIEW. The development environment also includes the FPGA R-Series Multifunction RIO - PXI-7854R card (illustration in Figure 1) with VIRTEX-5 LX110 [11].

Figure 1. Field-Programmable Gate Array (FPGA) PXI-7854R illustration [11].

The PXI-7854R FPGA-based card has a few dozen Input/Output (I/O) resources, which include analog/digital converters (ADCs), digital/analog converters (DACs) and digital I/O lines. The design tool LabVIEW accesses the FPGA PXI-7854R device through the bus interfaces (PXI Triggers and PXI BUS). This connection makes possible timing, triggering, processing and custom I/O measurements, based on FPGA target programming, and most of it is available for demanded functions. Those required functions may use varied amounts of logic, besides using I/O resources. We assume that all control techniques require the same I/O resources, such as signals controlling transistors, or input faults. Therefore, the number of FPGA resources, used for logical and arithmetic operations, indicates the algorithmic complexity and it will be considered as one essential parameter to evaluate the control techniques implemented in FPGA.

2.2. PWM Modulator Algorithm

To control the converter switches the PWM modulator is used. Figure 2 shows the main flowchart of the PWM modulator. For better visualization, we divide all operations into three (III) parts. In part I, the user may set the main control parameters: the average switching frequency f_{sw} and duty cycle of the PWM signal D. Then, the program calculates the required number of clock ticks for the period (N) and the duration of high state (N_d) according to the Equations (1) and (2).

$$N = f_{FPGA} / (f_{sw} \cdot SC_{TL}) \tag{1}$$

$$N_d = D \cdot N \tag{2}$$

where:
f_{FPGA}—onboard available clock frequency, f_{sw}—switching frequency and SC_{TL}—single cycle timed loop (in clock ticks).

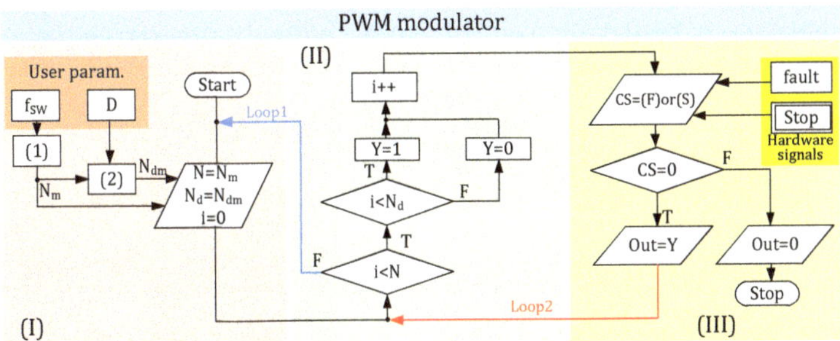

Figure 2. The main flowchart of the PWM modulator, providing parameters (I), counter ramp (II), and generation of output signal (III).

The index m (in Figure 2) means that the N_{dm} and N_m value is taken and recalculated for the m'th PWM period. The presented modulator can, therefore, cooperate with an external voltage controller and dynamically change the factor D. The duration of the period, represented by N_m, may be constant for deterministic modulation (DetM) or may be changed randomly for RanM case. Part II illustrated the carrier function counter ramp represented by the variable i. Parameters N and N_d are compared with the variable i to determine the value of output signal Y and control the loop I execution. Part III presents the generation of control (output) signal and the interaction with a CS function. In the presence of hardware signals of fault (F) or the stop button (S), the function CS returns 1, and the PWM operation is interrupted. The fault signal is typically generated by converter hardware in any emergency, while the stop signal is from the user. Figure 3 shows the operation of the modulator.

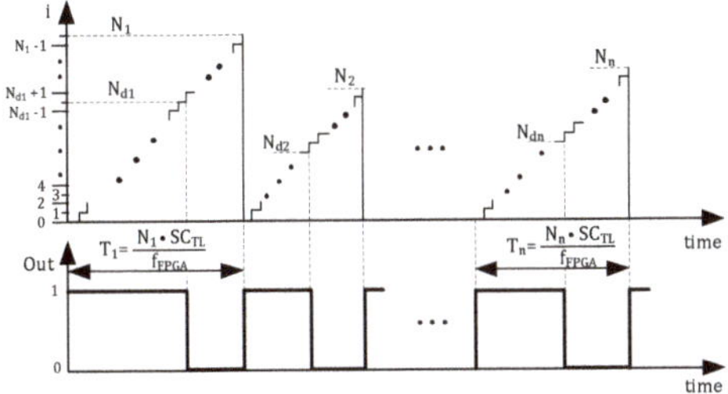

Figure 3. The principle of the PWM operation.

2.3. Random Number Generator

The main advantage of RanM instead DetM is highlighted in the literature [8,10,12–15] as the EMI noise reduction related to the f_{sw} and its harmonics. The random number generator is crucial to the operation of RanM control algorithms. Since we consider FPGA features connected with the fixed-point operation, we propose to generate the random stream by a linear congruential generator (LCG). The LCG is a modular arithmetic algorithm, which provides a stream of pseudo-randomized numbers calculated with a Equation (3).

$$RN_m = (\alpha \cdot RN_{m-1} + c) \, mod \, n \tag{3}$$

where:

RN_m—the m'th random number provided by LCG, α, c, n—LCG coefficients.

The values of coefficients configured for the FPGA implementation were $\alpha = 17$ and $RN_0 = 17$, $c = 0$ and $n = 2^{32}$. The modulo operation is made automatically due to the use of long integer variables with 32 bits of storage. For RanM, the generation of a random stream is done in a loop to get a new pseudo-random number for each PWM signal period. The presented LCG generates pseudo-random numbers with a period of 2^{31} and a normal distribution. To show the property of a random stream, Figure 4 shows the 10,000 numbers RN_m generated by LCG and their histogram/distribution.

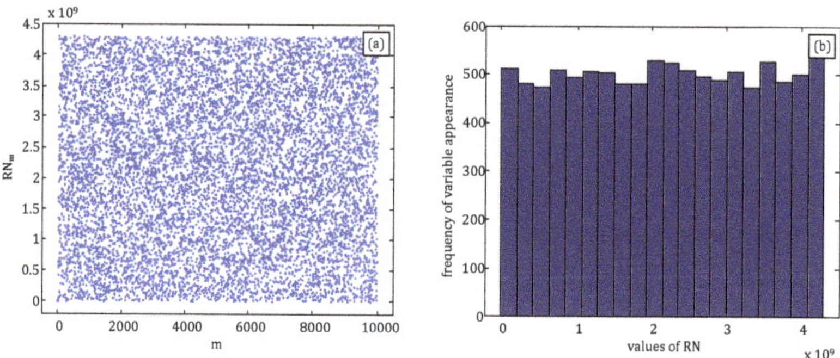

Figure 4. RN_m distribution (**a**), and histogram (**b**).

2.4. FPGA Implementation

This section presents the practical implementation of DetM and RandM in FPGA systems utilizing LabVIEW software. Additionaly, for RanM probability density function PDF analysis in MATLAB software will be shown.

Traditionally, during LabVIEW programming, we use modules with palettes of structures and functions. The number of such functions in LabVIEW FPGA Module is much smaller than for typical control devices with a microprocessor. Further, available functions can only operate on integer variables. Despite these drawbacks, the LabVIEW program Implementation in FPGA systems allows full control of program execution and high speed calculation. For instance, the LabVIEW FPGA Module has access to Single-Cycle Timed Loops (SCTL). The SCTL is a unique loop structure that executes all functions inside within one fixed time period (SC_{TL}) [11]. The SC_{TL} may be defined by the user as a time period or in ticks of the FPGA clock. In the control board used, the maximum clock frequency is 40 MHz.

To realize the algorithm from Figure 2 in LabVIEW for FPGA, the While Loop structure with a sub-system For Loop was chosen. The For Loop corresponds to the loop II in Figure 2 and is executed N times. The While Loop corresponds to the loop I in Figure 2. The While Loop is performed until the CS function is activated. For DetM we consider the $SC_{TL} = 1$ tick, and the number $N = N_m$ is constant.

2.4.1. Single Randomization

For RandM the number N_m is randomly changed for each PWM period using a random stream from the LCG (described in previous section). However, due to the use of fixed-point operations in FPGA, the resulting random number must be scaled (reduction of bit precision) to make it suitable for N_m calculation. Figure 5 illustrates in the part I how to scale the RN_m using the Reshaped Array Function (R_{AF}) which is available in the LabVIEW environment.

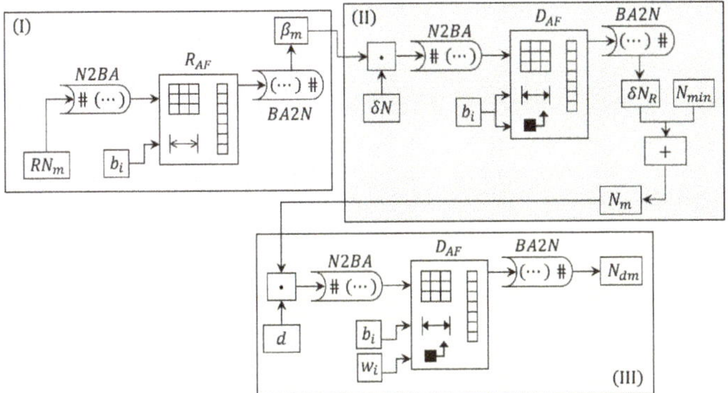

Figure 5. The scheme of the RN_m scaling process, led to lower bit precision (**I**), and for calculating the random stream of N_m (**II**) and N_{dm} (**III**).

Figure 5 shows in part I the parameter b_i, which is responsible for providing the length of the output array and must be numeric. Before resizing, the number RN_m is converted to a Boolean Array by the $N2BA$ function. After resizing, the Boolean Array is converted to a number again by the $BA2N$ function. As a result, a random number, β_m, from 0 to $2^{b_i} - 1$ is obtained. The next task is to calculate N_m and N_{dm} (from Figure 2), to change randomly according to the random β_m number for each PWM period. The changes of the N_m and N_{dm} should also take place within the assumed range from the N maximum and minimum values. These operations should be performed with the appropriate precision in fixed-point arithmetic. However, LabVIEW software for FPGA does not allow direct actions of the mathematical division. The solution to the stated problem for N_m calculation is shown in part II of the Figure 5. The δN describes the maximum assumed changes in the PWM period and may be calculated as:

$$\delta N = N_{max} - N_{min} + 1. \tag{4}$$

The δN can be also represented relative to the average value of N, e.g., $\delta N = 50\% \cdot N_{AV} + 1$. Thus, for the first interaction illustrated in Figure 5, part II, consider the multiplication of the β_m by δN. Since a divide operation is not available, the Delete from Array Function (D_{AF}) is applied to provide δN_R, a random stream that corresponds to the range between the 0 and the value of $\delta N - 1$. In this step, the N_m value is calculated using Equation (5). The final formula for calculating N_m according to the concept from Figure 5, part II, is presented in Equation (6).

$$\delta N_R = (((RN_m \gg (32 - b_i)) \cdot \delta N) \gg b_i) \tag{5}$$

$$N_m = \delta N_R + N_{min} \tag{6}$$

Although Equation (6) accurately describes random changes in the N value (which corresponds to the PWM period—T_{PWM}), it does not provide information on the average value. Therefore, later in the article, discussing RanM properties, we will give the average value of $N - N_{AV}$, and δN, so the variation range of N is equal to $N_{AV} \pm \delta N/2$. Choosing the right δN is not apparent and should be the subject of a broader analysis. Despite, this topic will not be fully discussed in this article. However, we will show the basic challenges that face when choosing parameters for RanM.

We can take δN value based on expected frequency randomization. For instance, if we assume that $f_{sw} = 80$ kHz and this frequency may change $\pm 50\%$, we will obtain $max(f_{sw}) = 80$ kHz $+ 50\% = 120$ kHz, $min(f_{sw}) = 80$ kHz $- 50\% = 40$ kHz. The N_{min} value will be related to 120 kHz and N_{max} to 40 kHz (based on Equation (1)), so $\delta N = 1000 - 333 + 1 = 668$. However, the value of $N_{AV} = 667$ which does not correspond to a frequency of 80 kHz. Figure 6 shows the histogram of randomized N_m values and the histogram of randomized frequencies f_{sw} related to N_m, for $N_{AV} = 667$ with $\delta N = 668$. Figure 6b shows that switching frequency f_{sw} varies within the assumed range from 40 to 120 kHz. Despite this, the frequency distribution is not uniform and the average frequency is not equal to 80 kHz. Adopting the δN in such a way is therefore not particularly useful. Another possible approach to selecting δN value may be made based on a relative change in time. Figure 7 shows the histogram of N_m values and histogram of f_{sw}, for $N_{AV} = 500$ (value related to $f_{sw} = 80$ kHz) with $\delta N = 334$ so $N \in < N_{AV} \pm 30\% >$. In such a case, the average value of the frequency is closer to that intended, but the frequency distribution is still strongly non-linear.

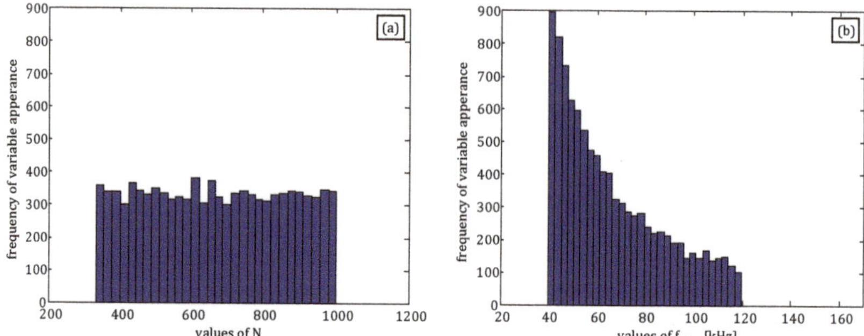

Figure 6. Histogram of N_m ticks distribution (**a**), and histogram of frequency distribution (**b**), for $N_{AV} = 667$ and $\delta N = 668$.

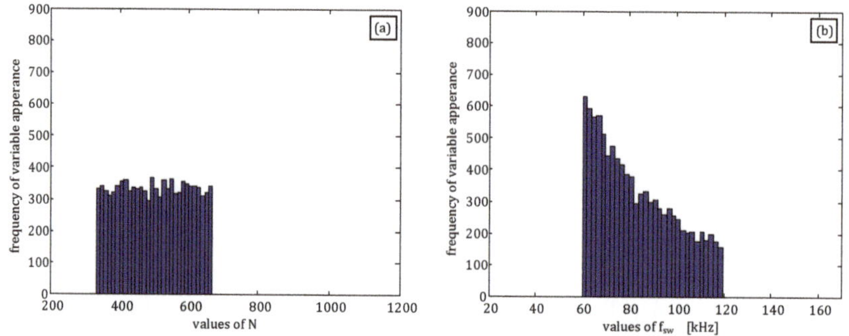

Figure 7. Histogram of N_m ticks distribution (**a**) and histogram of frequency distribution (**b**), for $N_{AV} = 500$ and $\delta N = 333$.

Figure 5 shows in part III the D_{AF} function with consideration of the set up of length and index of array, to provide calculation of N_{dm}. We assumed that the coefficient d, represented as an 8 bit integer will be proportional to duty cycle factor D. The FPGA implementation considers the index value (w_i) equal to 8, and the b_i equal to 23. Thus, the final value of N_{dm} is configurable, maintaining the proportionality with N_m, within a range defined by d, as follows in Equation (7).

$$N_{dm} = D \cdot N_m = ((d \cdot N_m) >> w_i) = \frac{d \cdot N_m}{2^{w_i}} \tag{7}$$

To calculate the period of PWM signal and its Duty cycle for RanM we use Equations (6) and (7), respectively. These equations are computed in the fixed time of a one loop execution—SC_{TL}. We may consider the situation, in which the time SC_{TL} is also changed randomly (RSC_{TL}).

2.4.2. RanM with RSC_{TL}—Additional Randomization

When we randomly change both the number of periods of the For Loop (N_m) and the duration of this loop, we can talk about an additional randomization (RanM with RSC_{TL}). Basically, for RSC_{TL} generation, we assume the same principle, as illustrated in Figure 5, part I and part II, however instead of $SC_{TL} = 1$ tick we generate random value from 7 to 13. The m'th PWM period T_{PWMm} is equal to $(N_m \cdot SC_{TLm})/f_{FPGA}$. Figure 8 shows the histogram of the T_{PWMm} and histogram of related f_{sw}, for $N = 50$, $\delta N = 34$ and $RSC_{TL} \in< 7 : 13 >$. Figure 8 shows that when we consider N_m and RSC_{TL}, the density distribution in both cases, (a) and (b), is changed. This approach of N_m and RSC_{TL} randomization gives us the possibility of shaping the density distribution. The frequency distribution is closer to the Gaussian distribution, which should be more favourable in terms of the average frequency and converter losses.

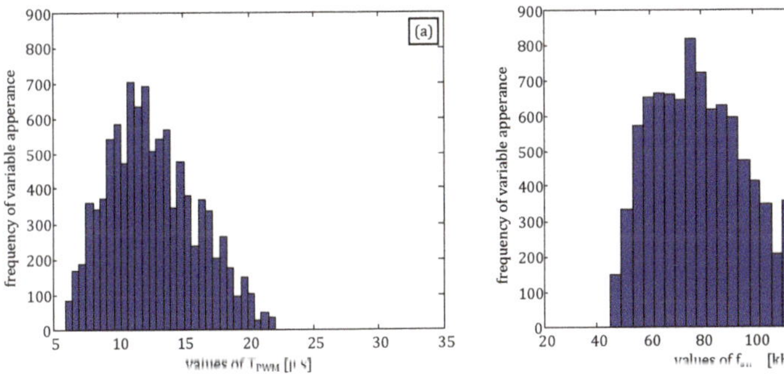

Figure 8. Histogram of T_{PWMm} distribution (**a**) and histogram of related f_{sw} distribution (**b**), for $N_{AV} = 50$ with $\delta N = 34$, and $RSC_{TL} \in< 7 : 13 >$.

2.4.3. RanM2—Split Distribution of Variable

The other method to shape the T_{PWMm} and f_{sw} distributions is to split the N_m distribution to a few sub-ranges in the entire Nm range. For the presented FPGA implementation, we propose to use two predefined random sequence ranges. Then we consider the use of one random bit, digital 1 and 0 levels to choose the range of N_m. The FPGA implementation is executed with a particular function (SF), which is available in the LabVIEW environment. The parameter S determines whether the SF returns the value wired to T or F. Thus, for each parameter, F or T, we assign one of the two predefined random sequence ranges, both provided by Nm. Figure 9 illustrating the proposed FPGA implementation. We will denote random modulation with such a distribution as RanM2.

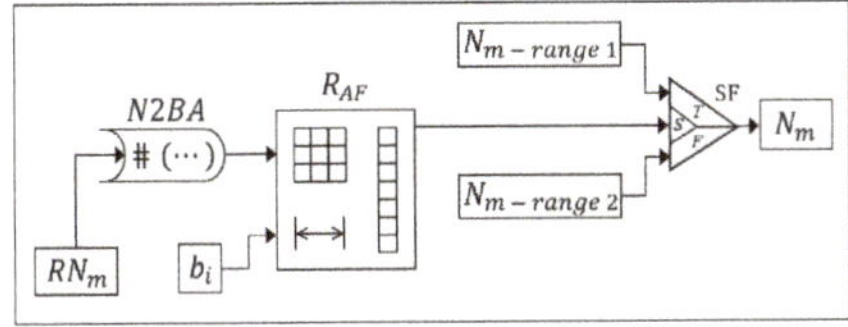

Figure 9. Illustration of N_m generation in proposed RanM2.

Figure 10 shows the histograms of Nm values and their corresponding f_{sw} values, for $N_{AV1} = 750$ with $\delta N_1 = 500$ and $N_{AV2} = 416$ with $\delta N_2 = 167$.

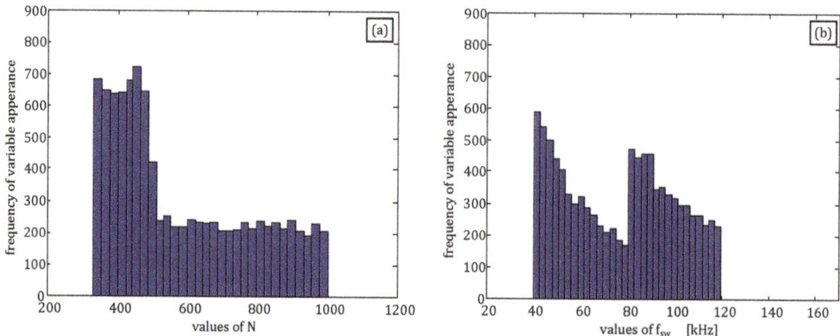

Figure 10. Histogram of N_m ticks distribution (**a**) and histogram of frequency distribution (**b**), for RanM2 with parameters: $N_{AV1} = 750$ with $\delta N_1 = 500$ and $N_{AV2} = 416$ with $\delta N_2 = 167$.

As one can see, the use of two probability distributions for a PWM period (which is proportional to N_m) produces a more equal alignment of the frequency distribution. Therefore, in such a manner there is a possibility to create, with obvious limitations, the distribution of frequency.

2.4.4. RanM2 with RSC_{TL}

The presented concept of two distributions of N_m (RanM2) may be linked with the concept of additional randomization RSC_{TL}. Figure 11 considers such a case with $RSC_{TL} \in\ <7 : 13>$ and for $N_{AV1} = 75$ with $\delta N_1 = 50$ and $N_{AV2} = 42$ with $\delta N_2 = 17$. The obtained histograms are more smooth than histograms in Figure 10, which can be an advantage. However, the use of RSC_{TL} increases the frequency spread.

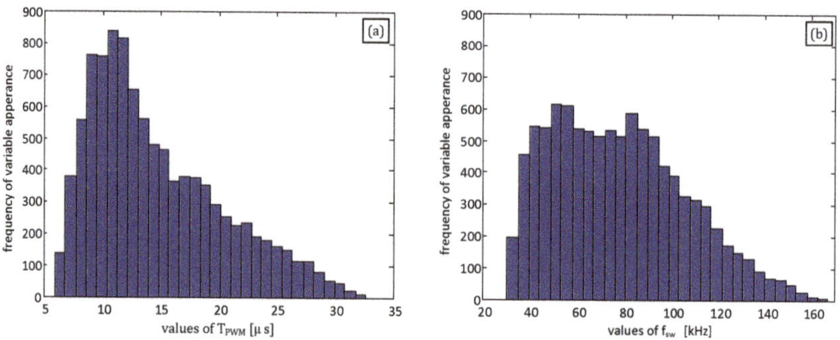

Figure 11. Histogram of N_m ticks distribution (**a**) and frequency distribution (**b**), for RanM2 with RSC_{TL} with parameters: $RSC_{TL} \in\ <7 : 13>$, $N_{AV1} = 75$, $\delta N_1 = 50$, $N_{AV2} = 42$ and $\delta N_2 = 17$.

Figure 12 shows the LabVIEW general program implemented in FPGA. In Figure 12, the parts corresponding to (I) and (II) refer to the basic modulator configuration (Figure 2). Therefore, it is applicable to both DetM and RanM. Part (III) presents the random number generator with a number scaling block, corresponding to part I of (Figure 5) and is used only for RanM, for both approaches of randomization discussed in Section 2.4 Parts (IV) and (V) correspond to part II and part III of (Figure 5), respectively. Since the FPGA needs to execute predefined random sequence ranges (the concept of additional randomization), the D_{AF} function proportionally increases.

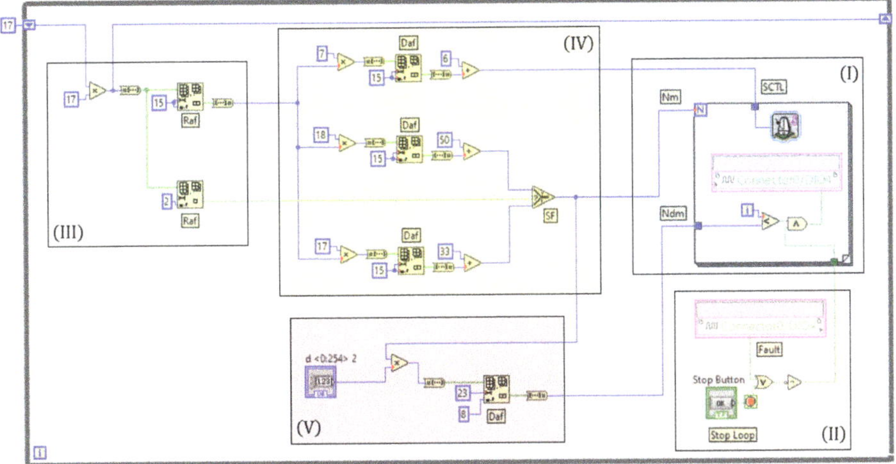

Figure 12. LabVIEW general program implemented in FPGA, for loop (counter ramp) and output signal generation (I), conditions for stopping the program II), random number generator (III), N_m and SC_{TL} calculation (IV), and N_{dm} calculation (V).

3. Experimental Results

This section provides the results of an experimental system. All presented analyses and measurements concern a buck-converter topology, with a C2-class high speed insulated-gate bipolar transistor (IGBT), and a hardware interface for signal and ground to the R-Series Multifunction RIO (FPGA PXI-7854R). The control signal output (RanM or DetM) is provided, at the hardware level, by the NI SCB-68A shielded connector block. Combined with the shielded cables, the SCB-68A provides rugged, very low-noise signal termination to the transistor gate drive. Figure 13 illustrates the scheme of the measuring testbed.

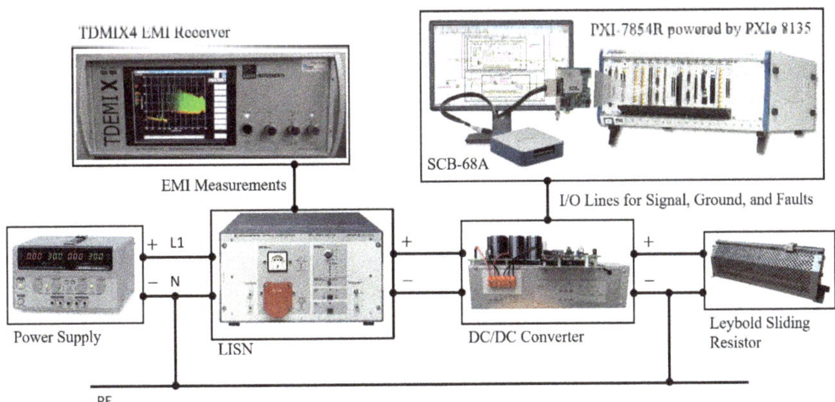

Figure 13. Schematic diagram of measuring testbed.

According to the schematic diagram illustrated in Figure 13, the buck-converter topology is powered by a regulated laboratory power supply. Additionally, the FPGA control board power is controlled by the PXIe 8135 to prevent additional couplings through the power source. A Leybold sliding resistor 320 Ω, 1.5 A was connected as load for the buck converter output. The EMI measurement was performed with the 50 Ω/50 μH Line Impedance Stabilization Network (LISN). The important parameters of the buck-converter and the testbed are summarized in Table 1.

Table 1. The main parameters of buck-converter topology.

Component/Function	Specification
Transistors type	IXGH40N60C2D1
I_C (max)	40 A
t_{on}	40 ns
t_{off}	180 ns
Transistor Gate Drivers	HCPL-316J
Converter Power	1800 W (max)
DC capacitors	1500 μF
Max DC voltage	450 V
Load	sliding resistor 320 Ω (max), 1.5 A (max)

Figure 14 shows the measurements for all cases (DetM and RanM) presented in Section 2. The results have been obtained using the TDMI X6 EMI receiver, which provides a 3D spectrogram for Quasi Peak (QP) detector, which is required by EMC standards in CISPR A frequency band.

The Figure 14a refers to the measurement for DetM with $N_m = N_{AV} = 500$ and $\delta N = 0$. The first harmonic magnitude (occurring at 80 kHz) is the most significant in the whole frequency spectrum. The magnitude of this harmonic is equal to 93.76 dBμV. The Figure 14b refers to measurement for RanM with $N_{AV} = 667$, $\delta N = 668$ and $SC_{TL} = 1$. As expected, the maximal harmonic magnitude is not connected with the $f_{sw} = 80$ kHz. Despite this, the frequency varied within the assumed range (Figure 5), and the spectrum level is lowered to value 74.24 dBμV. The Figure 14c presents the measurement results for RanM with $N_{AV} = 500$, $\delta N = 330$ and $SC_{TL} = 1$. As expected, the maximal harmonic magnitude is lowered and its frequency is more connected with the $f_{sw} = 80$ kHz. The maximum amplitude of the disturbances is equal in this case to 73.43 dBμV, and despite the smaller range of f_{sw}, variation is lower than in the case of RanM from Figure 14b. Therefore, we can conclude that increasing the δN range does not always lead to a lowering of the spectrum level. The Figure 14d refers to the measurement for RanM with $N_{AV} = 50$, $\delta N = 34$ and $RSC_{TL} \in <7:13>$. The maximal harmonic magnitude is connected with the $f_{sw} = 80$ kHz, and a little EMI noise reduction is provided, whether compared with Figure 14b,c. The Figure 14e refers to the measurement for RanM2 for parameters: $N_{AV1} = 750$ with $\delta N_1 = 500$, $N_{AV2} = 416$ with $\delta N_2 = 167$, and SC_{TL}. As expected, two extremes are visible in the spectrum. The Figure 14f shows the result of measurement for RanM2 with RSC_{TL} (concept of additional randomization). The parameters of modulator are: $RSC_{TL} \in <7:13>$, $N_{AV1} = 75$, $\delta N_1 = 50$, $N_{AV2} = 42$ and $\delta N_2 = 17$. The EMI noise is spread with a better shape between all RanM proposed. Unfortunately, the spread of f_{sw} value is the largest of the analysed cases (Figure 11). Based on the measurement, it is difficult in this case to determine the main/dominant frequency of disturbances.

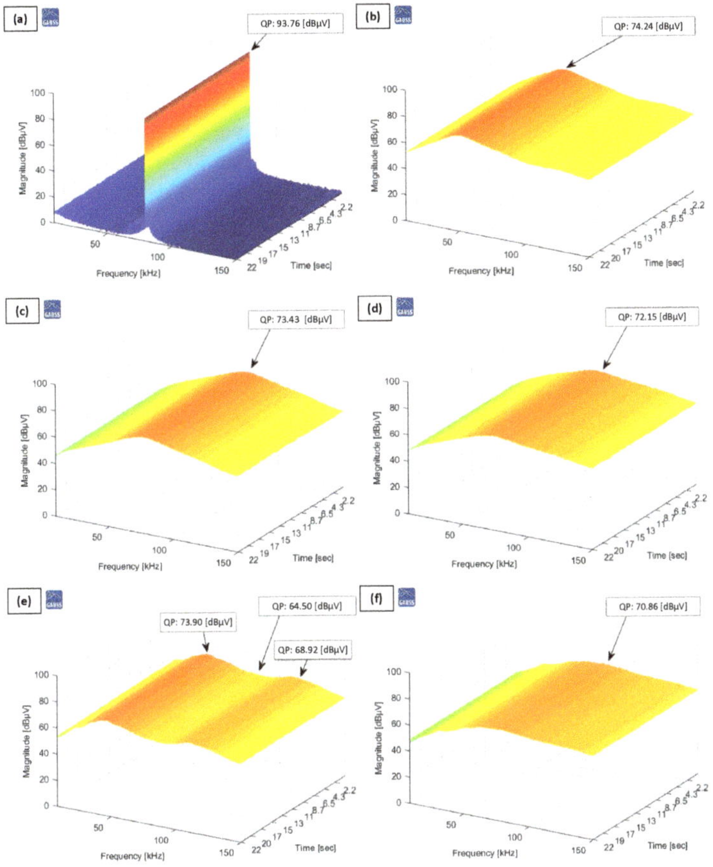

Figure 14. The electromagnetic interference (EMI) spectrum for DetM with $N = 500$ (**a**), RanM with $N_{AV} = 667$, $\delta N = 668$ (**b**), RanM with $N_{AV} = 500$, $\delta N = 330$ (**c**), RanM with RSC_{TL} for $N_{AV} = 50$, $\delta N = 34$ and $RSC_{TL} \in\, < 7:13 >$ (**d**), RanM2 with $N_{AV1} = 750$, $\delta N_1 = 500$, $N_{AV2} = 416$, $\delta N_2 = 167$ (**e**), RanM2 $RSC_{TL} \in\, < 7:13 >$, $N_{AV1} = 75$, $\delta N_1 = 50$, $N_{AV2} = 42$ and $\delta N_2 = 17$ (**f**).

4. Discussion and Analysis of Results

According to Figure 14a, the maximum harmonic magnitude in the spectrum (observed at 80 kHz) for DetM is about 93.76 dBμV. Furthermore, Figure 14d shows the maximum harmonic magnitude in the spectrum (observed at 80 kHz) for RanM with RSC_{TL}, which is about 72.15 dBμV. Therefore, RanM with RSC_{TL} provides 21.61 dBμV for QP detector, with lower EMI levels than DetM. Applying the concept of additional randomization with RSC_{TL} and sectional distribution of Nm, RanM2 obtained the best results—Figure 14f. The results are about 22.90 dBμV for the QP detector, being lower than DetM. According to the literature, the evaluated harmonic reduction could also be provided by the Harmonic Spread Factor (HSF) [16,17]. The HSF is an accurate evaluation index of any waveform for testing its harmonic spreading effects, and is defined as follows in Equations (8) and (9).

$$HSF = \sqrt{\frac{1}{N}\sum_{j=1}^{N}(H_j - H_o)^2} \tag{8}$$

$$H_0 = \frac{1}{N} \sum_{j=1}^{N} (H_j) \tag{9}$$

where: H_j is the amplitude of the jth harmonics and H_0 is the average value of all N harmonics.

The ideally spread spectrum must be near zero, i.e., white noise, presenting an HSF equal to zero. In this manuscript, the HSF analysis provides in absolute value, the harmonic reduction into CISPR A frequency band, for QP detector measurement and additionally for AV detector. Figure 15 shows the results of HSF calculation.

It is possible to observe that there are almost no differences between all HSF provided by RanM and RanM2 (for both QP and AV detectors). Of course, there is a significant difference between RanM2 and DetM. RanM2 based on the concept of additional randomization with SC_{TL} presents the best HSF. However, whether compared with RanM2 based on the concept of additional randomization with RSC_{TL} the difference is only 0.3% for QP and 0.1% for AV. In terms of EMI noise reduction, the difference between RanM2 with SC_{TL} and with RSC_{TL} is not too big for both detectors (less than 4 dB). Despite this, the frequency distribution was different in each case.

Additionally, hardware resources were also analysed. The hardware resources on an FPGA are indicated by the number of slices. The DetM code after the compilation process presents total slices of 3.0% from all available slices in the VIRTEX-5 LX110 FPGA. On the other hand, among all RM codes, the concept of additional randomization, RanM2 with RSC_{TL} presents greater use of the number of slices, with 5.5%.

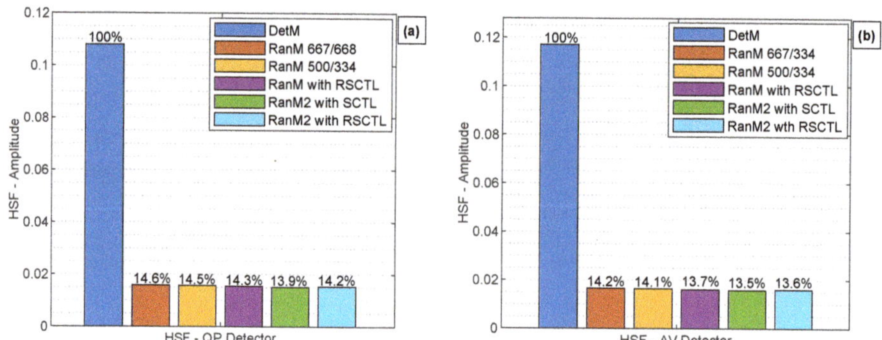

Figure 15. Harmonic Spread Factor (HSF) calculations: (**a**) detector Quasi Peak (QP) and (**b**) detector Average (AV).

5. Conclusions

This manuscript demonstrates how to design RanM, and DetM oriented for FPGA implementation with the LabVIEW engineering software. Since there are some FPGA software limitations (fixed-point operation, and a lack of basic arithmetic functions), it may be challenging to complete the RanM and the DetM implementation. Therefore in the article, we have highlighted how to do some calculations required for PWM modulator implementation in FPGA. We have also shown that the realization of RanM in FPGA should consider the distribution of randomized time and frequency parameters of the PWM signal. One should pay the primary attention to the range of switching frequency f_{sw} changes and the average value of this frequency.

The presented algorithms have been implemented in FPGA - R-Series Multifunction RIO (PXI-7854R), with VIRTEX-5 LX110. The most extensive version of the algorithm (RanM2 with RSC_{TL}) used only 5.5% of FPGA resources. Presented algorithms are weary simple, and they can be easily expanded. Likewise, implementing the modulator to a control system inside the same FPGA is also possible. Therefore, the proposed solutions can be used in all DC/DC converters applications. However, one may obtain the most significant benefits in automotive and lighting applications where there are direct limits on interference emission in the CISPR A band, and proposed control methods help fulfill these requirements.

Proposed modulators were tested experimentally. All measurements were carried out according to EMC standards in the frequency domain for CISPR A frequency band. The EMI emission (for design RanM) was significantly reduced compared to the DetM. We have achieved a reduction of the maximum EMI level value by over 20 dB. One should also remember, according to [8], that although pseudo-random modulations reduce the maximum level of interference, they do not change its aggregate power. Considering experimental research and implementation in FPGA, we assess that offered solutions have reached level 7 of Technology readiness levels (TRLs).

The differences between the EMI emission for different RanM were not significant. However, presented modulators differed in f_{sw} distribution. The minimum value of f_{sw} affects the operating conditions of the filters and and maximum amplitude of output current ripple. The maximum value of f_{sw} frequency will be significant with the restrictions of the transistors. Keeping the average value of this frequency unchanged, we will not change the overall losses, and converter efficiency will be consistent. Therefore, by using the proposed methods (with appropriate parameters), one can achieve energy neutrality for the converter. In the future research, we are planning to investigate which shape of the frequency distribution is preferred in respect of EMI level emissions and other operation of the converter.

Author Contributions: Conceptualization, H.L., P.L., R.S. D.N. and W.S.; methodology, H.L., P.L. and R.S.; validation, formal analysis, visualization H.L. and P.L.; software, investigation, H.L.; writing–original draft preparation, H.L., P.L. and W.S.; writing–review and editing, H.L., P.L. and R.S.; supervision, project administration, funding acquisition, R.S. All authors have read and agreed to the published version of the manuscript.

Funding: This paper is part of a project that has received funding from the European Union's Horizon 2020 research and innovation program under the Marie Skłodowska-Curie grant agreement No 812391 – SCENT.

Conflicts of Interest: The authors declare no conflict of interest.

Abbreviations

The following abbreviations are used in this manuscript:

AV	Average
DetM	Deterministic Modulation
DSP	Digital Signal Processors
EMC	ElectroMagnetic Compatibility
EMI	ElectroMagnetic Interference
FPGA	Filed-Programmable Gate Array
HSF	Harmonic Spread Factor
LGG	Linear Congruential Generator
LISN	Line Impedance Stabilization Network
PDF	Probability Density Function
QP	Quasi Peak
PWM	Pulse-Width Modulation
RanM	Pseudo-Random Modulator

References

1. Estrela, V.V.; Saotome, O.; Loschi, H.J.; Hemanth, J.; Farfan, W.S.; Aroma, J.; Saravanan, C.; Grata, E.G. Emergency response cyber-physical framework for landslide avoidance with sustainable electronics. *Technologies* **2018**, *6*, 42. [CrossRef]
2. Smoleński, R. Selected conducted electromagnetic interference issues in distributed power systems. *Bull. Pol. Acad. Sci. Tech. Sci.* **2009**, *57*, 383–393. [CrossRef]
3. Peng, Q.; Jiang, Q.; Yang, Y.; Liu, T.; Wang, H.; Blaabjerg, F. On the Stability of Power Electronics-Dominated Systems: Challenges and Potential Solutions. *IEEE Trans. Ind. Appl.* **2019**, *55*, 7657–7670. [CrossRef]

4. Yuan, B.; Liu, M.X.; Ng, W.T.; Lai, X.Q. Hybrid Buck Converter With Constant Mode Changing Point and Smooth Mode Transition for High-Frequency Applications. *IEEE Trans. Ind. Electron.* **2019**, *67*, 1466–1474. [CrossRef]

5. Arunkumari, T.; Indragandhi, V. An overview of high voltage conversion ratio DC-DC converter configurations used in DC micro-grid architectures. *Renew. Sustain. Energy Rev.* **2017**, *77*, 670–687. [CrossRef]

6. Loschi, H.; Ferreira, L.; Nascimento, D.; Cardoso, P.; Carvalho, S.; Conte, F. EMC Evaluation of Off-Grid and Grid-Tied Photovoltaic Systems for the Brazilian Scenario. *J. Clean Energy Technol.* **2018**, *6*, 125–133. [CrossRef]

7. Bojarski, J.; Smolenski, R.; Lezynski, P.; Sadowski, Z. Diophantine equation based model of data transmission errors caused by interference generated by DC-DC converters with deterministic modulation. *Bull. Pol. Acad. Sci. Tech. Sci.* **2016**, *64*, 575–580. [CrossRef]

8. Lezynski, P. Random modulation in inverters with respect to electromagnetic compatibility and power quality. *IEEE J. Emerg. Sel. Top. Power Electron.* **2017**, *6*, 782–790. [CrossRef]

9. Kubík, Z.; Skála, J. Electromagnetic Interference from DC/DC Converter of Photovoltaic System. In Proceedings of the International Conference on Applied Electronics (AE), Pilsen, Czech Republic, 15 April 2019; pp. 1–4.

10. Lezynski, P.; Smolenski, R.; Loschi, H.; Thomas, D.; Moonen, N. A novel method for EMI evaluation in random modulated power electronic converters. *Measurement* **2020**, *151*, 107098. [CrossRef]

11. National Instruments Corporation, Multifunction RIO, NI R Series Multifunction RIO User Manual NI 781xR, NI 783xR, Ni 784xR, and NI 785xR Devices. Technical Report 370489G-01, June 2009. Available online: https://www.ni.com/pdf/manuals/370489g.pdf (accessed on 8 May 2020).

12. Lai, Y.S.; Chen, B.Y. New random PWM technique for a full-bridge DC/DC converter with harmonics intensity reduction and considering efficiency. *IEEE Trans. Power Electron.* **2013**, *28*, 5013–5023. [CrossRef]

13. Chatterjee, S.; Garg, A.K.; Chatterjee, K.; Kumar, H. Chaotic PWM spread spectrum scheme for conducted noise mitigation in DC-DC converters. In Proceedings of the International Conference on Energy, Power and Environment: Towards Sustainable Growth, Shillong, India, 12 June 2015; pp. 1–6.

14. Marsala, G.; Ragusa, A. Spread spectrum in random PWM DC-DC converters by PSO&GA optimized randomness levels. In Proceedings of the 5th International Symposium on Electromagnetic Compatibility (EMC-Beijing), Beijing, China, 28–31 October 2017; pp. 1–6.

15. Gamoudi, R.; Chariag, D.E.; Sbita, L. A Review of Spread-Spectrum-Based PWM Techniques—A Novel Fast Digital Implementation. *IEEE Trans. Power Electron.* **2018**, *33*, 10292–10307. [CrossRef]

16. Liou, W.R.; Villaruza, H.M.; Yeh, M.L.; Roblin, P. A digitally controlled low-EMI SPWM generation method for inverter applications. *IEEE Trans. Ind. Inform.* **2013**, *10*, 73–83. [CrossRef]

17. Jing, H.; Weiying, Z.; Jinhong, L. Study on improving EMC of APFC converter with chaotic spread-spectrum technique. In Proceedings of the 2nd Advanced Information Technology, Electronic and Automation Control Conference, Chongqing, China, 25–26 March 2017; pp. 433–437.

Article

Dielectric Barrier Discharge Systems with HV Generators and Discharge Chambers for Surface Treatment and Decontamination of Organic Products

Jan Mucko [1,*], Robert Dobosz [2] and Ryszard Strzelecki [3]

[1] Faculty of Telecommunications, Computer Science and Electrical Engineering, UTP University of Science and Technology in Bydgoszcz, 85-796 Bydgoszcz, Poland
[2] Institute of Mathematics and Physics, UTP University of Science and Technology in Bydgoszcz, 85-796 Bydgoszcz, Poland; robertd@utp.edu.pl
[3] Faculty of Electrical and Control Engineering, Gdańsk University of Technology, 80-233 Gdańsk, Poland; ryszard.strzelecki@pg.edu.pl
* Correspondence: mucko@utp.edu.pl; Tel.: +48-52-340-8511

Received: 3 September 2020; Accepted: 30 September 2020; Published: 5 October 2020

Abstract: The article presents applications of systems with power electronic converters, high voltage transformers, and discharge chambers used for nonthermal, dielectric barrier discharge plasma treatment of a plastic surface and decontamination of organic loose products. In these installations, the inductance of the high voltage transformers and the capacitances of the electrode sets form resonant circuits that are excited by inverters. The article presents characteristic features of the installations and basic mathematical relationships as well as the impact of individual parameters of system components. These converters with their output installations were designed, built, and tested by the authors. Some of the converters developed by the authors are manufactured and used in the industry.

Keywords: resonant inverter; dielectric barrier discharge; nonthermal plasma; treatment of plastic surface; decontamination of organic loose products

---

1. Introduction

Plasma systems and plasma treated materials are now commonly used. The cold, nonthermal plasma (NTP) is produced usually by high voltage (HV) electrical discharges. In nonthermal plasma, most of the electric energy is used to produce high-energy electrons, not to heat the gas. The electrons themselves do not treat the surface. The high-energy electrons (1–2 eV) excite electronic states of molecules, vibrational states, and provide molecular dissociation (oxygen at the most). Namely atomic oxygen and electronically excited molecules contribute to the surface treatment and pollution control. Cold plasma applications are very diverse. The main applications of NTP include surface modification of plastics [1–3], ozone generation [4,5], surface decontamination [6–11], sterilization of wounds and soil [12,13], toxic and harmful gas and sewage decomposition [14–20]. One of the methods of producing cold plasma is the dielectric barrier discharge (DBD). A number of studies present theoretical foundations of barrier discharges, including models of discharge chambers and their substitute schemes, analytical description of current and voltage waveforms, voltage-charge characteristics (the Lissajous figures) [21–24]. The article [21] additionally includes a review of the applications of DBD to high power CO_2 lasers, excimer based ultraviolet and fluorescent lamps and flat large-area plasma displays. Another important application of barrier discharges and corona discharges (CD) is the investigation of the ionic wind generation and examination of its results for the development of propulsion [25]. The use of both barrier and corona discharges when supplying a set of several electrodes with alternating and direct voltage enables the creation of curtains from nonthermal plasma with a relatively large width of the air gap [26].

Many articles are focused on the development of high voltage generators that are components of DBD plasma systems. Depending on the application, the power of supplies ranges from single watts to hundreds of kilowatts. The choice of power and feed method is important for the operation of the DBD reactor and for the intensity of the reaction. The most common voltage waveform used to power DBD plasma reactors is the high voltage alternating current (AC) wave. In DBD reactors and in pulsed corona discharge reactors (PCD, without a dielectric layer) unipolar and bipolar voltage waves are used, sometimes a discontinuous wave with a prepolarization.

Already in 1857, Werner von Siemens reported on first experimental investigations with DBD. He applied a mechanical pulser ("Wagnerscher Hammer") interrupting the primary winding circuit of the HV transformer as the high-voltage generator [27,28]. Another simple high-voltage generator for barrier discharges can consist of an autotransformer and a high-voltage transformer fed directly from the power grid of voltage frequency 50 or 60 Hz [3]. However, the high voltage delivered to the electrodes with a frequency of 50 or 60 Hz has a much higher value than the voltage with the increased frequency generated by an inverter. Very popular solutions are voltage source inverters (VSI), with unregulated or regulated input voltage and with high voltage transformers connected to the output [22,26,29–35]. These inverters can often have a full or half-bridge structure. When supplying a DBD system from an inverter the selection of the supply frequency is associated with the resonant frequency of the transformer inductances (and additional inductances if present) and electrode set capacities. Other generator designs are also used. Low power generators can be made as fly-back converters [22,25]. Bridge converters can be equipped with snubber circuits that reduce du/dt or di/dt when transistors are switching (on/off). An interesting solution is the full-bridge in which only one branch uses the LDR (coil, diode, resistor) circuit for di/dt reduction [36]. This solution may be useful if a pulse width modulation with phase shift (PS-PWM) control is used. An example of the multiresonant generator is presented in [37]. The transformer inductances and capacitances of the electrode set (together with an additional capacitor) form one resonant circuit. Another resonant circuit provides conditions for zero current switching (ZCS) off transistors. The paper [38] describes variations of a diagonal half-bridge resonant converter topology (with four diodes and two transistors), which can be used to produce a single-period AC sinusoidal waveform. The method allows power regulation within very wide limits and makes possible the precharge pulse generation for transformer magnetization and gap voltage symmetrization. The construction of a HV generator [39,40] that allows the production of voltage waves with many different shapes is also very interesting. It consists of a 24-level cascaded H-bridge inverter and works without an HV transformer. A transformerless HV generator for DBD plasma producing is described in [41]. In the abovementioned example, the HV generator uses the phenomenon of voltage increase in a circuit under serial resonance conditions. The topology of the inverters with additional AC intermediate resonant circuits has been presented in [42]. Depending on the AC intermediate circuits these generators are characterized by properties of the current or voltage sources. In case of operation as the current source, the DBD discharges were very stable and the system was insured against arc discharges in a natural way.

The basic control methods of high voltage alternating current (HVAC) generators that are used to generate DBD plasma are described in [29,30,43]. These are pulse amplitude modulation (PAM), pulse width modulation (PWM), phase shift-pulse width modulation (PS-PWM, PSC), pulse density modulation (PDM), and pulse frequency modulation (PFM). In [31,32], the hybrid control of PDM and PFM is described.

While the above articles provide a good overview of the fundamentals of DBD discharge, their applications and HV generators design, they do not discuss the impact of individual components parameters and control variables on the power of DBD discharges. The aim of the paper at hand is to fill the gap. The discussion focusses on the following parameters: inverter input voltage, inverter output voltage frequency, DBD discharge ignition voltage, resonant circuit parameters together with the discharge chamber model parameters and transformer ratio. This article presents DBD generators and plasma reactors used for surface treatment of plastics and decontamination of loose organic materials, developed under the supervision of the authors. Generators are presented as well as entire technological devices.

The HV generators, which the authors designed, manufactured, and tested consist of power electronics converters: rectifier, DC/DC converter (if any), voltage-source inverter, and HV transformer. Transformer leakage inductances (and additional inductances if used) form a series resonant circuit with the capacities of the electrode sets. The resonant circuits create the conditions for soft switching, which is almost lossless. Thanks to the soft switching, the converters have high efficiency and generate little radio interference. Soft switching may appear as zero current switching (ZCS) or zero voltage switching (ZVS). The ZVS will be preferred in the analyzed systems, which is characterized by lower losses in the frequency and power range that the authors are interested in (up to about 100 kHz by 0.5 kW and about 25 kHz by 10 kW of rated power).

1.1. Power Electronics Converter Topology and Their Control Methods

The topology of the used power electronics converter is presented in Figure 1. Depending on the inverter control method and the assumed power and frequency range the step-down converter and/or HF resonant choke may be omitted. Then, only the transformer leakage inductances will perform the function of a resonant choke.

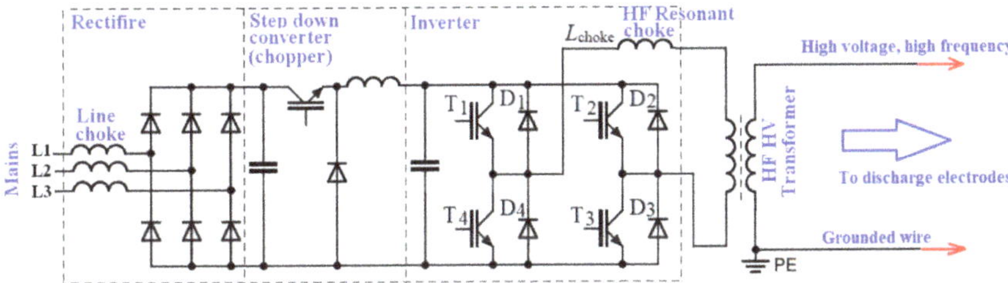

Figure 1. Schematic diagram of the power converter circuit used in the developed technological devices.

To control the output power of the converter, the following adjustment methods were considered [35]:

- pulse width modulation (PWM, with the constant inverter input voltage, the chopper can be omitted),
- changing the frequency of the inverter output (PFM, at constant inverter input voltage, the chopper can be omitted),
- changing the inverter input voltage (PAM, the chopper is necessary),
- PDM modulation (at constant inverter input voltage, the chopper can be omitted),
- combinations of the above methods.

Figure 2 shows the examples of the output current and voltage waveforms of the inverter with a series resonant circuit at the output for various analyzed control methods.

1.1.1. Output Power Control by Pulse Width Modulation, with Constant Inverter Input Voltage

Earlier studies and implementations carried out by one of the authors concerned the regulation of generator power using PWM. The chopper can be omitted to simplify the generator main circuit. This method is not recommended in the frequency range above several dozen (or even several) kHz and powers above a few kW. Turning off each of the transistors can be done "softly" (with appropriate transistors control) in the ZVS technique. However, switching on must be done "hard". Hard commutation, at high switching frequency, causes significant losses and current stress caused by the sum of the load current and the reverse current of the inverter's diodes (Figure 2a). In each half-period of the output current, a reverse diode conducts then a transistor and then again reverse diode. There are six switching operations during the inverter operation period. This causes the inverter output voltage to oscillate at the frequency three times greater than the current wave. In the previously tested inverters, the transistors had to be oversized and the inverter was a strong source of radio frequency interference, also for its own control circuits.

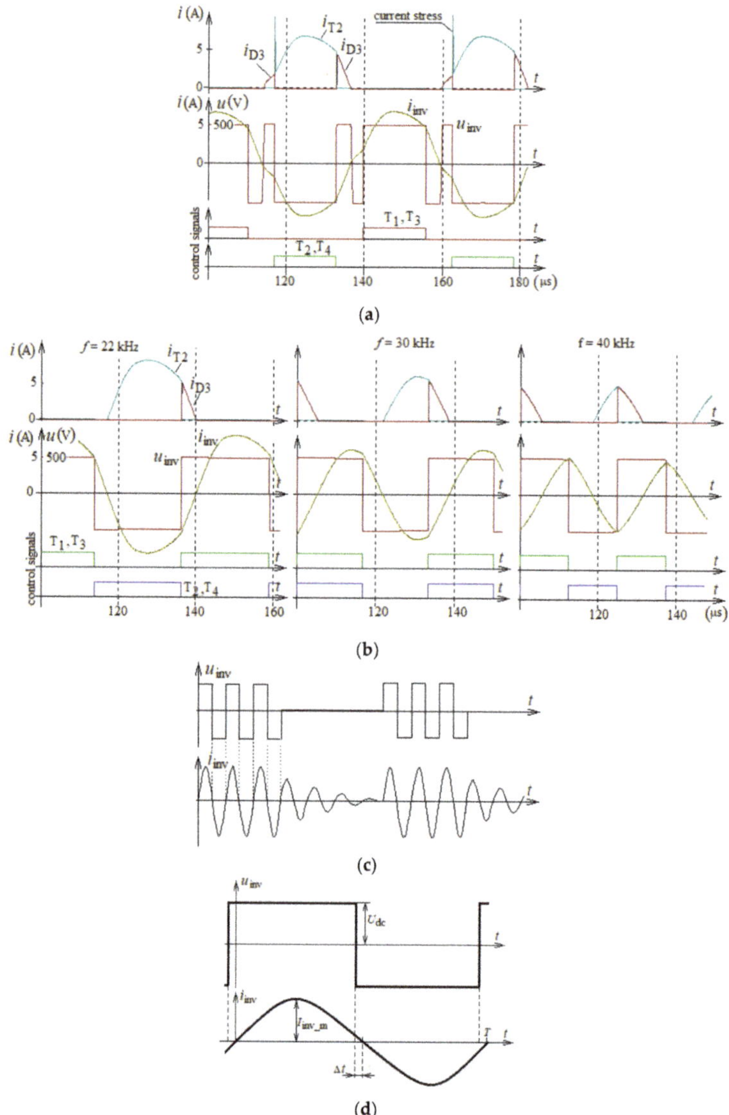

Figure 2. Waveforms of the inverter output voltage and current with different control methods, reproduced from Przegląd Elektrotechniczny [35]: (**a**) PWM, (**b**) PFM, (**c**) PDM, (**d**) PAM: i_T, i_D—transistor and diode current; i_{inv}—inverter output current; u_{inv}—inverter output voltage; for PWM, PDM, and PAM modulation it was assumed that the switching frequency is approximately equal to the resonant frequency and the voltage waveform is synchronized with the current in conditions for operation with ZVS switches.

Another kind of PWM modulation is the phase-shift pulse width modulation (PS-PWM, phase-shift control, PSC). In PS-PWM, transistors are conductive for half of the period and during diode conduction, the inverter output voltage is zero. This modulation has some advantages compared to PWM but also does not eliminate the hard switching in a wide range of power control [33,36].

1.1.2. Power Control by Pulse Frequency Modulation, with Constant Inverter Input Voltage

In this arrangement, the output power of the system is adjusted by changing the transistors switching frequency (Figure 2b). When operating below the resonant frequency, current stress in transistors occurs due to reverse currents of diodes. Thus, the authors do not recommend operating below the resonant frequency due to increased commutation losses. This method was previously used in series resonant inverters made in the thyristor technique. At the resonant frequency, the system works with maximum power. System operation (and power control) should take place at frequencies above resonance. Then ZVS conditions are created for the transistor's operation. The converter power circuit is relatively simple due to the unregulated DC voltage (no chopper). The inverter operation frequency should be limited both above and below so as not to go beyond the frequency range accepted in a given production process.

1.1.3. Power Control by Changing the Voltage at the Inverter Input

Earlier implementation carried out by one of the authors also considered regulation of generator power by changing the DC voltage supplying the inverter. This DC voltage source can be a controlled or semicontrolled thyristor rectifier, a noncontrolled rectifier with PFC converter or a transistor chopper (Figure 1). The system, although more complex, has a number of advantages. To enable the inverter transistors to work as ZVS switches, the transistors are turned off before the output current reaches zero. At the same time, if the switching frequency is close to the resonance frequency then the transistors turn off the low current. This is quasi-ZCS switching (Figure 2d). When ZVS and at the same time quasi-ZCS switching occurs, the inverter works in the most favorable conditions. Commutation losses are eliminated, and voltage and current steepness are limited. To ensure these optimal conditions it is necessary to automatically adjust the inverter switching frequency employing PLL. All power control processes take place in the chopper. Independent control of the inverter (PLL) and chopper is provided. Under these conditions, the time intervals at which energy returns to the DC source are very small. Then the amplitude, RMS, and average of the inverter output current (and transistor current) will be the lowest at the given output power. Assuming a sinusoidal current waveform and that $\Delta t << T_s$ (Figure 2d) one gets an approximate equation for the output power of the bridge inverters (Equation (1)):

$$P \approx \frac{1}{T_s/2} \int_0^{T_s/2} U_{dc} I_{inv\_m} \sin(\omega_s t) \quad dt = \frac{2 U_{dc} I_{inv\_m}}{\pi} \approx 0.9 U_{dc} I_{inv\_RMS} \tag{1}$$

where: U_{dc}—DC inverter input voltage, I_{inv_m}, I_{inv_RMS}—maximum and RMS value of the inverter output current, T_s, f_s, ω_s—period, switching, and angular frequency.

1.1.4. Power Control by PDM Modulation (at Constant Inverter Input Voltage)

PDM modulation ensures even distribution of discharges over the entire length of the electrodes with a wide range of changes in average process power. The power regulation by means of PDM consists of sending in "packets" the maximum power with the frequency of a modulating generator with regulated filling [29,30,43]. The transistors in the inverter can work with ZVS soft commutation at a switching frequency greater than the resonant frequency. For maximum use of components (minimum transistors current in relation to the transferred power), the system should work with a frequency close to resonance. The general working principle is shown in Figure 2c. Turning off the "packet" is accomplished by switching on of two transistors T_1 and T_2 or T_3 and T_4. Then the oscillations in the resonant circuit go out automatically. The generator power circuit is relatively simple due to the unregulated DC voltage. The control system should also ensure that the transformer does not saturate, regardless of the length of the "packet" of power pulses and the pause time [43]. In particular, an even number of half-waves of the inverter output voltage should be maintained. During a break in power transfer, one must remember the switching frequency to which the system tuned during power flow. Assuming a sinusoidal current waveform and that $\Delta t << T_s$ (as in Figure 2c,d), one obtains the equation for the output power of the bridge inverter:

$$P \approx \frac{2U_{dc}I_{inv\_m}}{\pi}D_{PDM} \approx 0.9U_{dc}I_{inv\_RMS}D_{PDM} \tag{2}$$

where: D_{PDM}—duty cycle of PDM.

1.1.5. Author's Method Based on Simultaneous PDM and PFM Modulation

The author's method [24,44] based on simultaneous PDM and PFM modulation differs from the methods described in [31,32]. This control method was used in DBD discharge generators manufactured and used at the Institute of Polymer Materials and Dyes Engineering (IMPIB, Toruń, Poland) [45]. The essence of this control method is the combination of PDM and PFM modulation while the inverter operation is not stopped (switching off all transistors or short-circuit state of the load) but there is a periodic increase in the switching frequency (Figure 3). The range of frequency changes is limited from above and below according to the assumed maximum and minimum power. The PDM modulating signal can have a rectangular shape with variable (Figure 3a) or with fixed (Figure 3b) filling. Limiting the maximum power protects against damage to the processed material or arcing caused by a dielectric breakdown. This control method ensures uniform discharge in a wide range of power control (about 10–100% of nominal power) and has additional advantages over the classic PDM method [31,32]. The control system does not require the use of an additional system that remembers the frequency from the moment before the inverter stops and does not require a system counting the number of half-waves of the output voltage. Another advantage is the ease of modification of existing PFM control systems to work according to the proprietary method [24].

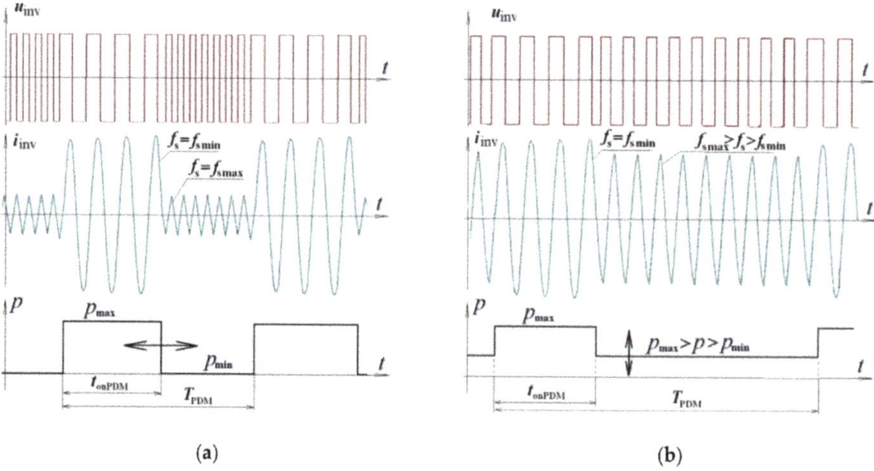

(a) (b)

Figure 3. Examples of inverter output voltage and current waveforms using the PDM-PFM method developed by the authors: (**a**) with variable duty cycle of PDM signal, (**b**) at a constant duty cycle of PDM signal.

1.1.6. Choice of Control Method

In further projects and implementations, the PWM method was abandoned, because the inverter transistors could not work with ZVS soft commutation. The following methods were used to control HV generators for barrier discharges:

1. Power regulation by input voltage changes of the inverter (PAM, system with chopper) and switching of the inverter transistors with a frequency slightly higher than the resonant frequency. The switching frequency was tuned using the PLL loop. The inverter output voltage was ahead of the output current wave by approx. 2–3 µs. The inverter transistors switched under ZVS and almost ZCS conditions with very low commutation losses. This method of regulation ensured uniform discharges over the entire length of the electrodes in the range of 20–100% of the nominal power (P_N) of the device.

2. Power regulation by means of frequency modulation, above the resonant frequency (PFM, system without chopper), ensured the correct generation of DBD in the range of 20–100% of P_N. For a power lower than 20% of P_N, the system switched to PDM + PFM modulation according to the method developed by one of the authors. In this way, the power could be adjusted in the range of about 5–100% of P_N. This type of control ensured the operation of the inverter transistors in ZVS conditions.

2. Matching of HV Generators and DBD Reactors

Dielectric Barrier Discharges—Theoretical Basics

Figure 4 shows a simplified model of the discharge chamber. Figure 4a shows the construction ideas and Figure 4b presents simplified equivalent diagrams of the generator and the chamber. Figure 4b also contains the simplified characteristic of the barrier discharge in the air [21–24].

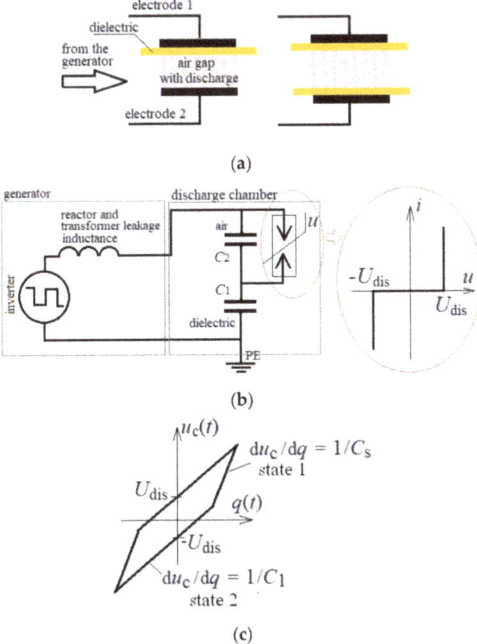

Figure 4. The idea of the construction of discharge chambers (**a**), simplified diagrams of the generator and the discharge chamber (**b**), and the trajectory of the voltage on the electrodes as a function of the charge supplied to these electrodes (**c**).

Capacities C_1, C_2 and discharge (and ignition) voltage U_{dis} depend on the shape and size of the electrodes, dielectric thickness, and its type and width of the air gap [46]. Figure 4c shows the trajectory of the voltage on the electrodes as a function of the charge supplied to these electrodes. This trajectory allows determining the capacities of the equivalent diagram and ignition voltage. To design a device for generating DBD discharges, one needs to know the parameters of the discharge chamber, transformer ratio, voltage and frequency range of the inverter output voltage, inductance in the resonant circuit.

According to the simplified scheme and the $u_C(q)$ trajectory (Figure 4b,c) one can distinguish two states in chamber operation (Figure 4c): state 1, wherein there are no discharges and state 2 in which DBD discharges occur. In state 1 the increase rate du_C/dq on the chamber terminals depends on the substitute capacity C_S made up of series-connected capacitors C_1 and C_2 (3). In state 2 the capacitor C_2 voltage does not change and the du_C/dq depends on the C_1 capacity (4). Parameters of the discharge chamber can be experimentally determined based on $u_C(q)$ trajectory (Figure 4c).

$$\frac{du_C}{dq} = \frac{1}{C_S} = \frac{C_1 + C_2}{C_1 \quad C_2} \tag{3}$$

$$\frac{du_C}{dq} = \frac{1}{C_1} \tag{4}$$

The chamber parameters related to the primary side of transformer are $C'_1 \dot= \vartheta^2 C_1$, $C'_2 \dot= \vartheta^2 C_2$, $C'_S \dot= \vartheta^2 C_S$, $U_{dis} \dot= U_{dis}/\vartheta$, where U_{dis}—discharge (and ignition) voltage, ϑ—transformation ratio. The frequency f_{syn} at which the inverter output voltage and current are synchronized is in the range $f_{r\_max} > f_{syn} > f_{r\_min}$ (Equations (5) and (6)) [34]. The synchronization frequency is the boundary of the transistors' abilities to work as ZVS or ZCS switches. The synchronization frequency f_{syn} at the rectangular inverter output voltage is only approximately equal to the resonant frequency [47].

$$f_{r\_max} = \frac{1}{2\pi \sqrt{L_r C_S \vartheta^2}} \tag{5}$$

$$f_{r\_min} = \frac{1}{2\pi \sqrt{L_r C_1 \vartheta^2}} \tag{6}$$

where $L_r = L_{choke} + L_\sigma$, L_{choke}—the additional choke (Figure 1) between the inverter output and the HV transformer, L_σ—the leakage inductance of the transformer seen from the low voltage side.

For the inverter voltage and frequency at which the capacitor C_2 voltage does not reach the U_{dis}, there are no discharges. In such conditions, the discharge chamber is a linear load. This creates a capacitor with a capacity of C_S. Capacitor C_S together with L_r creates a resonant circuit with low-pass filter properties. The shapes of electrode current and voltage are sinusoidal and the classical ac analysis can be used.

The amplitude of the capacitor C_2 voltage, which is referred to the first harmonic amplitude of the inverter output voltage is described by Equation (7), wherein U_{C2_1m}—the amplitude of the capacitor C_2 voltage, U_{inv_1m}—the first harmonic amplitude of the inverter output voltage (in the full bridge topology and a maximum duty cycle), $\omega_s = 2\pi f_s$—circular frequency of the inverter output voltage, f_s—transistors switching frequency [24,34].

$$\frac{U_{C2\_1m}/\vartheta}{U_{inv\_1m}} = \left| \frac{1}{\omega_s^2 L_r \quad \vartheta^2 C_S - 1} \cdot \frac{C_S}{C_2} \right| \tag{7}$$

where $U_{inv\_1m} = \frac{4}{\pi} U_{dc}$, $U_{C2m} = U_{dis} \approx U_{C2\_1m}$.

Equation (7) determines when the amplitude of the capacitor C_2 voltage reaches U_{dis}. The limit values of the switching frequencies at which the discharges appear, are determined based on Equations (8) and (9):

$$f_{s\_lim\_1} = \frac{1}{2\pi} \sqrt{\frac{1}{L_r \vartheta^2 C_2} \left(\frac{C_2}{C_S} - \frac{4}{\pi} \frac{U_{dc}}{U_{dis}/\vartheta} \right)} \tag{8}$$

$$f_{s\_lim\_2} = \frac{1}{2\pi} \sqrt{\frac{1}{L_r \vartheta^2 C_2} \left(\frac{C_2}{C_S} + \frac{4}{\pi} \frac{U_{dc}}{U_{dis}/\vartheta} \right)} \tag{9}$$

For PWM modulation, this equation is modified as shown in [29]. The discharges occur when $f_{s\_lim\_1} < f_s < f_{s\_lim\_2}$. The frequency limits depend on the capacity of the electrodes, the inductance of L_r, the discharge voltage, the inverter output voltage, and transformer winding ratio. The ratio of transformer winding has an impact on the operating frequency range and power of the device. By reducing the inverter output voltage these frequency limits approach to f_{r_max} (Equations (5) and (9)).

Figure 5a shows characteristics of discharges power as functions of frequency and inverter input voltage. Figure 5b illustrates power, current, and voltage characteristics as functions of frequency at a constant inverter input voltage (300 Vdc). Figure 5b illustrates the frequency limits according to the Equations (5), (6), (8), and (9). The characteristics from Figure 5a,b have been determined by assuming a constant value of the inductance L_r and transformer winding ratio. The following parameters of the real system (for treatment of plastic foil surface) were assumed in the simulation model: power $P_N = 3$ kW: $U_{dc} \approx 300$ V, two rotating electrodes: length 1700 mm, diameter 100 mm, 2 mm silicone insulation; two immovable electrodes: length 1600, width 36 mm toothed profile; air gap of approx. 2–4 mm (teeth); capacitance $C'_1 \approx 1.59$ nF, $C'_2 \approx 0.794$ nF; $L_r \approx 1.3$ mH; $\vartheta = 9.17$.

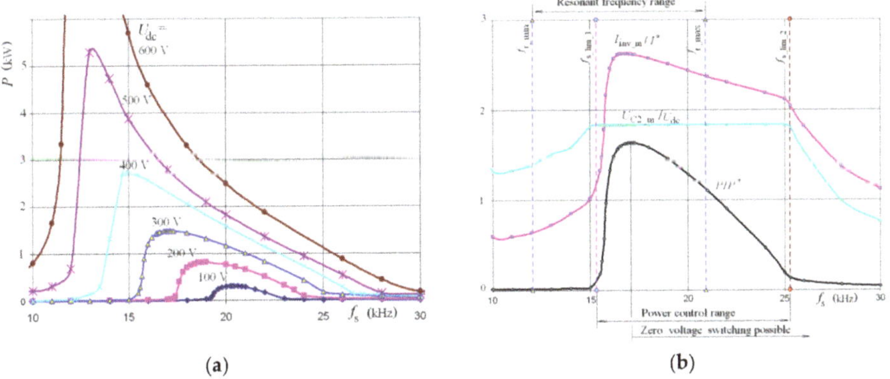

(a)   (b)

Figure 5. Characteristics of DBD discharges: (**a**) as a function of inverter output voltage and frequency, (**b**) as a function of inverter output frequency and constant inverter input voltage $U_{dc} = 300$ V, reproduced from Przegląd Elektrotechniczny [34]; the simulation results; base values: $I^* = U_{dc}/(L_r C_1')^{1/2}$, $P^* = U_{dc}^2/(L_r C_1')^{1/2}$.

Figure 6 shows the impact of the inductance and the transformation ratio on the frequency limits that define the range of switching frequency at the PFM modulation. The characteristics are derived by mathematical analysis (Equations (8) and (9)) for the above data. The same frequency limits were obtained by simulation. This is illustrated by the points on the curves in Figure 6. It is worth noting that experimentally measured frequencies did not diverge more than a few hundred Hz from those obtained by calculation and simulation.

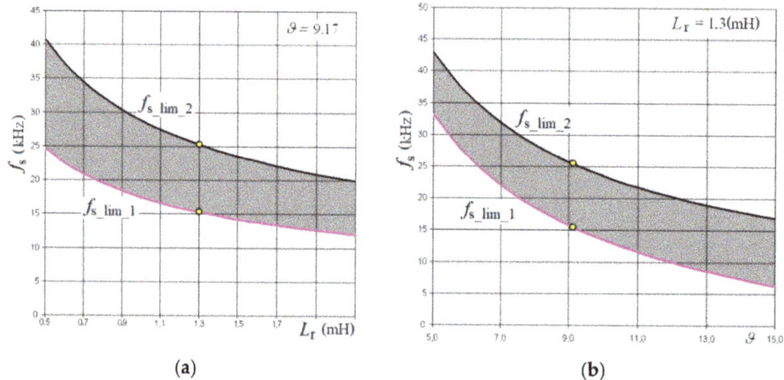

Figure 6. The range of the inverter output frequency at which the discharges appear (derived by mathematical analysis) as a function of: (**a**) inductance L_r; (**b**) transformer windings ratio, reproduced from Przegląd Elektrotechniczny [34].

Increasing demand for processing different kinds of materials with different sizes generates the need to examine the impact of the transformer windings ratio and the inductance in the treatment process. Figure 7 presents the power control characteristics for the device with the same parameters as described above. The electrodes capacitance and voltage U_{dis} are fixed. Figure 7 shows that with a change of the transformation ratio the electrode capacities and voltage U_{dis} (referred to the inverter side) also change.

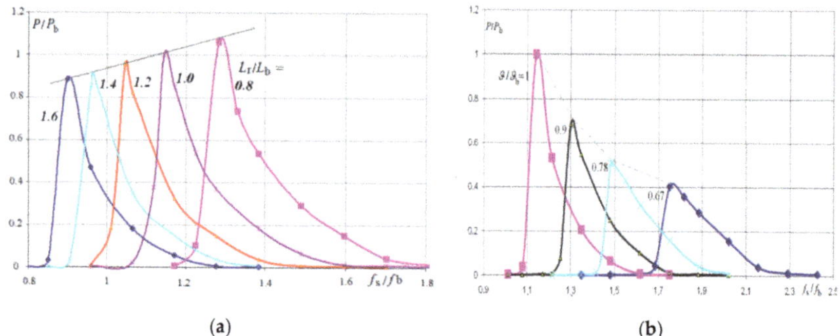

Figure 7. Relative discharge power as a function of frequency for: (**a**) various values of L_r; (**b**) various values of ϑ, reproduced from Przegląd Elektrotechniczny [34]; base values: L_b = 1 mH, ϑ_b = 9.17, $f_b = 1/(2\pi(C_1 L_b)^{1/2})$.

3. Developed Prototype and Industrial Systems

3.1. Systems with Resonant Inverters for Surface Treatment (Activation) of Plastics

In order to modify the surface of plastics during printing, laminating, and gluing the DBD discharges (so-called corona treatment) are used. To achieve the desired level of adhesion the discharge energy in the range of 0.65–1.3 kJ/m^2 should be delivered. Parameters of HV generators for plastics surface treatment are generally in the range of power—0.5–10 kVA; frequency—5–50 kHz; voltage—4–20 kV. The schematic diagram of the power converter circuit used in the developed technological devices is shown in Figure 1. The idea of construction and the equivalent circuit of discharge chambers are presented in Figure 4a,b.

Figure 8 shows the construction principle of discharge chamber for foil processing, the trajectory $u(q)$ and waveforms of current, voltage, charge, and instantaneous power of the electrode set obtained experimentally. Discharges occur between the cylindrical (rotating) and rod electrode (Figure 8a). Capacitors assembly consists of the electrodes and two dielectric layers (silicon, quartz glass or ceramics, and air). The treated plastic makes the third layer of dielectric. Capacities of silicone and treated plastic foil are analyzed as one capacitor. Capacities of the electrodes and the leakage of transformer and additional choke inductances create a resonant circuit. The selection of transformer winding ratio and choke inductance allows for operating of the system in a given frequency range and assumed output power (Figures 5–7). The density of energy E/s (J/m^2) supplied to the plastic surface for the device as in Figure 8a can be determined from the equation:

$$E/s = P/(vd), \tag{10}$$

where P—DBD discharge power, v—speed of the foil, d—width of discharges (electrodes).

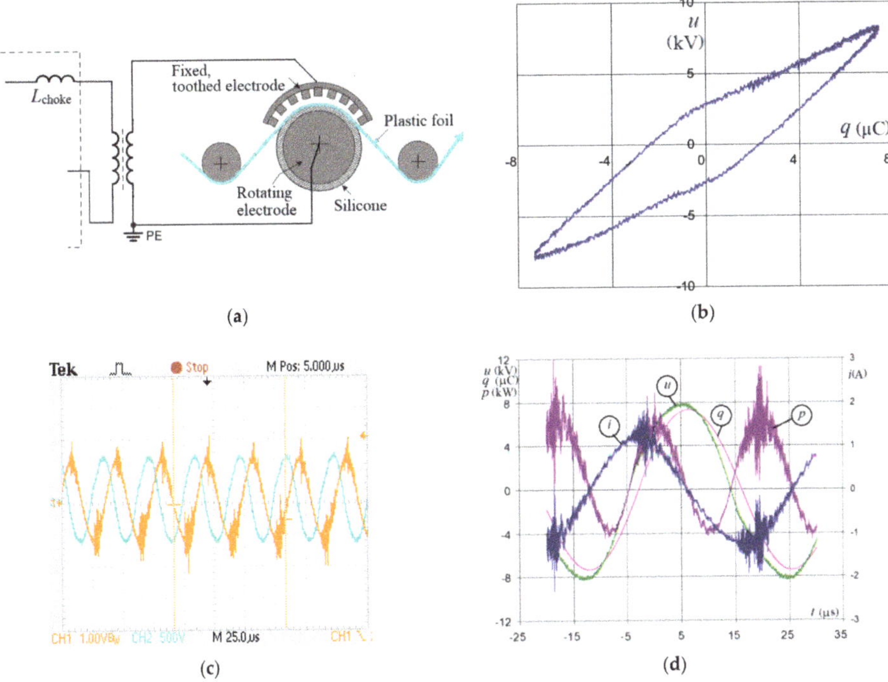

Figure 8. The discharge chamber for foil processing: (**a**) construction principle; (**b**) the trajectory $u(q)$ obtained experimentally at the frequency 26.8 kHz of the inverter output voltage and power 1.5 kW; (**c**) oscillogram of electrode current and voltage at the frequency of 26.8 kHz, CH1 1 A/div., CH2 6.25 kV/div.; (**d**) waveforms of current, voltage, charge, and instantaneous power of the electrode set obtained from the data from the oscillogram.

The waveforms of currents and voltages presented in Figure 8c were recorded using measuring devices: oscilloscope Tektronix TDS2024, current probe PA-622, high voltage differential probe P5200 with an additional voltage divider (1/12.5) at the input. The recorded data (from Figure 8c) were used to determine the trajectoryfrom Figure 8b and the waveforms from Figure 8d. Excel was used for this purpose. The capacities of C_1 and C_2 were determined on the basis of the trajectory from Figure 8b and Equations (3) and (4). In order to determine the inductance $L_r = (L_{choke} + L_\sigma)$, the secondary winding of the HV transformer was shorted and the additional choke and the transformer were powered from the inverter at reduced voltage. The rectangular voltage wave and the triangular current waveform at the inverter output were recorded. The inductance was determined on the basis of the relationship $L_r = U_{dc}(\Delta t/\Delta i)$ where Δt is half of the period of the inverter output voltage and Δi is the current increase during this time. The determined parameters values were used during the simulation, the results of which are shown in Figures 5–7. The dimensions of the discharge chamber and the determined parameters values can be found in the description of Figure 5.

The trajectorypresented in Figure 8b prove that the model adopted for the analysis and simulation is correct for the averaged values of voltage and current of the electrodes. During the analysis and simulation with the use of this model, the electrodes current and power do not experience high-frequency oscillations visible in Figure 8c, d. This model can be used in the design and simplified analysis of the phenomena occurring in the discharge chamber. On the other hand, the oscillograms in Figure 8c, d show that many ignition and extinguishing processes occur simultaneously.

The generators for surface treatment of plastics by DBD discharge, which are described in this article, are produced now based on documentation and under the supervision of one of the authors at the Institute of Polymer Materials and Dyes Engineering (IMPiB, formerly Metalchem) in Torun, Poland [45]. Figure 9 presents the exemplary generator and discharge electrodes. The nominal powers of these generators in the range from 0.5 to 10 kW are produced.

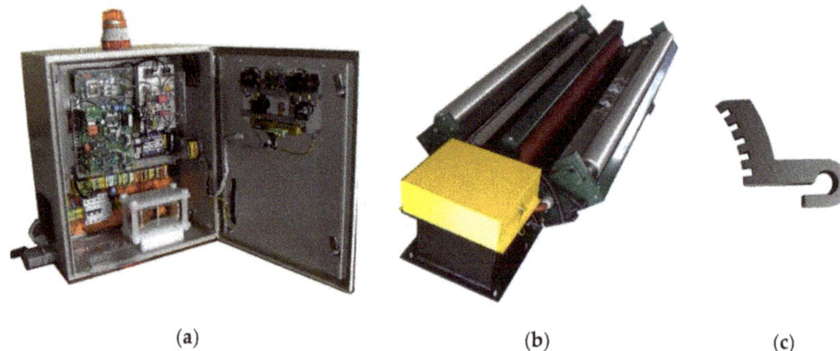

(a)   (b)   (c)

Figure 9. Construction of the devices for processing plastic film using DBD discharges: (**a**) power electronics generator with additional choke; (**b**) HV transformer and electrodes assembly, reproduced from web page IMPiB [45]; (**c**) part of the electrode.

3.2. Systems with Resonant Inverters for Decontamination of Loose Organic Material

Dielectric barrier discharges and ozone produced in this process can be used to decontaminate products such as seeds or ground dried plants. The use of plasma technologies in the food industry and agriculture has been described many times in literature [6,48–51]. However, these articles usually did not describe the construction details of plasma generators and reactors. Descriptions of some reactor designs can be found in patents [10,11]. The description of the DBD generation is analogous to the generation for surface treatment of plastics. However, the constructions of the discharge chambers are different. Figure 10 presents an equivalent diagram of part of the generator with HV transformer and construction of two types of discharge chambers for decontamination of loose organic material. The prototypes according to Figure 10b,c were built and tested under the supervision of one of the authors [7]. The first chamber (Figure 10b) has one fixed electrode, and the other in the form of a movable trolley that performs reciprocating movements transporting the treated organic material. The second version (Figure 10c) has two rotary electrodes in the form of cylinders between which the processed material is decontaminated.

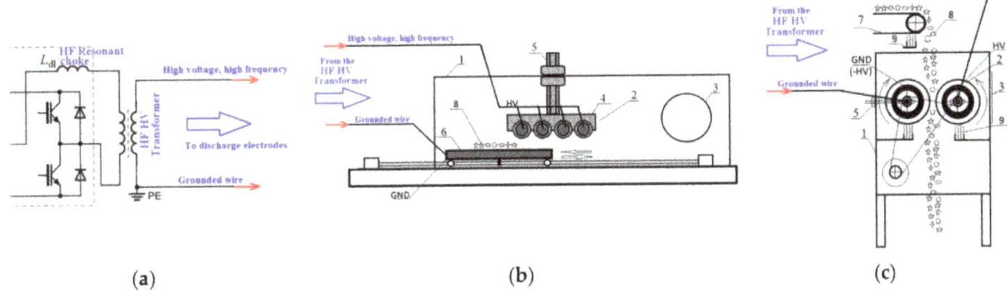

Figure 10. Construction of the devices for decontamination of loose organic material using DBD discharges: (**a**) equivalent diagram of part of a generator with transformer; (**b**) discharge chamber with a sliding electrode; (**c**) discharge chamber with rotating electrodes; 1—discharge chamber, 2—electrodes assembly, 3—suction hole, 4—insulating support, 5—electrodes gap adjustment knob, 6—transport trolley, 7—belt conveyor, 8—processed material, 9—sweeper.

New reactor designs were developed to increase the discharge power and thus to reduce the plasma exposure time and speed up the technological process. Plasma processing time is short compared with other known solutions [10]. Figure 11 shows two types of prototypes of devices for decontamination of crushed dried plants.

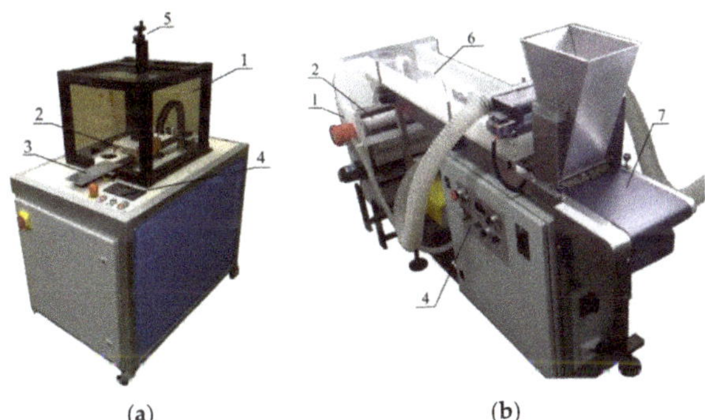

Figure 11. View of devices for decontamination of loose organic materials using DBD discharges developed and tested by the authors: (**a**) device with a sliding electrode; (**b**) device with rotating electrodes; 1—discharge chamber, 2—electrodes assembly, 3—transport trolley, 4—operator panel, 5—electrodes gap adjustment knob, 6—ozone chamber, 7—belt conveyor.

Figure 11b shows a solution that can be part of a technological line. This design is equipped with a support decontamination system which uses ozone generated in the discharge chamber. The conveyor speed determines the remaining time of the processed material in the ozone chamber. Electrodes in the form of rotating cylinders provide better cooling conditions than fixed electrodes. To increase the plasma operating time, the chamber may consist of several electrode assemblies. The implementation of such devices for the food industry is envisaged.

The power of discharges was regulated in the range of 200–1000 W by PFM or PDM + PFM modulation. The PDM + PFM modulation was used in the power range of 200–300 W to ensure even discharges over the entire length of the electrodes at low power. The decontamination efficiency of these prototypes was tested at the Faculty of Agriculture and Biotechnology of the UTP University in Bydgoszcz. The tests [52] confirm the effectiveness of DBD plasma and ozone in reducing microbial contamination of dried fragmented plants.

4. Conclusions

The article considers the most common problems concerning a proper matching of HV generators and DBD reactors It focused on parameters of electrodes sets (equivalent capacitance, discharge voltage), generator parameters (frequency and output voltage, modulation methods), transformer parameters (transformation ratio, leakage inductances), and the resonant circuit choke inductance. Thus, the article contributes knowledge to designing equipment for surface treatment of plastics and for decontamination by DBD method.

A common feature of the presented systems is that the transistors of the inverters work with the ZVS soft commutation in the whole range of power regulation. In the case of PAM + PFM modulation, the transistors work with ZVS and "almost" ZCS commutation, which radically reduces switching losses. Resonant converters created in this way had better parameters than similar systems in which transistors operated with hard commutation. This concerned parameters such as efficiency and generation of radioelectric disturbances.

The innovative solutions presented in the article are the inverters for DBD plasma generators, which use the proprietary PDM + PFM modulation method. This method ensures the extension of the power regulation range and maintaining the uniformity of discharges in DBD devices for plastic surface treatment and decontamination. The generators were built using the theoretical considerations presented in this article. New designs of discharge chambers for decontamination have been developed. They have been reserved in the European patent office. The innovative solution of the first structure, with a movable GND electrode, is the ability to precisely select the dose of energy and decontamination conditions by adjusting the discharge power, the distance between the electrodes, the speed of the trolley with the GND electrode, the number of runs of the trolley during processing. An innovative solution of the second design is, among others, the use of rotating cylindrical electrodes, which improves cooling conditions and the use of ozone produced in the process for initial decontamination.

Author Contributions: Conceptualization, J.M.; methodology, J.M.; validation, J.M., R.S.; formal analysis, J.M., R.S.; investigation, J.M., R.D.; writing—original draft preparation, J.M.; writing—review and editing, R.S., R.D.; funding acquisition, J.M. All authors have read and agreed to the published version of the manuscript.

Funding: The mechanical design and construction of equipment for decontamination of loose plant products was financed by the UTP and UKW University consortium under the "Inkubator Innowacyjności Plus" project. Project application no. 17/02/2018/UTP. Other presented research received no external funding.

Acknowledgments: The authors would like to thank the employees of the Institute of Polymer Materials and Dyes Engineering (IMPIB) in Torun for their help and sharing of electrode sets and HV transformers. The authors also thank the staff of the Faculty of Agriculture and Biotechnology of the UTP University in Bydgoszcz for their help in evaluating the concept of devices for decontamination of loose organic materials.

Conflicts of Interest: The authors declare no conflict of interest.

References

1. Mittal, K.L.; Lyons, C.S. *Plasma Surface Modification of Polymers: Relevance to Adhesion*; Taylor & Francis Group: Abingdon, UK, 2019; p. 290. ISBN1 0367449463. ISBN2 9780367449469.
2. Nastuta, A.V.; Rusu, G.B.; Topala, I.; Chiper, A.; Popa, G. Surface modifications of polymer induced by atmospheric DBD plasma in different configurations. *J. Optoelectron. Adv. Mater.* **2008**, *10*, 2038–2042.
3. Santos, A.L.; Botelho, E.; Kostov, K.; Nascente, P.A.P.; Da Silva, L.L.G. Atmospheric plasma treatment of carbon fibers for enhancement of their adhesion properties. *IEEE Trans. Plasma Sci.* **2013**, *41*, 319–324. [CrossRef]
4. Kinnares, V.; Hothongkham, P. Circuit analysis and modeling of a phase-shifted pulsewidth modulation full-bridge-inverter-fed ozone generator with constant applied electrode voltage. *IEEE Trans. Power Electron.* **2010**, *25*, 1739–1752. [CrossRef]
5. Francke, K.-P.; Rudolph, R.; Miessner, H. Design and operating characteristics of a simple and reliable DBD reactor for use with atmospheric air. *Plasma Chem. Plasma Process.* **2003**, *23*, 47–57. [CrossRef]
6. Grabowski, M.; Holub, M.; Balcerak, M.; Kalisiak, S.; Dąbrowski, W. Black pepper powder microbiological quality improvement using DBD systems in atmospheric pressure. *Eur. Phys. J. Appl. Phys.* **2015**, *71*, 6. [CrossRef]

7.   Mucko, J.; Panka, D.; Domanowski, P.; Bujnowski, S. Non-Thermal Plasma Reactor of Barrier Discharge for Disinfection and/or Sterilization of Organic Products. European. Patent Patent Application No. 20460005.0-1211, 29 January 2020.

8.   Lin, C.M.; Chu, Y.C.; Hsiao, C.P.; Wu, J.S.; Hsieh, C.W.; Hou, C.Y. The optimization of plasma-activated water treatments to inactivate salmonella enteritidis (ATCC 13076) on shell eggs. *Foods* **2019**, *8*, 520. [CrossRef]

9.   Ponomarev, A.V.; Gusev, A.I.; Pedos, M.S.; Sergeev, A.G.; Ustyuzhanin, A.V.; Podymova, A.S. High-Frequency Generator Based on Pulsed Excitation of the Oscillating Circuit for Biological Decontamination. In Proceedings of the IEEE International Conference on Plasma Science (ICOPS), San Francisco, CA, USA, 16–21 June 2013; p. 5. [CrossRef]

10. Grabowski, M.; Dąbrowski, W.; Balcerek, M.; Marcin, H.; Kalisiak, A.; Jakubowski, T.; Chyla, M. Reaktor Plazmy Nietermicznej do Sterylizacji Produktów Pochodzenia Organicznego. (Non-Thermal Plasma Reactor for Sterilizing Products of Organic Origin). U.S. Patent 2,307,98, 22 April 2015.

11. Heindl, A.; Müller, J.; Schulz, A.; Stroth, U.; Walker, M. Plasmaentkeimung. U.S. Patent DE 1,020,090,258,64A1, 14 October 2010.

12. Fridman, G.; Peddinghaus, M.; Fridman, A.; Balasubramanian, M.; Gutsol, A.; Friedman, G. Use of Non-Thermal Atmospheric Pressure Plasma Discharge for Coagulation and Sterilization of Surface Wounds. In Proceedings of the 17th International Conference on Plasma Chemistry, Toronto, ON, Canada, 7–12 August 2005; Paper No. 665. p. 6. [CrossRef]

13. Takayama, M.; Ebihara, K.; Stryczewska, H.; Ikegami, T.; Gyoutoku, Y.; Kubo, K.; Tachibana, M. *Ozone Generation by Dielectric Barrier Discharge for Soil Sterilization*; Elsevier: Amsterdam, The Netherlands, 2006; Volume 506–507, pp. 396–399. [CrossRef]

14. Oda, T.; Takahashi, T.; Nakano, H.; Masuda, S. Decomposition of fluorocarbon gaseous contaminants by surface discharge–induced plasma chemical processing. *IEEE Trans. Ind. Appl.* **1993**, *29*, 787–792. [CrossRef]

15. Yoshida, K. Aftertreatment of carbon particles emitted by diesel engine using a combination of corona and dielectric barrier discharge. *IEEE Trans. Ind. Appl.* **2019**, *55*, 5261–5268. [CrossRef]

16. Jakubowski, T.; Kalisiak, S.; Holub, M.; Palka, R.; Borkowski, T.; Myskow, J. New Resonant Inverter Topology with Active Energy Recovery in PDM Mode for DBD Plasma Reactor Supply. In Proceedings of the IEEE 14th European Conference on Power Electronics and Applications, Birmingham, UK, 30 August–1 September 2011; p. 8.

17. Du, Z.; Lin, X. Research Progress in Application of Low Temperature Plasma Technology for Wastewater Treatment. In Proceedings of the IOP Conference Series Earth and Environmental Science, Hangzhou, China, 17–19 April 2020; p. 7. [CrossRef]

18. Baloul, Y.; Aubry, O.; Rabat, H.; Colas, C.; Hong, D. Paracetamol degradation in aqueous solution by non-thermal plasma. *Eur. Phys. J. Appl. Phys.* **2017**, *79*. [CrossRef]

19. Holub, M.; Brandenburg, R.; Grosch, H.; Weinmann, S.; Hansel, B. Plasma supported odour removal from waste air in water treatment plants: An industrial case study. *Aerosol Air Qual. Res.* **2014**, *14*, 697–707. [CrossRef]

20. Holub, M.; Borkowski, T.; Jakubowski, T.; Kalisiak, S.; Myskow, J. Experimental results of a combined dbd reactor-catalyst assembly for a direct marine diesel-engine exhaust treatment. *IEEE Trans. Plasma Sci.* **2013**, *41*, 1562–1569. [CrossRef]

21. Kogelschatz, U. Dielectric-barrier discharges: Their history, discharge physics, and industrial applications. *Plasma Chem. Plasma Process.* **2003**, *23*, 1–46. [CrossRef]

22. Meißer, M. Resonant Behaviour of Pulse Generators for the Efficient Drive of Optical Radiation Sources Based on Dielectric Barrier Discharges. Ph.D. Thesis, Karlsruher Institut für Technologie, Karlsruhe, Germany, 2013. [CrossRef]

23. Rosenthal, L.A.; Davis, D.A. Electrical characterization of a corona discharge for surface treatment. *IEEE Trans. Ind. Appl.* **1975**, *11*, 328–335. [CrossRef]

24. Mucko, J. Corona Treatment System with Resonant Inverter-Selected Proprieties. In Proceedings of the 13th International Power Electronics and Motion Control Conference (EPE-PEMC), Poznan, Poland, 1–3 September 2008; pp. 1339–1343. [CrossRef]

25. Mehmood, A.; Jamal, H. Analysis on the Propulsion of Ionic Wind During Corona Discharge in Various Electrode Configuration with High Voltage Sources. In Proceedings of the IEEE International Conference on Applied and Engineering Mathematics (ICAEM), Taxila, Pakistan, 27–29 August 2019. [CrossRef]

26. Sosa, R.; Grondona, D.; Márquez, A.; Artana, G.; Kelly, H. On the induced gas flow by a trielectrode plasma curtain at atmospheric pressure. *IEEE Trans. Ind. Appl.* **2010**, *46*, 1132–1137. [CrossRef]

27. Siemens, W. Ueber die elektrostatische induction und die verzögerung des stroms in flaschendrähten. *Ann. Phys.* **1857**, *178*, 66–122. [CrossRef]

28. Pekarek, S. DC corona ozone generation enhanced by TiO2 photocatalyst. *Eur. Phys. J. D* **2008**, *50*, 171–175. [CrossRef]

29. Tsai, M.T.; Ke, C.W. Design and Implementation of a High-Voltage High-Frequency Pulse Power Generation System for Plasma Applications. In Proceedings of the International Conference on Power Electronics and Drive Systems (PEDS), Taipei, Taiwan, 2–5 November 2009; pp. 1556–1560. [CrossRef]

30. Tsai, M.T.; Chu, C.L. Power Control Strategies Evaluation of a Series Resonant Inverter for Atmosphere Plasma Applications. In Proceedings of the IEEE International Symposium on Industrial Electronics (ISIE), Seoul, Korea, 5–8 July 2009; pp. 632–637. [CrossRef]

31. Liu, Y.; He, X. A Series Resonant Inverter System with PDM and PFM hybrid Control for Plastic Film Surface Treatment. In Proceedings of the Industry Applications Conference (IAS), Hong Kong, China, 2–6 October 2005; pp. 1700–1704. [CrossRef]

32. Liu, Y.; He, X. PDM and PFM hybrid control of a series-resonant inverter for corona surface treatment. *IEE Proc. Electr. Power Appl.* **2005**, *152*, 1445. [CrossRef]

33. Luan, S.; Tang, X.; Wang, L.; Chen, S. Research of Phase-Shift Series-Load Resonant Power Supply for DBD-Ozone Generator. In Proceedings of the 6th International Conference on Power Electronics Systems and Applications (PESA), Hong Kong, China, 15–17 December 2015; p. 7. [CrossRef]

34. Mucko, J. Influence of the inverter output circuit on the dielectric barrier discharges during the surface treatment of plastics. *Prz. Elektrotechn. (Electrotech. Rev.)* **2015**, *91*, 95–98. [CrossRef]

35. Mucko, J. Aktywator folii z falownikiem rezonansowym - właściwości, metody i układy sterowania. (Corona treatment with resonant inverter-propriety, control methods and circuits). *Prz. Elektrotech. (Electrotech. Rev.)* **2005**, *81*, 42–49.

36. Changsheng, H.; Yushui, H.; Liqiao, W.; Zhongchao, Z. A Closed-loop Control for the Power Source of the Ozonizer. In Proceedings of the 35th Annual IEEE Power Electronics Specialists Conference, Aachen, Germany, 20–25 June 2004; pp. 3984–3987. [CrossRef]

37. Kalisiak, S.; Hołub, M.; Jakubowski, T. Resonant Inverter with Output Voltage Pulse-Phase-Shift Control for DBD Plasma Reactor Supply. In Proceedings of the 13th European Conference on Power Electronics and Applications, Barcelona, Spain, 8–10 September 2009; p. 9.

38. Kalisiak, S.; Jakubowski, T.; Holub, M. Single-Period AC Source with Energy Transfer Control for Dielectric Barrier Discharge Plasma Applications. In Proceedings of the 14th International Power Electronics and Motion Control Conference, EPE-PEMC, Ohrid, Macedonia, 6–8 September 2010; pp. 103–109. [CrossRef]

39. Dragonas, F.A.; Grandi, G.; Neretti, G. High-Voltage High-Frequency Arbitrary Waveform Multilevel Generator for Dielectric Barrier Discharge. In Proceedings of the International Symposium on Power Electronics, Electrical Drives, Automation and Motion, Ischia, Italy, 18–20 June 2014; pp. 57–61. [CrossRef]

40. Dragonas, F.A.; Neretti, G.; Sanjeevikumar, P.; Grandi, G. High-voltage high-frequency arbitrary waveform multilevel generator for dbd plasma actuators. *IEEE Trans. Ind. Appl.* **2015**, *51*, 3334–3342. [CrossRef]

41. Burány, N.; Huber, L.; Pejovic, P. Corona discharge surface treater without high voltage transformer. *IEEE Trans. Power Electron.* **2008**, *23*, 993–1002. [CrossRef]

42. Mućko, J.; Strzelecki, R.; Kozakiewicz, J.; Lutomirski, S. Resonant Inverters with Improved Output Characteristics in Application for Corona Discharge Treatment. In Proceedings of the European Conference on Power Electronics and Applications (EPE'99), Lausanne, Switzerland, 7–9 September 1999; pp. 1–6, Paper no. 857.

43. Fujita, H.; Ogasawara, S.; Akagi, H. An Approach to Broad Range of Power Control in Voltage-Source Series-Resonant Inverters for Corona Discharge Treatment. In Proceedings of the IEEE Power Electronics Specialists Conference (PESC), Saint Louis, MO, USA, 22–27 June 1997; pp. 1000–1006. [CrossRef]

44. Mućko, J. Control Method of Resonant Inverter for Application of a Foil Treatment (Sposób Sterowania Falownikiem Rezonansowym w Zastosowaniu Aktywatora Folii). Polish Patent No. 384865, 7 April 2008.

45. Instytut Inżynierii Materiałów Polimetowych i Barwników (Institute of Polymer Materials and Dyes Engineering), Torun, Poland, Selected Constructions of the Implemented Devices. Available online: http://www.torun.impib.pl/oferta/oferta-urzadzen-i-maszyn/aktywatory (accessed on 4 August 2020).

46. Meek, J.; Craggs, J. *Electrical Breakdown of Gases*; John Wiley & Sons: New York, NY, USA, 1978.

47. Mucko, J.; Strzelecki, R. Errors in the Analysis of Series Resonant Inverter/Converter Assuming Sinusoidal Waveforms of Voltage and Current. In Proceedings of the 10th International Conference on Compatibility, Power Electronics and Power Engineering (CPE-POWERENG), Bydgoszcz, Poland, 29 June–1 July 2016; pp. 369–374. [CrossRef]

48. Misra, N.N.; Schluter, O.; Cullen, P.J. *Cold Plasma in Food and Agriculture: Fundamentals and Applications*; Academic Press: Cambridge, MA, USA, 2016; p. 380. ISBN 9780128014899.

49. Santos, L.C.O.; Vieira, A.L.; Moecke, E.H.S.; Ribeiro, D.H.B.; Amante, E.R. Use of cold plasma to inactivate escherichia coli and physicochemical evaluation in pumpkin puree. *J. Food Prot.* **2018**, *81*, 1897–1905. [CrossRef] [PubMed]

50. Pankaj, S.; Wan, Z.; Keener, K.M. Effects of cold plasma on food quality: A review. *Foods* **2018**, *7*, 4. [CrossRef]

51. Han, L.; Patil, S.; Boehm, D.; Milosavljević, V.; Cullen, P.J.; Bourke, P. Mechanisms of inactivation by high-voltage atmospheric cold plasma differ for escherichia coli and staphylococcus aureus. *Appl. Environ. Microbiol.* **2016**, *82*, 450–458. [CrossRef]

52. Panka, D.; Jeske, M.; Lisiecki, K.; Mucko, J. Research on use of Cold Plasma for Disinfection of Plant Material. In Proceedings of the 4th World Summit & Expo on Food Technology and Probiotics, Bangkok, Thailand, 24–25 October 2019; p. 29.

MDPI

St. Alban-Anlage 66

4052 Basel

Switzerland

Tel. +41 61 683 77 34

Fax +41 61 302 89 18

www.mdpi.com

Energies Editorial Office

E-mail: energies@mdpi.com

www.mdpi.com/journal/energies